NASA STI Program...in Profile

Since its founding, the National Aeronautics and Space Administration (NASA) has been dedicated to the advancement of aeronautics and space science. The NASA Scientific and Technical Information (STI) program plays a key part in helping NASA maintain this important role.

The NASA STI program operates under the auspices of the Agency Chief Information Officer. It collects, organizes, provides for archiving, and disseminates NASA's STI. The NASA STI program provides access to the NASA Aeronautics and Space Database and its public interface, the NASA technical report server, thus providing one of the largest collections of aeronautical and space science STI in the world. Results are published in both non-NASA channels and by NASA in the NASA STI report series, which include the following report types:

- **Technical Publication:** Reports of completed research or a major significant phase of research that present the results of NASA programs and include extensive data or theoretical analysis. Includes compilations of significant scientific and technical data and information deemed to be of continuing reference value. NASA counterpart of peer-reviewed formal professional papers but has less stringent limitations on manuscript length and extent of graphic presentations.

- **Technical Memorandum:** Scientific and technical findings that are preliminary or of specialized interest, e.g., quick release reports, working papers, and bibliographies that contain minimal annotation. Does not contain extensive analysis.

- **Contractor Report:** Scientific and technical findings by NASA-sponsored contractors and grantees.

- **Conference Publication:** Collected papers from scientific and technical conferences, symposia, seminars, or other meetings sponsored or co-sponsored by NASA.

- **Special Publication:** Scientific, technical, or historical information from NASA programs, projects, and missions, often concerned with subjects having substantial public interest.

- **Technical Translation:** English-language translations of foreign scientific and technical material pertinent to NASA's mission.

Specialized services also include creating custom thesauri, building customized databases, and organizing and publishing research results.

For more information about the NASA STI program, see the following:

- Access the NASA STI program home page at www.sti.nasa.gov

- E-mail your question via the Internet to help@sti.nasa.gov

- Fax your question to the NASA STI help desk at 301-621-0134

- Phone the NASA STI help desk at 301-621-0390

- Write to:

NASA STI Help Desk
NASA Center for AeroSpace Information
7115 Standard Drive
Hanover, MD 21076-1320

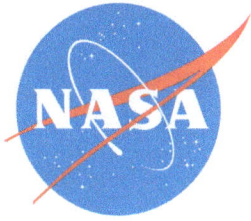

NASA/SP-2007-6105 Rev1

Systems Engineering Handbook

National Aeronautics and Space Administration
NASA Headquarters
Washington, D.C. 20546

December 2007

To request print or electronic copies or provide comments,
contact the Office of the Chief Engineer via
SP6105rev1SEHandbook@nasa.gov

Electronic copies are also available from
NASA Center for AeroSpace Information
7115 Standard Drive
Hanover, MD 21076-1320
at
http://ntrs.nasa.gov/

Table of Contents

Figures

Table of Contents

Boxes

Preface

Since the writing of NASA/SP-6105 in 1995, systems engineering at the National Aeronautics and Space Administration (NASA), within national and international standard bodies, and as a discipline has undergone rapid evolution. Changes include implementing standards in the International Organization for Standardization (ISO) 9000, the use of Carnegie Mellon Software Engineering Institute's Capability Maturity Model® Integration (CMMI®) to improve development and delivery of products, and the impacts of mission failures. Lessons learned on systems engineering were documented in reports such as those by the NASA Integrated Action Team (NIAT), the Columbia Accident Investigation Board (CAIB), and the follow-on Diaz Report. Out of these efforts came the NASA Office of the Chief Engineer (OCE) initiative to improve the overall Agency systems engineering infrastructure and capability for the efficient and effective engineering of NASA systems, to produce quality products, and to achieve mission success. In addition, Agency policy and requirements for systems engineering have been established. This handbook update is a part of the OCE-sponsored Agencywide systems engineering initiative.

In 1995, SP-6105 was initially published to bring the fundamental concepts and techniques of systems engineering to NASA personnel in a way that recognizes the nature of NASA systems and the NASA environment. This revision of SP-6105 maintains that original philosophy while updating the Agency's systems engineering body of knowledge, providing guidance for insight into current best Agency practices, and aligning the handbook with the new Agency systems engineering policy.

The update of this handbook was twofold: a top-down compatibility with higher level Agency policy and a bottom-up infusion of guidance from the NASA practitioners in the field. The approach provided the opportunity to obtain best practices from across NASA and bridge the information to the established NASA systems engineering process. The attempt is to communicate principles of good practice as well as alternative approaches rather than specify a particular way to accomplish a task. The result embodied in this handbook is a top-level implementation approach on the practice of systems engineering unique to NASA. The material for updating this handbook was drawn from many different sources, including NASA procedural requirements, field center systems engineering handbooks and processes, as well as non-NASA systems engineering textbooks and guides.

This handbook consists of six core chapters: (1) systems engineering fundamentals discussion, (2) the NASA program/project life cycles, (3) systems engineering processes to get from a concept to a design, (4) systems engineering processes to get from a design to a final product, (5) crosscutting management processes in systems engineering, and (6) special topics relative to systems engineering. These core chapters are supplemented by appendices that provide outlines, examples, and further information to illustrate topics in the core chapters. The handbook makes extensive use of boxes and figures to define, refine, illustrate, and extend concepts in the core chapters without diverting the reader from the main information.

The handbook provides top-level guidelines for good systems engineering practices; it is not intended in any way to be a directive.

NASA/SP-2007-6105 Rev1 supersedes SP-6105, dated June 1995.

Acknowledgments

Primary points of contact: Stephen J. Kapurch, Office of the Chief Engineer, NASA Headquarters, and Neil E. Rainwater, Marshall Space Flight Center.

The following individuals are recognized as contributing practitioners to the content of this handbook revision:

- ■ Core Team Member (or Representative) from Center, Directorate, or Office
- ◆ Integration Team Member
- ● Subject Matter Expert Team Champion
- ▲ Subject Matter Expert

Arden Acord, NASA/Jet Propulsion Laboratory ▲

Danette Allen, NASA/Langley Research Center ▲

Deborah Amato, NASA/Goddard Space Flight Center ▲●

Jim Andary, NASA/Goddard Space Flight Center ▲◆

Tim Beard, NASA/Ames Research Center ▲

Jim Bilbro, NASA/Marshall Space Flight Center ▲

Mike Blythe, NASA/Headquarters ■

Linda Bromley, NASA/Johnson Space Center ◆●▲■

Dave Brown, Defense Acquisition University ▲

John Brunson, NASA/Marshall Space Flight Center ▲●

Joe Burt, NASA/Goddard Space Flight Center ▲

Glenn Campbell, NASA/Headquarters ▲

Joyce Carpenter, NASA/Johnson Space Center ▲●

Keith Chamberlin, NASA/Goddard Space Flight Center ▲

Peggy Chun, NASA/NASA Engineering and Safety Center ◆●▲■

Cindy Coker, NASA/Marshall Space Flight Center ▲

Nita Congress, Graphic Designer ◆

Catharine Conley, NASA/Headquarters ▲

Shelley Delay, NASA/Marshall Space Flight Center ▲

Rebecca Deschamp, NASA/Stennis Space Center ▲

Homayoon Dezfuli, NASA/Headquarters ▲●

Olga Dominguez, NASA/Headquarters ▲

Rajiv Doreswamy, NASA/Headquarters ■

Larry Dyer, NASA/Johnson Space Center ▲

Nelson Eng, NASA/Johnson Space Center ▲

Patricia Eng, NASA/Headquarters ▲

Amy Epps, NASA/Marshall Space Flight Center ▲

Chester Everline, NASA/Jet Propulsion Laboratory ▲

Karen Fashimpaur, Arctic Slope Regional Corporation ◆▲

Orlando Figueroa, NASA/Goddard Space Flight Center ■

Stanley Fishkind, NASA/Headquarters ■

Brad Flick, NASA/Dryden Flight Research Center ■

Marton Forkosh, NASA/Glenn Research Center ■

Dan Freund, NASA/Johnson Space Center ▲

Greg Galbreath, NASA/Johnson Space Center ▲

Louie Galland, NASA/Langley Research Center ▲

Yuri Gawdiak, NASA/Headquarters ■●▲

Theresa Gibson, NASA/Glenn Research Center ▲

Ronnie Gillian, NASA/Langley Research Center ▲

Julius Giriunas, NASA/Glenn Research Center ▲

Ed Gollop, NASA/Marshall Space Flight Center ▲

Lee Graham, NASA/Johnson Space Center ▲

Larry Green, NASA/Langley Research Center ▲

Owen Greulich, NASA/Headquarters ■

Ben Hanel, NASA/Ames Research Center ▲

Gena Henderson, NASA/Kennedy Space Center ●▲

Amy Hemken, NASA/Marshall Space Flight Center ▲

Bob Hennessy, NASA/NASA Engineering and Safety Center ▲

Ellen Herring, NASA/Goddard Space Flight Center ●▲

Renee Hugger, NASA/Johnson Space Center ▲

Brian Hughitt, NASA/Headquarters ▲

Eric Isaac, NASA/Goddard Space Flight Center ■

Tom Jacks, NASA/Stennis Space Center ▲

Ken Johnson, NASA/NASA Engineering and Safety Center ▲

Ross Jones, NASA/Jet Propulsion Laboratory ■

John Juhasz, NASA/Johnson Space Center ▲

Stephen Kapurch, NASA/Headquarters ■◆●

Jason Kastner, NASA/Jet Propulsion Laboratory ▲

Kristen Kehrer, NASA/Kennedy Space Center ▲

John Kelly, NASA/Headquarters ▲

Kriss Kennedy, NASA/Johnson Space Center ▲

Acknowledgments

Steven Kennedy, NASA/Kennedy Space Center ▲

Tracey Kickbusch, NASA/Kennedy Space Center ■

Casey Kirchner, NASA/Stennis Space Center ▲

Kenneth Kumor, NASA/Headquarters ▲

Janne Lady, SAITECH/CSC ▲

Jerry Lake, Systems Management international ▲

Kenneth W. Ledbetter, NASA/Headquarters ■

Steve Leete, NASA/Goddard Space Flight Center ▲

William Lincoln, NASA/Jet Propulsion Laboratory ▲

Dave Littman, NASA/Goddard Space Flight Center ▲

John Lucero, NASA/Glenn Research Center ▲

Paul Luz, NASA/Marshall Space Flight Center ▲

Todd MacLeod, NASA/Marshall Space Flight Center ▲

Roger Mathews, NASA/Kennedy Space Center ▲●

Bryon Maynard, NASA/Stennis Space Center ▲

Patrick McDuffee, NASA/Marshall Space Flight Center ▲

Mark McElyea, NASA/Marshall Space Flight Center ▲

William McGovern, Defense Acquisition University ◆

Colleen McGraw, NASA/Goddard Space Flight Center ▲◆●

Melissa McGuire, NASA/Glenn Research Center ▲

Don Mendoza, NASA/Ames Research Center ▲

Leila Meshkat, NASA/Jet Propulsion Laboratory ▲

Elizabeth Messer, NASA/Stennis Space Center ▲●

Chuck Miller, NASA/Headquarters ▲

Scott Mimbs, NASA/Kennedy Space Center ▲

Steve Newton, NASA/Marshall Space Flight Center ▲

Tri Nguyen, NASA/Johnson Space Center ▲

Chuck Niles, NASA/Langley Research Center ▲●

Cynthia Null, NASA/NASA Engineering and Safety Center ▲

John Olson, NASA/Headquarters ▲

Tim Olson, QIC, Inc. ▲

Sam Padgett, NASA/Johnson Space Center ▲

Christine Powell, NASA/Stennis Space Center ◆●▲■

Steve Prahst, NASA/Glenn Research Center ▲

Pete Prassinos, NASA/Headquarters ■

Mark Prill, NASA/Marshall Space Flight Center ▲

Neil Rainwater, NASA/Marshall Space Flight Center ■◆

Ron Ray, NASA/Dryden Flight Research Center ▲

Gary Rawitscher, NASA/Headquarters ▲

Joshua Reinert, ISL Inc. ▲

Norman Rioux, NASA/Goddard Space Flight Center ▲

Steve Robbins, NASA/Marshall Space Flight Center ▲●

Dennis Rohn, NASA/Glenn Research Center ▲◆

Jim Rose, NASA/Jet Propulsion Laboratory ▲

Arnie Ruskin,* NASA/Jet Propulsion Laboratory ▲●

Harry Ryan, NASA/Stennis Space Center ▲

George Salazar, NASA/Johnson Space Center ▲

Nina Scheller, NASA/Ames Research Center ■

Pat Schuler, NASA/Langley Research Center ▲●

Randy Seftas, NASA/Goddard Space Flight Center ▲

Joey Shelton, NASA/Marshall Space Flight Center ▲●

Robert Shishko, NASA/Jet Propulsion Laboratory ▲◆

Burton Sigal, NASA/Jet Propulsion Laboratory ▲

Sandra Smalley, NASA/Headquarters ▲

Richard Smith, NASA/Kennedy Space Center ▲

John Snoderly, Defense Acquisition University ▲

Richard Sorge, NASA/Glenn Research Center ▲

Michael Stamatelatos, NASA/Headquarters ■

Tom Sutliff, NASA/Glenn Research Center ▲●

Todd Tofil, NASA/Glenn Research Center ▲

John Tinsley, NASA/Headquarters ▲

Rob Traister, Graphic Designer ◆

Clayton Turner, NASA/Langley Research Center ■

Paul VanDamme, NASA/Jet Propulsion Laboratory ▲

Karen Vaner, NASA/Stennis Space Center ▲

Lynn Vernon, NASA/Johnson Space Center ▲

Linda Voss, Technical Writer ◆

Britt Walters, NASA/Johnson Space Center ■

Tommy Watts, NASA/Marshall Space Flight Center ▲

Richard Weinstein, NASA/Headquarters ▲

Katie Weiss, NASA/Jet Propulsion Laboratory ▲●

Martha Wetherholt, NASA/Headquarters ▲

Becky Wheeler, NASA/Jet Propulsion Laboratory ▲

Cathy White, NASA/Marshall Space Flight Center ▲

Reed Wilcox, NASA/Jet Propulsion Laboratory ▲

Barbara Woolford, NASA/Johnson Space Center ▲●

Felicia Wright, NASA/Langley Research Center ▲

Robert Youngblood, ISL Inc. ▲

Tom Zang, NASA/Langley Research Center ▲

_____

*In memory of.

1.0 Introduction

1.1 Purpose

This handbook is intended to provide general guidance and information on systems engineering that will be useful to the NASA community. It provides a generic description of Systems Engineering (SE) as *it should be* applied throughout NASA. A goal of the handbook is to increase awareness and consistency across the Agency and advance the practice of SE. This handbook provides perspectives relevant to NASA and data particular to NASA.

This handbook should be used as a companion for implementing *NPR 7123.1, Systems Engineering Processes and Requirements*, as well as the Center-specific handbooks and directives developed for implementing systems engineering at NASA. It provides a companion reference book for the various systems engineering related courses being offered under NASA's auspices.

1.2 Scope and Depth

The coverage in this handbook is limited to general concepts and generic descriptions of processes, tools, and techniques. It provides information on systems engineering best practices and pitfalls to avoid. There are many Center-specific handbooks and directives as well as textbooks that can be consulted for in-depth tutorials.

This handbook describes systems engineering as it should be applied to the development and implementation of large and small NASA programs and projects. NASA has defined different life cycles that specifically address the major project categories, or product lines, which are: Flight Systems and Ground Support (FS&GS), Research and Technology (R&T), Construction of Facilities (CoF), and Environmental Compliance and Restoration (ECR). The technical content of the handbook provides systems engineering best practices that should be incorporated into all NASA product lines. (Check the NASA On-Line Directives Information System (NODIS) electronic document library for applicable NASA directives on topics such as product lines.) For simplicity this handbook uses the FS&GS product line as an example. The specifics of FS&GS can be seen in the description of the life cycle and the details of the milestone reviews. Each product line will vary in these two areas; therefore, the reader should refer to the applicable NASA procedural requirements for the specific requirements for their life cycle and reviews. The engineering of NASA systems requires a systematic and disciplined set of processes that are applied recursively and iteratively for the design, development, operation, maintenance, and closeout of systems throughout the life cycle of the programs and projects.

The handbook's scope properly includes systems engineering functions regardless of whether they are performed by a manager or an engineer, in-house, or by a contractor.

2.0 Fundamentals of Systems Engineering

Systems engineering is a methodical, disciplined approach for the design, realization, technical management, operations, and retirement of a system. A "system" is a construct or collection of different elements that together produce results not obtainable by the elements alone. The elements, or parts, can include people, hardware, software, facilities, policies, and documents; that is, all things required to produce system-level results. The results include system-level qualities, properties, characteristics, functions, behavior, and performance. The value added by the system as a whole, beyond that contributed independently by the parts, is primarily created by the relationship among the parts; that is, how they are interconnected.[1] It is a way of looking at the "big picture" when making technical decisions. It is a way of achieving stakeholder functional, physical, and operational performance requirements in the intended use environment over the planned life of the systems. In other words, systems engineering is a logical way of thinking.

Systems engineering is the art and science of developing an operable system capable of meeting requirements within often opposed constraints. Systems engineering is a holistic, integrative discipline, wherein the contributions of structural engineers, electrical engineers, mechanism designers, power engineers, human factors engineers, and many more disciplines are evaluated and balanced, one against another, to produce a coherent whole that is not dominated by the perspective of a single discipline.[2]

Systems engineering seeks a safe and balanced design in the face of opposing interests and multiple, sometimes conflicting constraints. The systems engineer must develop the skill and instinct for identifying and focusing efforts on assessments to optimize the overall design

and not favor one system/subsystem at the expense of another. The art is in knowing when and where to probe. Personnel with these skills are usually tagged as "systems engineers." They may have other titles—lead systems engineer, technical manager, chief engineer—but for this document, we will use the term systems engineer.

The exact role and responsibility of the systems engineer may change from project to project depending on the size and complexity of the project and from phase to phase of the life cycle. For large projects, there may be one or more systems engineers. For small projects, sometimes the project manager may perform these practices. But, whoever assumes those responsibilities, the systems engineering functions must be performed. The actual assignment of the roles and responsibilities of the named systems engineer may also therefore vary. The lead systems engineer ensures that the system technically fulfills the defined needs and requirements and that a proper systems engineering approach is being followed. The systems engineer oversees the project's systems engineering activities as performed by the technical team and directs, communicates, monitors, and coordinates tasks. The systems engineer reviews and evaluates the technical aspects of the project to ensure that the systems/subsystems engineering processes are functioning properly and evolves the system from concept to product. The entire technical team is involved in the systems engineering process.

The systems engineer will usually play the key role in leading the development of the system architecture, defining and allocating requirements, evaluating design tradeoffs, balancing technical risk between systems, defining and assessing interfaces, providing oversight of verification and validation activities, as well as many other tasks. The systems engineer will usually have the prime responsibility in developing many of the project documents, including the Systems Engineering Management Plan (SEMP), requirements/specification documents, verification and validation documents, certification packages, and other technical documentation.

---

[1] Rechtin, *Systems Architecting of Organizations: Why Eagles Can't Swim.*

[2] Comments on systems engineering throughout Chapter 2.0 are extracted from the speech "System Engineering and the Two Cultures of Engineering" by Michael D. Griffin, NASA Administrator.

In summary, the systems engineer is skilled in the art and science of balancing organizational and technical interactions in complex systems. However, since the entire team is involved in the systems engineering approach, in some ways everyone is a systems engineer. Systems engineering is about tradeoffs and compromises, about generalists rather than specialists. Systems engineering is about looking at the "big picture" and not only ensuring that they get the design right (meet requirements) but that they get the right design.

To explore this further, put SE in the context of project management. As discussed in *NPR 7120.5, NASA Space Flight Program and Project Management Requirements*, project management is the function of planning, overseeing, and directing the numerous activities required to achieve the requirements, goals, and objectives of the customer and other stakeholders within specified cost, quality, and schedule constraints. Project management can be thought of as having two major areas of emphasis, both of equal weight and importance. These areas are systems engineering and project control. Figure 2.0-1 is a notional graphic depicting this concept. Note that there are areas where the two cornerstones of project management overlap. In these areas, SE provides the technical aspects or inputs; whereas project control provides the programmatic, cost, and schedule inputs.

This document will focus on the SE side of the diagram. These practices/processes are taken from *NPR 7123.1,*

NASA Systems Engineering Processes and Requirements. Each will be described in much greater detail in subsequent chapters of this document, but an overview is given below.

2.1 The Common Technical Processes and the SE Engine

There are three sets of common technical processes in *NPR 7123.1, NASA Systems Engineering Processes and Requirements:* system design, product realization, and technical management. The processes in each set and their interactions and flows are illustrated by the NPR systems engineering "engine" shown in Figure 2.1-1. The processes of the SE engine are used to develop and realize the end products. This chapter provides the application context of the 17 common technical processes required in NPR 7123.1. The system design processes, the product realization processes, and the technical management processes are discussed in more details in Chapters 4.0, 5.0, and 6.0, respectively. Steps 1 through 9 indicated in Figure 2.1-1 represent the tasks in execution of a project. Steps 10 through 17 are crosscutting tools for carrying out the processes.

- **System Design Processes:** The four system design processes shown in Figure 2.1-1 are used to define and baseline stakeholder expectations, generate and baseline technical requirements, and convert the technical requirements into a design solution that will satisfy the baselined stakeholder expectations. These processes are applied to each product of the system structure from the top of the structure to the bottom until the lowest products in any system structure branch are defined to the point where they can be built, bought, or reused. All other products in the system structure are realized by integration. Designers not only develop the design solutions to the products intended to perform the operational functions of the system, but also

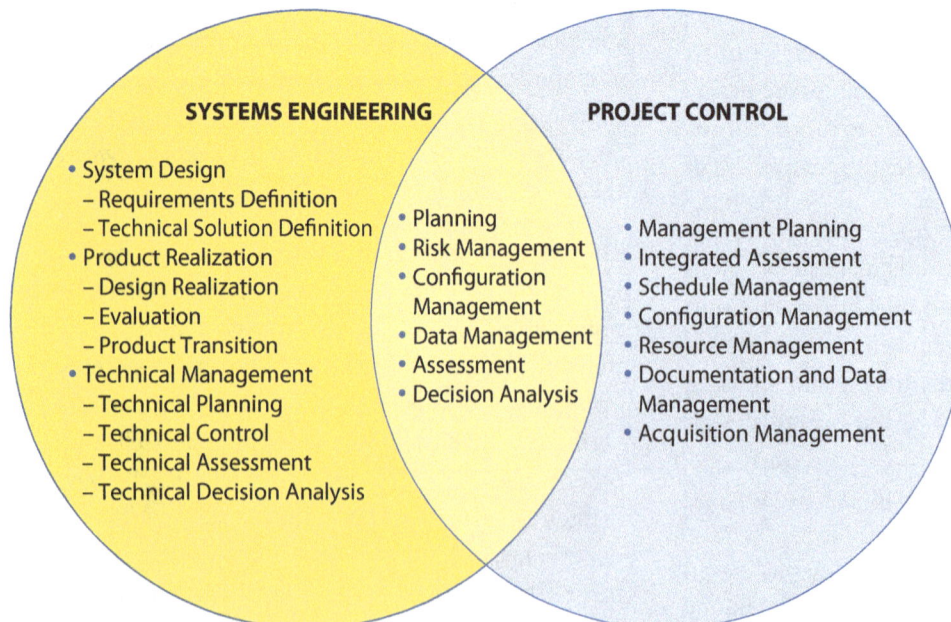

SYSTEMS ENGINEERING

- System Design
  - Requirements Definition
  - Technical Solution Definition
- Product Realization
  - Design Realization
  - Evaluation
  - Product Transition
- Technical Management
  - Technical Planning
  - Technical Control
  - Technical Assessment
  - Technical Decision Analysis

- Planning
- Risk Management
- Configuration Management
- Data Management
- Assessment
- Decision Analysis

PROJECT CONTROL

- Management Planning
- Integrated Assessment
- Schedule Management
- Configuration Management
- Resource Management
- Documentation and Data Management
- Acquisition Management

Figure 2.0-1 SE in context of overall project management

Requirements flow down
from level above

Realized products
to level above

**TECHNICAL MANAGEMENT
PROCESSES**

**SYSTEM
DESIGN
PROCESSES**

**PRODUCT
REALIZATION
PROCESSES**

Technical Planning
Process

10. Technical Planning

Requirements Definition
Processes

1. Stakeholder Expectations
Definition
2. Technical Requirements
Definition

Technical Control
Processes

11. Requirements Management
12. Interface Management
13. Technical Risk Management
14. Configuration Management
15. Technical Data Management

Product Transition Process

9. Product Transition

Evaluation Processes

7. Product Verification
8. Product Validation

Technical Solution
Definition Processes

3. Logical Decomposition
4. Design Solution Definition

Technical Assessment
Process

16. Technical Assessment

Design Realization
Processes

5. Product Implementation
6. Product Integration

Technical Decision Analysis
Process

17. Decision Analysis

Requirements flow down
to level below

Realized products
from level below

System design processes -
applied to each work breakdown
structure model down and
across system structure

Product realization processes -
applied to each product -
up and across -
system structure -

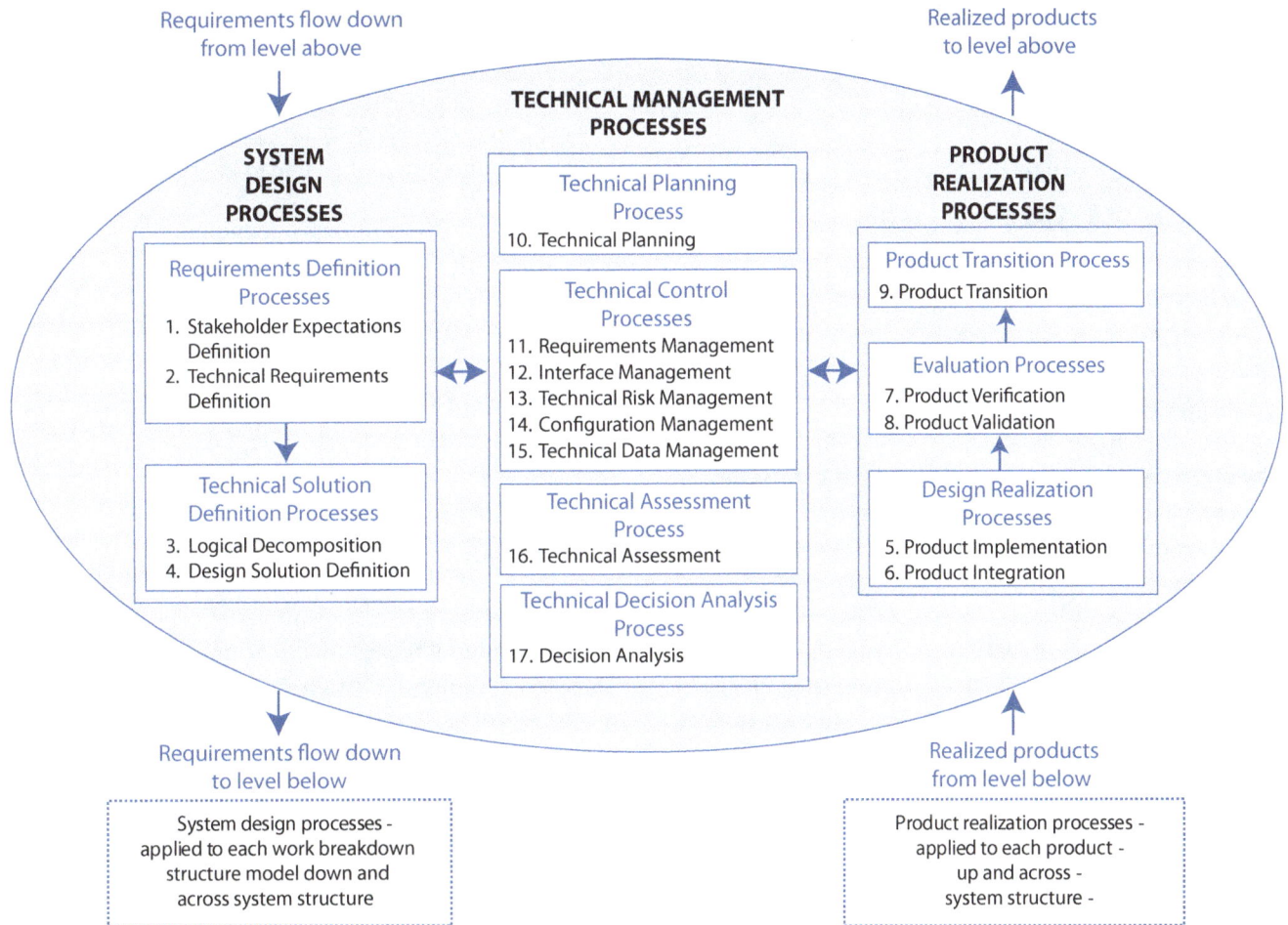

Figure 2.1-1 The systems engineering engine

establish requirements for the products and services that enable each operational/mission product in the system structure.

- **Product Realization Processes:** The product realization processes are applied to each operational/mission product in the system structure starting from the lowest level product and working up to higher level integrated products. These processes are used to create the design solution for each product (e.g., by the Product Implementation or Product Integration Process) and to verify, validate, and transition up to the next hierarchical level products that satisfy their design solutions and meet stakeholder expectations as a function of the applicable life-cycle phase.

- **Technical Management Processes:** The technical management processes are used to establish and evolve technical plans for the project, to manage communication across interfaces, to assess progress against the plans and requirements for the system products or services, to control technical execution of

the project through to completion, and to aid in the decisionmaking process.

The processes within the SE engine are used both iteratively and recursively. As defined in NPR 7123.1, "iterative" is the "application of a process to the same product or set of products to correct a discovered discrepancy or other variation from requirements," whereas "recursive" is defined as adding value to the system "by the repeated application of processes to design next lower layer system products or to realize next upper layer end products within the system structure. This also applies to repeating application of the same processes to the system structure in the next life-cycle phase to mature the system definition and satisfy phase success criteria." The example used in Section 2.3 will further explain these concepts. The technical processes are applied recursively and iteratively to break down the initializing concepts of the system to a level of detail concrete enough that the technical team can implement a product from the information. Then the processes are applied recursively and iteratively to integrate the smallest product into greater and larger systems until the whole of the system has been assembled, verified, validated, and transitioned.

2.2 An Overview of the SE Engine by Project Phase

Figure 2.2-1 conceptually illustrates how the SE engine is used during each of the seven phases of a project. Figure 2.2-1 is a *conceptual* diagram. For all of the details, refer to the poster version of this figure, which accompanies this handbook.

The uppermost horizontal portion of this chart is used as a reference to project system maturity, as the project progresses from a feasible concept to an as-deployed system; phase activities; Key Decision Points (KDPs); and major project reviews.

The next major horizontal band shows the technical development processes (steps 1 through 9) in each project phase. The systems engineering engine cycles five times

from Pre-Phase A through Phase D. Please note that NASA's management has structured Phases C and D to "split" the technical development processes in half in Phases C and D to ensure closer management control. The engine is bound by a dashed line in Phases C and D.

Once a project enters into its operational state (Phase E) and closes with a closeout phase (Phase F), the technical work shifts to activities commensurate with these last two project phases.

The next major horizontal band shows the eight technical management processes (steps 10 through 17) in each project phase. The SE engine cycles the technical management processes seven times from Pre-Phase A through Phase F.

Each of the engine entries is given a 6105 paragraph label that is keyed to Chapters 4.0, 5.0, and 6.0 in this handbook. For example, in the technical development processes, "Get Stakeholder Expectations" discussions and details are in Section 4.1.

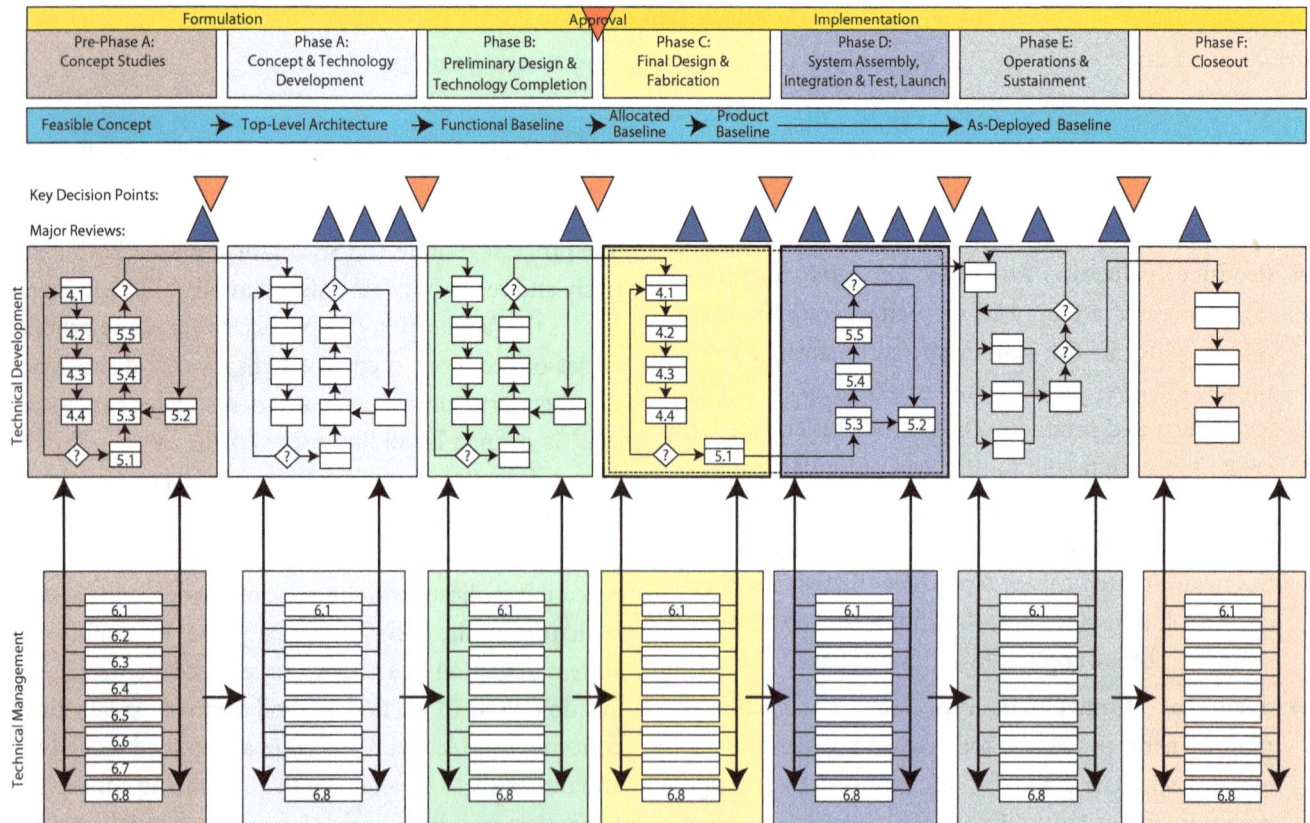

Figure 2.2-1   A miniaturized conceptualization of the poster-size NASA project life-cycle process flow for flight and ground systems accompanying this handbook

2.3 Example of Using the SE Engine

To help in understanding how the SE engine is applied, an example will be posed and walked through the processes. Pertinent to this discussion are the phases of the program and project life cycles, which will be discussed in greater depth in Chapter 3.0 of this document. As described in Chapter 3.0, NPR 7120.5 defines the life cycle used for NASA programs and projects. The life-cycle phases are described in Table 2.3-1.

Use of the different phases of a life cycle allows the various products of a project to be gradually developed and matured from initial concepts through the fielding of the product and to its final retirement. The SE engine shown in Figure 2.1-1 is used throughout all phases.

In Pre-Phase A, the SE engine is used to develop the initial concepts; develop a preliminary/draft set of key high-level requirements; realize these concepts through modeling, mockups, simulation, or other means; and verify and validate that these concepts and products would be able to meet the key high-level requirements. Note that this is not the formal verification and validation program that will be performed on the final product but is a methodical runthrough ensuring that the concepts that are being developed in this Pre-Phase A would be able to meet the likely requirements and expectations of the stakeholders. Concepts would be developed to the lowest level necessary to ensure that the concepts are feasible and to a level that will reduce the risk low enough to satisfy the project. Academically, this process could proceed down to the circuit board level for every system.

Table 2.3-1 Project Life-Cycle Phases

	Phase	Purpose	Typical Output
Formulation	Pre-Phase A Concept Studies	To produce a broad spectrum of ideas and alternatives for missions from which new programs/projects can be selected. Determine feasibility of desired system, develop mission concepts, draft system-level requirements, identify potential technology needs.	Feasible system concepts in the form of simulations, analysis, study reports, models, and mockups
	Phase A Concept and Technology Development	To determine the feasibility and desirability of a suggested new major system and establish an initial baseline compatibility with NASA's strategic plans. Develop final mission concept, system-level requirements, and needed system structure technology developments.	System concept definition in the form of simulations, analysis, engineering models, and mockups and trade study definition
	Phase B Preliminary Design and Technology Completion	To define the project in enough detail to establish an initial baseline capable of meeting mission needs. Develop system structure end product (and enabling product) requirements and generate a preliminary design for each system structure end product.	End products in the form of mockups, trade study results, specification and interface documents, and prototypes
Implementation	Phase C Final Design and Fabrication	To complete the detailed design of the system (and its associated subsystems, including its operations systems), fabricate hardware, and code software. Generate final designs for each system structure end product.	End product detailed designs, end product component fabrication, and software development
	Phase D System Assembly, Integration and Test, Launch	To assemble and integrate the products to create the system, meanwhile developing confidence that it will be able to meet the system requirements. Launch and prepare for operations. Perform system end product implementation, assembly, integration and test, and transition to use.	Operations-ready system end product with supporting related enabling products
	Phase E Operations and Sustainment	To conduct the mission and meet the initially identified need and maintain support for that need. Implement the mission operations plan.	Desired system
	Phase F Closeout	To implement the systems decommissioning/disposal plan developed in Phase E and perform analyses of the returned data and any returned samples.	Product closeout

However, that would involve a great deal of time and money. There may be a higher level or tier of product than circuit board level that would enable designers to accurately determine the feasibility of accomplishing the project (purpose of Pre-Phase A).

During Phase A, the recursive use of the SE engine is continued, this time taking the concepts and draft key requirements that were developed and validated during Pre-Phase A and fleshing them out to become the set of baseline system requirements and Concept of Operations (ConOps). During this phase, key areas of high risk might be simulated or prototyped to ensure that the concepts and requirements being developed are good ones and to identify verification and validation tools and techniques that will be needed in later phases.

During Phase B, the SE engine is applied recursively to further mature requirements for all products in the developing product tree, develop ConOps preliminary designs, and perform feasibility analysis of the verification and validation concepts to ensure the designs will likely be able to meet their requirements.

Phase C again uses the left side of the SE engine to finalize all requirement updates, finalize ConOps, develop the final designs to the lowest level of the product tree, and begin fabrication. Phase D uses the right side of the SE engine to recursively perform the final implementation, integration, verification, and validation of the end product, and at the final pass, transition the end product to the user. The technical management processes of the SE engine are used in Phases E and F to monitor performance; control configuration; and make decisions associated with the operations, sustaining engineering, and closeout of the system. Any new capabilities or upgrades of the existing system would reenter the SE engine as new developments.

2.3.1 Detailed Example

Since it is already well known, the NASA Space Transportation System (STS) will be used as an example to look at how the SE engine would be used in Phase A. This example will be simplified to illustrate the application of the SE processes in the engine, but will in no way be as detailed as necessary to actually build the highly complex vehicle. The SE engine is used recursively to drive out more and more detail with each pass. The icon shown in Figure 2.3-1 will be used to keep track of the applicable place in the SE engine. The numbers in the icon

correspond to the numbered processes within the SE engine as shown in Figure 2.1-1. The various layers of the product hierarchy will be called "tiers." Tiers are also called "layers," or "levels." But basically, the higher the number of the tier or level, the lower in the product hierarchy the product is going and the more detailed the product is becoming (e.g., going from boxes, to circuit boards, to components).

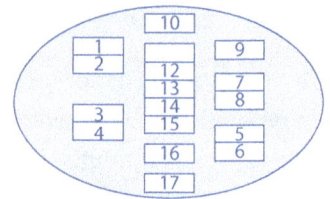

Figure 2.3-1 SE engine tracking icon

2.3.2 Example Premise

NASA decides that there is a need for a transportation system that will act like a "truck" to carry large pieces of equipment and crew into Low Earth Orbit (LEO). Referring back to the project life cycle, the project first enters the Pre-Phase A. During this phase, several concept studies are performed, and it is determined that it is feasible to develop such a "space truck." This is determined through combinations of simulations, mockups, analyses, or other like means. For simplicity, assume feasibility will be proven through concept models. The processes and framework of the SE engine will be used to design and implement these models. The project would then enter the Phase A activities to take the Pre-Phase A concepts and refine them and define the system requirements for the end product. The detailed example will begin in Phase A and show how the SE engine is used. As described in the overview, a similar process is used for the other project phases.

2.3.2.1 Example Phase A System Design Passes

First Pass

Taking the preliminary concepts and drafting key system requirements developed during the Pre-Phase A activities, the SE engine is entered at the first process and used to determine who the product (i.e., the STS) stakeholders are and what they want.

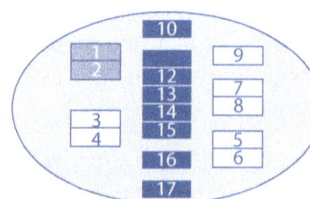

During Pre-Phase A these needs and expectations were pretty general ideas, probably just saying the Agency needs a "space truck" that will carry X tons of payload into LEO, accommodate a payload of so-and-so size, carry a crew of seven, etc. During this Phase A pass, these general concepts are detailed out and agreed to. The ConOps (sometimes referred to as operational concept) generated in Pre-Phase A is also detailed out and agreed to to ensure all stakeholders are in agreement as to what is really expected of the product—in this case the transportation system. The detailed expectations are then converted into good requirement statements. (For more information on what constitutes a good requirement, see Appendix C.) Subsequent passes and subsequent phases will refine these requirements into specifications that can actually be built. Also note that all of the technical management processes (SE engine processes numbered 10 through 17) are also used during this and all subsequent passes and activities. These ensure that all the proper planning, control, assessment, and decisions are used and maintained. Although for simplification they will not be mentioned in the rest of this example, they will *always* be in effect.

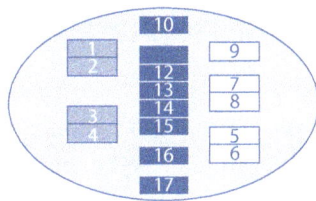

Next, using the requirements and the ConOps previously developed, logical decomposition models/diagrams are built up to help bring the requirements into perspective and to show their relationship. Finally, these diagrams, requirements, and ConOps documents are used to develop one or more feasible design solutions. Note that at this point, since this is only the first pass through the SE engine, these design solutions are not detailed enough to actually build anything. Consequently, the design solutions might be summarized as, "To accomplish this transportation system, the best option in our trade studies is a three-part system: a reusable orbiter for the crew and cargo, a large external tank to hold the propellants, and two solid rocket boosters to give extra power for liftoff that can be recovered, refurbished, and reused." (Of course, the actual design solution would be much more descriptive and detailed). So, for this first pass, the first tier of the product hierarchy might look like Figure 2.3-2. There would also be other enabling products that might appear in the product tree, but for simplicity only, the main products are shown in this example.

Now, obviously design solution is not yet at a detailed enough level to actually build the prototypes or models of any of these products. The requirements, ConOps, functional diagrams, and design solutions are still at a

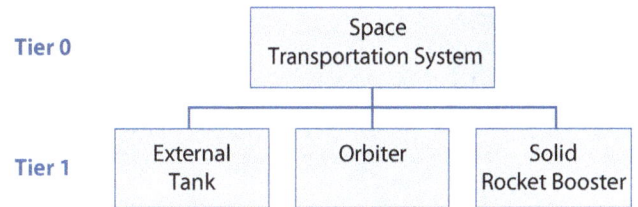

Figure 2.3-2 Product hierarchy, tier 1: first pass through the SE engine

pretty high, general level. Note that the SE processes on the right side (i.e., the product realization processes) of the SE engine have yet to be addressed. The design must first be at a level that something can actually be built, coded, or reused before that side of the SE engine can be used. So, a second pass of the left side of the SE engine will be started.

Second Pass

The SE engine is completely recursive. That is, each of the three elements shown in the tier 1 diagram can now be considered a product of its own and the SE engine is therefore applied to each of the three elements separately. For example, the external tank is considered an end product and the SE engine resets back to the first processes. So now, just focusing on the external tank, who are the stakeholders and what they expect of the external tank is determined. Of course, one of the main stakeholders will be the owners of the tier 1 requirements and the STS as an end product, but there will also be other new stakeholders. A new ConOps for how the external tank would operate is generated. The tier 1 requirements that are applicable (allocated) to the external tank would be "flowed down" and validated. Usually, some of these will be too general to implement into a design, so the requirements will have to be detailed out. To these derived requirements, there will also be added new requirements that are generated from the stakeholder expectations, and other applicable standards for workmanship, safety, quality, etc.

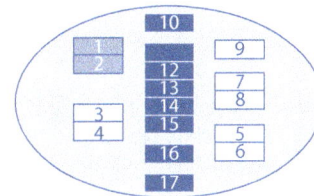

Next, the external tank requirements and the external tank ConOps are established, and functional diagrams are developed as was done in the first pass with the STS product. Finally, these diagrams, requirements, and ConOps documents are used to develop some feasible design solutions for the external tank. At this pass, there

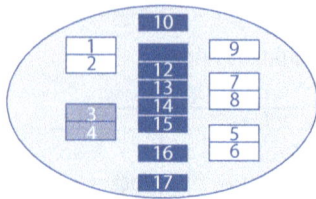

will also not be enough detail to actually build or prototype the external tank. The design solution might be summarized as, "To build this external tank, since our trade studies showed the best option was to use cryogenic propellants, a tank for the liquid hydrogen will be needed as will another tank for the liquid oxygen, instrumentation, and an outer structure of aluminum coated with foam." Thus, the tier 2 product tree for the external tank might look like Figure 2.3-3.

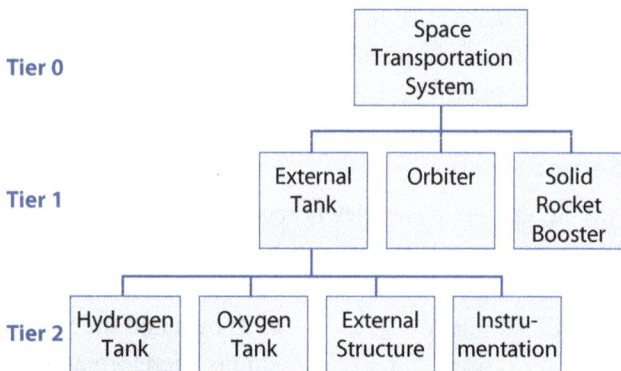

Figure 2.3-3 Product hierarchy, tier 2: external tank

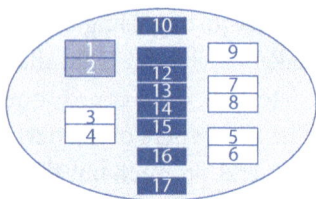

In a similar manner, the orbiter would also take another pass through the SE engine starting with identifying the stakeholders and their expectations, and generating a ConOps for the orbiter element. The tier 1 requirements that are applicable (allocated) to the orbiter would be "flowed down" and validated; new requirements derived from them and any additional requirements (including interfaces with the other elements) would be added.

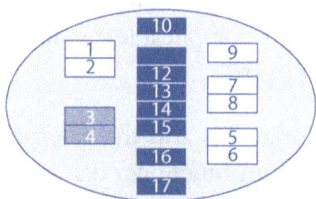

Next, the orbiter requirements and the ConOps are taken, functional diagrams are developed, and one or more feasible design solutions for the orbiter are generated. As with the external tank, at this pass, there will not be enough detail to actually build or do a complex model of the orbiter. The orbiter design solution might be summarized as, "To build this orbiter will require a winged vehicle with a thermal protection system; an avionics system; a guidance, navigation, and control system; a propulsion system; an environmental control system; etc." So the tier 2 product tree for the orbiter element might look like Figure 2.3-4.

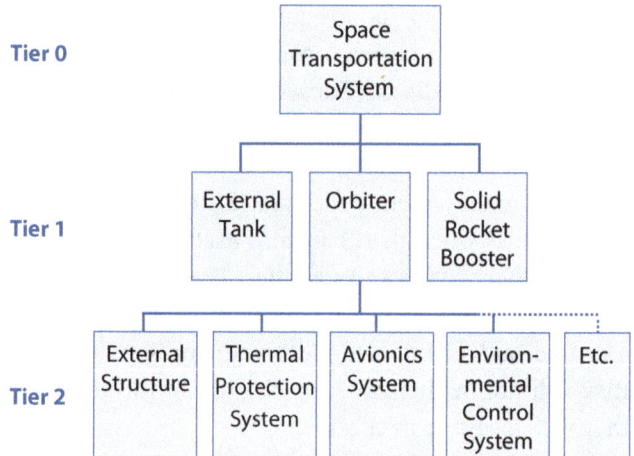

Figure 2.3-4 Product hierarchy, tier 2: orbiter

Likewise, the solid rocket booster would also be considered an end product, and a pass through the SE engine would generate a tier 2 design concept, just as was done with the external tank and the orbiter.

Third Pass

Each of the tier 2 elements is also considered an end product, and each undergoes another pass through the SE engine, defining stakeholders, generating ConOps, flowing down allocated requirements, generating new and derived requirements, and developing functional diagrams and design solution concepts. As an example of just the avionics system element, the tier 3 product hierarchy tree might look like Figure 2.3-5.

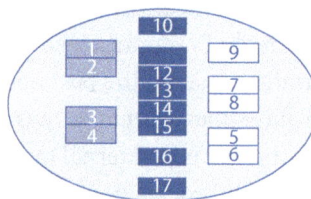

Passes 4 Through n

For this Phase A set of passes, this recursive process is continued for each product (model) on each tier down to the lowest level in the product tree. Note that in some projects it may not be feasible, given an estimated project cost and schedule, to perform this recursive process completely down to the smallest component during Phase A.

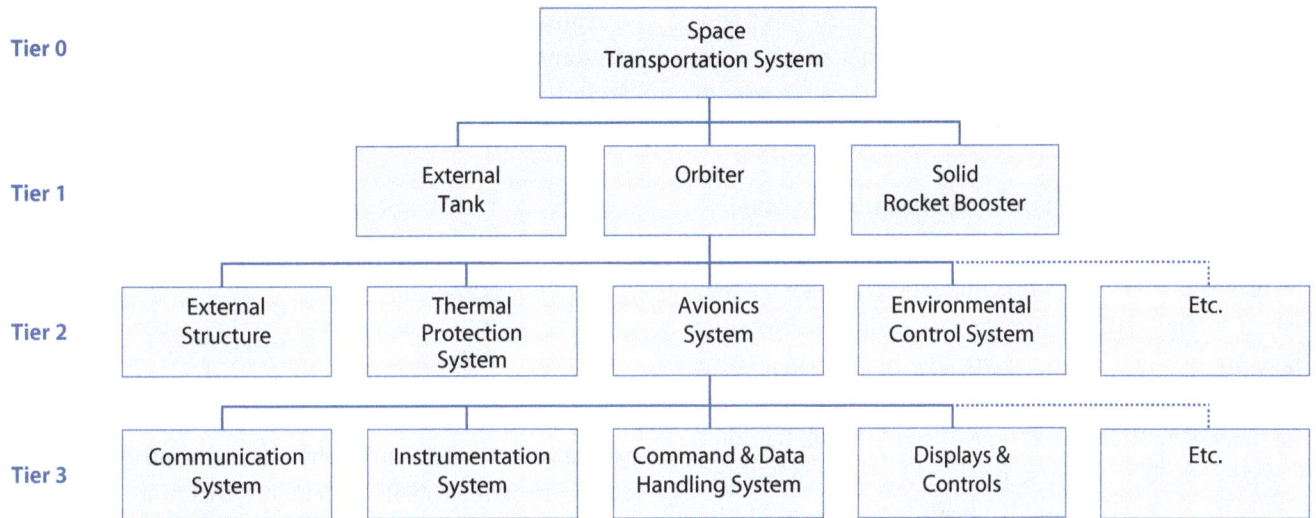

Figure 2.3-5 Product hierarchy, tier 3: avionics system

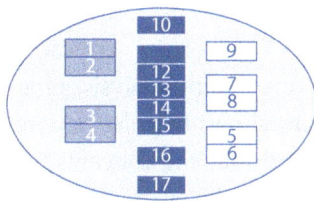

In these cases, engineering judgment must be used to determine what level of the product tier is feasible. Note that the lowest feasible level may occur at different tiers depending on the product-line complexity. For example, for one product line it may occur at tier 2; whereas, for a more complex product, it could occur at tier 8. This also means that it will take dif-ferent amounts of time to reach the bottom. Thus, for any given program or project, products will be at various stages of development. For this Phase A example, Figure 2.3-6 depicts the STS product hierarchy after completely passing through the system design processes side of the SE engine. At the end of this set of passes, system requirements, ConOps, and high-level conceptual functional and physical architectures for each product in the tree would exist. Note that these would not yet be the detailed or even preliminary designs for the end prod-

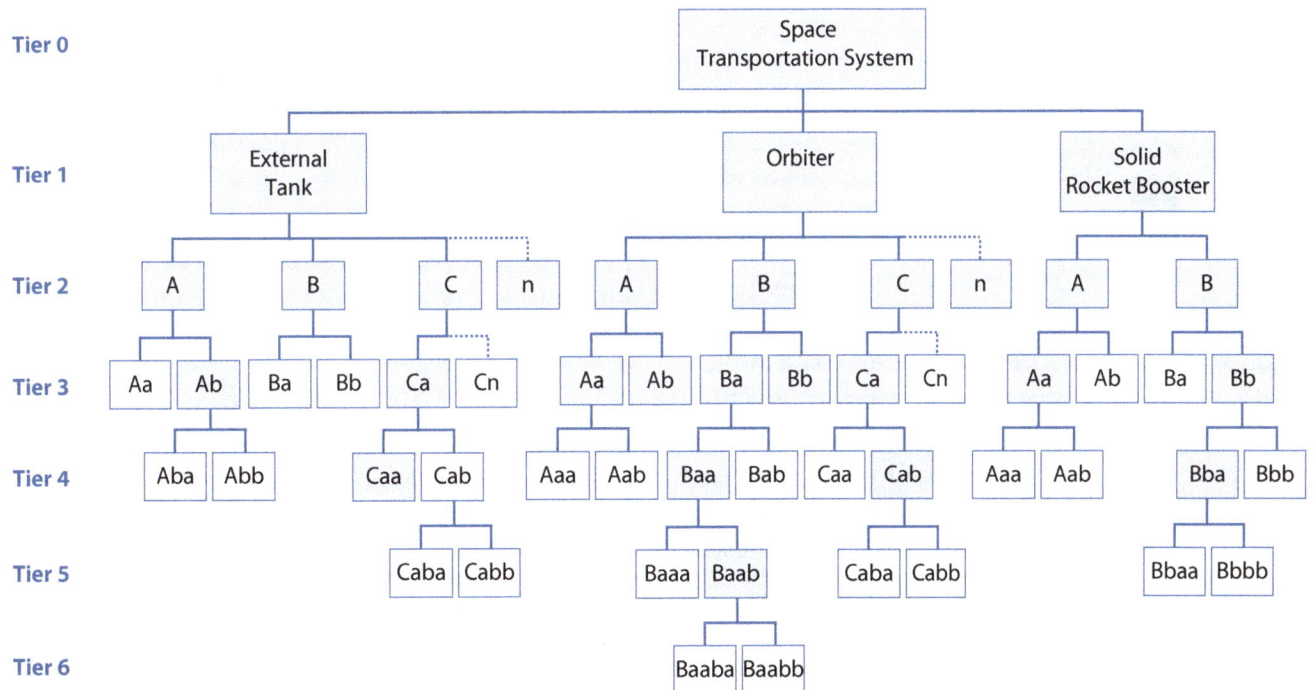

Figure 2.3-6 Product hierarchy: complete pass through system design processes side of the SE engine

Note: The unshaded boxes represent bottom-level phase products.

ucts. These will come later in the life cycle. At this point, enough conceptual design work has been done to ensure that at least the high-risk requirements are achievable as will be shown in the following passes.

2.3.2.2 Example Product Realization Passes

So now that the requirements and conceptual designs for the principal Phase A products have been developed, they need to be checked to ensure they are achievable. Note that there are two types of products. The first product is the "end product"—the one that will actually be delivered to the final user. The second type of product will be called a "phase product." A phase product is generated within a particular life-cycle phase that helps move the project toward delivering a final product. For example, while in Pre-Phase A, a foam-core mockup might be built to help visualize some of the concepts. Those mockups would not be the final "end product," but would be the "phase product." For this Phase A example, assume some computer models will be created and simulations performed of these key concepts to show that they are achievable. These will be the phase product for our example.

Now the focus shifts to the right side (i.e., product realization processes) of the SE engine, which will be applied recursively, starting at the bottom of the product hierarchy and moving upwards.

First Pass

Each of the phase products (i.e., our computer models) for the bottom-level product tier (ones that are unshaded in Figure 2.3-6) is taken individually and realized—that is, it is either bought, built, coded, or reused. For our example, assume the external tank product model Aa is a standard Commercial-Off-the-Shelf (COTS) product that is bought. Aba is a model that can be reused from another project, and product Abb is a model that will have to be developed with an in-house design that is to be built. Note that these models are parts of a larger model product that will be assembled or integrated on a subsequent runthrough of the SE engine. That is, to realize the model for product Ab of the external tank, models for products Aba and Abb must be first implemented and then later integrated together. This pass of the SE engine will be the realizing part. Likewise, each of the unshaded bottom-level model products

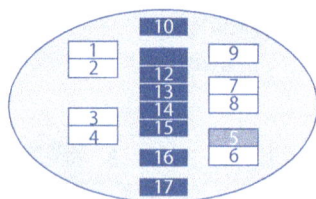

is realized in this first pass. The models will help us understand and plan the method to implement the final end product and will ensure the feasibility of the implemented method.

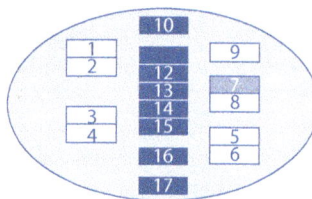

Next, each of the realized models (phase products) are used to verify that the end product would likely meet the requirements as defined in the Technical Requirements Definition Process during the system design pass for this product. This shows the product would likely meet the "shall" statements that were allocated, derived, or generated for it by method of test, analysis, inspection, or demonstration—that it was "built right." Verification is performed for each of the unshaded bottom-level model products. Note that during this Phase A pass, this process is not the formal verification of the final end product. However, using analysis, simulation, models, or other means shows that the requirements are good (verifiable) and that the concepts will most likely satisfy them. This also allows draft verification procedures of key areas to be developed. What can be formally verified, however, is that the phase product (the model) meets the requirements for the model.

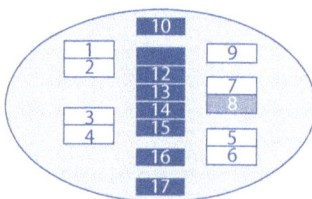

After the phase product (models) has been verified and used for planning the end product verification, the models are then used for validation. That is, additional test, analysis, inspection, or demonstrations are conducted to ensure that the proposed conceptual designs will likely meet the expectations of the stakeholders for this phase product and for the end product. This will track back to the ConOps that was mutually developed with the stakeholders during the Stakeholder Expectations Definition Process of the system design pass for this product. This will help ensure that the project has "built the right" product at this level.

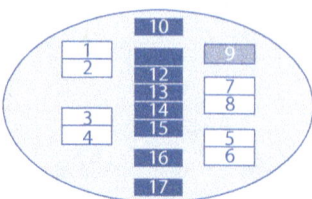

After verification and validation of the phase product (models) and using it for planning the verification and validation of the end product, it is time to prepare the model for transition to the next level up. Depending on complexity, where the model will be transitioned, security requirements, etc., transition may involve crating and shipment, transmitting over a network, or hand carrying over to the next lab. Whatever is appropriate, each model for the bottom-level product is prepared and handed to the next level up for further integration.

Second Pass

Now that all the models (phase products) for the bottom-level end products are realized, verified, validated, and transitioned, it is time to start integrating them into the next higher level product. For example, for the external tank, realized tier 4 models for product Aba and Abb are integrated to form the model for the tier 3 product Ab. Note that the Product Implementation Process only occurs at the bottommost product. All subsequent passes of the SE engine will employ the Product Integration Process since already realized products will be integrated to form the new higher level products. Integrating the lower tier phase products will result in the next-higher-tier phase product. This integration process can also be used for planning the integration of the final end products.

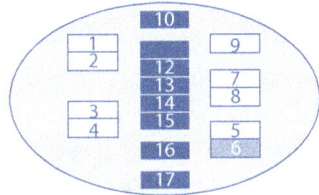

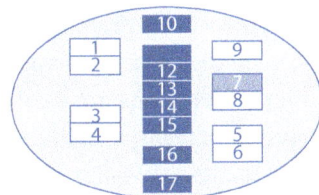

After the new integrated phase product (model) has been formed (tier 3 product Ab for example), it must now be proven that it meets its requirements. These will be the allocated, derived, or generated requirements developed during the Technical Requirements Definition Process during the system design pass for the model for this integrated product. This ensures that the integrated product was built (assembled) right. Note that just verifying the component parts (i.e., the individual models) that were used in the integration is not sufficient to assume that the integrated product will work right. There are many sources of problems that could occur—incomplete requirements at the interfaces, wrong assumptions during design, etc. The only sure way of knowing if an integrated product is good is to perform verification and validation at each stage. The knowledge gained from verifying this integrated phase product can also be used for planning the verification of the final end products.

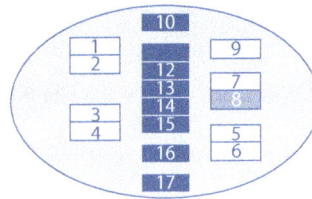

Likewise, after the integrated phase product is verified, it needs to be validated to show that it meets the expectations as documented in the ConOps for the model of the product at this level. Even though the component parts making up the integrated product will have been validated at this point, the only way to know that the project has built the "right" integrated product is to perform validation on the integrated product itself. Again, this information will help in the planning for the validation of the end products.

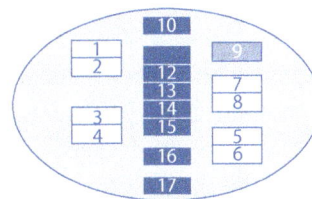

The model for the integrated phase product at this level (tier 3 product Ab for example) is now ready to be transitioned to the next higher level (tier 2 for the example). As with the products in the first pass, the integrated phase product is prepared according to its needs/requirements and shipped or handed over. In the example, the model for the external tank tier 3 integrated product Ab is transitioned to the owners of the model for the tier 2 product A. This effort with the phase products will be useful in planning for the transition of the end products.

Passes 3 Through n

In a similar manner as the second pass, the tier 3 models for the products are integrated together, realized, verified, validated, and transitioned to the next higher tier. For the example, the realized model for external tank tier 3 integrated phase product Ab is integrated with the model for tier 3 realized phase product Aa to form the tier 2 phase product A. Note that tier 3 product Aa is a bottom-tier product that has yet to go through the integration process. It may also have been realized some time ago and has been waiting for the Ab product line to become realized. Part of its transition might have been to place it in secure storage until the Ab product line became available. Or it could be that Aa was the long-lead item and product Ab had been completed some time ago and was waiting for the Aa purchase to arrive before they could be integrated together.

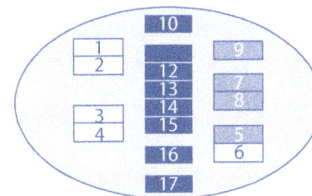

The length of the branch of the product tree does not necessarily translate to a corresponding length of time. This is why good planning in the first part of a project is so critical.

Final Pass

At some point, all the models for the tier 1 phase products will each have been used to ensure the system requirements and concepts developed during this Phase A cycle can be implemented, integrated, verified, validated, and transitioned. The elements are now defined as the external tank, the orbiter, and the solid rocket boosters. One final pass through the SE engine will show that they will likely be successfully implemented, integrated, verified, and validated. The final of these products—in the form of the baselined system requirements, ConOps, conceptual functional and physical designs—are made to provide inputs into the next life-cycle phase (B) where they will be further matured. In later phases, the products will actually be built into physical form. At this stage of the project, the key characteristics of each product are passed downstream in key SE documentation, as noted.

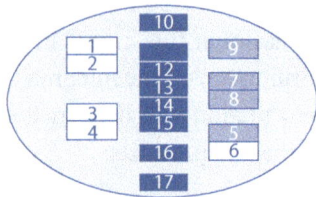

2.3.2.3 Example Use of the SE Engine in Phases B Through D

Phase B begins the preliminary design of the final end product. The recursive passes through the SE engine are repeated in a similar manner to that discussed in the detailed Phase A example. At this phase, the phase product might be a prototype of the product(s). Prototypes could be developed and then put through the planned verification and validation processes to ensure the design will likely meet all the requirements and expectations prior to the build of the final flight units. Any mistakes found on prototypes are much easier and less costly to correct than if not found until the flight units are built and undergoing the certification process.

Whereas the previous phases dealt with the final product in the form of analysis, concepts, or prototypes, Phases C and D work with the final end product itself. During Phase C, we recursively use the left side of the SE engine to develop the final design. In Phase D, we recursively use the right side of the SE engine to realize the final product and conduct the formal verification and validation of the

final product. As we come out of the last pass of the SE engine in Phase D, we have the final fully realized end product, the STS, ready to be delivered for launch.

2.3.2.4 Example Use of the SE Engine in Phases E and F

Even in Phase E (Operations and Sustainment) and Phase F (Closeout) of the life cycle, the technical management processes in the SE engine are still being used. During the operations phase of a project, a number of activities are still going on. In addition to the day-to-day use of the product, there is a need to monitor or manage various aspects of the system. This is where the key Technical Performance Measures (TPMs) that were defined in the early stages of development continue to play a part. (TPMs are described in Subsection 6.7.2.) These are great measures to monitor to ensure the product continues to perform as designed and expected. Configurations are still under control, still executing the Configuration Management Process. Decisions are still being made using the Decision Analysis Process. Indeed, all of the technical management processes still apply. For this discussion, the term "systems management" will be used for this aspect of operations. In addition to systems management and systems operation, there may also be a need for periodic refurbishment, repairing broken parts, cleaning, sparing, logistics, or other activities. Although other terms are used, for the purposes of this discussion the term "sustaining engineering" will be used for these activities. Again, all of the technical management processes still apply to these

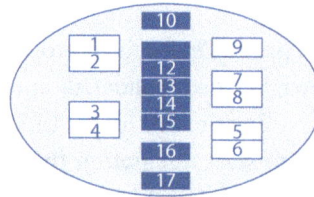

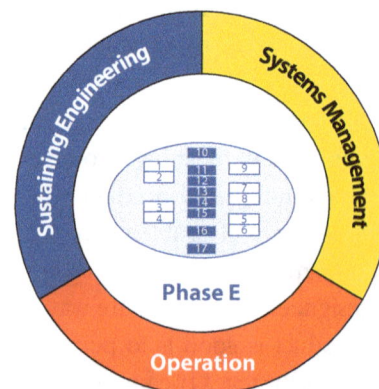

Figure 2.3-7 Model of typical activities during operational phase (Phase E) of a product

activities. Figure 2.3-7 represents these three activities occurring simultaneously and continuously throughout the operational lifetime of the final product. Some portions of the SE processes need to continue even after the system becomes nonoperational to handle retirement, decommissioning, and disposal. This is consistent with the basic SE principle of handling the full system life cycle from "cradle to grave."

However, if at any point in this phase a new product, a change that affects the design or certification of a product, or an upgrade to an existing product is needed, the development processes of the SE engine are reentered at the top. That is, the first thing that is done for an upgrade is to determine who the stakeholders are and what they expect. The entire SE engine is used just as for a newly developed product. This might be pictorially portrayed as in Figure 2.3-8. Note that in the figure although the SE engine is shown only once, it is used recursively down through the product hierarchy for upgraded products, just as described in our detailed example for the initial product.

2.4 Distinctions Between Product Verification and Product Validation

From a process perspective, the Product Verification and Product Validation Processes may be similar in nature, but the objectives are fundamentally different. Verifica-

tion of a product shows proof of compliance with requirements—that the product can meet each "shall" statement as proven though performance of a test, analysis, inspection, or demonstration. Validation of a product shows that the product accomplishes the intended purpose in the intended environment—that it meets the expectations of the customer and other stakeholders as shown through performance of a test, analysis, inspection, or demonstration.

Verification testing relates back to the approved requirements set and can be performed at different stages in the product life cycle. The approved specifications, drawings, parts lists, and other configuration documentation establish the configuration baseline of that product, which may have to be modified at a later time. Without a verified baseline and appropriate configuration controls, later modifications could be costly or cause major performance problems.

Validation relates back to the ConOps document. Validation testing is conducted under realistic conditions (or simulated conditions) on end products for the purpose of determining the effectiveness and suitability of the product for use in mission operations by typical users.

The selection of the verification or validation method is based on engineering judgment as to which is the most effective way to reliably show the product's conformance to requirements or that it will operate as intended and described in the ConOps.

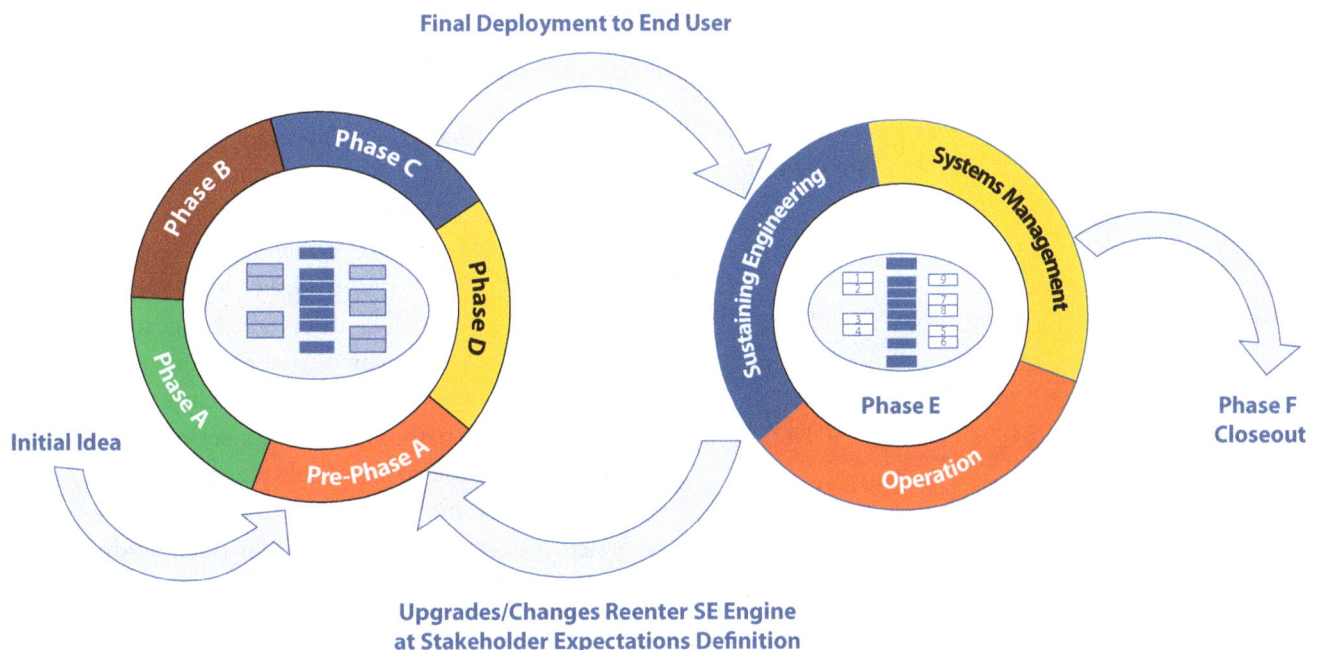

Figure 2.3-8 New products or upgrades reentering the SE engine

2.5 Cost Aspect of Systems Engineering

The objective of systems engineering is to see that the system is designed, built, and operated so that it accomplishes its purpose safely in the most cost-effective way possible considering performance, cost, schedule, and risk.

A cost-effective and safe system must provide a particular kind of balance between effectiveness and cost: the system must provide the most effectiveness for the resources expended, or equivalently, it must be the least expensive for the effectiveness it provides. This condition is a weak one because there are usually many designs that meet the condition. Think of each possible design as a point in the tradeoff space between effectiveness and cost. A graph plotting the maximum achievable effectiveness of designs available with current technology as a function of cost would, in general, yield a curved line such as the one shown in Figure 2.5-1. (In the figure, all the dimensions of effectiveness are represented by the ordinate (y axis) and all the dimensions of cost by the abscissa (x axis).) In other words, the curved line represents the envelope of the currently available technology in terms of cost-effectiveness.

Points above the line cannot be achieved with currently available technology; that is, they do not represent feasible designs. (Some of those points may be feasible in the future when further technological advances have been made.) Points inside the envelope are feasible, but are said to be dominated by designs whose combined cost and effectiveness lie on the envelope line. Designs represented by points on the envelope line are called cost-effective (or efficient or nondominated) solutions.

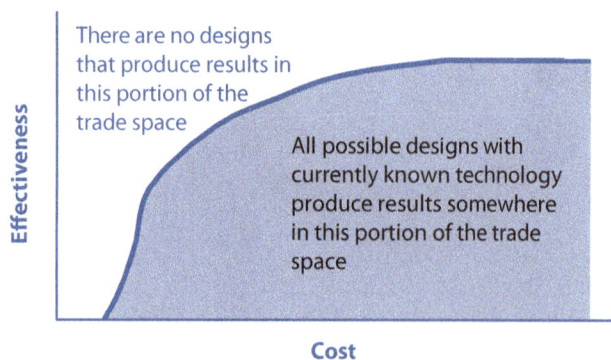

Figure 2.5-1 The enveloping surface of nondominated designs

> ### System Cost, Effectiveness, and Cost-Effectiveness
>
> - **Cost:** The cost of a system is the value of the resources needed to design, build, operate, and dispose of it. Because resources come in many forms—work performed by NASA personnel and contractors; materials; energy; and the use of facilities and equipment such as wind tunnels, factories, offices, and computers—it is convenient to express these values in common terms by using monetary units (such as dollars of a specified year).
> - **Effectiveness:** The effectiveness of a system is a quantitative measure of the degree to which the system's purpose is achieved. Effectiveness measures are usually very dependent upon system performance. For example, launch vehicle effectiveness depends on the probability of successfully injecting a payload onto a usable trajectory. The associated system performance attributes include the mass that can be put into a specified nominal orbit, the trade between injected mass and launch velocity, and launch availability.
> - **Cost-Effectiveness:** The cost-effectiveness of a system combines both the cost and the effectiveness of the system in the context of its objectives. While it may be necessary to measure either or both of those in terms of several numbers, it is sometimes possible to combine the components into a meaningful, single-valued *objective function* for use in design optimization. Even without knowing how to trade effectiveness for cost, designs that have lower cost and higher effectiveness are always preferred.

Design trade studies, an important part of the systems engineering process, often attempt to find designs that provide a better combination of the various dimensions of cost and effectiveness. When the starting point for a design trade study is inside the envelope, there are alternatives that either reduce costs with change to the overall effectiveness or alternatives that improve effectiveness without a cost increase (i.e., moving closer to the envelope curve). Then, the systems engineer's decision is easy. Other than in the sizing of subsystems, such "win-win" design trades are uncommon, but by no means rare. When the alternatives in a design trade study require trading cost for effectiveness, or even one dimension of effectiveness for another at the same cost (i.e., moving parallel to the envelope curve), the decisions become harder.

The process of finding the most cost-effective design is further complicated by uncertainty, which is shown in Figure 2.5-2. Exactly what outcomes will be realized by a particular system design cannot be known in advance with certainty, so the projected cost and effectiveness of a design are better described by a probability distribution than by a point. This distribution can be thought of as a cloud that is thickest at the most likely value and thinnest farthest away from the most likely point, as is shown for design concept A in the figure. Distributions resulting from designs that have little uncertainty are dense and highly compact, as is shown for concept B. Distributions associated with risky designs may have significant probabilities of producing highly undesirable outcomes, as is suggested by the presence of an additional low-effectiveness/high-cost cloud for concept C. (Of course, the envelope of such clouds cannot be a sharp line such as is shown in the figure, but must itself be rather fuzzy. The line can now be thought of as representing the envelope at some fixed confidence level, that is, a specific, numerical probability of achieving that effectiveness.)

Both effectiveness and cost may require several descriptors. Even the Echo balloons (circa 1960), in addition to their primary mission as communications satellites, obtained scientific data on the electromagnetic environment and atmospheric drag. Furthermore, Echo was the first satellite visible to the naked eye, an unquanti-

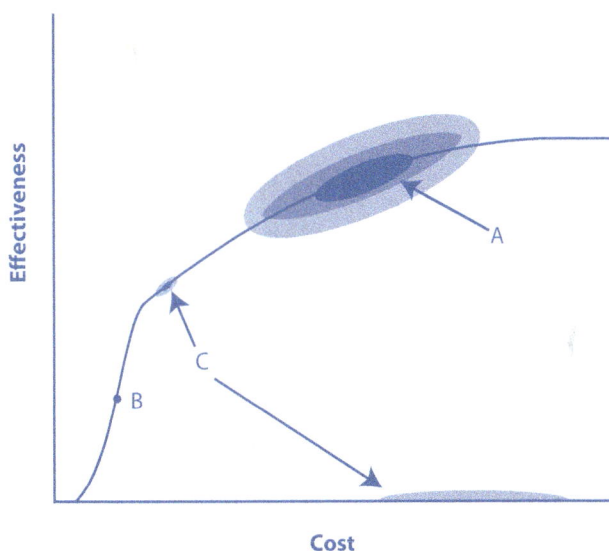

Figure 2.5-2 Estimates of outcomes to be obtained from several design concepts including uncertainty

Note: A, B, and C are design concepts with different risk patterns.

The Systems Engineer's Dilemma

At each cost-effective solution:

- To reduce cost at constant risk, performance must be reduced.
- To reduce risk at constant cost, performance must be reduced.
- To reduce cost at constant performance, higher risks must be accepted.
- To reduce risk at constant performance, higher costs must be accepted.

In this context, time in the schedule is often a critical resource, so that *schedule* behaves like a kind of *cost*.

---

fiable—but not unrecognized at the beginning of the space race—aspect of its effectiveness. Sputnik (circa 1957), for example, drew much of its effectiveness from the fact that it was a "first." Costs, the expenditure of limited resources, may be measured in the several dimensions of funding, personnel, use of facilities, and so on. Schedule may appear as an attribute of effectiveness or cost, or as a constraint. A mission to Mars that misses its launch window has to wait about two years for another opportunity—a clear schedule constraint.

In some contexts, it is appropriate to seek the most effectiveness possible within a fixed budget and with a fixed risk; in other contexts, it is more appropriate to seek the least cost possible with specified effectiveness and risk. In these cases, there is the question of what level of effectiveness to specify or what level of costs to fix. In practice, these may be mandated in the form of performance or cost requirements. It then becomes appropriate to ask whether a slight relaxation of requirements could produce a significantly cheaper system or whether a few more resources could produce a significantly more effective system.

The technical team must choose among designs that differ in terms of numerous attributes. A variety of methods have been developed that can be used to help uncover preferences between attributes and to quantify subjective assessments of relative value. When this can be done, trades between attributes can be assessed quantitatively. Often, however, the attributes seem to be truly incommensurate: decisions need to be made in spite of this multiplicity.

3.0 NASA Program/Project Life Cycle

One of the fundamental concepts used within NASA for the management of major systems is the program/project life cycle, which consists of a categorization of everything that should be done to accomplish a program or project into distinct phases, separated by Key Decision Points (KDPs). KDPs are the events at which the decision authority determines the readiness of a program/project to progress to the next phase of the life cycle (or to the next KDP). Phase boundaries are defined so that they provide more or less natural points for Go or No-Go decisions. Decisions to proceed may be qualified by liens that must be removed within an agreed to time period. A program or project that fails to pass a KDP may be allowed to "go back to the drawing board" to try again later—or it may be terminated.

All systems start with the recognition of a need or the discovery of an opportunity and proceed through various stages of development to a final disposition. While the most dramatic impacts of the analysis and optimization activities associated with systems engineering are obtained in the early stages, decisions that affect millions of dollars of value or cost continue to be amenable to the systems approach even as the end of the system lifetime approaches.

Decomposing the program/project life cycle into phases organizes the entire process into more manageable pieces. The program/project life cycle should provide managers with incremental visibility into the progress being made at points in time that fit with the management and budgetary environments.

NPR 7120.5, NASA Space Flight Program and Project Management Requirements defines the major NASA life-cycle phases as Formulation and Implementation. For Flight Systems and Ground Support (FS&GS) projects, the NASA life-cycle phases of Formulation and Implementation divide into the following seven incremental pieces. The phases of the project life cycle are:

- Pre-Phase A: Concept Studies (i.e., identify feasible alternatives)

- Phase A: Concept and Technology Development (i.e., define the project and identify and initiate necessary technology)
- Phase B: Preliminary Design and Technology Completion (i.e., establish a preliminary design and develop necessary technology)
- Phase C: Final Design and Fabrication (i.e., complete the system design and build/code the components)
- Phase D: System Assembly, Integration and Test, Launch (i.e., integrate components, and verify the system, prepare for operations, and launch)
- Phase E: Operations and Sustainment (i.e., operate and maintain the system)
- Phase F: Closeout (i.e., disposal of systems and analysis of data)

Figure 3.0-1 (NASA program life cycle) and Figure 3.0-2 (NASA project life cycle) identify the KDPs and reviews that characterize the phases. Sections 3.1 and 3.2 contain narrative descriptions of the purposes, major activities, products, and KDPs of the NASA program life-cycle phases. Sections 3.3 to 3.9 contain narrative descriptions of the purposes, major activities, products, and KDPs of the NASA project life-cycle phases. Section 3.10 describes the NASA budget cycle within which program/project managers and systems engineers must operate.

3.1 Program Formulation

The program Formulation phase establishes a cost-effective program that is demonstrably capable of meeting Agency and mission directorate goals and objectives. The program Formulation Authorization Document (FAD) authorizes a Program Manager (PM) to initiate the planning of a new program and to perform the analyses required to formulate a sound program plan. Major reviews leading to approval at KDP I are the P/SRR, P/SDR, PAR, and governing Program Management Council (PMC) review. (See full list of reviews in the program and project life cycle figures on the next page.) A summary of the required gate products for the pro-

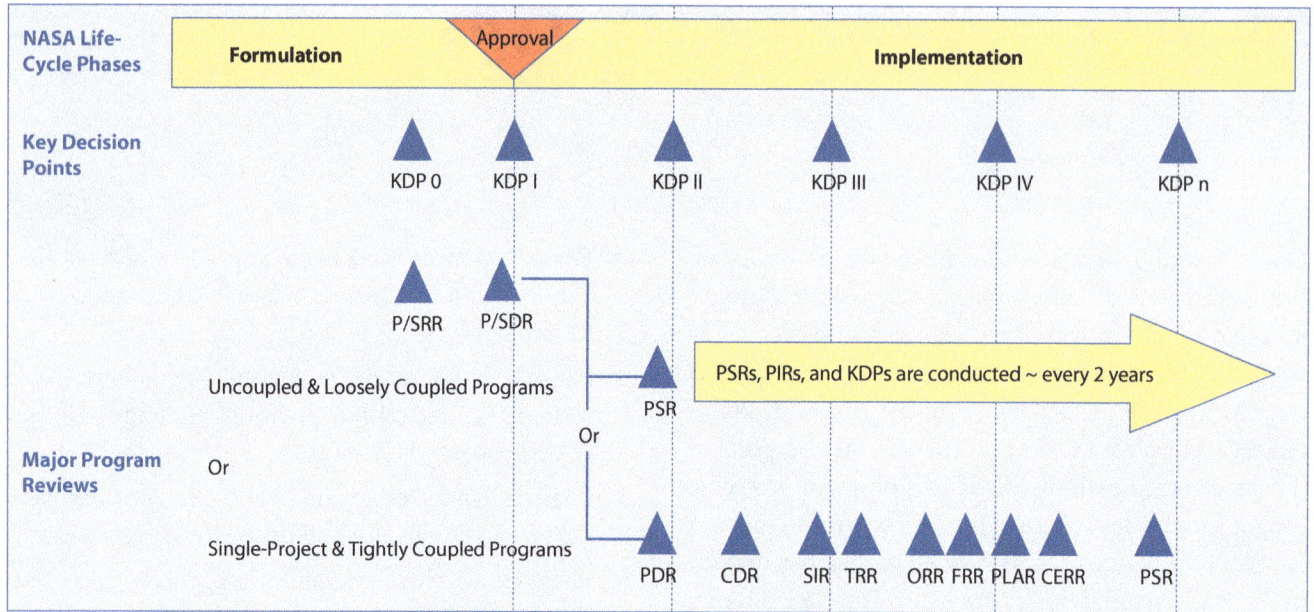

Figure 3.0-1 NASA program life cycle

CDR	Critical Design Review	PLAR	Post-Launch Assessment Review
CERR	Critical Events Readiness Review	PRR	Production Readiness Review
DR	Decommissioning Review	P/SDR	Program/System Definition Review
FRR	Flight Readiness Review	P/SRR	Program/System Requirements Review
KDP	Key Decision Point	PSR	Program Status Review
MCR	Mission Concept Review	SAR	System Acceptance Review
MDR	Mission Definition Review	SDR	System Definition Review
ORR	Operational Readiness Review	SIR	System Integration Review
PDR	Preliminary Design Review	SRR	System Requirements Review
PFAR	Post-Flight Assessment Review	TRR	Test Readiness Review
PIR	Program Implementation Review		

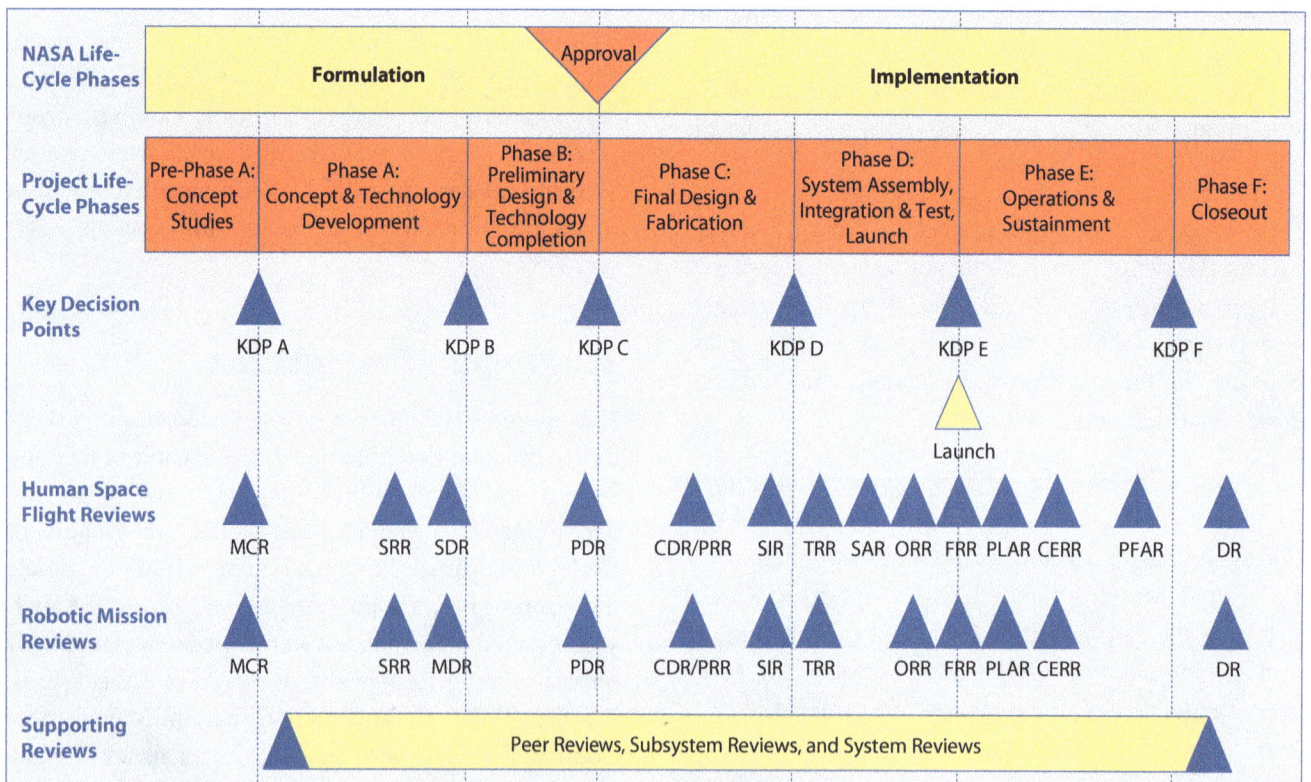

Figure 3.0-2 NASA project life cycle

Program Formulation

Purpose
To establish a cost-effective program that is demonstrably capable of meeting Agency and mission directorate goals and objectives

Typical Activities and Their Products
- Develop program requirements and allocate them to initial projects
- Define and approve program acquisition strategies
- Develop interfaces to other programs
- Start development of technologies that cut across multiple projects within the program
- Derive initial cost estimates and approve a program budget
- Perform required program Formulation technical activities defined in NPR 7120.5
- Satisfy program Formulation reviews' entrance/success criteria detailed in NPR 7123.1

Reviews
- P/SRR
- P/SDR

For uncoupled and loosely coupled programs, the Implementation phase only requires PSRs and PIRs to assess the program's performance and make a recommendation on its authorization at KDPs approximately every two years. Single-project and tightly coupled programs are more complex. For single-project programs, the Implementation phase program reviews shown in Figure 3.0-1 are synonymous (not duplicative) with the project reviews in the project life cycle (see Figure 3.0-2) through Phase D. Once in operations, these programs usually have biennial KDPs preceded by attendant PSRs/PIRs. Tightly coupled programs during implementation have program reviews tied to the project reviews to ensure the proper integration of projects into the larger system. Once in operations, tightly coupled programs also have biennial PSRs/PIRs/KDPs to assess the program's performance and authorize its continuation.

gram Formulation phase can be found in NPR 7120.5. Formulation for all program types is the same, involving one or more program reviews followed by KDP I where a decision is made approving a program to begin implementation. Typically, there is no incentive to move a program into implementation until its first project is ready for implementation.

3.2 Program Implementation

During the program Implementation phase, the PM works with the Mission Directorate Associate Administrator (MDAA) and the constituent project managers to execute the program plan cost effectively. Program reviews ensure that the program continues to contribute to Agency and mission directorate goals and objectives within funding constraints. A summary of the required gate products for the program Implementation phase can be found in NPR 7120.5. The program life cycle has two different implementation paths, depending on program type. Each implementation path has different types of major reviews.

Program Implementation

Purpose
To execute the program and constituent projects and ensure the program continues to contribute to Agency goals and objectives within funding constraints

Typical Activities and Their Products
- Initiate projects through direct assignment or competitive process (e.g., Request for Proposal (RFP), Announcement of Opportunity (AO))
- Monitor project's formulation, approval, implementation, integration, operation, and ultimate decommissioning
- Adjust program as resources and requirements change
- Perform required program Implementation technical activities from NPR 7120.5
- Satisfy program Implementation reviews' entrance/success criteria from NPR 7123.1

Reviews
- PSR/PIR (uncoupled and loosely coupled programs only)
- Reviews synonymous (not duplicative) with the project reviews in the project life cycle (see Figure 3.0-2) through Phase D (single-project and tightly coupled programs only)

3.3 Project Pre-Phase A: Concept Studies

The purpose of this phase, which is usually performed more or less continually by concept study groups, is to devise various feasible concepts from which new projects (programs) can be selected. Typically, this activity consists of loosely structured examinations of new ideas,

---

Pre-Phase A: Concept Studies

Purpose
To produce a broad spectrum of ideas and alternatives for missions from which new programs/projects can be selected

Typical Activities and Products
(Note: AO projects will have defined the deliverable products.)

- Identify missions and architecture consistent with charter
- Identify and involve users and other stakeholders
- Identify and perform tradeoffs and analyses
- Identify requirements, which include:
  - ▸ Mission,
  - ▸ Science, and
  - ▸ Top-level system.
- Define measures of effectiveness and measures of performance
- Identify top-level technical performance measures
- Perform preliminary evaluations of possible missions
- Prepare program/project proposals, which may include:
  - ▸ Mission justification and objectives;
  - ▸ Possible ConOps;
  - ▸ High-level WBSs;
  - ▸ Cost, schedule, and risk estimates; and
  - ▸ Technology assessment and maturation strategies.
- Prepare preliminary mission concept report
- Perform required Pre-Phase A technical activities from NPR 7120.5
- Satisfy MCR entrance/success criteria from NPR 7123.1

Reviews
- MCR
- Informal proposal review

---

usually without central control and mostly oriented toward small studies. Its major product is a list of suggested projects, based on the identification of needs and the discovery of opportunities that are potentially consistent with NASA's mission, capabilities, priorities, and resources.

Advanced studies may extend for several years and may be a sequence of papers that are only loosely connected. These studies typically focus on establishing mission goals and formulating top-level system requirements and ConOps. Conceptual designs are often offered to demonstrate feasibility and support programmatic estimates. The emphasis is on establishing feasibility and desirability rather than optimality. Analyses and designs are accordingly limited in both depth and number of options.

3.4 Project Phase A: Concept and Technology Development

During Phase A, activities are performed to fully develop a baseline mission concept and begin or assume responsibility for the development of needed technologies. This work, along with interactions with stakeholders, helps establish a mission concept and the program requirements on the project.

In Phase A, a team—often associated with a program or informal project office—readdresses the mission concept to ensure that the project justification and practicality are sufficient to warrant a place in NASA's budget. The team's effort focuses on analyzing mission requirements and establishing a mission architecture. Activities become formal, and the emphasis shifts toward establishing optimality rather than feasibility. The effort addresses more depth and considers many alternatives. Goals and objectives are solidified, and the project develops more definition in the system requirements, top-level system architecture, and ConOps. Conceptual designs are developed and exhibit more engineering detail than in advanced studies. Technical risks are identified in more detail, and technology development needs become focused.

In Phase A, the effort focuses on allocating functions to particular items of hardware, software, personnel, etc. System functional and performance requirements, along with architectures and designs, become firm as system tradeoffs and subsystem tradeoffs iterate back and forth

Phase A: Concept and Technology Development

Purpose
To determine the feasibility and desirability of a suggested new major system and establish an initial baseline compatibility with NASA's strategic plans

Typical Activities and Their Products
- Prepare and initiate a project plan
- Develop top-level requirements and constraints
- Define and document system requirements (hardware and software)
- Allocate preliminary system requirements to next lower level
- Define system software functionality description and requirements
- Define and document internal and external interface requirements
- Identify integrated logistics support requirements
- Develop corresponding evaluation criteria and metrics
- Document the ConOps
- Baseline the mission concept report
- Demonstrate that credible, feasible design(s) exist
- Perform and archive trade studies
- Develop mission architecture
- Initiate environmental evaluation/National Environmental Policy Act process
- Develop initial orbital debris assessment (NASA Safety Standard 1740.14)
- Establish technical resource estimates
- Define life-cycle cost estimates and develop system-level cost-effectiveness model
- Define the WBS
- Develop SOWs
- Acquire systems engineering tools and models
- Baseline the SEMP
- Develop system risk analyses
- Prepare and initiate a risk management plan
- Prepare and Initiate a configuration management plan
- Prepare and initiate a data management plan
- Prepare engineering specialty plans (e.g., contamination control plan, electromagnetic interference/electromagnetic compatibility control plan, reliability plan, quality control plan, parts management plan)
- Prepare a safety and mission assurance plan
- Prepare a software development or management plan (see NPR 7150.2)
- Prepare a technology development plan and initiate advanced technology development
- Establish human rating plan
- Define verification and validation approach and document it in verification and validation plans
- Perform required Phase A technical activities from NPR 7120.5
- Satisfy Phase A reviews' entrance/success criteria from NPR 7123.1

Reviews
- SRR
- MDR (robotic mission only)
- SDR (human space flight only)

in the effort to seek out more cost-effective designs. (Trade studies should precede—rather than follow—system design decisions.) Major products to this point include an accepted functional baseline for the system and its major end items. The effort also produces various engineering and management plans to prepare for managing the project's downstream processes, such as verification and operations, and for implementing engineering specialty programs.

3.5 Project Phase B: Preliminary Design and Technology Completion

During Phase B, activities are performed to establish an initial project baseline, which (according to NPR 7120.5 and NPR 7123.1) includes "a formal flow down of the project-level performance requirements to a complete set of system and subsystem design specifications for both flight and ground elements" and "corresponding preliminary designs." The technical requirements should be sufficiently detailed to establish firm schedule and cost estimates for the project. It also should be noted, especially for AO-driven projects, that Phase B is where the top-level requirements and the requirements flowed down to the next level are finalized and placed under configuration control. While the requirements should be baselined in Phase A, there are just enough changes resulting from the trade studies and analyses in late Phase A and early Phase B that changes are inevitable. However, by mid-Phase B, the top-level requirements should be finalized.

Actually, the Phase B baseline consists of a collection of evolving baselines covering technical and business aspects of the project: system (and subsystem) requirements and specifications, designs, verification and operations plans, and so on in the technical portion of the baseline, and schedules, cost projections, and management plans in the business portion. Establishment of baselines implies the implementation of configuration management procedures. (See Section 6.5.)

In Phase B, the effort shifts to establishing a functionally complete preliminary design solution (i.e., a functional baseline) that meets mission goals and objectives. Trade studies continue. Interfaces among the

---

Phase B: Preliminary Design and Technology Completion

Purpose

To define the project in enough detail to establish an initial baseline capable of meeting mission needs

Typical Activities and Their Products

- Baseline the project plan
- Review and update documents developed and baselined in Phase A
- Develop science/exploration operations plan based on matured ConOps
- Update engineering specialty plans (e.g., contamination control plan, electromagnetic interference/electromagnetic compatibility control plan, reliability plan, quality control plan, parts management plan)
- Update technology maturation planning
- Report technology development results
- Update risk management plan
- Update cost and schedule data
- Finalize and approve top-level requirements and flowdown to the next level of requirements
- Establish and baseline design-to specifications (hardware and software) and drawings, verification and validation plans, and interface documents at lower levels
- Perform and archive trade studies' results
- Perform design analyses and report results
- Conduct engineering development tests and report results
- Select a baseline design solution
- Baseline a preliminary design report
- Define internal and external interface design solutions (e.g., interface control documents)
- Define system operations as well as PI/contract proposal management, review, and access and contingency planning
- Develop appropriate level safety data package
- Develop preliminary orbital debris assessment
- Perform required Phase B technical activities from NPR 7120.5
- Satisfy Phase B reviews' entrance/success criteria from NPR 7123.1

Reviews

- PDR
- Safety review

major end items are defined. Engineering test items may be developed and used to derive data for further design work, and project risks are reduced by successful technology developments and demonstrations. Phase B culminates in a series of PDRs, containing the system-level PDR and PDRs for lower level end items as appropriate. The PDRs reflect the successive refinement of requirements into designs. (See the doctrine of successive refinement in Subsection 4.4.1.2 and Figure 4.4-2.) Design issues uncovered in the PDRs should be resolved so that final design can begin with unambiguous design-to specifications. From this point on, almost all changes to the baseline are expected to represent successive refinements, not fundamental changes. Prior to baselining, the system architecture, preliminary design, and ConOps must have been validated by enough technical analysis and design work to establish a credible, feasible design in greater detail than was sufficient for Phase A.

3.6 Project Phase C: Final Design and Fabrication

During Phase C, activities are performed to establish a complete design (allocated baseline), fabricate or produce hardware, and code software in preparation for integration. Trade studies continue. Engineering test units more closely resembling actual hardware are built and tested to establish confidence that the design will function in the expected environments. Engineering specialty analysis results are integrated into the design, and the manufacturing process and controls are defined and validated. All the planning initiated back in Phase A for the testing and operational equipment, processes and analysis, integration of the engineering specialty analysis, and manufacturing processes and controls is implemented. Configuration management continues to track and control design changes as detailed interfaces are defined. At each step in the successive refinement of the final design, corresponding integration and verification activities are planned in greater detail. During this phase, technical parameters, schedules, and budgets are closely tracked to ensure that undesirable trends (such as an unexpected growth in

spacecraft mass or increase in its cost) are recognized early enough to take corrective action. These activities focus on preparing for the CDR, PRR (if required), and the SIR.

Phase C contains a series of CDRs containing the system-level CDR and CDRs corresponding to the different levels of the system hierarchy. A CDR for each end item should be held prior to the start of fabrication/production for hardware and prior to the start of coding of deliverable software products. Typically, the sequence of CDRs reflects the integration process that will occur in the next phase—that is, from lower level CDRs to the system-level CDR. Projects, however, should tailor the sequencing of the reviews to meet the needs of the project. If there is a production run of products, a PRR will be performed to ensure the production plans, facilities, and personnel are ready to begin production. Phase C culminates with an SIR. The final product of this phase is a product ready for integration.

3.7 Project Phase D: System Assembly, Integration and Test, Launch

During Phase D, activities are performed to assemble, integrate, test, and launch the system. These activities focus on preparing for the FRR. Activities include assembly, integration, verification, and validation of the system, including testing the flight system to expected environments within margin. Other activities include the initial training of operating personnel and implementation of the logistics and spares planning. For flight projects, the focus of activities then shifts to prelaunch integration and launch. Although all these activities are conducted in this phase of a project, the planning for these activities was initiated in Phase A. The planning for the activities cannot be delayed until Phase D begins because the design of the project is too advanced to incorporate requirements for testing and operations. Phase D concludes with a system that has been shown to be capable of accomplishing the purpose for which it was created.

Phase C: Final Design and Fabrication

Purpose

To complete the detailed design of the system (and its associated subsystems, including its operations systems), fabricate hardware, and code software

Typical Activities and Their Products

- Update documents developed and baselined in Phase B
- Update interface documents
- Update mission operations plan based on matured ConOps
- Update engineering specialty plans (e.g., contamination control plan, electromagnetic interference/electromagnetic compatibility control plan, reliability plan, quality control plan, parts management plan)
- Augment baselined documents to reflect the growing maturity of the system, including the system architecture, WBS, and project plans
- Update and baseline production plans
- Refine integration procedures
- Baseline logistics support plan
- Add remaining lower level design specifications to the system architecture
- Complete manufacturing and assembly plans and procedures
- Establish and baseline build-to specifications (hardware and software) and drawings, verification and validation plans, and interface documents at all levels
- Baseline detailed design report
- Maintain requirements documents
- Maintain verification and validation plans
- Monitor project progress against project plans
- Develop verification and validation procedures
- Develop hardware and software detailed designs
- Develop the system integration plan and the system operation plan
- Develop the end-to-end information system design
- Develop spares planning
- Develop command and telemetry list
- Prepare launch site checkout and operations plans
- Prepare operations and activation plan
- Prepare system decommissioning/disposal plan, including human capital transition, for use in Phase F
- Finalize appropriate level safety data package
- Develop preliminary operations handbook
- Perform and archive trade studies
- Fabricate (or code) the product
- Perform testing at the component or subsystem level
- Identify opportunities for preplanned product improvement
- Baseline orbital debris assessment
- Perform required Phase C technical activities from NPR 7120.5
- Satisfy Phase C reviews' entrance/success criteria from NPR 7123.1

Reviews

- CDR
- PRR
- SIR
- Safety review

Phase D: System Assembly, Integration and Test, Launch

Purpose

To assemble and integrate the products and create the system, meanwhile developing confidence that it will be able to meet the system requirements; conduct launch and prepare for operations

Typical Activities and Their Products

- Integrate and verify items according to the integration and verification plans, yielding verified components and (sub-systems)
- Monitor project progress against project plans
- Refine verification and validation procedures at all levels
- Perform system qualification verifications
- Perform system acceptance verifications and validation(s) (e.g., end-to-end tests encompassing all elements (i.e., space element, ground system, data processing system)
- Perform system environmental testing
- Assess and approve verification and validation results
- Resolve verification and validation discrepancies
- Archive documentation for verifications and validations performed
- Baseline verification and validation report
- Baseline "as-built" hardware and software documentation
- Update logistics support plan
- Document lessons learned
- Prepare and baseline operator's manuals
- Prepare and baseline maintenance manuals
- Approve and baseline operations handbook
- Train initial system operators and maintainers
- Train on contingency planning
- Finalize and implement spares planning
- Confirm telemetry validation and ground data processing
- Confirm system and support elements are ready for flight
- Integrate with launch vehicle(s) and launch, perform orbit insertion, etc., to achieve a deployed system
- Perform initial operational verification(s) and validation(s)
- Perform required Phase D technical activities from NPR 7120.5
- Satisfy Phase D reviews' entrance/success criteria from NPR 7123.1

Reviews

- TRR (at all levels)
- SAR (human space flight only)
- ORR
- FRR
- System functional and physical configuration audits
- Safety review

3.8 Project Phase E: Operations and Sustainment

During Phase E, activities are performed to conduct the prime mission and meet the initially identified need and maintain support for that need. The products of the phase are the results of the mission. This phase encompasses the evolution of the system only insofar as that evolution does not involve major changes to the system architecture. Changes of that scope constitute new "needs," and

the project life cycle starts over. For large flight projects, there may be an extended period of cruise, orbit insertion, on-orbit assembly, and initial shakedown operations. Near the end of the prime mission, the project may apply for a mission extension to continue mission activities or attempt to perform additional mission objectives.

3.9 Project Phase F: Closeout

During Phase F, activities are performed to implement the systems decommissioning disposal planning and analyze any returned data and samples. The products of the phase are the results of the mission.

Phase F deals with the final closeout of the system when it has completed its mission; the time at which this occurs depends on many factors. For a flight system that returns to Earth with a short mission duration, closeout may require little more than deintegration of the hardware and its return to its owner. On flight projects of long duration, closeout may proceed according to established plans or may begin as a result of unplanned events, such as failures. Refer to *NPD 8010.3, Notification of Intent to Decommission or Terminate Operating Space Systems and Terminate Missions* for terminating an operating mission. Alternatively, technological advances may make it uneconomical to continue operating the system either in its current configuration or an improved one.

Phase E: Operations and Sustainment

Purpose
To conduct the mission and meet the initially identified need and maintain support for that need

Typical Activities and Their Products
- Conduct launch vehicle performance assessment
- Conduct in-orbit spacecraft checkout
- Commission and activate science instruments
- Conduct the intended prime mission(s)
- Collect engineering and science data
- Train replacement operators and maintainers
- Train the flight team for future mission phases (e.g., planetary landed operations)
- Maintain and approve operations and maintenance logs
- Maintain and upgrade the system
- Address problem/failure reports
- Process and analyze mission data
- Apply for mission extensions, if warranted, and conduct mission activities if awarded
- Prepare for deactivation, disassembly, decommissioning as planned (subject to mission extension)
- Complete post-flight evaluation reports
- Complete final mission report
- Perform required Phase E technical activities from NPR 7120.5
- Satisfy Phase E reviews' entrance/success criteria from NPR 7123.1

Reviews
- PLAR
- CERR
- PFAR (human space flight only)
- System upgrade review
- Safety review

Phase F: Closeout

Purpose
To implement the systems decommissioning/disposal plan developed in Phase C and analyze any returned data and samples

Typical Activities and Their Products
- Dispose of the system and supporting processes
- Document lessons learned
- Baseline mission final report
- Archive data
- Begin transition of human capital (if applicable)
- Perform required Phase F technical activities from NPR 7120.5
- Satisfy Phase F reviews' entrance/success criteria from NPR 7123.1

Reviews
- DR

To limit space debris, *NPR 8715.6, NASA Procedural Requirements for Limiting Orbital Debris* provides guidelines for removing Earth-orbiting robotic satellites from their operational orbits at the end of their useful life. For Low Earth Orbiting (LEO) missions, the satellite is usually deorbited. For small satellites, this is accomplished by allowing the orbit to slowly decay until the satellite eventually burns up in the Earth's atmosphere. Larger, more massive satellites and observatories must be designed to demise or deorbited in a controlled manner so that they can be safely targeted for impact in a remote area of the ocean. The Geostationary (GEO) satellites at 35,790 km above the Earth cannot be practically deorbited, so they are boosted to a higher orbit well beyond the crowded operational GEO orbit.

In addition to uncertainty as to when this part of the phase begins, the activities associated with safe closeout of a system may be long and complex and may affect the system design. Consequently, different options and strategies should be considered during the project's earlier phases along with the costs and risks associated with the different options.

3.10 Funding: The Budget Cycle

NASA operates with annual funding from Congress. This funding results, however, from a continuous rolling process of budget formulation, budget enactment, and finally, budget execution. NASA's *Financial Management Requirements (FMR) Volume 4* provides the concepts, the goals, and an overview of NASA's budget system of resource alignment referred to as Planning, Programming, Budgeting, and Execution (PPBE) and establishes guidance on the programming and budgeting phases of the PPBE process, which are critical to budget formulation for NASA. Volume 4 includes strategic budget planning and resources guidance, program review, budget development, budget presentation, and justification of estimates to the Office of Management and Budget (OMB) and to Congress. It also provides detailed descriptions of the roles and responsibilities for key players in each step of the process. It consolidates current legal, regulatory, and administrative policies and procedures applicable to NASA. A highly simplified representation of the typical NASA budget cycle is shown in Figure 3.10-1.

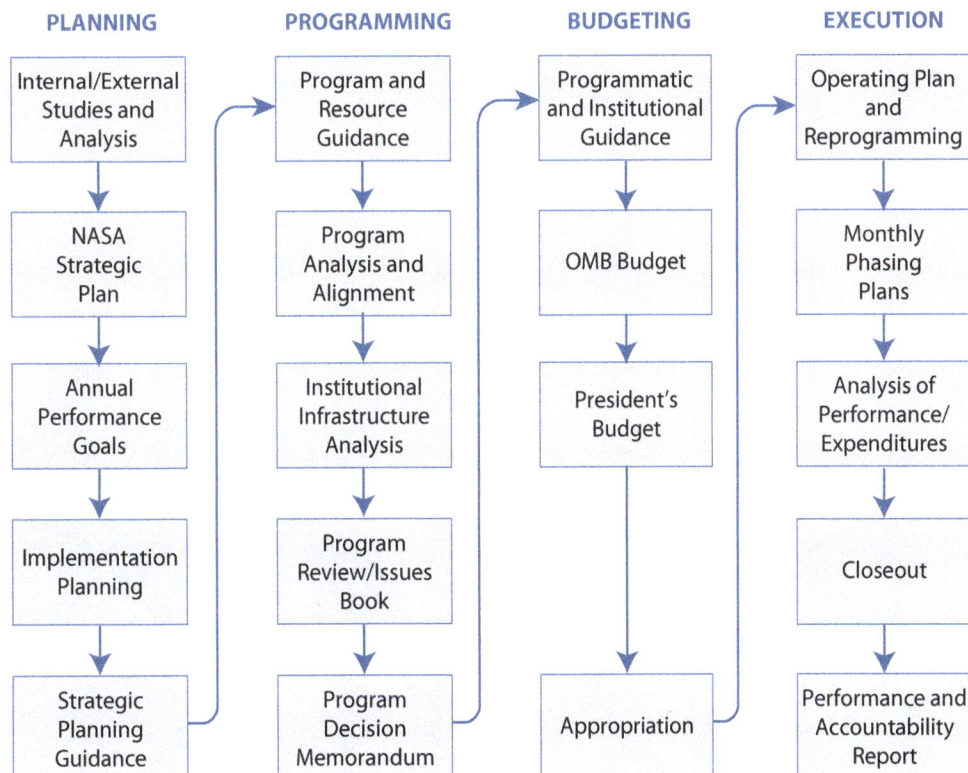

PLANNING
- Internal/External Studies and Analysis
- NASA Strategic Plan
- Annual Performance Goals
- Implementation Planning
- Strategic Planning Guidance

PROGRAMMING
- Program and Resource Guidance
- Program Analysis and Alignment
- Institutional Infrastructure Analysis
- Program Review/Issues Book
- Program Decision Memorandum

BUDGETING
- Programmatic and Institutional Guidance
- OMB Budget
- President's Budget
- Appropriation

EXECUTION
- Operating Plan and Reprogramming
- Monthly Phasing Plans
- Analysis of Performance/Expenditures
- Closeout
- Performance and Accountability Report

Figure 3.10-1 Typical NASA budget cycle

NASA typically starts developing its budget each February with economic forecasts and general guidelines as identified in the most recent President's budget. By late August, NASA has completed the planning, programming, and budgeting phases of the PPBE process and prepares for submittal of a preliminary NASA budget to the OMB. A final NASA budget is submitted to the OMB in September for incorporation into the President's budget transmittal to Congress, which generally occurs in January. This proposed budget is then subjected to congressional review and approval, culminating in the passage of bills authorizing NASA to obligate funds in accordance with congressional stipulations and appropriating those funds. The congressional process generally lasts through the summer. In recent years, however, final bills have often been delayed past the start of the fiscal year on October 1. In those years, NASA has operated on continuing resolution by Congress.

With annual funding, there is an implicit funding control gate at the beginning of every fiscal year. While these gates place planning requirements on the project and can make significant replanning necessary, they are not part of an orderly systems engineering process. Rather, they constitute one of the sources of uncertainty that affect project risks, and they are essential to consider in project planning.

4.0 System Design

This chapter describes the activities in the system design processes listed in Figure 2.1-1. The chapter is separated into sections corresponding to steps 1 to 4 listed in Figure 2.1-1. The processes within each step are discussed in terms of inputs, activities, and outputs. Additional guidance is provided using examples that are relevant to NASA projects. The system design processes are four interdependent, highly iterative and recursive processes, resulting in a validated set of requirements and a validated design solution that satisfies a set of stakeholder expectations. The four system design processes are to develop stakeholder expectations, technical requirements, logical decompositions, and design solutions.

Figure 4.0-1 illustrates the recursive relationship among the four system design processes. These processes start with a study team collecting and clarifying the stakeholder expectations, including the mission objectives, constraints, design drivers, operational objectives, and criteria for defining mission success. This set of stakeholder expectations and high-level requirements is used to drive an iterative design loop where a strawman architecture/design, the concept of operations, and derived requirements are developed. These three products must be consistent with each other and will require iterations and design decisions to achieve this consistency. Once consistency is achieved, analyses allow the project team to validate the design against the stakeholder expectations. A simplified validation asks the questions: Does the system work? Is the system safe and reliable? Is the system achievable within budget and schedule constraints? If the answer to any of these questions is no,

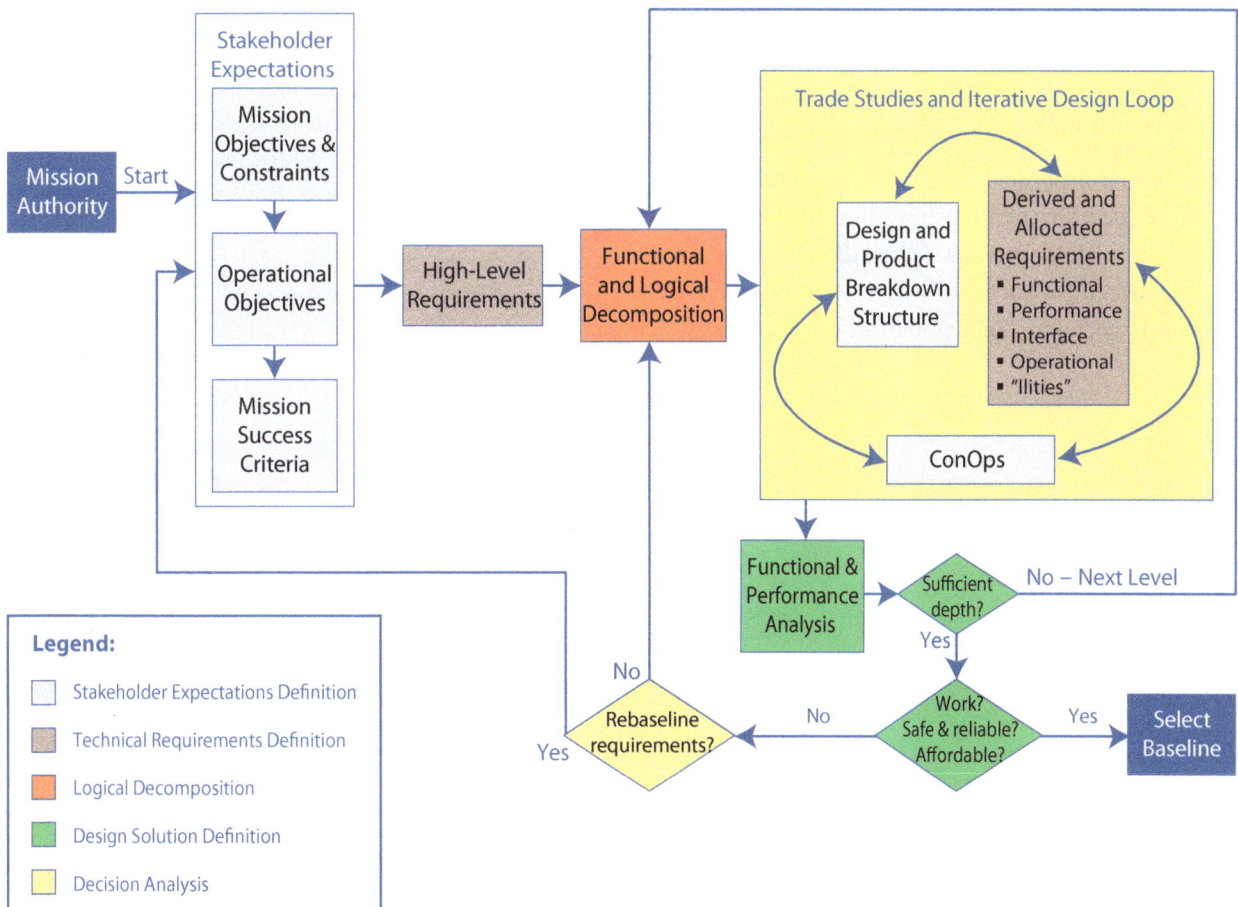

Figure 4.0-1 Interrelationships among the system design processes

then changes to the design or stakeholder expectations will be required, and the process started again. This process continues until the system—architecture, ConOps, and requirements—meets the stakeholder expectations.

The depth of the design effort must be sufficient to allow analytical verification of the design to the requirements. The design must be feasible and credible when judged by a knowledgeable independent review team and must have sufficient depth to support cost modeling.

Once the system meets the stakeholder expectations, the study team baselines the products and prepares for the next phase. Often, intermediate levels of decomposition are validated as part of the process. In the next level of decomposition, the baselined derived (and allocated) requirements become the set of high-level requirements for the decomposed elements and the process begins again. These system design processes are primarily applied in Pre-Phase A and continue through Phase C.

The system design processes during Pre-Phase A focus on producing a feasible design that will lead to Formulation approval. During Phase A, alternative designs and additional analytical maturity are pursued to optimize the design architecture. Phase B results in a preliminary design that satisfies the approval criteria. During Phase C, detailed, build-to designs are completed.

This has been a simplified description intended to demonstrate the recursive relationship among the system design processes. These processes should be used as guidance and tailored for each study team depending on the size of the project and the hierarchical level of the study team. The next sections describe each of the four system design processes and their associated products for a given NASA mission.

System Design Keys

- Successfully understanding and defining the mission objectives and operational concepts are keys to capturing the stakeholder expectations, which will translate into quality requirements over the life cycle of the project.

- Complete and thorough requirements traceability is a critical factor in successful validation of requirements.

- Clear and unambiguous requirements will help avoid misunderstanding when developing the overall system and when making major or minor changes.

- Document all decisions made during the development of the original design concept in the technical data package. This will make the original design philosophy and negotiation results available to assess future proposed changes and modifications against.

- The design solution verification occurs when an acceptable design solution has been selected and documented in a technical data package. The design solution is verified against the system requirements and constraints. However, the validation of a design solution is a continuing recursive and iterative process during which the design solution is evaluated against stakeholder expectations.

4.1 Stakeholder Expectations Definition

The Stakeholder Expectations Definition Process is the initial process within the SE engine that establishes the foundation from which the system is designed and the product is realized. The main purpose of this process is to identify who the stakeholders are and how they intend to use the product. This is usually accomplished through use-case scenarios, Design Reference Missions (DRMs), and ConOps.

4.1.1 Process Description

Figure 4.1-1 provides a typical flow diagram for the Stakeholder Expectations Definition Process and identifies typical inputs, outputs, and activities to consider in addressing stakeholder expectations definition.

4.1.1.1 Inputs

Typical inputs needed for the Stakeholder Expectations Definition Process would include the following:

- **Upper Level Requirements and Expectations:** These would be the requirements and expectations (e.g., needs, wants, desires, capabilities, constraints, external interfaces) that are being flowed down to a particular system of interest from a higher level (e.g., program, project, etc.).
- **Identified Customers and Stakeholders:** The organization or individual who has requested the product(s) and those who are affected by or are in some way accountable for the product's outcome.

4.1.1.2 Process Activities

Identifying Stakeholders

Advocacy for new programs and projects may originate in many organizations. These include Presidential directives, Congress, NASA Headquarters (HQ), the NASA Centers, NASA advisory committees, the National Academy of Sci-

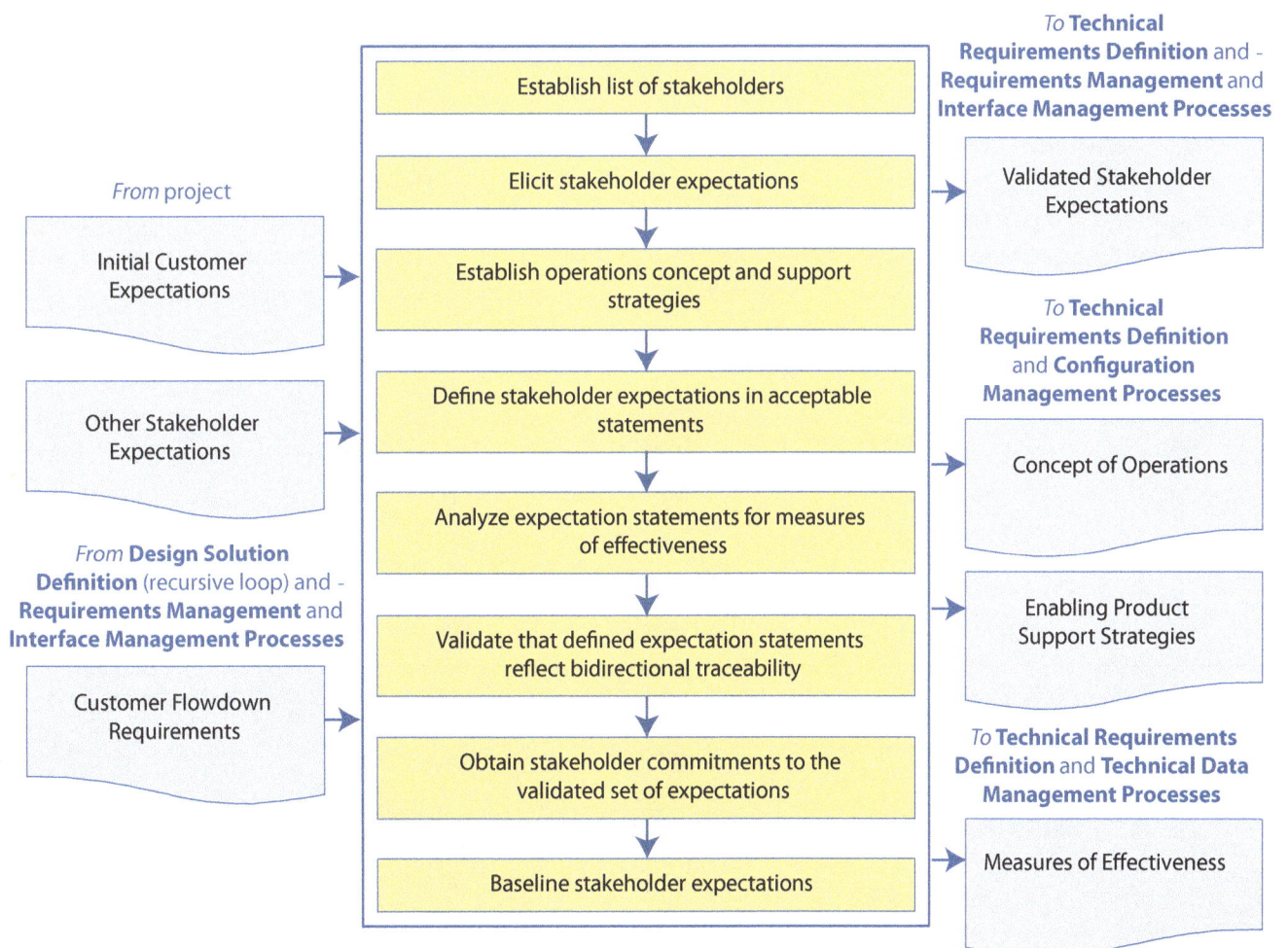

Figure 4.1-1 Stakeholder Expectations Definition Process

ences, the National Space Council, and many other groups in the science and space communities. These organizations are commonly referred to as stakeholders. A stakeholder is a group or individual who is affected by or is in some way accountable for the outcome of an undertaking.

Stakeholders can be classified as customers and other interested parties. Customers are those who will receive the goods or services and are the direct beneficiaries of the work. Examples of customers are scientists, project managers, and subsystems engineers.

Other interested parties are those who affect the project by providing broad, overarching constraints within which the customers' needs must be achieved. These parties may be affected by the resulting product, the manner in which the product is used, or have a responsibility for providing life-cycle support services. Examples include Congress, advisory planning teams, program managers, users, operators, maintainers, mission partners, and NASA contractors. It is important that the list of stakeholders be identified early in the process, as well as the primary stakeholders who will have the most significant influence over the project.

Identifying Stakeholder Expectations

Stakeholder expectations, the vision of a particular stakeholder individual or group, result when they specify what is desired as an end state or as an item to be produced and put bounds upon the achievement of the goals. These bounds may encompass expenditures (resources), time to deliver,

performance objectives, or other less obvious quantities such as organizational needs or geopolitical goals.

Figure 4.1-2 shows the type of information needed when defining stakeholder expectations and depicts how the information evolves into a set of high-level requirements. The yellow paths depict validation paths. Examples of the types of information that would be defined during each step are also provided.

Defining stakeholder expectations begins with the mission authority and strategic objectives that the mission is meant to achieve. *Mission authority* changes depending on the category of the mission. For example, science missions are usually driven by NASA Science Mission Directorate strategic plans; whereas the exploration missions may be driven by a Presidential directive.

An early task in defining stakeholder expectations is understanding the *objectives of the mission*. Clearly describing and documenting them helps ensure that the project team is working toward a common goal. These objectives form the basis for developing the mission, so they need to be clearly defined and articulated.

Defining the objectives is done by eliciting the needs, wants, desires, capabilities, external interfaces, assumptions, and constraints from the stakeholders. Arriving at an agreed-to set of objectives can be a long and arduous task. The proactive iteration with the stakeholders throughout the systems engineering process is the way

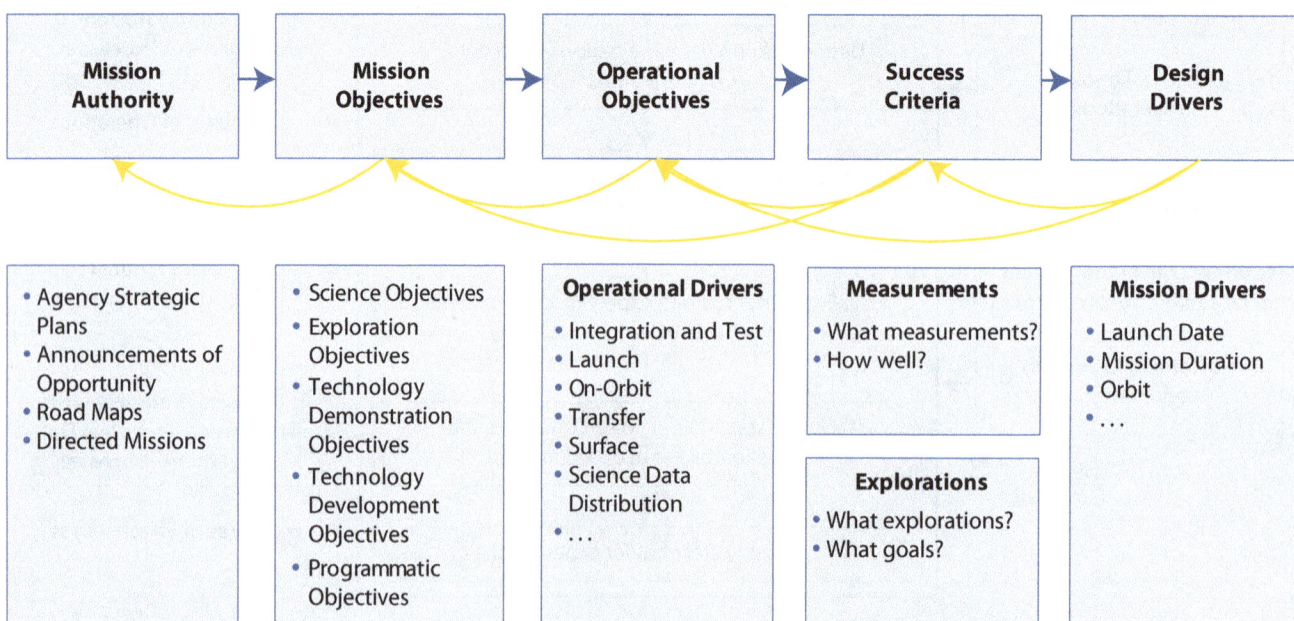

Figure 4.1-2 Product flow for stakeholder expectations

that all parties can come to a true understanding of what should be done and what it takes to do the job. It is important to know who the primary stakeholders are and who has the decision authority to help resolve conflicts.

The project team should also identify the *constraints* that may apply. A constraint is a condition that must be met. Sometimes a constraint is dictated by external factors such as orbital mechanics or the state of technology; sometimes constraints are the result of the overall budget environment. It is important to document the constraints and assumptions along with the mission objectives.

Operational objectives also need to be included in defining the stakeholder expectations. The operational objectives identify how the mission must be operated to achieve the mission objectives.

The *mission and operational success criteria* define what the mission must accomplish to be successful. This will be in the form of a measurement concept for science missions and exploration concept for human exploration missions. The success criteria also define how well the concept measurements or exploration activities must be accomplished. The success criteria capture the stakeholder expectations and, along with programmatic requirements and constraints, are used within the high-level requirements.

The *design drivers* will be strongly dependent upon the ConOps, including the operational environment, orbit, and mission duration requirements. For science missions, the design drivers may include, at a minimum, the mission launch date, duration, and orbit. If alternative orbits are to be considered, a separate concept is needed for each orbit. Exploration missions must consider the destination, the duration, the operational sequence (and system configuration changes), and the in situ exploration activities that allow the exploration to succeed.

The end result of this step is the discovery and delineation of the *system's goals*, which generally express the agree-

Note: It is extremely important to involve stakeholders in all phases of a project. Such involvement should be built in as a self-correcting feedback loop that will significantly enhance the chances of mission success. Involving stakeholders in a project builds confidence in the end product and serves as a validation and acceptance with the target audience.

ments, desires, and requirements of the eventual users of the system. The high-level requirements and success criteria are examples of the products representing the consensus of the stakeholders.

4.1.1.3 Outputs

Typical outputs for capturing stakeholder expectations would include the following:

- **Top-Level Requirements and Expectations:** These would be the top-level requirements and expectations (e.g., needs, wants, desires, capabilities, constraints, and external interfaces) for the product(s) to be developed.
- **ConOps:** This describes how the system will be operated during the life-cycle phases to meet stakeholder expectations. It describes the system characteristics from an operational perspective and helps facilitate an understanding of the system goals. Examples would be the ConOps document or a DRM.

4.1.2 Stakeholder Expectations Definition Guidance

4.1.2.1 Concept of Operations

The ConOps is an important component in capturing stakeholder expectations, requirements, and the architecture of a project. It stimulates the development of the requirements and architecture related to the user elements of the system. It serves as the basis for subsequent definition documents such as the operations plan, launch and early orbit plan, and operations handbook and provides the foundation for the long-range operational planning activities such as operational facilities, staffing, and network scheduling.

The ConOps is an important driver in the system requirements and therefore must be considered early in the system design processes. Thinking through the ConOps and use cases often reveals requirements and design functions that might otherwise be overlooked. A simple example to illustrate this point is adding system requirements to allow for communication during a particular phase of a mission. This may require an additional antenna in a specific location that may not be required during the nominal mission.

The ConOps is important for all projects. For science projects, the ConOps describes how the systems will be operated to achieve the measurement set required for a successful mission. They are usually driven by the data volume of the measurement set. The ConOps for exploration projects is likely to be more complex. There are typically more operational phases, more configuration changes, and additional communication links required for human interaction. For human spaceflight, functions and objectives must be clearly allocated between human operators and systems early in the project.

The ConOps should consider all aspects of operations including integration, test, and launch through disposal. Typical information contained in the ConOps includes a description of the major phases; operation timelines; operational scenarios and/or DRM; end-to-end communications strategy; command and data architecture; operational facilities; integrated logistic support (resupply, maintenance, and assembly); and critical events. The operational scenarios describe the dynamic view of the systems' operations and include how the system is perceived to function throughout the various modes and mode transitions, including interactions with external inter-

faces. For exploration missions, multiple DRMs make up a ConOps. The design and performance analysis leading to the requirements must satisfy all of them. Figure 4.1-3

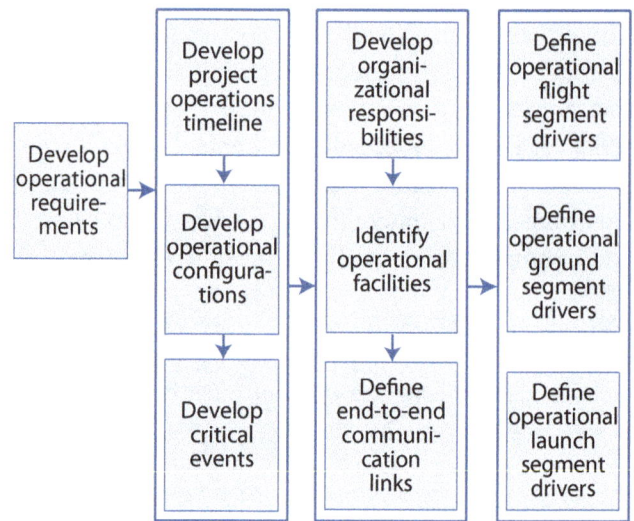

Figure 4.1-3 Typical ConOps development for a science mission

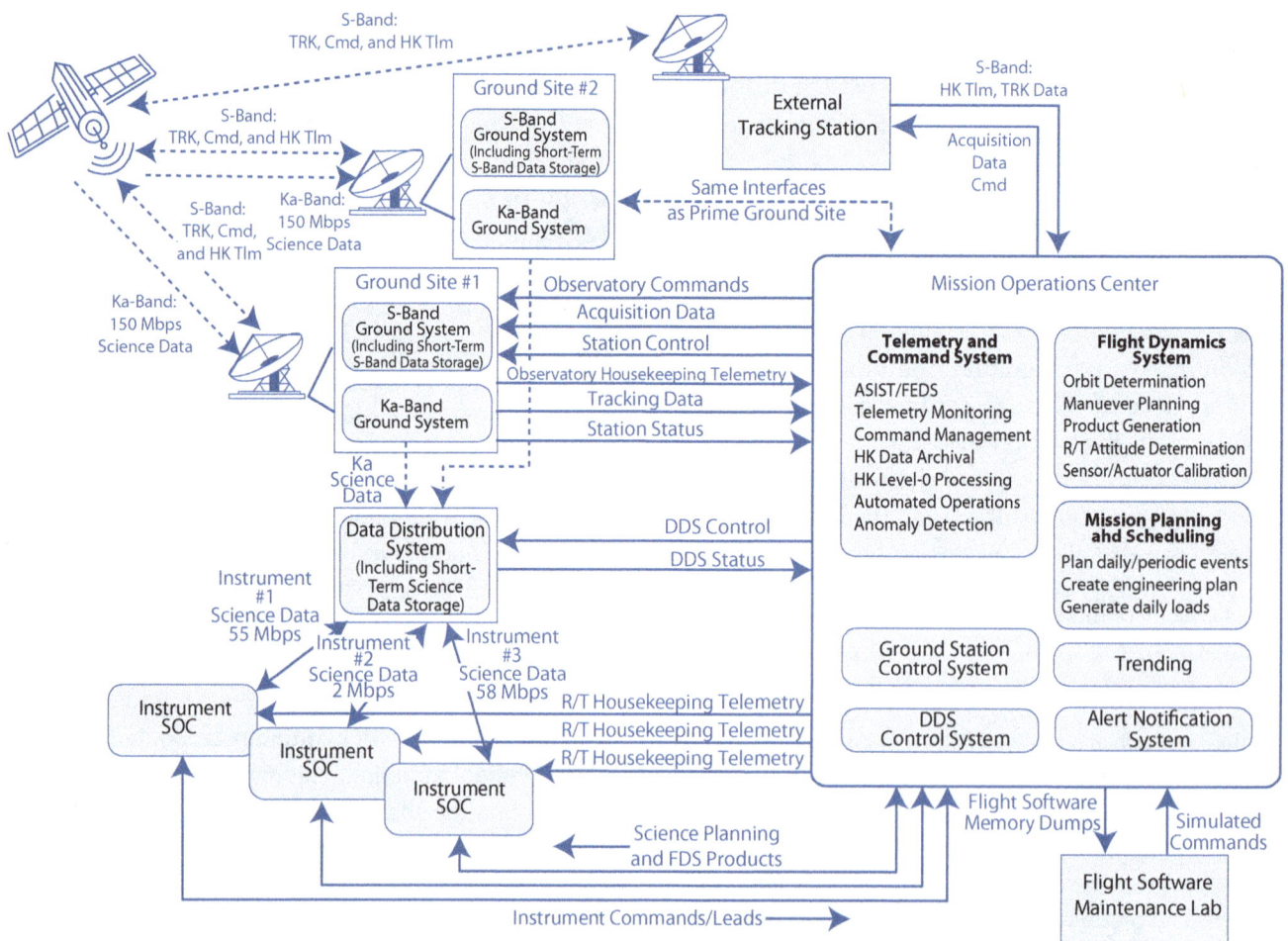

Figure 4.1-4 Example of an associated end-to-end operational architecture

illustrates typical information included in the ConOps for a science mission, and Figure 4.1-4 is an example of an end-to-end operational architecture. For more information about developing the ConOps, see ANSI/AIAA G-043-1992, *Guide for the Preparation of Operational Concept Documents.*

The operation timelines provide the basis for defining system configurations, operational activities, and other sequenced related elements necessary to achieve the mission objectives for each operational phase. It describes the activities, tasks, and other sequenced related elements necessary to achieve the mission objectives in each of the phases. Depending on the type of project (science, exploration, operational), the timeline could become quite complex.

The timeline matures along with the design. It starts as a simple time-sequenced order of the major events and matures into a detailed description of subsystem operations during all major mission modes or transitions. An example of a lunar sortie timeline and DRM early in the life cycle are shown in Figures 4.1-5a and b, respectively. An example of a more detailed, integrated time-

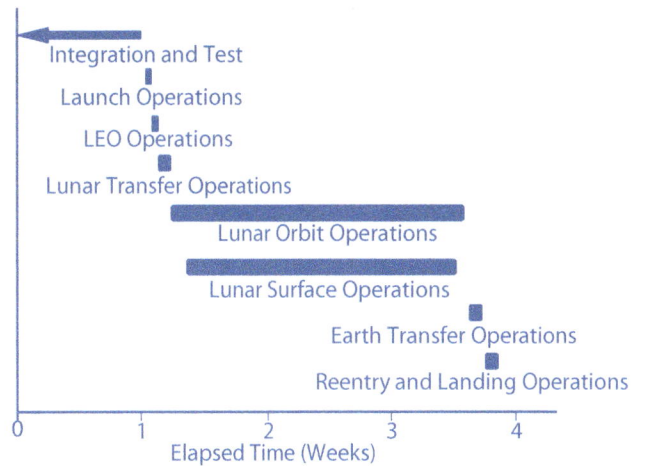

Figure 4.1-5a Example of a lunar sortie timeline developed early in the life cycle

line later in the life cycle for a science mission is shown in Figure 4.1-6.

An important part of the ConOps is defining the operational phases, which will span project Phases D, E, and F. The operational phases provide a time-sequenced

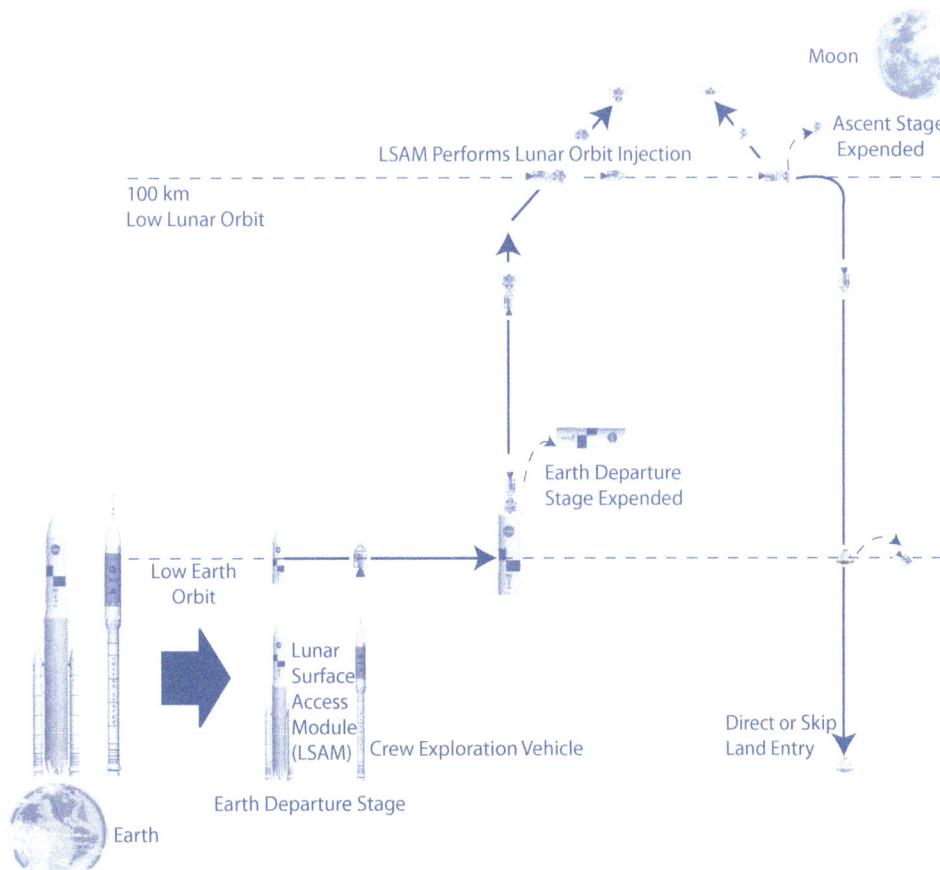

Figure 4.1-5b Example of a lunar sortie DRM early in the life cycle

structure for defining the configuration changes and operational activities needed to be carried out to meet the goals of the mission. For each of the operational phases, facilities, equipment, and critical events should also be included. Table 4.1-1 identifies some common examples of operational phases for a NASA mission.

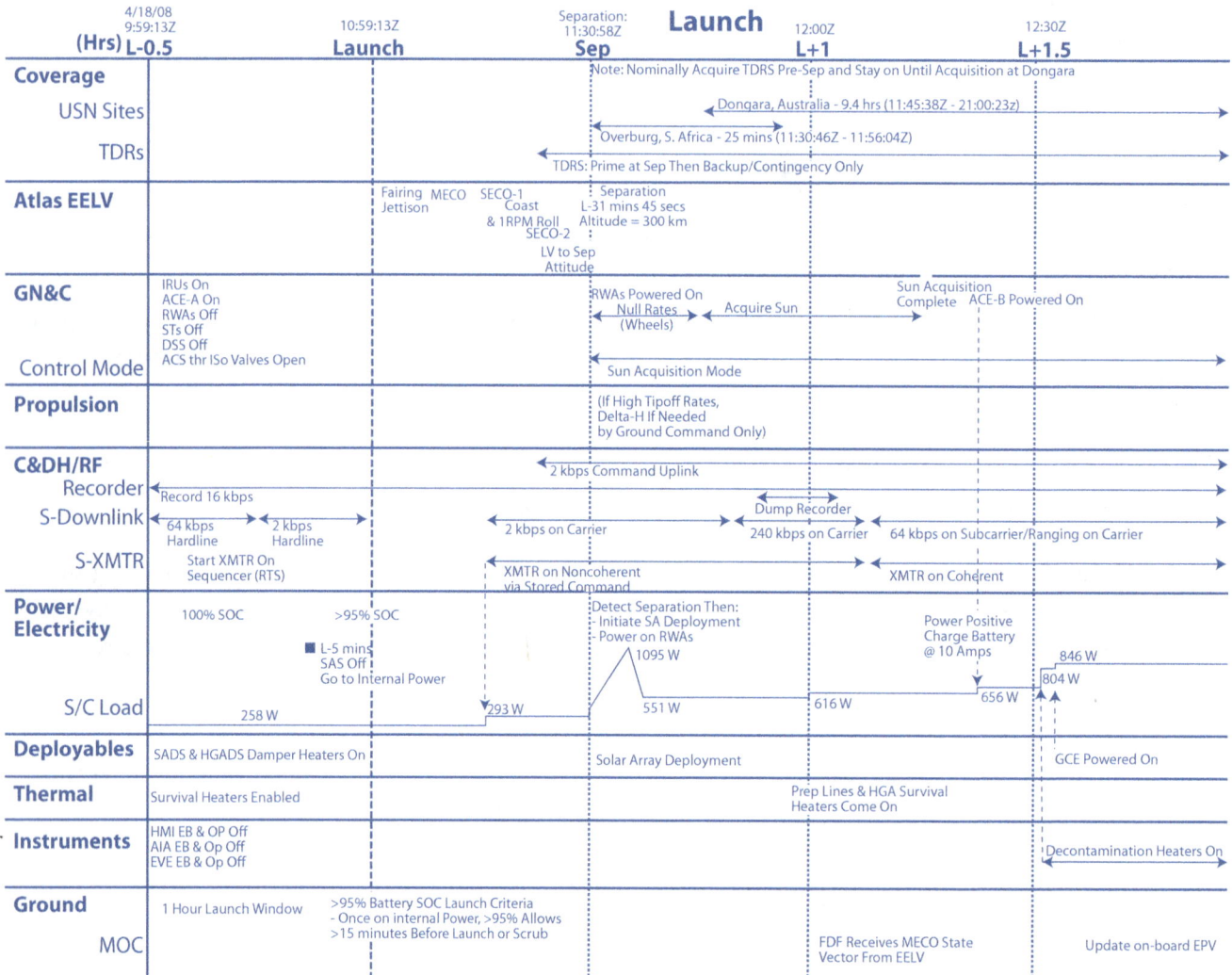

Figure 4.1-6 Example of a more detailed, integrated timeline later in the life cycle for a science mission

Table 4.1-1 Typical Operational Phases for a NASA Mission

Operational Phase	Description
Integration and test operations	**Project Integration and Test:** During the latter period of project integration and test, the system is tested by performing operational simulations during functional and environmental testing. The simulations typically exercise the end-to-end command and data system to provide a complete verification of system functionality and performance against simulated project operational scenarios.
	Launch Integration: The launch integration phase may repeat integration and test operational and functional verification in the launch-integrated configuration.
Launch operations	**Launch:** Launch operation occurs during the launch countdown, launch ascent, and orbit injection. Critical event telemetry is an important driver during this phase.
	Deployment: Following orbit injection, spacecraft deployment operations reconfigure the spacecraft to its orbital configuration. Typically, critical events covering solar array, antenna, and other deployments and orbit trim maneuvers occur during this phase.
	In-Orbit Checkout: In-orbit checkout is used to perform a verification that all systems are healthy. This is followed by on-orbit alignment, calibration, and parameterization of the flight systems to prepare for science operations.
Science operations	The majority of the operational lifetime is used to perform science operations.
Safe-hold operations	As a result of on-board fault detection or by ground command, the spacecraft may transition to a safe-hold mode. This mode is designed to maintain the spacecraft in a power positive, thermally stable state until the fault is resolved and science operations can resume.
Anomaly resolution and maintenance operations	Anomaly resolution and maintenance operations occur throughout the mission. They may require resources beyond established operational resources.
Disposal operations	Disposal operations occur at the end of project life. These operations are used to either provide a controlled reentry of the spacecraft or a repositioning of the spacecraft to a disposal orbit. In the latter case, the dissipation of stored fuel and electrical energy is required.

4.2 Technical Requirements Definition

The Technical Requirements Definition Process transforms the stakeholder expectations into a definition of the problem and then into a complete set of validated technical requirements expressed as "shall" statements that can be used for defining a design solution for the Product Breakdown Structure (PBS) model and related enabling products. The process of requirements definition is a recursive and iterative one that develops the stakeholders' requirements, product requirements, and lower level product/component requirements (e.g., PBS model products such as systems or subsystems and related enabling products such as external systems that provide or consume data). The requirements should enable the description of all inputs, outputs, and required relationships between inputs and outputs. The requirements documents organize and communicate requirements to the customer and other stakeholders and the technical community.

Technical requirements definition activities apply to the definition of all technical requirements from the program, project, and system levels down to the lowest level product/component requirements document.

> It is important to note that the team must not rely solely on the requirements received to design and build the system. Communication and iteration with the relevant stakeholders are essential to ensure a mutual understanding of each requirement. Otherwise, the designers run the risk of misunderstanding and implementing an unwanted solution to a different interpretation of the requirements.

4.2.1 Process Description

Figure 4.2-1 provides a typical flow diagram for the Technical Requirements Definition Process and identifies typical inputs, outputs, and activities to consider in addressing technical requirements definition.

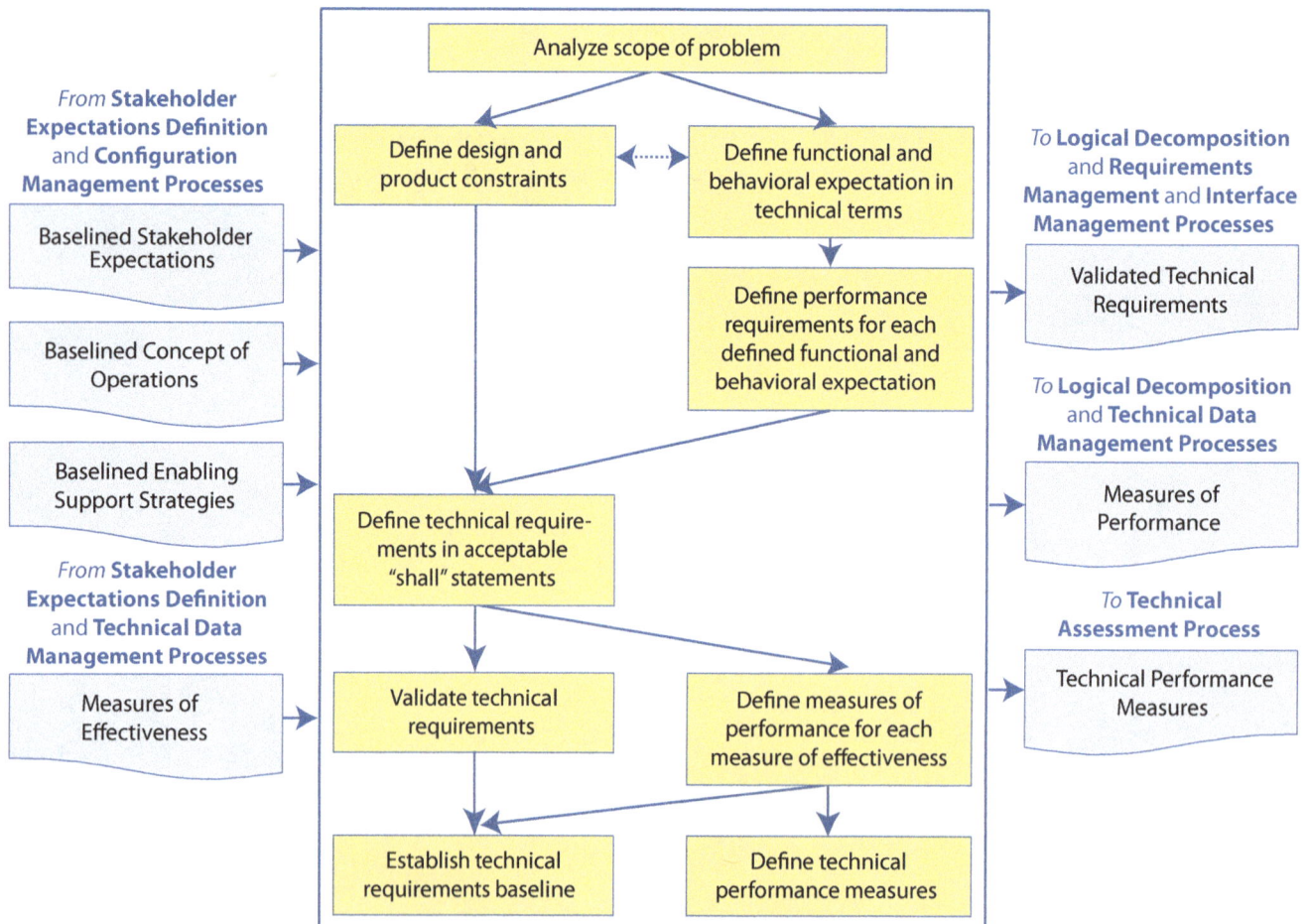

Figure 4.2-1 Technical Requirements Definition Process

4.2.1.1 Inputs

Typical inputs needed for the requirements process would include the following:

- **Top-Level Requirements and Expectations:** These would be the agreed-to top-level requirements and expectations (e.g., needs, wants, desires, capabilities, constraints, external interfaces) for the product(s) to be developed coming from the customer and other stakeholders.

- **Concept of Operations:** This describes how the system will be operated during the life-cycle phases to meet stakeholder expectations. It describes the system characteristics from an operational perspective and helps facilitate an understanding of the system goals. Examples would be a ConOps document or a DRM.

4.2.1.2 Process Activities

The top-level requirements and expectations are initially assessed to understand the technical problem to be solved and establish the design boundary. This boundary is typically established by performing the following activities:

- Defining constraints that the design must adhere to or how the system will be used. The constraints are typically not able to be changed based on tradeoff analyses.

- Identifying those elements that are already under design control and cannot be changed. This helps establish those areas where further trades will be performed to narrow potential design solutions.

- Establishing physical and functional interfaces (e.g., mechanical, electrical, thermal, human, etc.) with which the system must interact.

- Defining functional and behavioral expectations for the range of anticipated uses of the system as identified in the ConOps. The ConOps describes how the system will be operated and the possible use-case scenarios.

With an overall understanding of the constraints, physical/functional interfaces, and functional/behavioral expectations, the requirements can be further defined by establishing performance criteria. The performance is expressed as the quantitative part of the requirement to indicate how well each product function is expected to be accomplished.

Finally, the requirements should be defined in acceptable "shall" statements, which are complete sentences with a single "shall" per statement. See Appendix C for guidance on how to write good requirements and Appendix E for validating requirements. A well-written requirements document provides several specific benefits to both the stakeholders and the technical team, as shown in Table 4.2-1.

4.2.1.3 Outputs

Typical outputs for the Technical Requirements Definition Process would include the following:

- **Technical Requirements:** This would be the approved set of requirements that represents a complete description of the problem to be solved and requirements that have been validated and approved by the customer and stakeholders. Examples of documentation that capture the requirements are a System Requirements Document (SRD), Project Requirements Document (PRD), Interface Requirements Document (IRD), etc.

- **Technical Measures:** An established set of measures based on the expectations and requirements that will be tracked and assessed to determine overall system or product effectiveness and customer satisfaction. Common terms for these measures are Measures of Effectiveness (MOEs), Measures of Performance (MOPs), and Technical Performance Measures (TPMs). See Section 6.7 for further details.

4.2.2 Technical Requirements Definition Guidance

4.2.2.1 Types of Requirements

A complete set of project requirements includes the functional needs requirements (what functions need to be performed), performance requirements (how well these functions must be performed), and interface requirements (design element interface requirements). For space projects, these requirements are decomposed and allocated down to design elements through the PBS.

Functional, performance, and interface requirements are very important but do not constitute the entire set of requirements necessary for project success. The space segment design elements must also survive and continue to perform in the project environment. These environmental drivers include radiation, thermal, acoustic, mechanical loads, contamination, radio frequency, and others. In addition, reliability requirements drive design choices in design robustness, failure tolerance, and redundancy. Safety requirements drive design choices in providing diverse functional redundancy. Other spe-

Table 4.2-1 Benefits of Well-Written Requirements

Benefit	Rationale
Establish the basis for agreement between the stakeholders and the developers on what the product is to do	The complete description of the functions to be performed by the product specified in the requirements will assist the potential users in determining if the product specified meets their needs or how the product must be modified to meet their needs. During system design, requirements are allocated to subsystems (e.g., hardware, software, and other major components of the system), people, or processes.
Reduce the development effort because less rework is required to address poorly written, missing, and misunderstood requirements	The Technical Requirements Definition Process activities force the relevant stakeholders to consider rigorously all of the requirements before design begins. Careful review of the requirements can reveal omissions, misunderstandings, and inconsistencies early in the development cycle when these problems are easier to correct thereby reducing costly redesign, remanufacture, recoding, and retesting in later life-cycle phases.
Provide a basis for estimating costs and schedules	The description of the product to be developed as given in the requirements is a realistic basis for estimating project costs and can be used to evaluate bids or price estimates.
Provide a baseline for validation and verification	Organizations can develop their validation and verification plans much more productively from a good requirements document. Both system and subsystem test plans and procedures are generated from the requirements. As part of the development, the requirements document provides a baseline against which compliance can be measured. The requirements are also used to provide the stakeholders with a basis for acceptance of the system.
Facilitate transfer	The requirements make it easier to transfer the product to new users or new machines. Stakeholders thus find it easier to transfer the product to other parts of their organization, and developers find it easier to transfer it to new stakeholders or reuse it.
Serve as a basis for enhancement	The requirements serve as a basis for later enhancement or alteration of the finished product.

cialty requirements also may affect design choices. These may include producibility, maintainability, availability, upgradeability, human factors, and others. Unlike functional needs requirements, which are decomposed and allocated to design elements, these requirements are levied across major project elements. Designing to meet these requirements requires careful analysis of design alternatives. Figure 4.2-2 shows the characteristics of functional, operational, reliability, safety, and specialty requirements. Top-level mission requirements are generated from mission objectives, programmatic constraints, and assumptions. These are normally grouped into function and performance requirements and include the categories of requirements in Figure 4.2-2.

Functional Requirements

The functional requirements need to be specified for all intended uses of the product over its entire lifetime. Functional analysis is used to draw out both functional and performance requirements. Requirements are partitioned into groups, based on established criteria (e.g., similar functionality, performance, or coupling, etc.), to facilitate and focus the requirements analysis. Func-

> Functional requirements define what functions need to be done to accomplish the objectives.
>
> Performance requirements define how well the system needs to perform the functions.

tional and performance requirements are allocated to functional partitions and subfunctions, objects, people, or processes. Sequencing of time-critical functions is considered. Each function is identified and described in terms of inputs, outputs, and interface requirements from the top down so that the decomposed functions are recognized as part of larger functional groupings. Functions are arranged in a logical sequence so that any specified operational usage of the system can be traced in an end-to-end path to indicate the sequential relationship of all functions that must be accomplished by the system.

It is helpful to walk through the ConOps and scenarios asking the following types of questions: what functions need to be performed, where do they need to be performed, how often, under what operational and environ-

Figure 4.2-2 Characteristics of functional, operational, reliability, safety, and specialty requirements

mental conditions, etc. Thinking through this process often reveals additional functional requirements.

Performance Requirements

Performance requirements quantitatively define how well the system needs to perform the functions. Again, walking through the ConOps and the scenarios often draws out the performance requirements by asking the following types of questions: how often and how well, to what accuracy (e.g., how good does the measurement need to be), what is the quality and quantity of the output, under what stress (maximum simultaneous data

Example of Functional and Performance Requirements

Initial Function Statement

The Thrust Vector Controller (TVC) shall provide vehicle control about the pitch and yaw axes.

This statement describes a high-level function that the TVC must perform. The technical team needs to transform this statement into a set of design-to functional and performance requirements.

Functional Requirements with Associated Performance Requirements

- The TVC shall gimbal the engine a maximum of 9 degrees, ± 0.1 degree.
- The TVC shall gimbal the engine at a maximum rate of 5 degrees/second ± 0.3 degrees/second.
- The TVC shall provide a force of 40,000 pounds, ± 500 pounds.
- The TVC shall have a frequency response of 20 Hz, ± 0.1 Hz.

requests) or environmental conditions, for what duration, at what range of values, at what tolerance, and at what maximum throughput or bandwidth capacity.

Be careful not to make performance requirements too restrictive. For example, for a system that must be able to run on rechargeable batteries, if the performance requirements specify that the time to recharge should be less than 3 hours when a 12-hour recharge time would be sufficient, potential design solutions are eliminated. In the same sense, if the performance requirements specify that a weight must be within ±0.5 kg, when ±2.5 kg is sufficient, metrology cost will increase without adding value to the product.

Wherever possible, define the performance requirements in terms of (1) a threshold value (the minimum acceptable value needed for the system to carry out its mission) and (2) the baseline level of performance desired. Specifying performance in terms of thresholds and baseline requirements provides the system designers with trade space in which to investigate alternative designs.

All qualitative performance expectations must be analyzed and translated into quantified performance requirements. Trade studies often help quantify performance requirements. For example, tradeoffs can show whether a slight relaxation of the performance requirement could produce a significantly cheaper system or whether a few more resources could produce a significantly more effective system. The rationale for thresholds and goals should be documented with the requirements to understand the reason and origin for the performance requirement in case it must be changed. The performance requirements that can be quantified by or changed by tradeoff analysis should be identified. See Section 6.8, Decision Analysis, for more information on tradeoff analysis.

Interface Requirements

It is important to define all interface requirements for the system, including those to enabling systems. The external interfaces form the boundaries between the product and the rest of the world. Types of interfaces include: operational command and control, computer to computer, mechanical, electrical, thermal, and data. One useful tool in defining interfaces is the context diagram (see Appendix F), which depicts the product and all of its external interfaces. Once the product components are defined, a block diagram showing the major components, interconnections, and external interfaces of the system should be developed to define both the components and their interactions.

Interfaces associated with all product life-cycle phases should also be considered. Examples include interfaces with test equipment; transportation systems; Integrated Logistics Support (ILS) systems; and manufacturing facilities, operators, users, and maintainers.

As the technical requirements are defined, the interface diagram should be revisited and the documented interface requirements refined to include newly identified interfaces information for requirements both internal and external. More information regarding interfaces can be found in Section 6.3.

Environmental Requirements

Each space mission has a unique set of environmental requirements that apply to the flight segment elements. It is a critical function of systems engineering to identify the external and internal environments for the particular mission, analyze and quantify the expected environments, develop design guidance, and establish a margin philosophy against the expected environments.

The environments envelope should consider what can be encountered during ground test, storage, transportation, launch, deployment, and normal operations from beginning of life to end of life. Requirements derived from the mission environments should be included in the system requirements.

External and internal environment concerns that must be addressed include acceleration, vibration, shock, static loads, acoustic, thermal, contamination, crew-induced loads, total dose radiation/radiation effects, Single-Event Effects (SEEs), surface and internal charging, orbital debris, atmospheric (atomic oxygen) control and quality, attitude control system disturbance (atmospheric drag, gravity gradient, and solar pressure), magnetic, pressure gradient during launch, microbial growth, and radio frequency exposure on the ground and on orbit.

The requirements structure must address the specialty engineering disciplines that apply to the mission environments across project elements. These discipline areas levy requirements on system elements regarding Electromagnetic Interference, Electromagnetic Compatibility (EMI/EMC), grounding, radiation and other shielding, contamination protection, and reliability.

Reliability Requirements

Reliability can be defined as the probability that a device, product, or system will not fail for a given period of time under specified operating conditions. Reliability is an inherent system design characteristic. As a principal contributing factor in operations and support costs and in system effectiveness, reliability plays a key role in determining the system's cost-effectiveness.

Reliability engineering is a major specialty discipline that contributes to the goal of a cost-effective system. This is primarily accomplished in the systems engineering process through an active role in implementing specific design features to ensure that the system can perform in the predicted physical environments throughout the mission, and by making independent predictions of system reliability for design trades and for test program, operations, and integrated logistics support planning.

Reliability requirements ensure that the system (and subsystems, e.g., software and hardware) can perform in the predicted environments and conditions as expected throughout the mission and that the system has the ability to withstand certain numbers and types of faults, errors, or failures (e.g., withstand vibration, predicted data rates, command and/or data errors, single-event upsets, and temperature variances to specified limits). Environments can include ground (transportation and handling), launch, on-orbit (Earth or other), planetary, reentry, and landing, or they might be for software within certain modes or states of operation. Reliability addresses design and verification requirements to meet the requested level of operation as well as fault and/or failure tolerance for all expected environments and conditions. Reliability requirements cover fault/failure prevention, detection, isolation, and recovery.

Safety Requirements

NASA uses the term "safety" broadly to include human (public and workforce), environmental, and asset safety. There are two types of safety requirements—deterministic and risk-informed. A deterministic safety requirement is the qualitative or quantitative definition of a threshold of action or performance that must be met by a mission-related design item, system, or activity for that item, system, or activity to be acceptably safe. Examples of deterministic safety requirements are incorporation of safety devices (e.g., build physical hardware stops into the system to prevent the hydraulic lift/arm from extending past allowed safety height and length limits); limits on the range of values a system input variable is allowed to take on; and limit checks on input commands to ensure they are within specified safety limits or constraints for that mode or state of the system (e.g., the command to retract the landing gear is only allowed if the airplane is in the airborne state). For those components identified as "safety critical," requirements include functional redundancy or failure tolerance to allow the system to meet its requirements in the presence of one or more failures or to take the system to a safe state with reduced functionality (e.g., dual redundant computer processors, safe-state backup processor); detection and automatic system shutdown if specified values (e.g., temperature) exceed prescribed safety limits; use of only a subset that is approved for safety-critical software of a particular computer language; caution or warning devices; and safety procedures. A risk-informed safety requirement is a requirement that has been established, at least in part, on the basis of the consideration of safety-related TPMs and their associated uncertainty. An example of a risk-informed safety requirement is the Probability of Loss of Crew (P(LOC)) not exceeding a certain value "p" with a certain confidence level. Meeting safety requirements involves identification and elimination of hazards, reducing the likelihood of the accidents associated with hazards, or reducing

the impact from the hazard associated with these accidents to within acceptable levels. (For additional information concerning safety, see, for example, *NPR 8705.2, Human-Rating Requirements for Space Systems, NPR 8715.3, NASA General Safety Program Requirements*, and *NASA-STD-8719.13, Software Safety Standard*.)

4.2.2.2 Human Factors Engineering Requirements

In human spaceflight, the human—as operator and as maintainer—is a critical component of the mission and system design. Human capabilities and limitations must enter into designs in the same way that the properties of materials and characteristics of electronic components do. Human factors engineering is the discipline that studies human-system interfaces and interactions and provides requirements, standards, and guidelines to ensure the entire system can function as designed with effective accommodation of the human component.

Humans are initially integrated into systems through analysis of the overall mission. Mission functions are allocated to humans as appropriate to the system architecture, technical capabilities, cost factors, and crew capabilities. Once functions are allocated, human factors analysts work with system designers to ensure that human operators and maintainers are provided the equipment, tools, and interfaces to perform their assigned tasks safely and effectively.

NASA-STD-3001, NASA Space Flight Human System Standards Volume 1: Crew Health ensures that systems are safe and effective for humans. The standards focus on the human integrated with the system, the measures needed (rest, nutrition, medical care, exercise, etc.) to ensure that the human stays healthy and effective, the workplace environment, and crew-system physical and cognitive interfaces.

4.2.2.3 Requirements Decomposition, Allocation, and Validation

Requirements are decomposed in a hierarchical structure starting with the highest level requirements imposed by Presidential directives, mission directorates, program, Agency, and customer and other stakeholders. These high-level requirements are decomposed into functional and performance requirements and allocated across the system. These are then further decomposed and allocated among the elements and subsystems. This decomposition and allocation process continues until a complete set of design-to requirements is achieved. At each level of decomposition (system, subsystem, component, etc.), the total set of derived requirements must be validated against the stakeholder expectations or higher level parent requirements before proceeding to the next level of decomposition.

The traceability of requirements to the lowest level ensures that each requirement is necessary to meet the stakeholder expectations. Requirements that are not allocated to lower levels or are not implemented at a lower level result in a design that does not meet objectives and is, therefore, not valid. Conversely, lower level requirements that are not traceable to higher level requirements

result in an overdesign that is not justified. This hierarchical flowdown is illustrated in Figure 4.2-3.

Figure 4.2-4 is an example of how science pointing requirements are successively decomposed and allocated from the top down for a typical science mission. It is important to understand and document the relationship between requirements. This will reduce the possibility of misinterpretation and the possibility of an unsatisfactory design and associated cost increases.

Throughout Phases A and B, changes in requirements and constraints will occur. It is imperative that all changes be thoroughly evaluated to determine the impacts on both higher and lower hierarchical levels. All changes must be

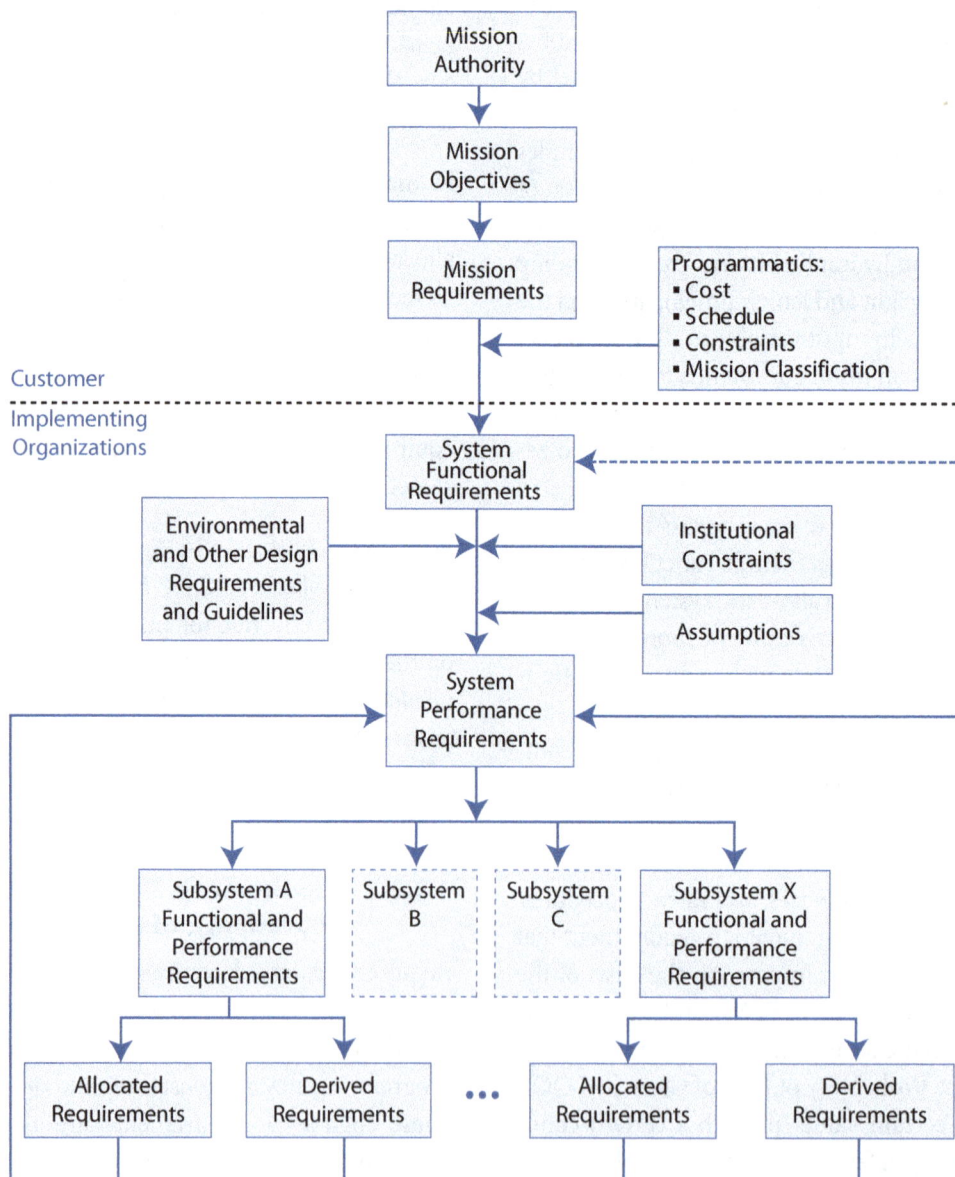

Figure 4.2-3 The flowdown of requirements

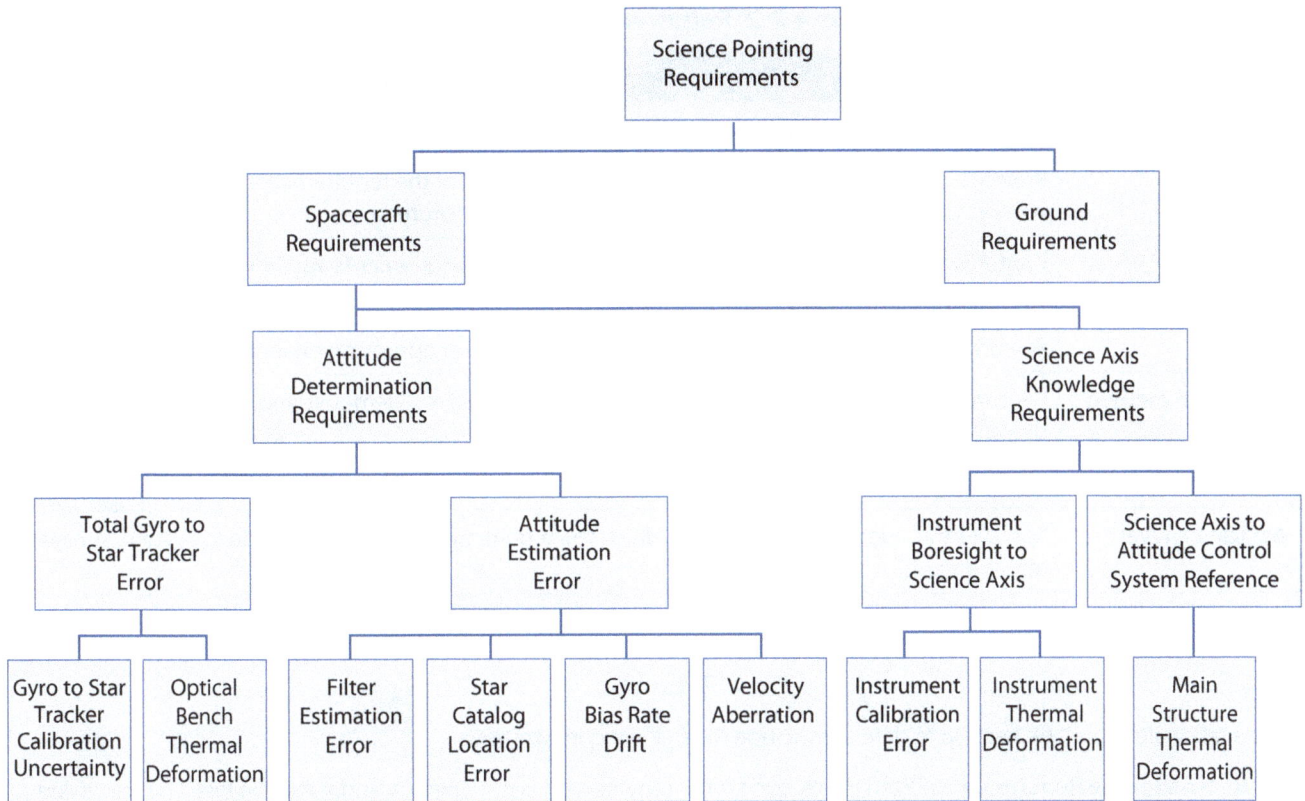

Figure 4.2-4 Allocation and flowdown of science pointing requirements

subjected to a review and approval cycle as part of a formal change control process to maintain traceability and to ensure the impacts of any changes are fully assessed for all parts of the system. A more formal change control process is required if the mission is very large and involves more than one Center or crosses other jurisdictional or organizational boundaries.

4.2.2.4 Capturing Requirements and the Requirements Database

At the time the requirements are written, it is important to capture the requirements statements along with the metadata associated with each requirement. The metadata is the supporting information necessary to help clarify and link the requirements.

The method of verification must also be thought through and captured for each requirement at the time it is developed. The verification method includes test, inspection, analysis, and demonstration. Be sure to document any new or derived requirements that are uncovered during determination of the verification method. An example is requiring an additional test port to give visibility to an internal signal during integration and test. If a requirement cannot be verified, then either it

should not be a requirement or the requirement statement needs to be rewritten. For example, the requirement to "minimize noise" is vague and cannot be verified. If the requirement is restated as "the noise level of the component X shall remain under Y decibels" then it is clearly verifiable. Examples of the types of metadata are provided in Table 4.2-2.

The requirements database is an extremely useful tool for capturing the requirements and the associated metadata and for showing the bidirectional traceability between requirements. The database evolves over time and could be used for tracking status information related to requirements such as To Be Determined (TBD)/To Be Resolved (TBR) status, resolution date, and verification status. Each project should decide what metadata will be captured. The database is usually in a central location that is made available to the entire project team. (See Appendix D for a sample requirements verification matrix.)

4.2.2.5 Technical Standards

Importance of Standards Application

Standards provide a proven basis for establishing common technical requirements across a program or

Table 4.2-2 Requirements Metadata

Item	Function
Requirement ID	Provides a unique numbering system for sorting and tracking.
Rationale	Provides additional information to help clarify the intent of the requirements at the time they were written. (See "Rationale" box below on what should be captured.)
Traced from	Captures the bidirectional traceability between parent requirements and lower level (derived) requirements and the relationships between requirements.
Owner	Person or group responsible for writing, managing, and/or approving changes to this requirement.
Verification method	Captures the method of verification (test, inspection, analysis, demonstration) and should be determined as the requirements are developed.
Verification lead	Person or group assigned responsibility for verifying the requirement.
Verification level	Specifies the level in the hierarchy at which the requirements will be verified (e.g., system, subsystem, element).

Rationale

The rationale should be kept up to date and include the following information:

- **Reason for the Requirement:** Often the reason for the requirement is not obvious, and it may be lost if not recorded as the requirement is being documented. The reason may point to a constraint or concept of operations. If there is a clear parent requirement or trade study that explains the reason, then reference it.

- **Document Assumptions:** If a requirement was written assuming the completion of a technology development program or a successful technology mission, document the assumption.

- **Document Relationships:** The relationships with the product's expected operations (e.g., expectations about how stakeholders will use a product). This may be done with a link to the ConOps.

- **Document Design Constraints:** Imposed by the results from decisions made as the design evolves. If the requirement states a method of implementation, the rationale should state why the decision was made to limit the solution to this one method of implementation.

project to avoid incompatibilities and ensure that at least minimum requirements are met. Common standards can also lower implementation cost as well as costs for inspection, common supplies, etc. Typically, standards (and specifications) are used throughout the product life cycle to establish design requirements and margins, materials and process specifications, test methods, and interface specifications. Standards are used as requirements (and guidelines) for design, fabrication, verification, validation, acceptance, operations, and maintenance.

Selection of Standards

NASA policy for technical standards is provided in *NPD 8070.6, Technical Standards*, which addresses selection, tailoring, application, and control of standards. In gen-

eral, the order of authority among standards for NASA programs and projects is as follows:

- Standards mandated by law (e.g., environmental standards),
- National or international voluntary consensus standards recognized by industry,
- Other Government standards,
- NASA policy directives, and
- NASA technical standards.

NASA may also designate mandatory or "core" standards that must be applied to all programs where technically applicable. Waivers to designated core standards must be justified and approved at the Agency level unless otherwise delegated.

4.3 Logical Decomposition

Logical Decomposition is the process for creating the detailed functional requirements that enable NASA programs and projects to meet the stakeholder expectations. This process identifies the "what" that must be achieved by the system at each level to enable a successful project. Logical decomposition utilizes functional analysis to create a system architecture and to decompose top-level (or parent) requirements and allocate them down to the lowest desired levels of the project.

The Logical Decomposition Process is used to:

- Improve understanding of the defined technical requirements and the relationships among the requirements (e.g., functional, behavioral, and temporal), and

- Decompose the parent requirements into a set of logical decomposition models and their associated sets of derived technical requirements for input to the Design Solution Definition Process.

4.3.1 Process Description

Figure 4.3-1 provides a typical flow diagram for the Logical Decomposition Process and identifies typical inputs, outputs, and activities to consider in addressing logical decomposition.

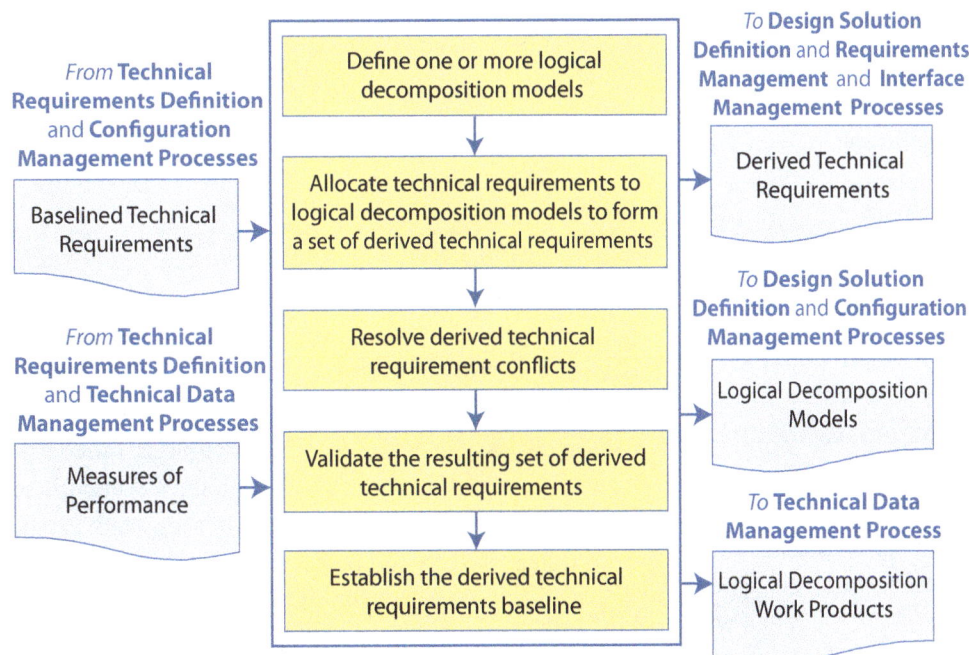

4.3.1.1 Inputs

Typical inputs needed for the Logical Decomposition Process would include the following:

- **Technical Requirements:** A validated set of requirements that represent a description of the problem to be solved, have been established by functional and performance analysis, and have been approved by the customer and other stakeholders. Examples of documentation that capture the requirements are an SRD, PRD, and IRD.

- **Technical Measures:** An established set of measures based on the expectations and requirements that will be tracked and assessed to determine overall system or product effectiveness and customer satisfaction. These measures are MOEs, MOPs, and a special subset of these called TPMs. See Subsection 6.7.2.2 for further details.

4.3.1.2 Process Activities

The key first step in the Logical Decomposition Process is establishing the system architecture model. The system architecture activity defines the underlying structure and relationships of hardware, software, communications, operations, etc., that provide for the implementation of Agency, mission directorate, program, project, and subsequent levels of the requirements. System architecture activities drive the partitioning of system elements and requirements to lower level functions and requirements to the point that design work can be accomplished. Interfaces and relationships between partitioned subsystems and elements are defined as well.

Once the top-level (or parent) functional requirements and constraints have been established, the system designer uses functional analysis to begin to formulate a conceptual system architecture. The system ar-

From **Technical Requirements Definition** and **Configuration Management Processes**

Baselined Technical Requirements

From **Technical Requirements Definition** and **Technical Data Management Processes**

Measures of Performance

- Define one or more logical decomposition models
- Allocate technical requirements to logical decomposition models to form a set of derived technical requirements
- Resolve derived technical requirement conflicts
- Validate the resulting set of derived technical requirements
- Establish the derived technical requirements baseline

To **Design Solution Definition** and **Requirements Management** and **Interface Management Processes**

Derived Technical Requirements

To **Design Solution Definition** and **Configuration Management Processes**

Logical Decomposition Models

To **Technical Data Management Process**

Logical Decomposition Work Products

Figure 4.3-1 Logical Decomposition Process

chitecture can be seen as the strategic organization of the functional elements of the system, laid out to enable the roles, relationships, dependencies, and interfaces between elements to be clearly defined and understood. It is strategic in its focus on the overarching structure of the system and how its elements fit together to contribute to the whole, instead of on the particular workings of the elements themselves. It enables the elements to be developed separately from each other while ensuring that they work together effectively to achieve the top-level (or parent) requirements.

Much like the other elements of functional decomposition, the development of a good system-level architecture is a creative, recursive, and iterative process that combines an excellent understanding of the project's end objectives and constraints with an equally good knowledge of various potential technical means of delivering the end products.

Focusing on the project's ends, top-level (or parent) requirements, and constraints, the system architect must develop at least one, but preferably multiple, concept architectures capable of achieving program objectives. Each architecture concept involves specification of the functional elements (what the pieces do), their relationships to each other (interface definition), and the ConOps, i.e., how the various segments, subsystems, elements, units, etc., will operate as a system when distributed by location and environment from the start of operations to the end of the mission.

The development process for the architectural concepts must be recursive and iterative, with feedback from stakeholders and external reviewers, as well as from subsystem designers and operators, provided as often as possible to increase the likelihood of achieving the program's ends, while reducing the likelihood of cost and schedule overruns.

In the early stages of the mission, multiple concepts are developed. Cost and schedule constraints will ultimately limit how long a program or project can maintain multiple architectural concepts. For all NASA programs, architecture design is completed during the Formulation phase. For most NASA projects (and tightly coupled programs), the selection of a single architecture will happen during Phase A, and the architecture and ConOps will be baselined during Phase B. Architectural changes at higher levels occasionally occur as decomposition to lower levels produces complications in design, cost, or schedule that necessitate such changes.

Aside from the creative minds of the architects, there are multiple tools that can be utilized to develop a system's architecture. These are primarily modeling and simulation tools, functional analysis tools, architecture frameworks, and trade studies. (For example, one way of doing architecture is the Department of Defense (DOD) Architecture Framework (DODAF). See box.) As each concept is developed, analytical models of the architecture, its elements, and their operations will be developed with increased fidelity as the project evolves. Functional decomposition, requirements development, and trade studies are subsequently undertaken. Multiple iterations of these activities feed back to the evolving architectural concept as the requirements flow down and the design matures.

Functional analysis is the primary method used in system architecture development and functional requirement decomposition. It is the systematic process of identifying, describing, and relating the functions a system must perform to fulfill its goals and objectives. Functional analysis identifies and links system functions, trade studies, interface characteristics, and rationales to requirements. It is usually based on the ConOps for the system of interest.

Three key steps in performing functional analysis are:

- Translate top-level requirements into functions that must be performed to accomplish the requirements.
- Decompose and allocate the functions to lower levels of the product breakdown structure.
- Identify and describe functional and subsystem interfaces.

The process involves analyzing each system requirement to identify all of the functions that must be performed to meet the requirement. Each function identified is described in terms of inputs, outputs, and interface requirements. The process is repeated from the top down so that subfunctions are recognized as part of larger functional areas. Functions are arranged in a logical sequence so that any specified operational usage of the system can be traced in an end-to-end path.

The process is recursive and iterative and continues until all desired levels of the architecture/system have been analyzed, defined, and baselined. There will almost cer-

DOD Architecture Framework

New ways, called architecture frameworks, have been developed in the last decade to describe and characterize evolving, complex system-of-systems. In such circumstances, architecture descriptions are very useful in ensuring that stakeholder needs are clearly understood and prioritized, that critical details such as interoperability are addressed upfront, and that major investment decisions are made strategically. In recognition of this, the U.S. Department of Defense has established policies that mandate the use of the DODAF in capital planning, acquisition, and joint capabilities integration.

An architecture can be understood as "the structure of components, their relationships, and the principles and guidelines governing their design and evolution over time."* To describe an architecture, the DODAF defines several views: operational, systems, and technical standards. In addition, a dictionary and summary information are also required. (See figure below.)

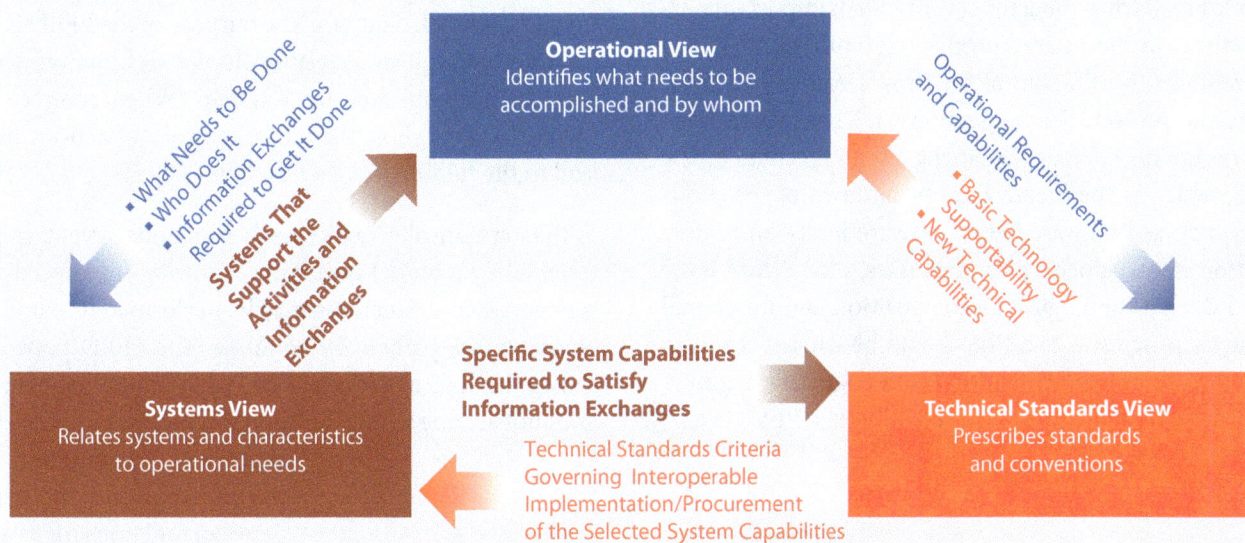

Within each of these views, DODAF contains specific *products*. For example, within the Operational View is a description of the operational nodes, their connectivity, and information exchange requirements. Within the Systems View is a description of all the systems contained in the operational nodes and their interconnectivity. Not all DODAF products are relevant to NASA systems engineering, but its underlying concepts and formalisms may be useful in structuring complex problems for the Technical Requirements Definition and Decision Analysis Processes.

*Definition based on Institute of Electrical and Electronics Engineers (IEEE) STD 610.12.

Source: DOD, *DOD Architecture Framework*.

tainly be alternative ways to decompose functions; therefore, the outcome is highly dependent on the creativity, skills, and experience of the engineers doing the analysis. As the analysis proceeds to lower levels of the architecture and system and the system is better understood, the systems engineer must keep an open mind and a willingness to go back and change previously established architecture and system requirements. These changes will then have to be decomposed down through the architecture and systems again, with the recursive process continuing until the system is fully defined, with all of the requirements understood and known to be viable, verifiable, and internally consistent. Only at that point should the system architecture and requirements be baselined.

4.3.1.3 Outputs

Typical outputs of the Logical Decomposition Process would include the following:

- **System Architecture Model:** Defines the underlying structure and relationship of the elements of the system (e.g., hardware, software, communications, operations, etc.) and the basis for the partitioning of requirements into lower levels to the point that design work can be accomplished.

- **End Product Requirements:** A defined set of make-to, buy-to, code-to, and other requirements from which design solutions can be accomplished.

4.3.2 Logical Decomposition Guidance

4.3.2.1 Product Breakdown Structure

The decompositions represented by the PBS and the Work Breakdown Structure (WBS) form important perspectives on the desired product system. The WBS is a hierarchical breakdown of the work necessary to complete the project. See Subsection 6.1.2.1 for further information on WBS development. The WBS contains the PBS, which is the hierarchical breakdown of the products such as hardware items, software items, and information items (documents, databases, etc.). The PBS is used during the Logical Decomposition and functional analysis processes. The PBS should be carried down to the lowest level for which there is a cognizant engineer or manager. Figure 4.3-2 is an example of a PBS.

4.3.2.2 Functional Analysis Techniques

Although there are many techniques available to perform functional analysis, some of the more popular are (1) Functional Flow Block Diagrams (FFBDs) to depict task sequences and relationships, (2) N2 diagrams (or N x N interaction matrix) to identify interactions or interfaces between major factors from a systems perspective, and (3) Timeline Analyses (TLAs) to depict the time sequence of time-critical functions.

Functional Flow Block Diagrams

The primary functional analysis technique is the functional flow block diagram. The purpose of the FFBD is to indicate the sequential relationship of all functions that must be accomplished by a system. When completed, these diagrams show the entire network of actions that lead to the fulfillment of a function.

FFBDs specifically depict each functional event (represented by a block) occurring following the preceding function. Some functions may be performed in parallel, or alternative paths may be taken. The FFBD network shows the logical sequence of "what" must happen; it does not ascribe a time duration to functions or between functions. The duration of the function and the time between functions may vary from a fraction of a second to many weeks. To understand *time-critical* requirements, a TLA is used. (See the TLA discussion later in this subsection.)

The FFBDs are function oriented, not equipment oriented. In other words, they identify "what" must happen and must not assume a particular answer to "how" a function will be performed. The "how" is then defined for each block at a given level by defining the "what" functions at the next lower level necessary to accomplish that block. In this way, FFBDs are developed from the top down,

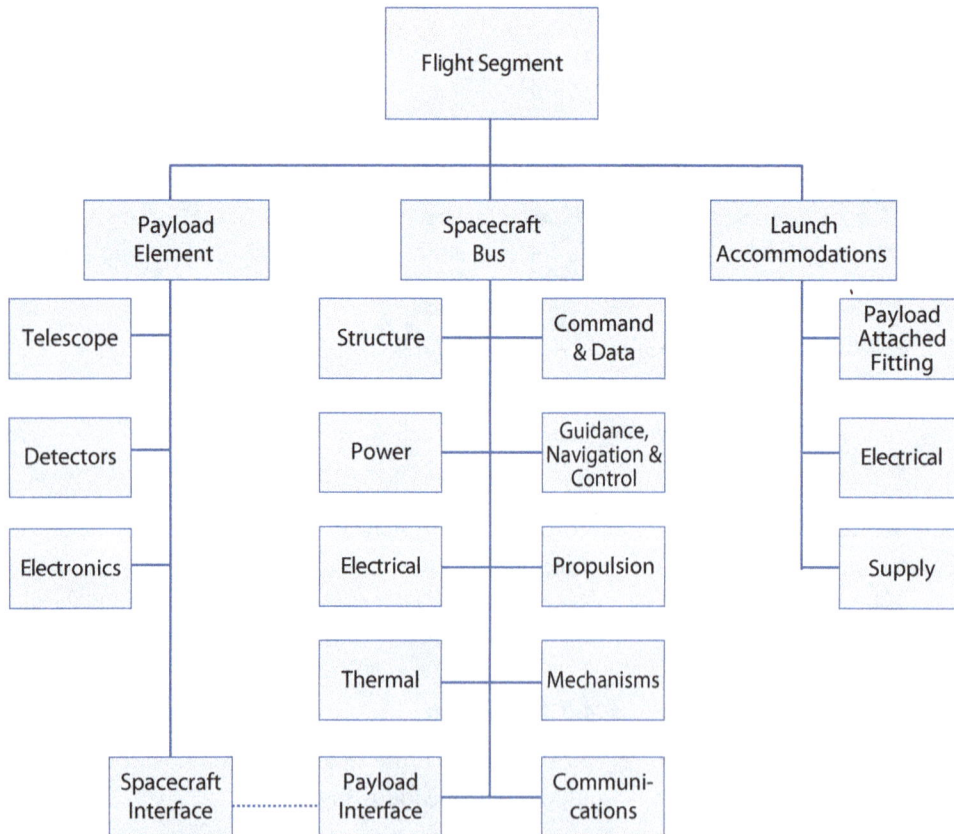

Figure 4.3-2 Example of a PBS

in a series of levels, with tasks at each level identified through functional decomposition of a single task at a higher level. The FFBD displays all of the tasks at each level in their logical, sequential relationship, with their required inputs and anticipated outputs (including metrics, if applicable), plus a clear link back to the single, higher level task.

An example of an FFBD is shown in Figure 4.3-3. The FFBD depicts the entire flight mission of a spacecraft.

Each block in the first level of the diagram is expanded to a series of functions, as shown in the second-level diagram for "Perform Mission Operations." Note that the diagram shows both input ("Transfer to OPS Orbit") and output ("Transfer to STS Orbit"), thus initiating the interface identification and control process. Each block in the second-level diagram can be progressively developed into a series of functions, as shown in the third-level diagram.

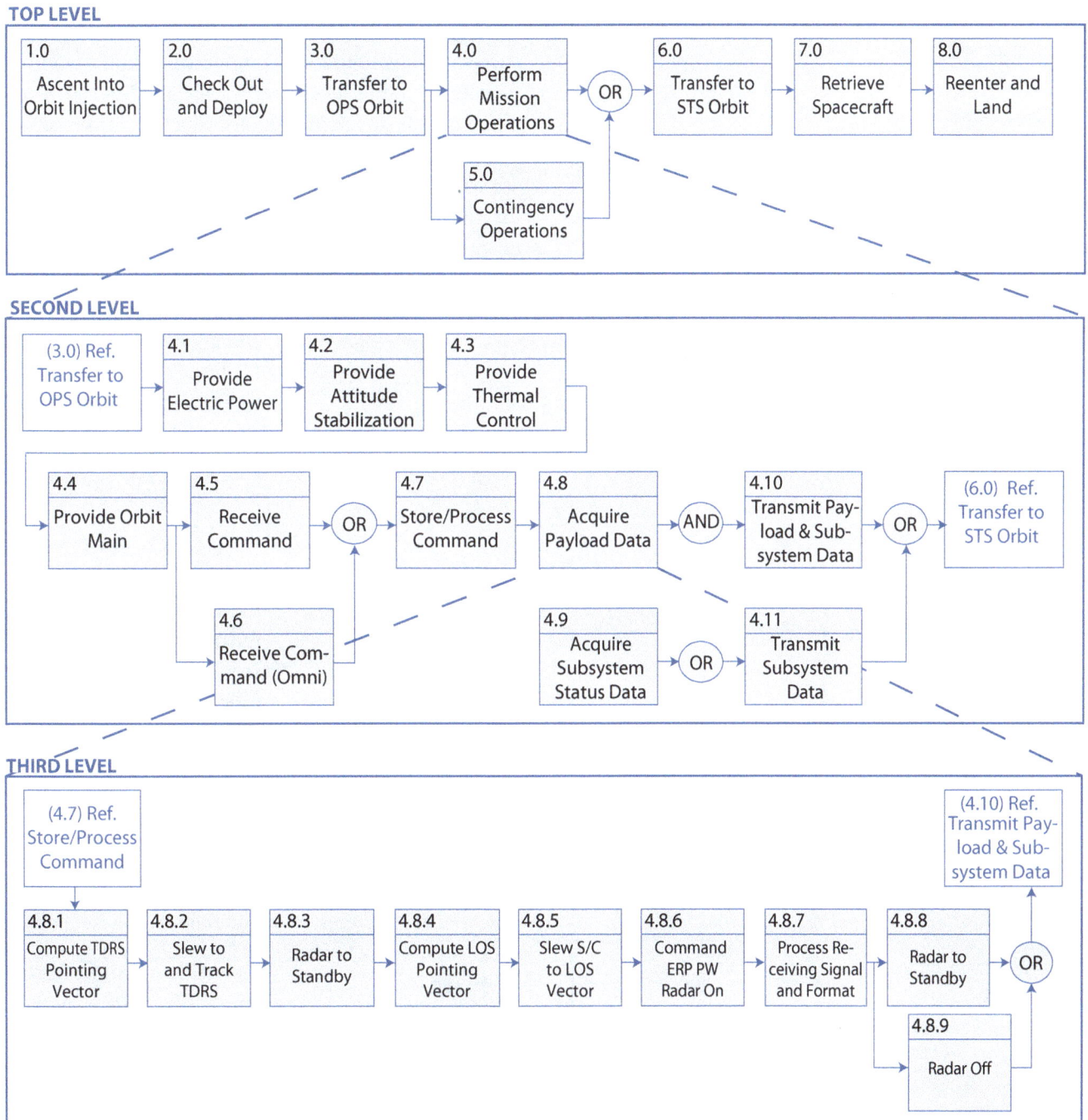

Figure 4.3-3 Example of a functional flow block diagram

FFBDs are used to develop, analyze, and flow down requirements, as well as to identify profitable trade studies, by identifying alternative approaches to performing each function. In certain cases, alternative FFBDs may be used to represent various means of satisfying a particular function until trade study data are acquired to permit selection among the alternatives.

The flow diagram also provides an understanding of the total operation of the system, serves as a basis for development of operational and contingency procedures, and pinpoints areas where changes in operational procedures could simplify the overall system operation.

N2 Diagrams

The N-squared (N2 or N^2) diagram is used to develop system interfaces. An example of an N2 diagram is shown in Figure 4.3-4. The system components or functions are placed on the diagonal; the remainder of

the squares in the N x N matrix represent the interface inputs and outputs. Where a blank appears, there is no interface between the respective components or functions. The N2 diagram can be taken down into successively lower levels to the component functional levels. In addition to defining the interfaces, the N2 diagram also pinpoints areas where conflicts could arise in interfaces, and highlights input and output dependency assumptions and requirements.

Timeline Analysis

TLA adds consideration of functional durations and is performed on those areas where time is critical to mission success, safety, utilization of resources, minimization of downtime, and/or increasing availability. TLA can be applied to such diverse operational functions as spacecraft command sequencing and launch; but for those functional sequences where time is not a critical factor, FFBDs or N2 diagrams are sufficient. The following areas are often categorized as time-critical: (1) functions affecting system reaction time, (2) mission turnaround time, (3) time countdown activities, and (4) functions for which optimum equipment and/or personnel utilization are dependent on the timing of particular activities.

Timeline Sheets (TLSs) are used to perform and record the analysis of time-critical functions and functional sequences. For time-critical functional sequences, the time requirements are specified with associated tolerances. Additional tools such as mathematical models and computer simulations may be necessary to establish the duration of each timeline.

For additional information on FFBD, N2 diagrams, timeline analysis, and other functional analysis methods, see Appendix F.

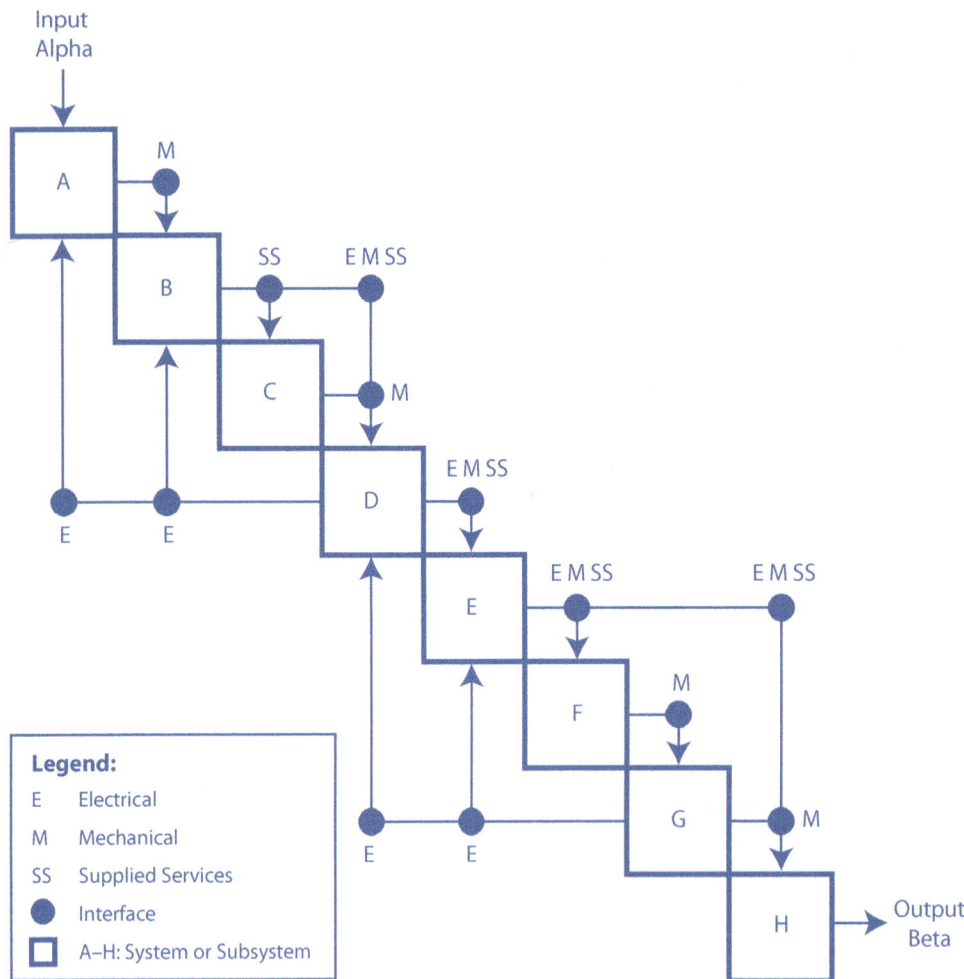

Figure 4.3-4 Example of an N2 diagram

4.4 Design Solution Definition

The Design Solution Definition Process is used to translate the high-level requirements derived from the stakeholder expectations and the outputs of the Logical Decomposition Process into a design solution. This involves transforming the defined logical decomposition models and their associated sets of derived technical requirements into alternative solutions. These alternative solutions are then analyzed through detailed trade studies that result in the selection of a preferred alternative. This preferred alternative is then fully defined into a final design solution that will satisfy the technical requirements. This design solution definition will be used to generate the end product specifications that will be used produce the product and to conduct product verification. This process may be further refined depending on whether there are additional subsystems of the end product that need to be defined.

4.4.1 Process Description

Figure 4.4-1 provides a typical flow diagram for the Design Solution Definition Process and identifies typical inputs, outputs, and activities to consider in addressing design solution definition.

4.4.1.1 Inputs

There are several fundamental inputs needed to initiate the Design Solution Definition Process:

- **Technical Requirements:** The customer and stakeholder needs that have been translated into a reason

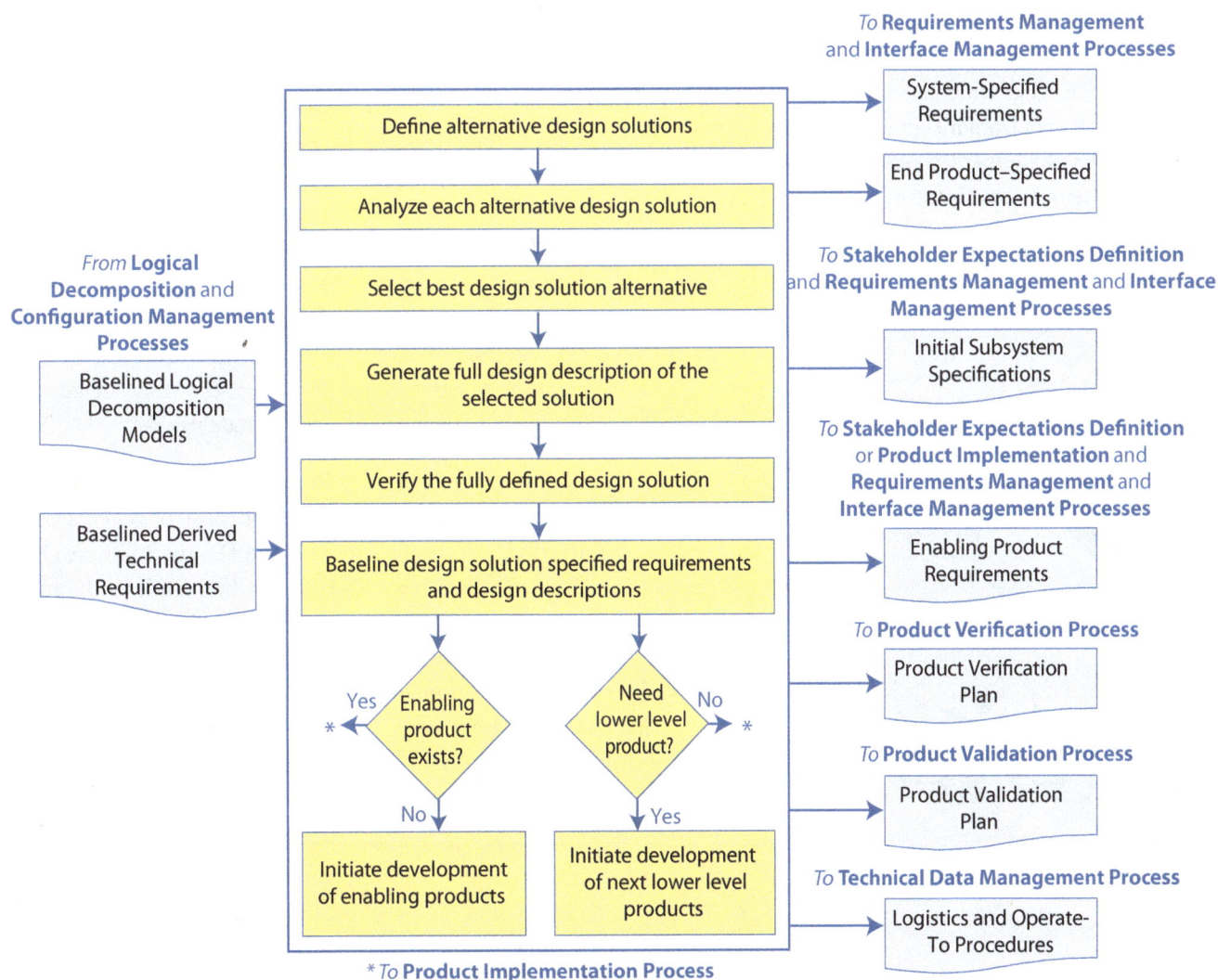

Figure 4.4-1 Design Solution Definition Process

ably complete set of validated requirements for the system, including all interface requirements.

- **Logical Decomposition Models:** Requirements decomposed by one or more different methods (e.g., function, time, behavior, data flow, states, modes, system architecture, etc.).

4.4.1.2 Process Activities

Define Alternative Design Solutions

The realization of a system over its life cycle involves a succession of decisions among alternative courses of action. If the alternatives are precisely defined and thoroughly understood to be well differentiated in the cost-effectiveness space, then the systems engineer can make choices among them with confidence.

To obtain assessments that are crisp enough to facilitate good decisions, it is often necessary to delve more deeply into the space of possible designs than has yet been done, as is illustrated in Figure 4.4-2. It should be realized, however, that this illustration represents neither the project life cycle, which encompasses the system development process from inception through disposal, nor the product development process by which the system design is developed and implemented.

Each create concepts step in Figure 4.4-2 involves a recursive and iterative design loop driven by the set of stake-

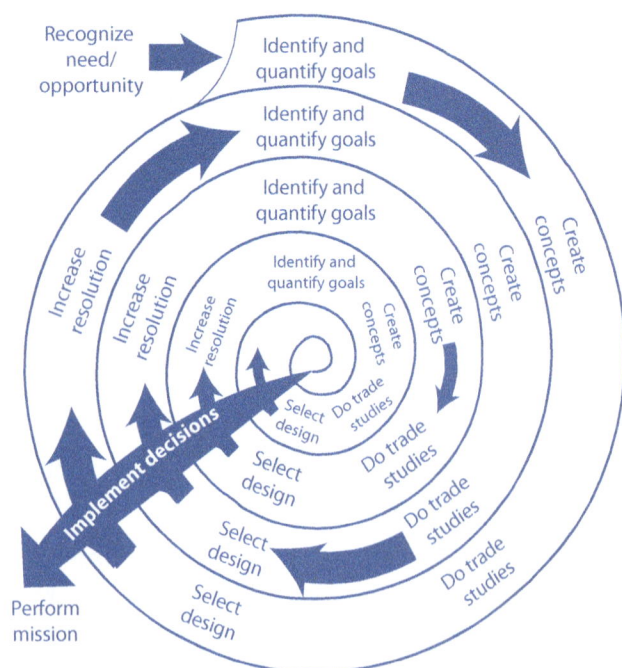

Figure 4.4-2 The doctrine of successive refinement

holder expectations where a strawman architecture/design, the associated ConOps, and the derived requirements are developed. These three products must be consistent with each other and will require iterations and design decisions to achieve this consistency. This recursive and iterative design loop is illustrated in Figure 4.0-1.

Each create concepts step also involves an assessment of potential capabilities offered by the continually changing state of technology and potential pitfalls captured through experience-based review of prior program/project lessons learned data. It is imperative that there be a continual interaction between the technology development process and the design process to ensure that the design reflects the realities of the available technology and that overreliance on immature technology is avoided. Additionally, the state of any technology that is considered enabling must be properly monitored, and care must be taken when assessing the impact of this technology on the concept performance. This interaction is facilitated through a periodic assessment of the design with respect to the maturity of the technology required to implement the design. (See Subsection 4.4.2.1 for a more detailed discussion of technology assessment.) These technology elements usually exist at a lower level in the PBS. Although the process of design concept development by the integration of lower level elements is a part of the systems engineering process, there is always a danger that the top-down process cannot keep up with the bottom-up process. Therefore, system architecture issues need to be resolved early so that the system can be modeled with sufficient realism to do reliable trade studies.

As the system is realized, its particulars become clearer—but also harder to change. The purpose of systems engineering is to make sure that the Design Solution Definition Process happens in a way that leads to the most cost-effective final system. The basic idea is that before those decisions that are hard to undo are made, the alternatives should be carefully assessed, particularly with respect to the maturity of the required technology.

Create Alternative Design Concepts

Once it is understood what the system is to accomplish, it is possible to devise a variety of ways that those goals can be met. Sometimes, that comes about as a consequence of considering alternative functional allocations and integrating available subsystem design options, all of which can have technologies at varying degrees of maturity. Ideally, as wide a range of plausible alternatives as is consistent with the design organization's charter should be defined, keeping in mind the current stage in the process of successive refinement. When the bottom-up process is operating, a problem for the systems engineer is that the designers tend to become fond of the designs they create, so they lose their objectivity; the systems engineer often must stay an "outsider" so that there is more objectivity. This is particularly true in the assessment of the technological maturity of the subsystems and components required for implementation. There is a tendency on the part of technology developers and project management to overestimate the maturity and applicability of a technology that is required to implement a design. This is especially true of "heritage" equipment. The result is that critical aspects of systems engineering are often overlooked.

On the first turn of the successive refinement in Figure 4.4-2, the subject is often general approaches or strategies, sometimes architectural concepts. On the next, it is likely to be functional design, then detailed design, and so on. The reason for avoiding a premature focus on a single design is to permit discovery of the truly best design. Part of the systems engineer's job is to ensure that the design concepts to be compared take into account all interface requirements. "Did you include the cabling?" is a characteristic question. When possible, each design concept should be described in terms of controllable design parameters so that each represents as wide a class of designs as is reasonable. In doing so, the systems engineer should keep in mind that the potentials for change may include organizational structure, schedules, procedures, and any of the other things that make up a system. When possible, constraints should also be described by parameters.

Analyze Each Alternative Design Solution

The technical team analyzes how well each of the design alternatives meets the system goals (technology gaps, effectiveness, cost, schedule, and risk, both quantified and otherwise). This assessment is accomplished through the use of trade studies. The purpose of the trade study process is to ensure that the system architecture and design decisions move toward the best solution that can be achieved with the available resources. The basic steps in that process are:

- Devise some alternative means to meet the functional requirements. In the early phases of the project life-cycle, this means focusing on system architectures; in later phases, emphasis is given to system designs.

- Evaluate these alternatives in terms of the MOEs and system cost. Mathematical models are useful in this step not only for forcing recognition of the relationships among the outcome variables, but also for helping to determine what the measures of performance must be quantitatively.

- Rank the alternatives according to appropriate selection criteria.

- Drop less promising alternatives and proceed to the next level of resolution, if needed.

The trade study process must be done openly and inclusively. While quantitative techniques and rules are used, subjectivity also plays a significant role. To make the process work effectively, participants must have open minds, and individuals with different skills—systems engineers, design engineers, specialty engineers, program analysts, decision scientists, and project managers—must cooperate. The right quantitative methods and selection criteria must be used. Trade study assumptions, models, and results must be documented as part of the project archives. The participants must remain focused on the functional requirements, including those for enabling products. For an in-depth discussion of the trade study process, see Section 6.8. The ability to perform these studies is enhanced by the development of system models that relate the design parameters to those assessments—but it does not depend upon them.

The technical team must consider a broad range of concepts when developing the system model. The model must define the roles of crew, hardware, and software in the system. It must identify the critical technologies required to implement the mission and must consider the entire life cycle, from fabrication to disposal. Evaluation criteria for selecting concepts must be established. Cost is always a limiting factor. However, other criteria, such as time to develop and certify a unit, risk, and reliability, also are critical. This stage cannot be accomplished without addressing the roles of operators and maintainers. These contribute significantly to life-cycle costs and to the system reliability. Reliability analysis should be performed based upon estimates of component failure rates for hardware. If probabilistic risk assessment models are applied, it may be necessary to include occurrence rates or probabilities for software faults or human error events. Assessments of the maturity of the required technology must be done and a technology development plan developed.

Controlled modification and development of design concepts, together with such system models, often permits the use of formal optimization techniques to find regions of the design space that warrant further investigation.

Whether system models are used or not, the design concepts are developed, modified, reassessed, and compared against competing alternatives in a closed-loop process that seeks the best choices for further development. System and subsystem sizes are often determined during the trade studies. The end result is the determination of bounds on the relative cost-effectiveness of the design alternatives, measured in terms of the quantified system goals. (Only bounds, rather than final values, are possible because determination of the final details of the design is intentionally deferred.) Increasing detail associated with the continually improving resolution reduces the spread between upper and lower bounds as the process proceeds.

Select the Best Design Solution Alternative

The technical team selects the best design solution from among the alternative design concepts, taking into account subjective factors that the team was unable to quantify as well as estimates of how well the alternatives meet the quantitative requirements; the maturity of the available technology; and any effectiveness, cost, schedule, risk, or other constraints.

The Decision Analysis Process, as described in Section 6.8, should be used to make an evaluation of the alternative design concepts and to recommend the "best" design solution.

When it is possible, it is usually well worth the trouble to develop a mathematical expression, called an "objective function," that expresses the values of combinations of possible outcomes as a single measure of cost-effectiveness, as illustrated in Figure 4.4-3, even if both cost and effectiveness must be described by more than one measure.

The objective function (or "cost function") assigns a real number to candidate solutions or "feasible solutions" in the alternative space or "search space." A feasible solution that minimizes (or maximizes, if that is the goal) the objective function is called an "optimal solution." When

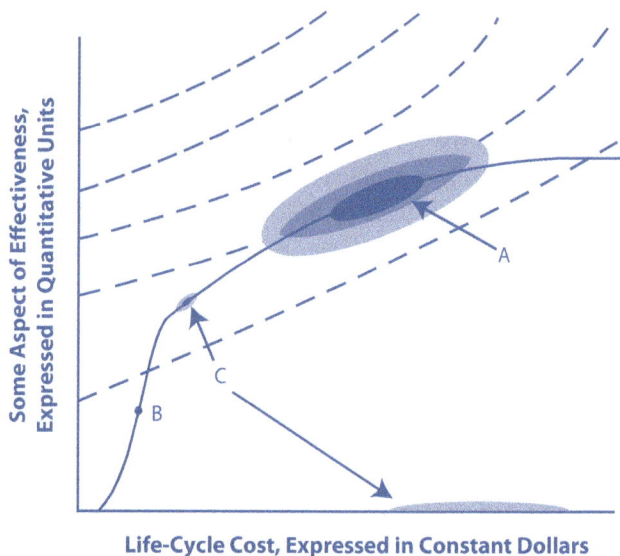

Figure 4.4-3 A quantitative objective function, dependent on life-cycle cost and all aspects of effectiveness

Note: The different shaded areas indicate different levels of uncertainty. Dashed lines represent constant values of objective function (cost-effectiveness). Higher values of cost-effectiveness are achieved by moving toward upper left. A, B, and C are design concepts with different risk patterns.

achievement of the goals can be quantitatively expressed by such an objective function, designs can be compared in terms of their value. Risks associated with design concepts can cause these evaluations to be somewhat nebulous (because they are uncertain and are best described by probability distributions).

In Figure 4.4-3, the risks are relatively high for design concept A. There is little risk in either effectiveness or cost for concept B, while the risk of an expensive failure is high for concept C, as is shown by the cloud of probability near the x axis with a high cost and essentially no effectiveness. Schedule factors may affect the effectiveness and cost values and the risk distributions.

The mission success criteria for systems differ significantly. In some cases, effectiveness goals may be much more important than all others. Other projects may demand low costs, have an immutable schedule, or require minimization of some kinds of risks. Rarely (if ever) is it possible to produce a combined quantitative measure that relates all of the important factors, even if it is expressed as a vector with several components. Even when that can be done, it is essential that the underlying factors and relationships be thoroughly revealed to and understood by the systems engineer. The systems engineer must weigh the importance of the unquantifiable factors along with the quantitative data.

Technical reviews of the data and analyses, including technology maturity assessments, are an important part of the decision support packages prepared for the technical team. The decisions that are made are generally entered into the configuration management system as changes to (or elaborations of) the system baseline. The supporting trade studies are archived for future use. An essential feature of the systems engineering process is that trade studies are performed before decisions are made. They can then be baselined with much more confidence.

Increase the Resolution of the Design

The successive refinement process of Figure 4.4-2 illustrates a continuing refinement of the system design. At each level of decomposition, the baselined derived (and allocated) requirements become the set of high-level requirements for the decomposed elements, and the process begins again. One might ask, "When do we stop refining the design?" The answer is that the design effort precedes to a depth that is sufficient to meet several needs: the design must penetrate sufficiently to allow analytical validation of the design to the requirements; it must also have sufficient depth to support cost modeling and to convince a review team of a feasible design with performance, cost, and risk margins.

The systems engineering engine is applied again and again as the system is developed. As the system is realized, the issues addressed evolve and the particulars of the activity change. Most of the major system decisions (goals, architecture, acceptable life-cycle cost, etc.) are made during the early phases of the project, so the successive refinements do not correspond precisely to the phases of the system life cycle. Much of the system architecture can be seen even at the outset, so the successive refinements do not correspond exactly to development of the architectural hierarchy, either. Rather, they correspond to the successively greater resolution by which the system is defined.

It is reasonable to expect the system to be defined with better resolution as time passes. This tendency is formalized at some point (in Phase B) by defining a baseline system definition. Usually, the goals, objectives, and constraints are baselined as the requirements portion of the baseline. The entire baseline is then subjected to configuration control in an attempt to ensure that any subsequent changes are indeed justified and affordable.

At this point in the systems engineering process, there is a logical branch point. For those issues for which the process of successive refinement has proceeded far enough, the next step is to implement the decisions at that level of resolution. For those issues that are still insufficiently resolved, the next step is to refine the development further.

Fully Describe the Design Solution

Once the preferred design alternative has been selected and the proper level of refinement has been completed, then the design is fully defined into a final design solution that will satisfy the technical requirements. The design solution definition will be used to generate the end product specifications that will be used to produce the product and to conduct product verification. This process may be further refined depending on whether there are additional subsystems of the end product that need to be defined.

The scope and content of the full design description must be appropriate for the product life-cycle phase, the phase success criteria, and the product position in the PBS (system structure). Depending on these factors, the form of the design solution definition could be simply a simulation model or a paper study report. The technical data package evolves from phase to phase, starting with conceptual sketches or models and ending with complete drawings, parts list, and other details needed for product implementation or product integration. Typical output definitions from the Design Solution Definition Process are shown in Figure 4.4-1 and are described in Subsection 4.4.1.3.

Verify the Design Solution

Once an acceptable design solution has been selected from among the various alternative designs and documented in a technical data package, the design solution must next be verified against the system requirements and constraints. A method to achieve this verification is by means of a peer review to evaluate the resulting design solution definition. Guidelines for conducting a peer review are discussed in Section 6.7.

In addition, peer reviews play a significant role as a detailed technical component of higher level technical and programmatic reviews. For example, the peer review of a component battery design can go into much more technical detail on the battery than the integrated power subsystem review. Peer reviews can cover the components of a subsystem down to the level appropriate for verifying the design against the requirements. Concerns raised at the peer review might have implications on the power subsystem design and verification and therefore must be reported at the next higher level review of the power subsystem.

The verification must show that the design solution definition:

- Is realizable within the constraints imposed on the technical effort;
- Has specified requirements that are stated in acceptable statements and have bidirectional traceability with the derived technical requirements, technical requirements, and stakeholder expectations; and
- Has decisions and assumptions made in forming the solution consistent with its set of derived technical requirements, separately allocated technical requirements, and identified system product and service constraints.

This design solution verification is in contrast to the verification of the end product described in the end product verification plan which is part of the technical data package. That verification occurs in a later life-cycle phase and is a result of the Product Verification Process (see Section 5.3) applied to the realization of the design solution as an end product.

Validate the Design Solution

The validation of the design solution is a recursive and iterative process as shown in Figure 4.0-1. Each alternative design concept is validated against the set of stakeholder expectations. The stakeholder expectations drive the iterative design loop in which a strawman architecture/design, the ConOps, and the derived requirements are developed. These three products must be consistent with each other and will require iterations and design decisions to achieve this consistency. Once consistency is achieved, functional analyses allow the study team to validate the design against the stakeholder expectations. A simplified validation asks the questions: Does the system work? Is the system safe and reliable? Is the system affordable? If the answer to any of these questions is no, then changes to the design or stakeholder expec-

tations will be required, and the process is started over again. This process continues until the system—architecture, ConOps, and requirements—meets the stakeholder expectations.

This design solution validation is in contrast to the validation of the end product described in the end product validation plan, which is part of the technical data package. That validation occurs in a later life-cycle phase and is a result of the Product Validation Process (see Section 5.4) applied to the realization of the design solution as an end product.

Identify Enabling Products

Enabling products are the life-cycle support products and services (e.g., production, test, deployment, training, maintenance, and disposal) that facilitate the progression and use of the operational end product through its life cycle. Since the end product and its enabling products are interdependent, they are viewed as a system. Project responsibility thus extends to responsibility for acquiring services from the relevant enabling products in each life-cycle phase. When a suitable enabling product does not already exist, the project that is responsible for the end product also can be responsible for creating and using the enabling product.

Therefore, an important activity in the Design Solution Definition Process is the identification of the enabling products that will be required during the life cycle of the selected design solution and then initiating the acquisition or development of those enabling products. Need dates for the enabling products must be realistically identified on the project schedules, incorporating appropriate schedule slack. Then firm commitments in the form of contracts, agreements, and/or operational plans must be put in place to ensure that the enabling products will be available when needed to support the product-line life-cycle phase activities. The enabling product requirements are documented as part of the technical data package for the Design Solution Definition Process.

An environmental test chamber would be an example of an enabling product whose use would be acquired at an appropriate time during the test phase of a space flight system.

Special test fixtures or special mechanical handling devices would be examples of enabling products that would have to be created by the project. Because of long development times as well as oversubscribed facilities, it is important to identify enabling products and secure the commitments for them as early in the design phase as possible.

Baseline the Design Solution

As shown earlier in Figure 4.0-1, once the selected system design solution meets the stakeholder expectations, the study team baselines the products and prepares for the next life-cycle phase. Because of the recursive nature of successive refinement, intermediate levels of decomposition are often validated and baselined as part of the process. In the next level of decomposition, the baselined requirements become the set of high-level requirements for the decomposed elements, and the process begins again.

Baselining a particular design solution enables the technical team to focus on one design out of all the alternative design concepts. This is a critical point in the design process. It puts a stake in the ground and gets everyone on the design team focused on the same concept. When dealing with complex systems, it is difficult for team members to design their portion of the system if the system design is a moving target. The baselined design is documented and placed under configuration control. This includes the system requirements, specifications, and configuration descriptions.

While baselining a design is beneficial to the design process, there is a danger if it is exercised too early in the Design Solution Definition Process. The early exploration of alternative designs should be free and open to a wide range of ideas, concepts, and implementations. Baselining too early takes the inventive nature out of the concept exploration. Therefore baselining should be one of the last steps in the Design Solution Definition Process.

4.4.1.3 Outputs

Outputs of the Design Solution Definition Process are the specifications and plans that are passed on to the product realization processes. They contain the design-to, build-to, and code-to documentation that complies with the approved baseline for the system.

As mentioned earlier, the scope and content of the full design description must be appropriate for the product-line life-cycle phase, the phase success criteria, and the product position in the PBS.

Outputs of the Design Solution Definition Process include the following:

- **The System Specification:** The system specification contains the functional baseline for the system that is the result of the Design Solution Definition Process. The system design specification provides sufficient guidance, constraints, and system requirements for the design engineers to execute the design.

- **The System External Interface Specifications:** The system external interface specifications describe the functional baseline for the behavior and characteristics of all physical interfaces that the system has with the external world. These include all structural, thermal, electrical, and signal interfaces, as well as the human-system interfaces.

- **The End-Product Specifications:** The end-product specifications contain the detailed build-to and code-to requirements for the end product. They are detailed, exact statements of design particulars, such as statements prescribing materials, dimensions, and quality of work to build, install, or manufacture the end product.

- **The End-Product Interface Specifications:** The end-product interface specifications contain the detailed build-to and code-to requirements for the behavior and characteristics of all logical and physical interfaces that the end product has with external elements, including the human-system interfaces.

- **Initial Subsystem Specifications:** The end-product subsystem initial specifications provide detailed information on subsystems if they are required.

- **Enabling Product Requirements:** The requirements for associated supporting enabling products provide details of all enabling products. Enabling products are the life-cycle support products and services that facilitate the progression and use of the operational end product through its life cycle. They are viewed as part of the system since the end product and its enabling products are interdependent.

- **Product Verification Plan:** The end-product verification plan provides the content and depth of detail necessary to provide full visibility of all verification activities for the end product. Depending on the scope of the end product, the plan encompasses qualification, acceptance, prelaunch, operational, and disposal verification activities for flight hardware and software.

- **Product Validation Plan:** The end-product validation plan provides the content and depth of detail necessary to provide full visibility of all activities to validate the realized product against the baselined stakeholder expectations. The plan identifies the type of validation, the validation procedures, and the validation environment that are appropriate to confirm that the realized end product conforms to stakeholder expectations.

- **Logistics and Operate-to Procedures:** The applicable logistics and operate-to procedures for the system describe such things as handling, transportation, maintenance, long-term storage, and operational considerations for the particular design solution.

4.4.2 Design Solution Definition Guidance

4.4.2.1 Technology Assessment

As mentioned in the process description (Subsection 4.4.1), the creation of alternative design solutions involves assessment of potential capabilities offered by the continually changing state of technology. A continual interaction between the technology development process and the design process ensures that the design reflects the realities of the available technology. This interaction is facilitated through periodic assessment of the design with respect to the maturity of the technology required to implement the design.

After identifying the technology gaps existing in a given design concept, it will frequently be necessary to undertake technology development in order to ascertain viability. Given that resources will always be limited, it will be necessary to pursue only the most promising technologies that are required to enable a given concept.

If requirements are defined without fully understanding the resources required to accomplish needed technology developments then the program/project is at risk. Technology assessment must be done iteratively until requirements and available resources are aligned within an acceptable risk posture. Technology development plays a far greater role in the life cycle of a program/project than has been traditionally considered, and it is the role of the systems engineer to develop an understanding of the extent of program/project impacts—maximizing benefits and minimizing adverse effects. Traditionally, from a program/project perspective, technology development has been associated with the development and incor-

poration of any "new" technology necessary to meet requirements. However, a frequently overlooked area is that associated with the modification of "heritage" systems incorporated into different architectures and operating in different environments from the ones for which they were designed. If the required modifications and/or operating environments fall outside the realm of experience, then these too should be considered technology development.

To understand whether or not technology development is required—and to subsequently quantify the associated cost, schedule, and risk—it is necessary to systematically assess the maturity of each system, subsystem, or component in terms of the architecture and operational environment. It is then necessary to assess what is required in the way of development to advance the maturity to a point where it can successfully be incorporated within cost, schedule, and performance constraints. A process for accomplishing this assessment is described in Appendix G. Because technology development has the potential for such significant impacts on a program/project, technology assessment needs to play a role throughout the design and development process from concept development through Preliminary Design Review (PDR). Lessons learned from a technology development point of view should then be captured in the final phase of the program.

4.4.2.2 Integrating Engineering Specialties into the Systems Engineering Process

As part of the technical effort, specialty engineers in cooperation with systems engineering and subsystem designers often perform tasks that are common across disciplines. Foremost, they apply specialized analytical techniques to create information needed by the project manager and systems engineer. They also help define and write system requirements in their areas of expertise, and they review data packages, Engineering Change Requests (ECRs), test results, and documentation for major project reviews. The project manager and/or systems engineer needs to ensure that the information and products so generated add value to the project commensurate with their cost. The specialty engineering technical effort should be well integrated into the project. The roles and responsibilities of the specialty engineering disciplines should be summarized in the SEMP.

The specialty engineering disciplines included in this handbook are safety and reliability, Quality Assurance

(QA), ILS, maintainability, producibility, and human factors. An overview of these specialty engineering disciplines is provided to give systems engineers a brief introduction. It is not intended to be a handbook for any of these discipline specialties.

Safety and Reliability

Overview and Purpose

A reliable system ensures mission success by functioning properly over its intended life. It has a low and acceptable probability of failure, achieved through simplicity, proper design, and proper application of reliable parts and materials. In addition to long life, a reliable system is robust and fault tolerant, meaning it can tolerate failures and variations in its operating parameters and environments.

Safety and Reliability in the System Design Process

A focus on safety and reliability throughout the mission life cycle is essential for ensuring mission success. The fidelity to which safety and reliability are designed and built into the system depends on the information needed and the type of mission. For human-rated systems, safety and reliability is the primary objective throughout the design process. For science missions, safety and reliability should be commensurate with the funding and level of risk a program or project is willing to accept. Regardless of the type of mission, safety and reliability considerations must be an intricate part of the system design processes.

To realize the maximum benefit from reliability analysis, it is essential to integrate the risk and reliability analysts within the design teams. The importance of this cannot be overstated. In many cases, the reliability and risk analysts perform the analysis on the design after it has been formulated. In this case, safety and reliability features are added on or outsourced rather than designed in. This results in unrealistic analysis that is not focused on risk drivers and does not provide value to the design.

Risk and reliability analyses evolve to answer key questions about design trades as the design matures. Reliability analyses utilize information about the system, identify sources of risk and risk drivers, and provide an important input for decisionmaking. *NASA-STD-8729.1, Planning, Developing, and Maintaining an Effective Reliability and Maintainability (R&M) Program* outlines engineering activities that should be tailored for each specific project. The concept is to choose an effective set of reliability and maintainability engineering activities to ensure that the systems designed, built, and deployed will operate successfully for the required mission life cycle.

In the early phases of a project, risk and reliability analyses help designers understand the interrelationships of requirements, constraints, and resources, and uncover key relationships and drivers so they can be properly considered. The analyst must help designers go beyond the requirements to understand implicit dependencies that emerge as the design concept matures. It is unrealistic to assume that design requirements will correctly capture all risk and reliability issues and "force" a reliable design. The systems engineer should develop a system strategy mapped to the PBS on how to allocate and coordinate reliability, fault tolerance, and recovery between systems both horizontally and vertically within the architecture to meet the total mission requirements. System impacts of designs must play a key role in the design. Making designers aware of impacts of their decisions on overall mission reliability is key.

As the design matures, preliminary reliability analysis occurs using established techniques. The design and concept of operations should be thoroughly examined for accident initiators and hazards that could lead to mishaps. Conservative estimates of likelihood and consequences of the hazards can be used as a basis for applying design resources to reduce the risk of failures. The team should also ensure that the goals can be met and failure modes are considered and take into account the entire system.

During the latter phases of a project, the team uses risk assessments and reliability techniques to verify that the design is meeting its risk and reliability goals and to help develop mitigation strategies when the goals are not met or discrepancies/failures occur.

Analysis Techniques and Methods

This subsection provides a brief summary of the types of analysis techniques and methods.

- Event sequence diagrams/event trees are models that describe the sequence of events and responses to off-nominal conditions that can occur during a mission.
- Failure Modes and Effects Analyses (FMEAs) are bottom-up analyses that identify the types of failures

that can occur within a system and identify the causes, effects, and mitigating strategies that can be employed to control the effects of the failures.

- Qualitative top-down logic models identify how failures within a system can combine to cause an undesired event.

- Quantitative logic models (probabilistic risk assessment) extend the qualitative models to include the likelihood of failure. These models involve developing failure criteria based on system physics and system success criteria, and employing statistical techniques to estimate the likelihood of failure along with uncertainty.

- Reliability block diagrams are diagrams of the elements to evaluate the reliability of a system to provide a function.

- Preliminary Hazard Analysis (PHA) is performed early based on the functions performed during the mission. Preliminary hazard analysis is a "what if" process that considers the potential hazard, initiating event scenarios, effects, and potential corrective measures and controls. The objective is to determine if the hazard can be eliminated, and if not, how it can be controlled.

- Hazard analysis evaluates the completed design. Hazard analysis is a "what if" process that considers the potential hazard, initiating event, effects, and potential corrective measures and controls. The objective is to determine if the hazard can be eliminated, and if not, how it can be controlled.

- Human reliability analysis is a method to understand how human failures can lead to system failure and estimate the likelihood of those failures.

- Probabilistic structural analysis provides a way to combine uncertainties in materials and loads to evaluate the failure of a structural element.

- Sparing/logistics models provide a means to estimate the interactions of systems in time. These models include ground-processing simulations and mission campaign simulations.

Limitations on Reliability Analysis

The engineering design team must understand that reliability is expressed as the probability of mission success. Probability is a mathematical measure expressing the likelihood of occurrence of a specific event. Therefore, probability estimates should be based on engineering and historical data, and any stated probabilities should include some measure of the uncertainty surrounding that estimate.

Uncertainty expresses the degree of belief analysts have in their estimates. Uncertainty decreases as the quality of data and understanding of the system improve. The initial estimates of failure rates or failure probability might be based on comparison to similar equipment, historical data (heritage), failure rate data from handbooks, or expert elicitation.

In summary,

- Reliability estimates express probability of success.

- Uncertainty should be included with reliability estimates.

- Reliability estimates combined with FMEAs provide additional and valuable information to aid in the decisionmaking process.

Quality Assurance

Even with the best designs, hardware fabrication and testing are subject to human error. The systems engineer needs to have some confidence that the system actually produced and delivered is in accordance with its functional, performance, and design requirements. QA provides an independent assessment to the project manager/systems engineer of the items produced and processes used during the project life cycle. The project manager/systems engineer must work with the quality assurance engineer to develop a quality assurance program (the extent, responsibility, and timing of QA activities) tailored to the project it supports.

QA is the mainstay of quality as practiced at NASA. *NPD 8730.5, NASA Quality Assurance Program Policy* states that NASA's policy is "to comply with prescribed requirements for performance of work and to provide for independent assurance of compliance through implementation of a quality assurance program." The quality function of Safety and Mission Assurance (SMA) ensures that both contractors and other NASA functions do what they say they will do and say what they intend to do. This ensures that end product and program quality, reliability, and overall risk are at the level planned.

The Systems Engineer's Relationship to QA

As with reliability, producibility, and other characteristics, quality must be designed as an integral part of any system. It is important that the systems engineer understands SMA's safeguarding role in the broad context of total risk and supports the quality role explicitly and vigorously. All of this is easier if the SMA quality function is actively included and if quality is designed in with buy-in by all roles, starting at concept development. This will help mitigate conflicts between design and quality requirements, which can take on the effect of "tolerance stacking."

Quality is a vital part of risk management. Errors, variability, omissions, and other problems cost time, program resources, taxpayer dollars, and even lives. It is incumbent on the systems engineer to know how quality affects their projects and to encourage best practices to achieve the quality level.

Rigid adherence to procedural requirements is necessary in high-risk, low-volume manufacturing. In the absence of large samples and long production runs, compliance to these written procedures is a strong step toward ensuring process, and, thereby, product consistency. To address this, NASA requires QA programs to be designed to mitigate risks associated with noncompliance to those requirements.

There will be a large number of requirements and procedures thus created. These must be flowed down to the supply chain, even to lowest tier suppliers. For circumstances where noncompliance can result in loss of life or loss of mission, there is a requirement to insert into procedures Government Mandatory Inspection Points (GMIPs) to ensure 100 percent compliance with safety/mission-critical attributes. Safety/mission-critical attributes include hardware characteristics, manufacturing process requirements, operating conditions, and functional performance criteria that, if not met, can result in loss of life or loss of mission. There will be in place a Program/Project Quality Assurance Surveillance Plan (PQASP) as mandated by Federal Acquisition Regulation (FAR) Subpart 46.4. Preparation and content for PQASPs are outlined in *NPR 8735.2, Management of Government Quality Assurance Functions for NASA Contracts.* This document covers quality assurance requirements for both low-risk and high-risk acquisitions and includes functions such as document review, product examination, process witnessing, quality system evaluation, nonconformance reporting and corrective action, planning for quality assurance and surveillance, and GMIPs. In addition, most NASA projects are required to adhere to either ISO 9001 (noncritical work) or AS9100 (critical work) requirements for management of quality systems. Training in these systems is mandatory for most NASA functions, so knowledge of their applicability by the systems engineer is assumed. Their texts and intent are strongly reflected in NASA's quality procedural documents.

Integrated Logistics Support

The objective of ILS activities within the systems engineering process is to ensure that the product system is supported during development (Phase D) and operations (Phase E) in a cost-effective manner. ILS is particularly important to projects that are reusable or serviceable. Projects whose primary product does not evolve over its operations phase typically only apply ILS to parts of the project (for example, the ground system) or to some of the elements (for example, transportation). ILS is primarily accomplished by early, concurrent consideration of supportability characteristics; performing trade studies on alternative system and ILS concepts; quantifying resource requirements for each ILS element using best practices; and acquiring the support items associated with each ILS element. During operations, ILS activities support the system while seeking improvements in cost-effectiveness by conducting analyses in response to actual operational conditions. These analyses continually reshape the ILS system and its resource requirements. Neglecting ILS or poor ILS decisions invariably have adverse effects on the life-cycle cost of the resultant system. Table 4.4-1 summarizes the ILS disciplines.

ILS planning should begin early in the project life cycle and should be documented. This plan should address the elements above including how they will be considered, conducted, and integrated into the systems engineering process needs.

Maintainability

Maintainability is defined as the measure of the ability of an item to be retained in or restored to specified conditions when maintenance is performed by personnel having specified skill levels, using prescribed procedures and resources, at each prescribed level of maintenance. It is the inherent characteristics of a design or installation that contribute to the ease, economy, safety, and accuracy with which maintenance actions can be performed.

Table 4.4-1 ILS Technical Disciplines

Technical Discipline	Definition
Maintenance support planning	Ongoing and iterative planning, organization, and management activities necessary to ensure that the logistics requirements for any given program are properly coordinated and implemented
Design interface	The interaction and relationship of logistics with the systems engineering process to ensure that supportability influences the definition and design of the system so as to reduce life-cycle cost
Technical data and technical publications	The recorded scientific, engineering, technical, and cost information used to define, produce, test, evaluate, modify, deliver, support, and operate the system
Training and training support	Encompasses all personnel, equipment, facilities, data/documentation, and associated resources necessary for the training of operational and maintenance personnel
Supply support	Actions required to provide all the necessary material to ensure the system's supportability and usability objectives are met
Test and support equipment	All tools, condition-monitoring equipment, diagnostic and checkout equipment, special test equipment, metrology and calibration equipment, maintenance fixtures and stands, and special handling equipment required to support operational maintenance functions
Packaging, handling, storage, and transportation	All materials, equipment, special provisions, containers (reusable and disposable), and supplies necessary to support the packaging, safety and preservation, storage, handling, and transportation of the prime mission-related elements of the system, including personnel, spare and repair parts, test and support equipment, technical data computer resources, and mobile facilities
Personnel	Involves identification and acquisition of personnel with skills and grades required to operate and maintain a system over its lifetime
Logistics facilities	All special facilities that are unique and are required to support logistics activities, including storage buildings and warehouses and maintenance facilities at all levels
Computer resources support	All computers, associated software, connecting components, networks, and interfaces necessary to support the day-to-day flow of information for all logistics functions

Source: Blanchard, *System Engineering Management.*

Role of the Maintainability Engineer

Maintainability engineering is another major specialty discipline that contributes to the goal of a supportable system. This is primarily accomplished in the systems engineering process through an active role in implementing specific design features to facilitate safe and effective maintenance actions in the predicted physical environments, and through a central role in developing the ILS system. Example tasks of the maintainability engineer include: developing and maintaining a system maintenance concept, establishing and allocating maintainability requirements, performing analysis to quantify the system's maintenance resource requirements, and verifying the system's maintainability requirements.

Producibility

Producibility is a system characteristic associated with the ease and economy with which a completed design can be transformed (i.e., fabricated, manufactured, or coded) into a hardware and/or software realization. While major NASA systems tend to be produced in small quantities, a particular producibility feature can be crit-

ical to a system's cost-effectiveness, as experience with the shuttle's thermal tiles has shown. Factors that influence the producibility of a design include the choice of materials, simplicity of design, flexibility in production alternatives, tight tolerance requirements, and clarity and simplicity of the technical data package.

Role of the Production Engineer

The production engineer supports the systems engineering process (as a part of the multidisciplinary product development team) by taking an active role in implementing specific design features to enhance producibility and by performing the production engineering analyses needed by the project. These tasks and analyses include:

- Performing the manufacturing/fabrication portion of the system risk management program. This is accomplished by conducting a rigorous production risk assessment and by planning effective risk mitigation actions.

- Identifying system design features that enhance producibility. Efforts usually focus on design simplification, fabrication tolerances, and avoidance of hazardous materials.

- Conducting producibility trade studies to determine the most cost-effective fabrication/manufacturing process.
- Assessing production feasibility within project constraints. This may include assessing contractor and principal subcontractor production experience and capability, new fabrication technology, special tooling, and production personnel training requirements.
- Identifying long-lead items and critical materials.
- Estimating production costs as a part of life-cycle cost management.
- Supporting technology readiness assessments.
- Developing production schedules.
- Developing approaches and plans to validate fabrication/manufacturing processes.

The results of these tasks and production engineering analyses are documented in the manufacturing plan with a level of detail appropriate to the phase of the project. The production engineer also participates in and contributes to major project reviews (primarily PDR and Critical Design Review (CDR)) on the above items, and to special interim reviews such as the PRR.

Prototypes

Experience has shown that prototype systems can be effective in enabling efficient producibility even when building only a single flight system. Prototypes are built early in the life cycle and they are made as close to the flight item in form, fit, and function as is feasible at that stage of the development. The prototype is used to "wring out" the design solution so that experience gained from the prototype can be fed back into design changes that will improve the manufacture, integration, and maintainability of a single flight item or the production run of several flight items. Unfortunately, prototypes are often deleted from projects to save cost. Along with that decision, the project accepts an increased risk in the development phase of the life cycle. Fortunately, advancements in computer-aided design and manufacturing have mitigated that risk somewhat by enabling the designer to visualize the design and "walk through" the integration sequence to uncover problems before they become a costly reality.

Human Factors Engineering

Overview and Purpose

Consideration of human operators and maintainers of systems is a critical part of the design process. Human factors engineering is the discipline that studies the human-system interfaces and provides requirements, standards, and guidelines to ensure the human component of the integrated system is able to function as intended. Human roles include operators (flight crews and ground crews), designers, manufacturers, ground support, maintainers, and passengers. Flight crew functions include system operation, troubleshooting, and in-flight maintenance. Ground crew functions include spacecraft and ground system manufacturing, assembly, test, checkout, logistics, ground maintenance, repair, refurbishment, launch control, and mission control.

Human factors are generally considered in four categories. The first is anthropometry and biomechanics—the physical size, shape, and strength of the humans. The second is sensation and perception—primarily vision and hearing, but senses such as touch are also important. The environment is a third factor—ambient noise and lighting, vibration, temperature and humidity, atmospheric composition, and contaminants. Psychological factors comprise memory; information processing components such as pattern recognition, decisionmaking, and signal detection; and affective factors—e.g., emotions, cultural patterns, and habits.

Human Factors Engineering in the System Design Process

- **Stakeholder Expectations:** The operators, maintainers, and passengers are all stakeholders in the system. The human factors specialist identifies roles and responsibilities that can be performed by humans and scenarios that exceed human capabilities. The human factors specialist ensures that system operational concept development includes task analysis and human/system function allocation. As these are refined, function allocation distributes operator roles and responsibilities for subtasks to the crew, external support teams, and automation. (For example, in aviation, tasks may be allocated to crew, air traffic controllers, or autopilots. In spacecraft, tasks may be performed by crew, mission control, or onboard systems.)

- **Requirements Definition:** Human factors requirements for spacecraft and space habitats are program/project dependent, derived from *NASA-STD-3001, NASA Space Flight Human System Standard Volume 1: Crew Health*. Other human factors requirements of other missions and Earth-based activities for human space flight missions are derived from human factors standards such as *MIL-STD-1472, Human Engineering*; NUREG-0700, *Human-System Interface Design Review Guidelines*; and the Federal Aviation Administration's *Human Factors Design Standard*.

- **Technical Solution:** Consider the human as a central component when doing logical decomposition and developing design concepts. The users—operators or maintainers—will not see the entire system as the designer does, only as the system interfaces with them. In engineering design reviews, human factors specialists promote the usability of the design solution. With early involvement, human factors assessments may catch usability problems at very early stages. For example, in one International Space Station payload design project, a human factors assessment of a very early block diagram of the layout of stowage and hardware identified problems that would have made operations very difficult. Changes were made to the conceptual design at negligible cost—i.e., rearranging conceptual block diagrams based on the sequence in which users would access items.

- **Usability Evaluations of Design Concepts:** Evaluations can be performed easily using rapid prototyping tools for hardware and software interfaces, standard human factors engineering data-gathering and analysis tools, and metrics such as task completion time and number of errors. Systematically collected subjective reports from operators also provide useful data. New technologies provide detailed objective information—e.g., eye tracking for display and control layout assessment. Human factors specialists provide assessment capabilities throughout the iterative design process.

- **Verification:** As mentioned, verification of requirements for usability, error rates, task completion times, and workload is challenging. Methods range from tests with trained personnel in mockups and simulators, to models of human performance, to inspection by experts. As members of the systems engineering team, human factors specialists provide verification guidance from the time requirements are first developed.

Human Factors Engineering Analyses Techniques and Methods

Example methods used to provide human performance data, predict human-system performance, and evaluate human-system designs include:

- **Task Analysis:** Produces a detailed description of the things a person must do in a system to accomplish a task, with emphasis on requirements for information presentation, decisions to be made, task times, operator actions, and environmental conditions.

- **Timeline Analysis:** Follows from task analysis. Durations of tasks are identified in task analyses, and the times at which these tasks occur are plotted in graphs, which also show the task sequences. The purpose is to identify requirements for simultaneous incompatible activities and activities that take longer than is available. Timelines for a given task can describe the activities of multiple operators or crewmembers.

- **Modeling and Simulation:** Models or mockups to make predictions about system performance, compare configurations, evaluate procedures, and evaluate alternatives. Simulations can be as simple as positioning a graphical human model with realistic anthropometric dimensions with a graphical model of an operator station, or they can be complex stochastic models capturing decision points, error opportunities, etc.

- **Usability Testing:** Based on a task analysis and preliminary design, realistic tasks are carried out in a controlled environment with monitoring and recording equipment. Objective measures such as performance time and number of errors are evaluated; subjective ratings are collected. The outputs systematically report on strengths and weaknesses of candidate design solutions.

- **Workload Assessment:** Measurement on a standardized scale such as the NASA-TLX or the Cooper-Harper rating scales of the amount and type of work. It assesses operator and crew task loading, which determines the ability of a human to perform the required tasks in the desired time with the desired accuracy.

- **Human Error and Human Reliability Assessment:** Top-down (fault tree analyses) and bottom-up (human factors process failure modes and effects analysis) analyses. The goal is to promote human reliability by creating a system that can tolerate and recover from human errors. Such a system must also support the human role in adding reliability to the system.

Roles of the Human Factors Specialist

The human factors specialist supports the systems engineering process by representing the users' and maintainers' requirements and capabilities throughout the design, production, and operations stages. Human factors specialists' roles include:

- Identify applicable requirements based on Agency standards for human-system integration during the requirements definition phase.
- Support development of mission concepts by providing information on human performance capabilities and limitations.
- Support task analysis and function allocation with information on human capabilities and limitations.
- Identify system design features that enhance usability. This integrates knowledge of human performance capabilities and design features.

- Support trade studies by providing data on effects of alternative designs on time to complete tasks, workload, and error rates.
- Support trade studies by providing data on effects of alternative designs on skills and training required to operate the system.
- Support design reviews to ensure compliance with human-systems integration requirements.
- Conduct evaluations using mockups and prototypes to provide detailed data on user performance.
- Support development of training and maintenance procedures in conjunction with hardware designers and mission planners.
- Collect data on human-system integration issues during operations to inform future designs.

5.0 Product Realization

This chapter describes the activities in the product realization processes listed in Figure 2.1-1. The chapter is separated into sections corresponding to steps 5 through 9 listed in Figure 2.1-1. The processes within each step are discussed in terms of the inputs, the activities, and the outputs. Additional guidance is provided using examples that are relevant to NASA projects.

The product realization side of the SE engine is where the rubber meets the road. In this portion of the engine, five interdependent processes result in systems that meet the design specifications and stakeholder expectations. These products are produced, acquired, reused, or coded; integrated into higher level assemblies; verified against design specifications; validated against stakeholder expectations; and transitioned to the next level of the system. As has been mentioned in previous sections, products can be models and simulations, paper studies or proposals, or hardware and software. The type and level of product depends on the phase of the life cycle and the product's specific objectives. But whatever the product, all must effectively use the processes to ensure the system meets the intended operational concept.

This effort starts with the technical team taking the output from the system design processes and using the appropriate crosscutting functions, such as data and configuration management, and technical assessments to make, buy, or reuse subsystems. Once these subsystems are realized, they must be integrated to the appropriate level as designated by the appropriate interface requirements. These products are then verified through the Technical Assessment Process to ensure they are consistent with the technical data package and that "the product was built right." Once consistency is achieved, the technical team will validate the products against the stakeholder expectations that "the right product was built." Upon successful completion of validation, the products are transitioned to the next level of the system. Figure 5.0-1 illustrates these processes.

This is an iterative and recursive process. Early in the life cycle, paper products, models, and simulations are run through the five realization processes. As the system matures and progresses through the life cycle, hardware and software products are run through these processes. It is important to catch errors and failures at the lowest level of integration and early in the life cycle so that changes can be made through the design processes with minimum impact to the project.

The next sections describe each of the five product realization processes and their associated products for a given NASA mission.

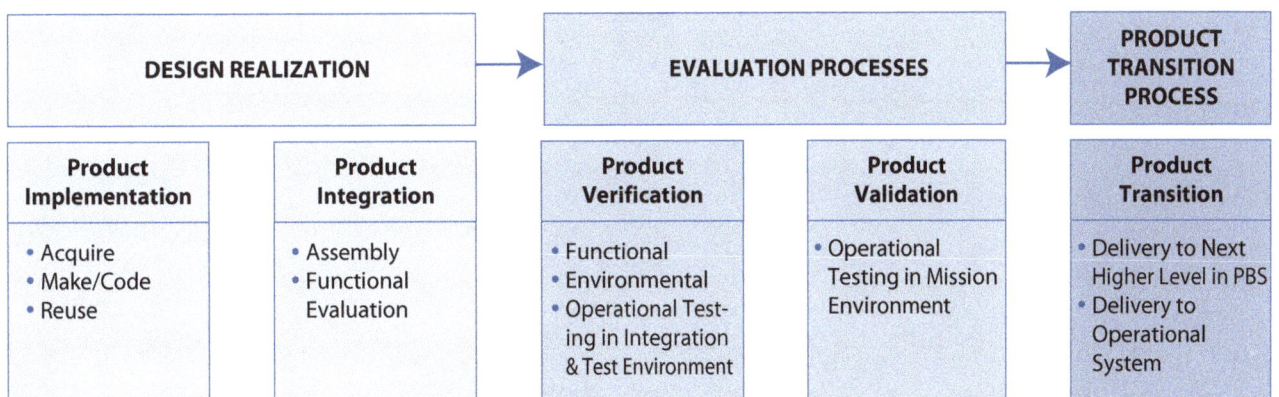

Figure 5.0-1 Product realization

Product Realization Keys

- Generate and manage requirements for off-the-shelf hardware/software products as for all other products.
- Understand the differences between verification testing and validation testing.
    - **Verification Testing:** Verification testing relates back to the approved requirements set (such as a System Requirements Document (SRD)) and can be performed at different stages in the product life cycle. Verification testing includes: (1) any testing used to assist in the development and maturation of products, product elements, or manufacturing or support processes; and/or (2) any engineering-type test used to verify status of technical progress, to verify that design risks are minimized, to substantiate achievement of contract technical performance, and to certify readiness for initial validation testing. Verification tests use instrumentation and measurements, and are generally accomplished by engineers, technicians, or operator-maintainer test personnel in a controlled environment to facilitate failure analysis.
    - **Validation Testing:** Validation relates back to the ConOps document. Validation testing is conducted under realistic conditions (or simulated conditions) on any end product for the purpose of determining the effectiveness and suitability of the product for use in mission operations by typical users; and the evaluation of the results of such tests. Testing is the detailed quantifying method of both verification and validation. However, testing is required to validate final end products to be produced and deployed.
- Consider all customer, stakeholder, technical, programmatic, and safety requirements when evaluating the input necessary to achieve a successful product transition.
- Analyze for any potential incompatibilities with interfaces as early as possible.
- Completely understand and analyze all test data for trends and anomalies.
- Understand the limitations of the testing and any assumptions that are made.
- Ensure that a reused product meets the verification and validation required for the relevant system in which it is to be used, as opposed to relying on the original verification and validation it met for the system of its original use. It would then be required to meet the same verification and validation as a purchased product or a built product. The "pedigree" of a reused product in its original application should not be relied upon in a different system, subsystem, or application.

5.1 Product Implementation

Product implementation is the first process encountered in the SE engine that begins the movement from the bottom of the product hierarchy up towards the Product Transition Process. This is where the plans, designs, analysis, requirements development, and drawings are realized into actual products.

Product implementation is used to generate a specified product of a project or activity through buying, making/coding, or reusing previously developed hardware, software, models, or studies to generate a product appropriate for the phase of the life cycle. The product must satisfy the design solution and its specified requirements.

The Product Implementation Process is the key activity that moves the project from plans and designs into realized products. Depending on the project and life-cycle phase within the project, the product may be hardware, software, a model, simulations, mockups, study reports, or other tangible results. These products may be realized through their purchase from commercial or other vendors, generated from scratch, or through partial or complete reuse of products from other projects or activities. The decision as to which of these realization strategies, or which combination of strategies, will be used for the

products of this project will have been made early in the life cycle using the Decision Analysis Process.

5.1.1 Process Description

Figure 5.1-1 provides a typical flow diagram for the Product Implementation Process and identifies typical inputs, outputs, and activities to consider in addressing product implementation.

5.1.1.1 Inputs

Inputs to the Product Implementation activity depend primarily on the decision as to whether the end product will be purchased, developed from scratch, or if the product will be formed by reusing part or all of products from other projects. Typical inputs are shown in Figure 5.1-1.

- **Inputs if Purchasing the End Product:** If the decision was made to purchase part or all of the products for this project, the end product design specifications are obtained from the configuration management system as well as other applicable documents such as the SEMP.

- **Inputs if Making/Coding the End Product:** For end products that will be made/coded by the technical

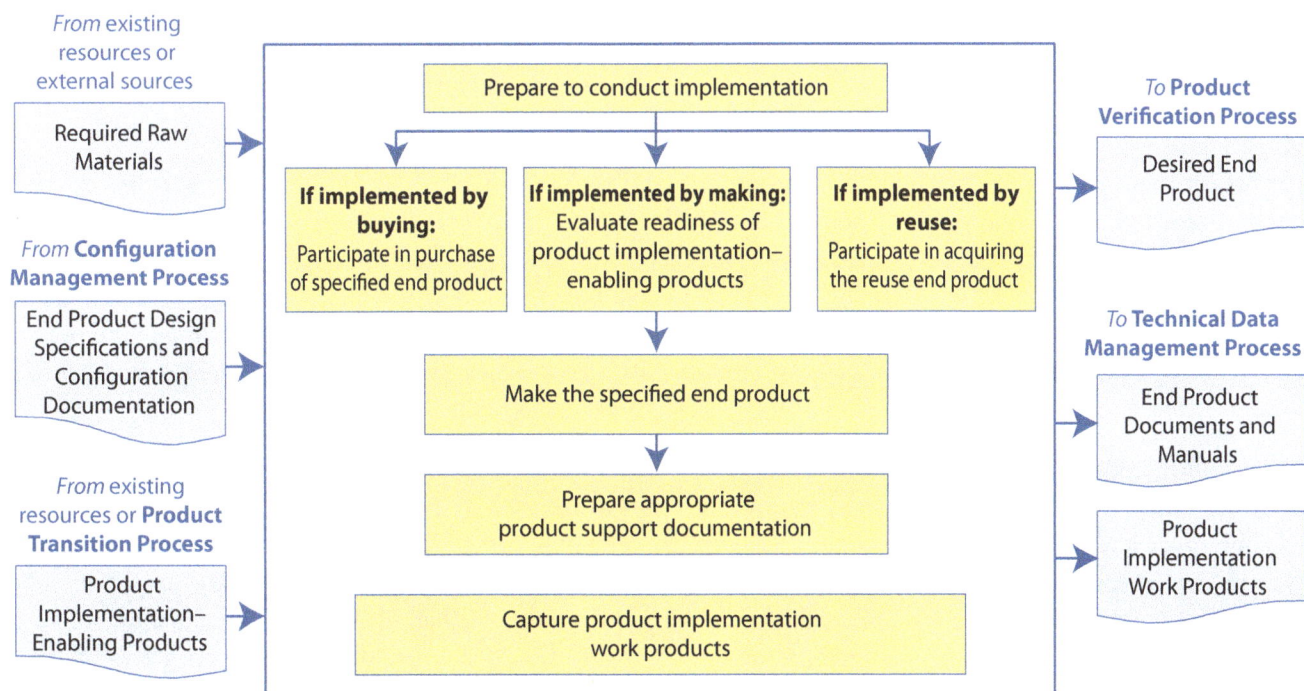

Figure 5.1-1 Product Implementation Process

team, the inputs will be the configuration controlled design specifications and raw materials as provided to or purchased by the project.

- **Inputs Needed if Reusing an End Product:** For end products that will reuse part or all of products generated by other projects, the inputs may be the documentation associated with the product, as well as the product itself. Care must be taken to ensure that these products will indeed meet the specifications and environments for this project. These would have been factors involved in the Decision Analysis Process to determine the make/buy/reuse decision.

5.1.1.2 Process Activities

Implementing the product can take one of three forms:

- Purchase/buy,
- Make/code, or
- Reuse.

These three forms will be discussed in the following subsections. Figure 5.1-1 shows what kind of inputs, outputs, and activities are performed during product implementation regardless of where in the product hierarchy or life cycle it is. These activities include preparing to conduct the implementation, purchasing/making/reusing the product, and capturing the product implementation work product. In some cases, implementing a product may have aspects of more than one of these forms (such as a build-to-print). In those cases, the appropriate aspects of the applicable forms are used.

Prepare to Conduct Implementation

Preparing to conduct the product implementation is a key first step regardless of what form of implementation has been selected. For complex projects, implementation strategy and detailed planning or procedures need to be developed and documented. For less complex projects, the implementation strategy and planning will need to be discussed, approved, and documented as appropriate for the complexity of the project.

The documentation, specifications, and other inputs will also need to be reviewed to ensure they are ready and at an appropriate level of detail to adequately complete the type of implementation form being employed and for the product life-cycle phase. For example, if the "make" implementation form is being employed, the design specifications will need to be reviewed to ensure they are at a design-to level that will allow the product to be de-

veloped. If the product is to be bought as a pure Commercial-Off-the-Shelf (COTS) item, the specifications will need to be checked to make sure they adequately describe the vendor characteristics to narrow to a single make/model of their product line.

Finally, the availability and skills of personnel needed to conduct the implementation as well as the availability of any necessary raw materials, enabling products, or special services should also be reviewed. Any special training necessary for the personnel to perform their tasks needs to be performed by this time.

Purchase, Make, or Reuse the Product

Purchase the Product

In the first case, the end product is to be purchased from a commercial or other vendor. Design/purchase specifications will have been generated during requirements development and provided as inputs. The technical team will need to review these specifications and ensure they are in a form adequate for the contract or purchase order. This may include the generation of contracts, Statements of Work (SOWs), requests for proposals, purchase orders, or other purchasing mechanisms. The responsibilities of the Government and contractor team should have been documented in the SEMP. This will define, for example, whether NASA expects the vendor to provide a fully verified and validated product or whether the NASA technical team will be performing those duties. The team will need to work with the acquisition team to ensure the accuracy of the contract SOW or purchase order and to ensure that adequate documentation, certificates of compliance, or other specific needs are requested of the vendor.

For contracted purchases, as proposals come back from the vendors, the technical team should work with the contracting officer and participate in the review of the technical information and in the selection of the vendor that best meets the design requirements for acceptable cost and schedule.

As the purchased products arrive, the technical team should assist in the inspection of the delivered product and its accompanying documentation. The team should ensure that the requested product was indeed the one delivered, and that all necessary documentation, such as source code, operator manuals, certificates of compliance, safety information, or drawings have been received.

The technical team should also ensure that any enabling products necessary to provide test, operations, maintenance, and disposal support for the product also are ready or provided as defined in the contract.

Depending on the strategy and roles/responsibilities of the vendor, as documented in the SEMP, a determination/analysis of the vendor's verification and validation compliance may need to be reviewed. This may be done informally or formally as appropriate for the complexity of the product. For products that were verified and validated by the vendor, after ensuring that all work products from this phase have been captured, the product may be ready to enter the Product Transition Process to be delivered to the next higher level or to its final end user. For products that will be verified and validated by the technical team, the product will be ready to be verified after ensuring that all work products for this phase have been captured.

Make/Code the Product

If the strategy is to make or code the product, the technical team should first ensure that the enabling products are ready. This may include ensuring all piece parts are available, drawings are complete and adequate, software design is complete and reviewed, machines to cut the material are available, interface specifications are approved, operators are trained and available, procedures/processes are ready, software personnel are trained and available to generate code, test fixtures are developed and ready to hold products while being generated, and software test cases are available and ready to begin model generation.

The product is then made or coded in accordance with the specified requirements, configuration documentation, and applicable standards. Throughout this process, the technical team should work with the quality organization to review, inspect, and discuss progress and status within the team and with higher levels of management as appropriate. Progress should be documented within the technical schedules. Peer reviews, audits, unit testing, code inspections, simulation checkout, and other techniques may be used to ensure the made or coded product is ready for the verification process.

Reuse

If the strategy is to reuse a product that already exists, care must be taken to ensure that the product is truly applicable to this project and for the intended uses and the environment in which it will be used. This should have been a factor used in the decision strategy to make/buy/reuse.

The documentation available from the reuse product should be reviewed by the technical team to become completely familiar with the product and to ensure it will meet the requirements in the intended environment. Any supporting manuals, drawings, or other documentation available should also be gathered.

The availability of any supporting or enabling products or infrastructure needed to complete the fabrication, coding, testing, analysis, verification, validation, or shipping of the product needs to be determined. If any of these products or services are lacking, they will need to be developed or arranged for before progressing to the next phase.

Special arrangements may need to be made or forms such as nondisclosure agreements may need to be acquired before the reuse product can be received.

A reused product will frequently have to undergo the same verification and validation as a purchased product or a built product. Relying on prior verification and validation should only be considered if the product's verification and validation documentation meets the verification, validation, and documentation requirements of the current project and the documentation demonstrates that the product was verified and validated against equivalent requirements and expectations. The savings gained from reuse is not necessarily from reduced testing, but in a lower likelihood that the item will fail tests and generate rework.

Capture Work Products

Regardless of what implementation form was selected, all work products from the make/buy/reuse process should be captured, including design drawings, design documentation, code listings, model descriptions, procedures used, operator manuals, maintenance manuals, or other documentation as appropriate.

5.1.1.3 Outputs

- **End Product for Verification:** Unless the vendor performs verification, the made/coded, purchased, or reused end product, in a form appropriate for the life-cycle phase, is provided for the verification process. The form of the end product is a function of the life-cycle phase and the placement within the system structure (the form of the end product could be hardware, software, model, prototype, first article for test, or single operational article or multiple production article).

- **End Product Documents and Manuals:** Appropriate documentation is also delivered with the end product to the verification process and to the technical data management process. Documentation may include applicable design drawings; operation, user, maintenance, or training manuals; applicable baseline documents (configuration baseline, specifications, stakeholder expectations); certificates of compliance; or other vendor documentation.

The process is complete when the following activities have been accomplished:

- End product is fabricated, purchased, or reuse modules acquired.
- End products are reviewed, checked, and ready for verification.
- Procedures, decisions, assumptions, anomalies, corrective actions, lessons learned, etc., resulting from the make/buy/reuse are recorded.

5.1.2 Product Implementation Guidance

5.1.2.1 Buying Off-the-Shelf Products

Off-the-Shelf (OTS) products are hardware/software that has an existing heritage and usually originates from one of several sources, which include commercial, military, and NASA programs. Special care needs to be taken when purchasing OTS products for use in the space environment. Most OTS products were developed for use in the more benign environments of Earth and may not be suitable to endure the harsh space environments, including vacuum, radiation, extreme temperature ranges, extreme lighting conditions, zero gravity, atomic oxygen, lack of convection cooling, launch vibration or acceleration, and shock loads.

When purchasing OTS products, requirements should still be generated and managed. A survey of available OTS is made and evaluated as to the extent they satisfy the requirements. Products that meet all the requirements are a good candidate for selection. If no product can be found to meet all the requirements, a trade study needs to be performed to determine whether the requirements can be relaxed or waived, the OTS can be modi-

fied to bring it into compliance, or whether another option to build or reuse should be selected.

Several additional factors should be considered when selecting the OTS option:

- Heritage of the product;
- Critical or noncritical application;
- Amount of modification required and who performs it;
- Whether sufficient documentation is available;
- Proprietary, usage, ownership, warranty, and licensing rights;
- Future support for the product from the vendor/provider;
- Any additional validation of the product needed by the project; and
- Agreement on disclosure of defects discovered by the community of users of the product.

5.1.2.2 Heritage

"Heritage" refers to the original manufacturer's level of quality and reliability that is built into parts and which has been proven by (1) time in service, (2) number of units in service, (3) mean time between failure performance, and (4) number of use cycles. High-heritage products are from the original supplier, who has maintained the great majority of the original service, design, performance, and manufacturing characteristics. Low-heritage products are those that (1) were not built by the original manufacturer; (2) do not have a significant history of test and usage; or (3) have had significant aspects of the original service, design, performance, or manufacturing characteristics altered. An important factor in assessing the heritage of a COTS product is to ensure that the use/application of the product is relevant to the application for which it is now intended. A product that has high heritage in a ground-based application could have a low heritage when placed in a space environment.

The focus of a "heritage review" is to confirm the applicability of the component for the current application. Assessments must be made regarding not only technical interfaces (hardware and software) and performance, but also the environments to which the unit has been previously qualified, including electromagnetic compatibility, radiation, and contamination. The compatibility of the design with parts quality requirements must also be assessed. All noncompliances must be identified, documented, and addressed either by modification to bring the component into compliance or formal waivers/deviations for accepted deficiencies. This heritage review is commonly held closely after contract award.

When reviewing a product's applicability, it is important to consider the nature of the application. A "catastrophic" application is one where a failure could cause loss of life or vehicle. A "critical" application is one where failure could cause loss of mission. For use in these applications, several additional precautions should be taken, including ensuring the product will not be used near the boundaries of its performance or environmental envelopes. Extra scrutiny by experts should be applied during Preliminary Design Reviews (PDRs) and Critical Design Reviews (CDRs) to ensure the appropriateness of its use.

Modification of an OTS product may be required for it to be suitable for a NASA application. This affects the product's heritage, and therefore, the modified product should be treated as a new design. If the product is modified by NASA and not the manufacturer, it would be beneficial for the supplier to have some involvement in reviewing the modification. NASA modification may also require the purchase of additional documentation from the supplier such as drawings, code, or other design and test descriptions.

For additional information and suggested test and analysis requirements for OTS products, see JSC EA-WI-016 or MSFC MWI 8060.1 both titled *Off the Shelf Hardware Utilization in Flight Hardware Development* and G-118-2006e *AIAA Guide for Managing the Use of Commercial Off the Shelf (COTS) Software Components for Mission-Critical Systems.*

5.2 Product Integration

Product Integration is one of the SE engine product realization processes that make up the system structure. In this process, lower level products are assembled into higher level products and checked to make sure that the integrated product functions properly. It is an element of the processes that lead realized products from a level below to realized end products at a level above, between the Product Implementation, Verification, and Validation Processes.

The purpose of the Product Integration Process is to systematically assemble the higher level product from the lower level products or subsystems (e.g., product elements, units, components, subsystems, or operator tasks); ensure that the product, as integrated, functions properly; and deliver the product. Product integration is required at each level of the system hierarchy. The activities associated with product integrations occur throughout the entire product life cycle. This includes all of the incremental steps, including level-appropriate testing, necessary to complete assembly of a product and to enable the top-level product tests to be conducted. The Product Integration Process may include and often begins with analysis and simulations (e.g., various types of prototypes) and progresses through increasingly more realistic incremental functionality until the final product is achieved. In each successive build, prototypes are constructed, evaluated, improved, and reconstructed based upon knowledge gained in the evaluation process. The degree of virtual versus physical prototyping required depends on the functionality of the design tools and the complexity of the product and its associated risk. There is a high probability that the product, integrated in this

manner, will pass product verification and validation. For some products, the last integration phase will occur when the product is deployed at its intended operational site. If any problems of incompatibility are discovered during the product verification and validation testing phase, they are resolved one at a time.

The Product Integration Process applies not only to hardware and software systems but also to service-oriented solutions, requirements, specifications, plans, and concepts. The ultimate purpose of product integration is to ensure that the system elements will function as a whole.

5.2.1 Process Description

Figure 5.2-1 provides a typical flow diagram for the Product Integration Process and identifies typical inputs, outputs, and activities to consider in addressing product integration. The activities of the Product Integration Process are truncated to indicate the action and object of the action.

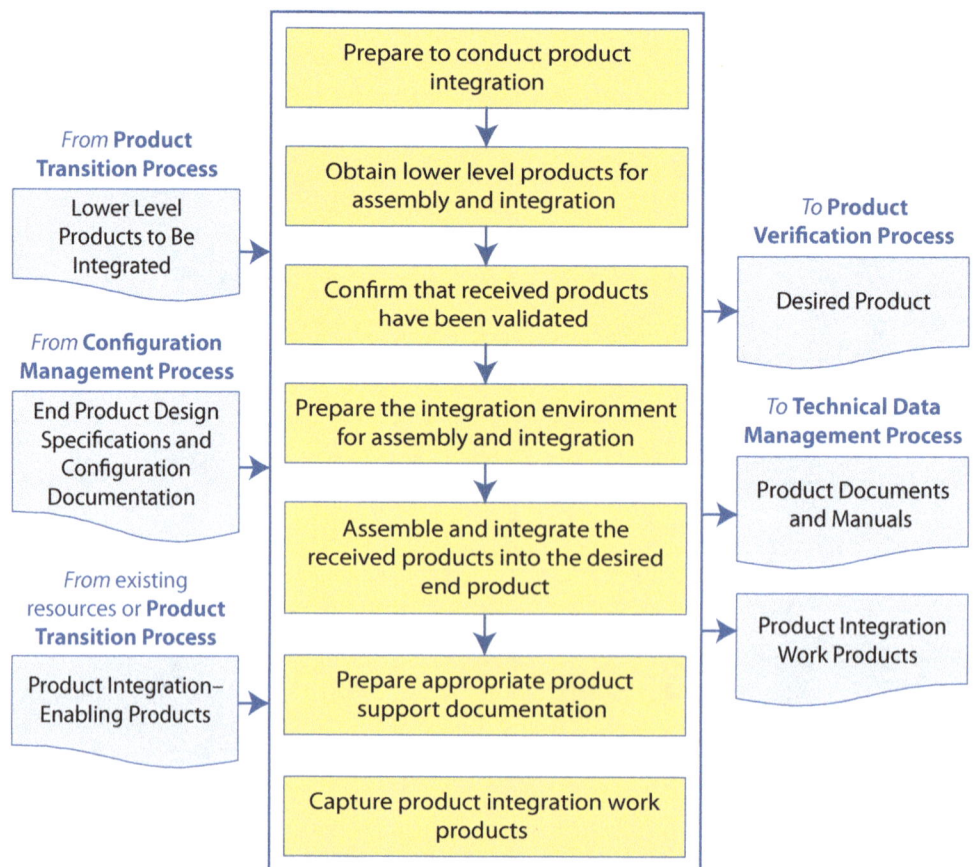

Figure 5.2-1 Product Integration Process

5.2.1.1 Inputs

Product Integration encompasses more than a one-time assembly of the lower level products and operator tasks at the end of the design and fabrication phase of the life cycle. An integration plan must be developed and documented. An example outline for an integration plan is provided in Appendix H. Product Integration is conducted incrementally, using a recursive process of assembling lower level products and operator tasks; evaluating them through test, inspection, analysis, or demonstration; and then assembling more lower level products and operator tasks. Planning for Product Integration should be initiated during the concept formulation phase of the life cycle. The basic tasks that need to be established involve the management of internal and external interfaces of the various levels of products and operator tasks to support product integration and are as follows:

- Define interfaces;
- Identify the characteristics of the interfaces (physical, electrical, mechanical, etc.);
- Ensure interface compatibility at all defined interfaces by using a process documented and approved by the project;
- Ensure interface compatibility at all defined interfaces;
- Strictly control all of the interface processes during design, construction, operation, etc.;
- Identify lower level products to be assembled and integrated (from the Product Transition Process);
- Identify assembly drawings or other documentation that show the complete configuration of the product being integrated, a parts list, and any assembly instructions (e.g., torque requirements for fasteners);
- Identify end-product, design-definition-specified requirements (specifications), and configuration documentation for the applicable work breakdown structure model, including interface specifications, in the form appropriate to satisfy the product-line life-cycle phase success criteria (from the Configuration Management Process); and
- Identify Product Integration–enabling products (from existing resources or the Product Transition Process for enabling product realization).

5.2.1.2 Process Activities

This subsection addresses the approach to the top-level implementation of the Product Integration Process, including the activities required to support the process,

The project would follow this approach throughout its life cycle.

The following are typical activities that support the Product Integration Process:

- Prepare to conduct Product Integration by (1) preparing a product integration strategy, detailed planning for the integration, and integration sequences and procedures and (2) determining whether the product configuration documentation is adequate to conduct the type of product integration applicable for the product-line life-cycle phase, location of the product in the system structure, and management phase success criteria.
- Obtain lower level products required to assemble and integrate into the desired product.
- Confirm that the received products that are to be assembled and integrated have been validated to demonstrate that the individual products satisfy the agreed-to set of stakeholder expectations, including interface requirements.
- Prepare the integration environment in which assembly and integration will take place, including evaluating the readiness of the product integration–enabling products and the assigned workforce.
- Assemble and integrate the received products into the desired end product in accordance with the specified requirements, configuration documentation, interface requirements, applicable standards, and integration sequencing and procedures.
- Conduct functional testing to ensure that assembly is ready to enter verification testing and ready to be integrated into the next level.
- Prepare appropriate product support documentation such as special procedures for performing product verification and product validation.
- Capture work products and related information generated while performing the product integration process activities.

5.2.1.3 Outputs

The following are typical outputs from this process and destinations for the products from this process:

- Integrated product(s) in the form appropriate to the product-line life-cycle phase and to satisfy phase success criteria (to the Product Verification Process).
- Documentation and manuals in a form appropriate for satisfying the life-cycle phase success criteria, in-

cluding as-integrated product descriptions and operate-to and maintenance manuals (to the Technical Data Management Process).

● Work products, including reports, records, and non-deliverable outcomes of product integration activities (to support the Technical Data Management Process); integration strategy document; assembly/check area drawings; system/component documentation sequences and rationale for selected assemblies; interface management documentation; personnel requirements; special handling requirements; system documentation; shipping schedules; test equipment and drivers' requirements; emulator requirements; and identification of limitations for both hardware and software.

5.2.2 Product Integration Guidance

5.2.2.1 Integration Strategy

An integration strategy is developed, as well as supporting documentation, to identify optimal sequence of receipt, assembly, and activation of the various components that make up the system. This strategy should use business as well as technical factors to ensure an assembly, activation, and loading sequence that minimizes cost and assembly difficulties. The larger or more complex the system or the more delicate the element, the more critical the proper sequence becomes, as small changes can cause large impacts on project results.

The optimal sequence of assembly is built from the bottom up as components become subelements, elements, and subsystems, each of which must be checked prior to fitting into the next higher assembly. The sequence will encompass any effort needed to establish and equip the assembly facilities (e.g., raised floor, hoists, jigs, test equipment, input/output, and power connections). Once established, the sequence must be periodically reviewed to ensure that variations in production and delivery schedules have not had an adverse impact on the sequence or compromised the factors on which earlier decisions were made.

5.2.2.2 Relationship to Product Implementation

As previously described, Product Implementation is where the plans, designs, analysis, requirements development, and drawings are realized into actual products. Product Integration concentrates on the control of the interfaces and the verification and validation to achieve the correct product to meet the requirements. Product Integration can be thought of as released or phased deliveries. Product Integration is the process that pulls together new and existing products and ensures that they all combine properly into a complete product without interference or complications. If there are issues, the Product Integration Process documents the exceptions, which can then be evaluated to determine if the product is ready for implementation/operations.

Integration occurs at every stage of a project's life cycle. In the Formulation phase, the decomposed requirements need to be integrated into a complete system to verify that nothing is missing or duplicated. In the Implementation phase, the design and hardware need to be integrated into an overall system to verify that they meet the requirements and that there are no duplications or omissions.

The emphasis on the recursive, iterative, and integrated nature of systems engineering highlights how the product integration activities are not only integrated across all of the phases of the entire life cycle in the initial planning stages of the project, but also used recursively across all of the life-cycle phases as the project product proceeds through the flow down and flow up conveyed by the SE engine. This ensures that when changes occur to requirements, design concepts, etc.—usually in response to updates from stakeholders and results from analysis, modeling, or testing—that adequate course corrections are made to the project. This is accomplished through re-evaluation by driving through the SE engine, enabling all aspects of the product integration activities to be appropriately updated. The result is a product that meets all of the new modifications approved by the project and eliminates the opportunities for costly and time-consuming modifications in the later stages of the project.

5.2.2.3 Product/Interface Integration Support

There are several processes that support the integration of products and interfaces. Each process allows either the integration of products and interfaces or the validation that the integrated products meet the needs of the project.

The following is a list of typical example processes and products that support the integration of products and interfaces and that should be addressed by the project in the overall approach to Product Integration: requirements documents; requirements reviews; design reviews; design drawings and specifications; integration and test plans; hardware configuration control documentation; quality assurance records; interface control requirements/documents; ConOps documents; verification requirement documents; verification reports/analysis; NASA, military, and industry standards; best practices; and lessons learned.

5.2.2.4 Product Integration of the Design Solution

This subsection addresses the more specific implementation of Product Integration related to the selected design solution.

Generally, system/product designs are an aggregation of subsystems and components. This is relatively obvious for complex hardware and/or software systems. The same holds true for many service-oriented solutions. For example, a solution to provide a single person access to the Internet involves hardware, software, and a communications interface. The purpose of Product Integration is to ensure that combination of these elements achieves the required result (i.e., works as expected). Consequently, internal and external interfaces must be considered in the design and evaluated prior to production.

There are a variety of different testing requirements to verify product integration at all levels. Qualification testing and acceptance testing are examples of two of these test types that are performed as the product is integrated. Another type of testing that is important to the design and ultimate product integration is a planned test process in which development items are tested under actual or simulated mission profile environments to disclose design deficiencies and to provide engineering information on failure modes and mechanisms. If accomplished with development items, this provides early insight into any issues that may otherwise only be observed at the late stages of product integration where it becomes costly to incorporate corrective actions. For large, complex system/products, integration/verification efforts are accomplished using a prototype.

5.2.2.5 Interface Management

The objective of the interface management is to achieve functional and physical compatibility among all interrelated system elements. Interface management is defined in more detail in Section 6.3. An interface is any boundary between one area and another. It may be cognitive, external, internal, functional, or physical. Interfaces occur within the system (internal) as well as between the system and another system (external) and may be functional or physical (e.g., mechanical, electrical) in nature. Interface requirements are documented in an Interface Requirements Document (IRD). Care should be taken to define interface requirements and to avoid specifying design solutions when creating the IRD. In its final form, the Interface Control Document (ICD) describes the detailed implementation of the requirements contained in the IRD. An interface control plan describes the management process for IRDs and ICDs. This plan provides the means to identify and resolve interface incompatibilities and to determine the impact of interface design changes.

5.2.2.6 Compatibility Analysis

During the program's life, compatibility and accessibility must be maintained for the many diverse elements. Compatibility analysis of the interface definition demonstrates completeness of the interface and traceability records. As changes are made, an authoritative means of controlling the design of interfaces must be managed with appropriate documentation, thereby avoiding the situation in which hardware or software, when integrated into the system, fails to function as part of the system as intended. Ensuring that all system pieces work together is a complex task that involves teams, stakeholders, contractors, and program management from the end of the initial concept definition stage through the operations and support stage. Physical integration is accomplished during Phase D. At the finer levels of resolution, pieces must be tested, assembled and/or integrated, and tested again. The systems engineer role includes performance of the delegated management duties such as configuration control and overseeing the integration, verification, and validation processes.

5.2.2.7 Interface Management Tasks

The interface management tasks begin early in the development effort, when interface requirements can be influenced by all engineering disciplines and applicable interface standards can be invoked. They continue through design and checkout. During design, emphasis is on ensuring that interface specifications are documented and communicated. During system element checkout, both prior to assembly and in the assembled configuration, emphasis is on verifying the implemented interfaces. Throughout the product integration process activities, interface baselines are controlled to ensure that changes in the design of system elements have minimal impact on other elements with which they interface. During testing or other validation and verification activities, multiple system elements are checked out as integrated subsystems or systems. The following provides more details on these tasks.

Define Interfaces

The bulk of integration problems arise from unknown or uncontrolled aspects of interfaces. Therefore, system and subsystem interfaces are specified as early as possible in the development effort. Interface specifications address logical, physical, electrical, mechanical, human, and environmental parameters as appropriate. Intra-system interfaces are the first design consideration for developers of the system's subsystems. Interfaces are used from previous development efforts or are developed in accordance with interface standards for the given discipline or technology. Novel interfaces are constructed only for compelling reasons. Interface specifications are verified against interface requirements. Typical products include interface descriptions, ICDs, interface requirements, and specifications.

Verify Interfaces

In verifying the interfaces, the systems engineer must ensure that the interfaces of each element of the system or subsystem are controlled and known to the developers. Additionally, when changes to the interfaces are needed, the changes must at least be evaluated for possible impact on other interfacing elements and then communicated to the affected developers. Although all affected developers are part of the group that makes changes, such changes need to be captured in a readily accessible place so that the current state of the interfaces can be known to all. Typical products include ICDs and exception reports.

The use of emulators for verifying hardware and software interfaces is acceptable where the limitations of the emulator are well characterized and meet the operating environment characteristics and behavior requirements for interface verification. The integration plan should specifically document the scope of use for emulators.

Inspect and Acknowledge System and Subsystem Element Receipt

Acknowledging receipt and inspecting the condition of each system or subsystem element is required prior to

assembling the system in accordance with the intended design. The elements are checked for quantity, obvious damage, and consistency between the element description and a list of element requirements. Typical products include acceptance documents, delivery receipts, and checked packing list.

Verify System and Subsystem Elements

System and subsystem element verification confirms that the implemented design features of developed or purchased system elements meet their requirements. This is intended to ensure that each element of the system or subsystem functions in its intended environment, including those elements that are OTS for other environments. Such verifications may be by test (e.g., regression testing as a tool or subsystem/elements are combined), inspection, analysis (deficiency or compliance reports), or demonstration and may be executed either by the organization that will assemble the system or subsystem or by the producing organization. A method of discerning the elements that "passed" verification from those elements that "failed" needs to be in place. Typical products include verified system features and exception reports.

Verify Element Interfaces

Verification of the system element interfaces ensures that the elements comply with the interface specification prior to assembly in the system. The intent is to ensure that the interface of each element of the system or subsystem is verified against its corresponding interface specification. Such verification may be by test, inspection, analysis, or demonstration and may be executed by the organization that will assemble the system or subsystem or by another organization. Typical products include verified system element interfaces, test reports, and exception reports.

Integrate and Verify

Assembly of the elements of the system should be performed in accordance with the established integration strategy. This ensures that the assembly of the system elements into larger or more complex assemblies is conducted in accordance with the planned strategy. To ensure that the integration has been completed, a verification of the integrated system interfaces should be performed. Typical products include integration reports, exception reports, and an integrated system.

5.3 Product Verification

The Product Verification Process is the first of the verification and validation processes conducted on a realized end product. As used in the context of the systems engineering common technical processes, a realized product is one provided by either the Product Implementation Process or the Product Integration Process in a form suitable for meeting applicable life-cycle phase success criteria. Realization is the act of verifying, validating, and transitioning the realized product for use at the next level up of the system structure or to the customer. Simply put, the Product Verification Process answers the critical question, Was the end product realized right? The Product Validation Process addresses the equally critical question, Was the right end product realized?

Verification proves that a realized product for any system model within the system structure conforms to the build-to requirements (for software elements) or realize-to specifications and design descriptive documents (for hardware elements, manual procedures, or composite products of hardware, software, and manual procedures).

Distinctions Between Product Verification and Product Validation

From a process perspective, product verification and validation may be similar in nature, but the objectives are fundamentally different.

It is essential to confirm that the realized product is in conformance with its specifications and design description documentation (i.e., verification). Such specifications and documents will establish the configuration baseline of that product, which may have to be modified at a later time. Without a verified baseline and appropriate configuration controls, such later modifications could be costly or cause major performance problems. However, from a customer point of view, the interest is in whether the end product provided will do what the customer intended within the environment of use (i.e., validation). When cost effective and warranted by analysis, the expense of validation testing alone can be mitigated by combining tests to perform verification and validation simultaneously.

The outcome of the Product Verification Process is confirmation that the "as-realized product," whether achieved by implementation or integration, conforms

to its specified requirements, i.e., verification of the end product. This subsection discusses the process activities, inputs, outcomes, and potential deficiencies.

5.3.1 Process Description

Figure 5.3-1 provides a typical flow diagram for the Product Verification Process and identifies typical inputs, outputs, and activities to consider in addressing product verification.

5.3.1.1 Inputs

Key inputs to the process are the product to be verified, verification plan, specified requirements baseline, and any enabling products needed to perform the Product Verification Process (including the ConOps, mission needs and goals, requirements and specifications, in-

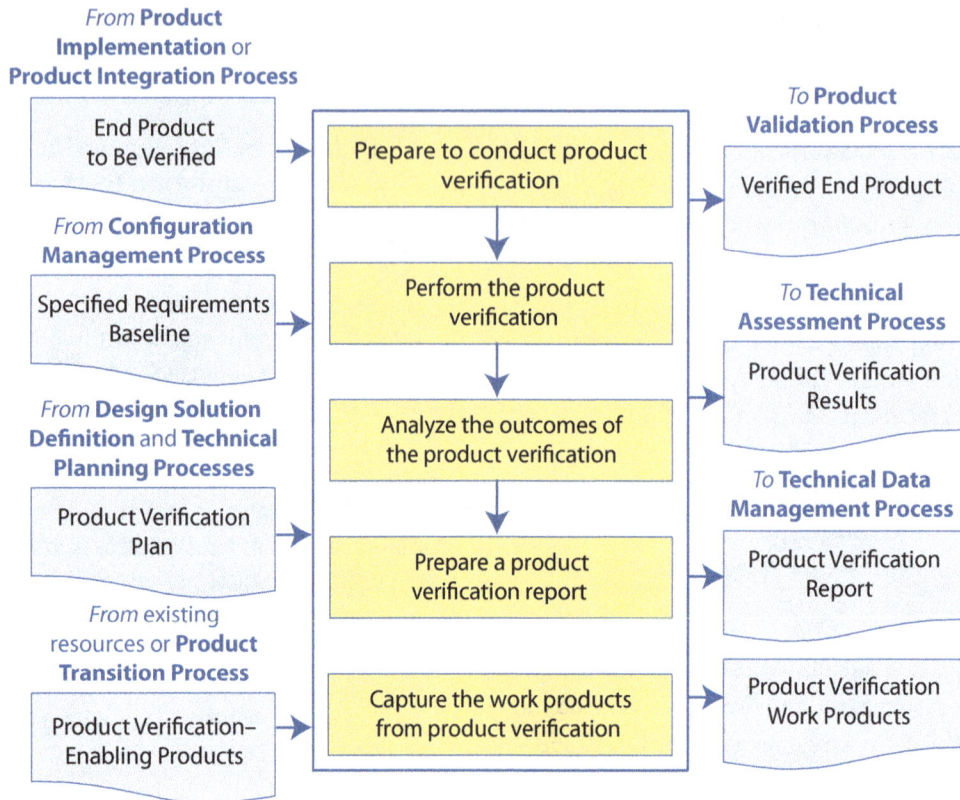

Figure 5.3-1 Product Verification Process

terface control drawings, testing standards and policies, and Agency standards and policies).

5.3.1.2 Process Activities

There are five major steps in the Product Verification Process: (1) verification planning (prepare to implement the verification plan); (2) verification preparation (prepare for conducting verification); (3) conduct verification (perform verification); (4) analyze verification results; and (5) capture the verification work products.

The objective of the Product Verification Process is to generate evidence necessary to confirm that end products, from the lowest level of the system structure to the highest, conform to the specified requirements (specifications and descriptive documents) to which they were realized whether by the Product Implementation Process or by the Product Integration Process.

Product Verification is usually performed by the developer that produced (or "realized") the end product, with participation of the end user and customer. Product Verification confirms that the as-realized product, whether it was achieved by Product Implementation or Product Integration, conforms to its specified require-

ments (specifications and descriptive documentation) used for making or assembling and integrating the end product. Developers of the system, as well as the users, are typically involved in verification testing. The customer and Quality Assurance (QA) personnel are also critical in the verification planning and execution activities.

Product Verification Planning

Planning to conduct the product verification is a key first step. From relevant specifications and product form, the type of verification (e.g., analysis, demonstration, inspection, or test) should be established based on the life-cycle phase, cost, schedule, resources, and the position of the end product within the system structure. The verification plan should be reviewed (an output of the Technical Planning Process, based on design solution outputs) for any specific procedures, constraints, success criteria, or other verification requirements. (See Appendix I for a sample verification plan outline.)

Verification Plan and Methods

The task of preparing the verification plan includes establishing the type of verification to be performed, de-

---

Types of Testing

There are many different types of testing that can be used in verification of an end product. These examples are provided for consideration:

- Aerodynamic
- Burn-in
- Drop
- Environmental
- High-/Low-Voltage Limits
- Leak Rates
- Nominal
- Parametric
- Pressure Limits
- Security Checks
- Thermal Limits

- Acceptance
- Characterization
- Electromagnetic Compatibility
- G-loading
- Human Factors Engineering/ Human-in-the-Loop Testing
- Lifetime/Cycling
- Off-Nominal
- Performance
- Qualification Flow
- System
- Thermal Vacuum

- Acoustic
- Component
- Electromagnetic Interference
- Go or No-Go
- Integration
- Manufacturing/Random Defects
- Operational
- Pressure Cycling
- Structural Functional
- Thermal Cycling
- Vibration

---

pendent on the life-cycle phase; position of the product in the system structure; the form of the product used; and related costs of verification of individual specified requirements. The types of verification include analyses, inspection, demonstration, and test or some combination of these four. The verification plan, typically written at a detailed technical level, plays a pivotal role in bottom-up product realization.

> Note: Close alignment of the verification plan with the project's SEMP is absolutely essential.

Verification can be performed recursively throughout the project life cycle and on a wide variety of product forms. For example:

- Simulated (algorithmic models, virtual reality simulator);
- Mockup (plywood, brass board, breadboard);
- Concept description (paper report);
- Prototype (product with partial functionality);
- Engineering unit (fully functional but may not be same form/fit);
- Design verification test units (form, fit, and function is the same, but they may not have flight parts);
- Qualification units (identical to flight units but may be subjected to extreme environments); and

- Flight units (end product that is flown, including proto-flight units).

Any of these types of product forms may be in any of these states:

- Produced (built, fabricated, manufactured, or coded);
- Reused (modified internal nondevelopmental products or OTS product); and
- Assembled and integrated (a composite of lower level products).

The conditions and environment under which the product is to be verified should be established and the verification planned based on the associated entrance/success criteria identified. The Decision Analysis Process should be used to help finalize the planning details.

Procedures should be prepared to conduct verification based on the type (e.g., analysis, inspection, demonstration, or test) planned. These procedures are typically developed during the design phase of the project life cycle and matured as the design is matured. Operational use

> Note: The final, official verification of the end product should be for a controlled unit. Typically, attempting to "buy off" a "shall" on a prototype is not acceptable; it is usually completed on a qualification, flight, or other more final, controlled unit.

Types of Verification

- **Analysis:** The use of mathematical modeling and analytical techniques to predict the suitability of a design to stakeholder expectations based on calculated data or data derived from lower system structure end product verifications. Analysis is generally used when a prototype; engineering model; or fabricated, assembled, and integrated product is not available. Analysis includes the use of modeling and simulation as analytical tools. A model is a mathematical representation of reality. A simulation is the manipulation of a model.

- **Demonstration:** Showing that the use of an end product achieves the individual specified requirement. It is generally a basic confirmation of performance capability, differentiated from testing by the lack of detailed data gathering. Demonstrations can involve the use of physical models or mockups; for example, a requirement that all controls shall be reachable by the pilot could be verified by having a pilot perform flight-related tasks in a cockpit mockup or simulator. A demonstration could also be the actual operation of the end product by highly qualified personnel, such as test pilots, who perform a one-time event that demonstrates a capability to operate at extreme limits of system performance, an operation not normally expected from a representative operational pilot.

- **Inspection:** The visual examination of a realized end product. Inspection is generally used to verify physical design features or specific manufacturer identification. For example, if there is a requirement that the safety arming pin has a red flag with the words "Remove Before Flight" stenciled on the flag in black letters, a visual inspection of the arming pin flag can be used to determine if this requirement was met.

- **Test:** The use of an end product to obtain detailed data needed to verify performance, or provide sufficient information to verify performance through further analysis. Testing can be conducted on final end products, breadboards, brass boards or prototypes. Testing produces data at discrete points for each specified requirement under controlled conditions and is the most resource-intensive verification technique. As the saying goes, "Test as you fly, and fly as you test." (See Subsection 5.3.2.5.)

---

scenarios are thought through so as to explore all possible verification activities to be performed.

Outcomes of verification planning include the following:

- The verification type that is appropriate for showing or proving the realized product conforms to its specified requirements is selected.

- The product verification procedures are clearly defined based on: (1) the procedures for each type of verification selected, (2) the purpose and objective of each procedure, (3) any pre-verification and post-ver-

> Note: Verification planning is begun early in the project life cycle during the requirements development phase. (See Section 4.2.) Which verification approach to use should be included as part of the requirements development to plan for the future activities, establish special requirements derived from verification-enabling products identified, and to ensure that the technical statement is a verifiable requirement. Updates to verification planning continue throughout logical decomposition and design development, especially as design reviews and simulations shed light on items under consideration. (See Section 6.1.)

ification actions, and (4) the criteria for determining the success or failure of the procedure.

- The verification environment (e.g., facilities, equipment, tools, simulations, measuring devices, personnel, and climatic conditions) in which the verification procedures will be implemented is defined.

- As appropriate, project risk items are updated based on approved verification strategies that cannot duplicate fully integrated test systems, configurations, and/or target operating environments. Rationales, trade space, optimization results, and implications of the approaches are documented in the new or revised risk statements as well as references to accommodate future design, test, and operational changes to the project baseline.

Product Verification Preparation

In preparation for verification, the specified requirements (outputs of the Design Solution Process) are collected and confirmed. The product to be verified is obtained (output from implementation or integration), as are any enabling products and support resources that are necessary for verification (requirements identified and acquisition initiated by design solution definition activities). The final element of verification preparation includes the preparation of the verification environment (e.g., facilities, equipment, tools, simulations, measuring devices, personnel, and climatic conditions). Identification of the environmental requirements is necessary and the implications of those requirements must be carefully considered.

> Note: Depending on the nature of the verification effort and the life-cycle phase the program is in, some type of review to assess readiness for verification (as well as validation later) is typically held. In earlier phases of the life cycle, these reviews may be held informally; in later phases of the life cycle, this review becomes a formal event called a Test Readiness Review. TRRs and other technical reviews are an activity of the Technical Assessment Process.
>
> On most projects, a number of TRRs with tailored entrance/success criteria are held to assess the readiness and availability of test ranges; test facilities; trained testers; instrumentation; integration labs; support equipment; and other enabling products; etc.
>
> Peer reviews are additional reviews that may be conducted formally or informally to ensure readiness for verification (as well as the results of the verification process).

Outcomes of verification preparation include the following:

- The preparations for performing the verification as planned are completed;
- An appropriate set of specified requirements and supporting configuration documentation is available and on hand;
- Articles/models to be used for verification are on hand, assembled, and integrated with the verification environment according to verification plans and schedules;
- The resources needed to conduct the verification are available according to the verification plans and schedules; and
- The verification environment is evaluated for adequacy, completeness, readiness, and integration.

Conduct Planned Product Verification

The actual act of verifying the end product is conducted as spelled out in the plans and procedures and confor-

mance established to each specified verification requirement. The responsible engineer should ensure that the procedures were followed and performed as planned, the verification-enabling products were calibrated correctly, and the data were collected and recorded for required verification measures.

The Decision Analysis Process should be used to help make decisions with respect to making needed changes in the verification plans, environment, and/or conduct.

Outcomes of conducting verification include the following:

- A verified product is established with supporting confirmation that the appropriate results were collected and evaluated to show completion of verification objectives,
- A determination as to whether the realized end product (in the appropriate form for the life-cycle phase) complies with its specified requirements,
- A determination that the verification product was appropriately integrated with the verification environment and each specified requirement was properly verified, and
- A determination that product functions were verified both together and with interfacing products throughout their performance envelope.

Analyze Product Verification Results

Once the verification activities have been completed, the results are collected and analyzed. The data are analyzed for quality, integrity, correctness, consistency, and validity, and any verification anomalies, variations, and out-of-compliance conditions are identified and reviewed.

Variations, anomalies, and out-of-compliance conditions must be recorded and reported for followup action and closure. Verification results should be recorded in the requirements compliance matrix developed during the Technical Requirements Definition Process or other mechanism to trace compliance for each verification requirement.

System design and product realization process activities may be required to resolve anomalies not resulting from poor verification conduct, design, or conditions. If there are anomalies not resulting from the verification conduct, design, or conditions, and the mitigation of these anomalies results in a change to the product, the verification may need to be planned and conducted again.

Outcomes of analyzing the verification results include the following:

- End-product variations, anomalies, and out-of-compliance conditions have been identified;
- Appropriate replanning, redefinition of requirements, design and reverification have been accomplished for resolution for anomalies, variations, or out-of-compliance conditions (for problems not caused by poor verification conduct);
- Variances, discrepancies, or waiver conditions have been accepted or dispositioned;
- Discrepancy and corrective action reports have been generated as needed; and
- The verification report is completed.

Reengineering

Based on analysis of verification results, it could be necessary to re-realize the end product used for verification or to reengineer the end products assembled and integrated into the product being verified, based on where and what type of defect was found.

Reengineering could require the reapplication of the system design processes (Stakeholder Expectations Definition, Technical Requirements Definition, Logical Decomposition, and Design Solution Definition).

Verification Deficiencies

Verification test outcomes can be unsatisfactory for several reasons. One reason is poor conduct of the verification (e.g., procedures not followed, equipment not calibrated, improper verification environmental conditions, or failure to control other variables not involved in verifying a specified requirement). A second reason could be that the realized end product used was not realized correctly. Reapplying the system design processes could create the need for the following:

> Note: Nonconformances and discrepancy reports may be directly linked with the Technical Risk Management Process. Depending on the nature of the nonconformance, approval through such bodies as a material review board or configuration control board (which typically includes risk management participation) may be required.

- Reengineering products lower in the system structure that make up the product that were found to be defective (i.e., they failed to satisfy verification requirements) and/or
- Reperforming the Product Verification Process.

Pass Verification But Fail Validation?

Many systems successfully complete verification but then are unsuccessful in some critical phase of the validation process, delaying development and causing extensive rework and possible compromises with the stakeholder. Developing a solid ConOps in early phases of the project (and refining it through the requirements development and design phases) is critical to preventing unsuccessful validation. Communications with stakeholders helps to identify operational scenarios and key needs that must be understood when designing and implementing the end product. Should the product fail validation, redesign may be a necessary reality. Review of the understood requirements set, the existing design, operational scenarios, and support material may be necessary, as well as negotiations and compromises with the customer, other stakeholders, and/or end users to determine what, if anything, can be done to correct or resolve the situation. This can add time and cost to the overall project or, in some cases, cause the project to fail or be cancelled.

Capture Product Verification Work Products

Verification work products (inputs to the Technical Data Management Process) take many forms and involve many sources of information. The capture and recording of verification results and related data is a very important, but often underemphasized, step in the Product Verification Process.

Verification results, anomalies, and any corrective action(s) taken should be captured, as should all relevant results from the application of the Product Verification Process (related decisions, rationale for the decisions made, assumptions, and lessons learned).

Outcomes of capturing verification work products include the following:

- Verification of work products are recorded, e.g., type of verification, procedures, environments, outcomes, decisions, assumptions, corrective actions, lessons learned.

- Variations, anomalies, and out-of-compliance conditions have been identified and documented, including the actions taken to resolve them.

- Proof that the realized end product did or did not satisfy the specified requirements is documented.

- The verification report is developed, including:
  - ▶ Recorded test/verification results/data;
  - ▶ Version of the set of specified requirements used;
  - ▶ Version of the product verified;
  - ▶ Version or standard for tools, data, and equipment used;
  - ▶ Results of each verification including pass or fail declarations; and
  - ▶ Expected versus actual discrepancies.

5.3.1.3 Outputs

Key outputs from the process are:

- Discrepancy reports and identified corrective actions;

- Verified product to validation or integration; and

- Verification report(s) and updates to requirements compliance documentation (including verification plans, verification procedures, verification matrices, verification results and analysis, and test/demonstration/inspection/analysis records).

Success criteria include: (1) documented objective evidence of compliance (or waiver, as appropriate) with each system-of-interest requirement and (2) closure of all discrepancy reports. The Product Verification Process is not considered or designated complete until all discrepancy reports are closed (i.e., all errors tracked to closure).

5.3.2 Product Verification Guidance

5.3.2.1 Verification Program

A verification program should be tailored to the project it supports. The project manager/systems engineer must work with the verification engineer to develop a verification program concept. Many factors need to be considered in developing this concept and the subsequent verification program. These factors include:

- Project type, especially for flight projects. Verification methods and timing depend on:
  - ▶ The type of flight article involved (e.g., an experiment, payload, or launch vehicle).

- ▶ NASA payload classification (*NPR 8705.4, Risk Classification for NASA Payloads*). Guidelines are intended to serve as a starting point for establishment of the formality of test programs which can be tailored to the needs of a specific project based on the "A-D" payload classification.

- ▶ Project cost and schedule implications. Verification activities can be significant drivers of a project's cost and schedule; these implications should be considered early in the development of the verification program. Trade studies should be performed to support decisions about verification methods and requirements and the selection of facility types and locations. For example, a trade study might be made to decide between performing a test at a centralized facility or at several decentralized locations.

- ▶ Risk implications. Risk management must be considered in the development of the verification program. Qualitative risk assessments and quantitative risk analyses (e.g., a Failure Mode and Effects Analysis (FMECA)) often identify new concerns that can be mitigated by additional testing, thus increasing the extent of verification activities. Other risk assessments contribute to trade studies that determine the preferred methods of verification to be used and when those methods should be performed. For example, a trade might be made between performing a model test versus determining model characteristics by a less costly, but less revealing, analysis. The project manager/systems engineer must determine what risks are acceptable in terms of the project's cost and schedule.

- Availability of verification facilities/sites and transportation assets to move an article from one location to another (when needed). This requires coordination with the Integrated Logistics Support (ILS) engineer.

- Acquisition strategy (i.e., in-house development or system contract). Often, a NASA field center can shape a contractor's verification process through the project's SOW.

- Degree of design inheritance and hardware/software reuse.

5.3.2.2 Verification in the Life Cycle

The type of verification completed will be a function of the life-cycle phase and the position of the end product within the system structure. The end product must be verified and validated before it is transitioned to the next level up as part of the bottom-up realization process. (See Figure 5.3-2.)

While illustrated here as separate processes, there can be considerable overlap between some verification and validation events when implemented.

Quality Assurance in Verification

Even with the best of available designs, hardware fabrication, software coding, and testing, projects are subject to the vagaries of nature and human beings. The systems engineer needs to have some confidence that the system actually produced and delivered is in accordance with its functional, performance, and design requirements. QA provides an independent assessment to the project manager/systems engineer of the items produced and processes used during the project life cycle. The QA engineer typically acts as the systems engineer's eyes and ears in this context.

The QA engineer typically monitors the resolution and closeout of nonconformances and problem/failure reports; verifies that the physical configuration of the system conforms to the build-to (or code-to) documentation approved at CDR; and collects and maintains QA data for subsequent failure analyses. The QA engineer also participates in major reviews (primarily SRR, PDR, CDR, and FRR) on issues of design, materials, workmanship, fabrication and verification processes, and other characteristics that could degrade product system quality.

The project manager/systems engineer must work with the QA engineer to develop a QA program (the extent, responsibility, and timing of QA activities) tailored to the project it supports. In part, the QA program ensures verification requirements are properly specified, especially with respect to test environments, test configurations, and pass/fail criteria, and monitors qualification and acceptance tests to ensure compliance with verification requirements and test procedures to ensure that test data are correct and complete.

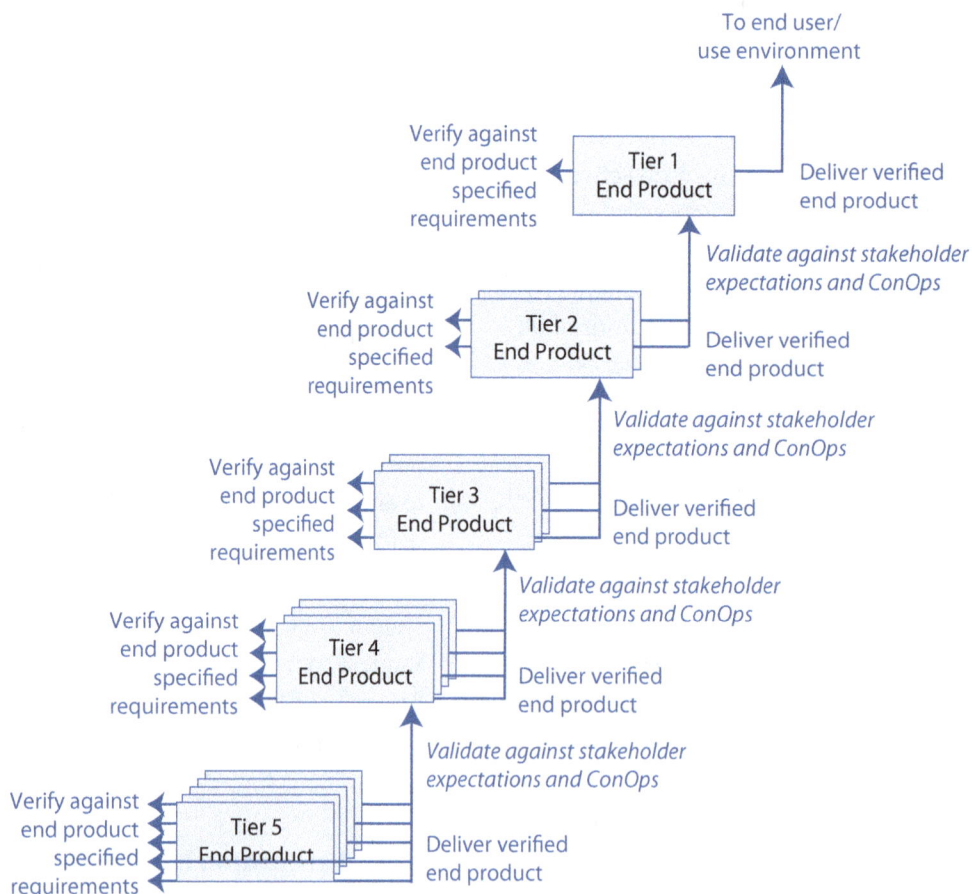

Figure 5.3-2 Bottom-up realization process

Configuration Verification

Configuration verification is the process of verifying that resulting products (e.g., hardware and software items) conform to the baselined design and that the baseline documentation is current and accurate. Configuration verification is accomplished by two types of control gate activity: audits and technical reviews.

Qualification Verification

Qualification-stage verification activities begin after completion of development of the flight/operations hardware designs and include analyses and testing to ensure that the flight/operations or flight-type hardware (and software) will meet functional and performance requirements in anticipated environmental conditions. During this stage, many performance requirements are verified, while analyses and models are updated as test data are acquired. Qualification tests generally are designed to subject the hardware to worst-case loads and environmental stresses plus a defined level of margin. Some of the verifications performed to ensure hardware compliance are vibration/acoustic, pressure limits, leak rates, thermal vacuum, thermal cycling, Electromagnetic Interference and Electromagnetic Compatibility (EMI/EMC), high- and low-voltage limits, and lifetime/cycling. Safety requirements, defined by hazard analysis reports, may also be satisfied by qualification testing.

Qualification usually occurs at the component or subsystem level, but could occur at the system level as well. A project deciding against building dedicated qualification hardware—and using the flight/operations hardware itself for qualification purposes—is termed "protoflight." Here, the requirements being verified are typically less than that of qualification levels but higher than that of acceptance levels.

Qualification verification verifies the soundness of the design. Test levels are typically set with some margin above expected flight/operations levels, including the maximum number of cycles that can be accumulated during acceptance testing. These margins are set to address design safety margins in general, and care should be exercised not to set test levels so that unrealistic failure modes are created.

Acceptance Verification

Acceptance-stage verification activities provide the assurance that the flight/operations hardware and software are in compliance with all functional, performance, and design requirements and are ready for shipment to the launch site. The acceptance stage begins with the acceptance of each individual component or piece part for assembly into the flight/operations article, continuing through the System Acceptance Review (SAR). (See Subsection 6.7.2.1.)

Some verifications cannot be performed after a flight/operations article, especially a large one, has been assembled and integrated (e.g., due to inaccessibility). When this occurs, these verifications are to be performed *during* fabrication and integration, and are known as "in-process" tests. In this case, acceptance testing begins with in-process testing and continues through functional testing, environmental testing, and end-to-end compatibility testing. Functional testing normally begins at the component level and continues at the systems level, ending with all systems operating simultaneously.

When flight/operations hardware is unavailable, or its use is inappropriate for a specific test, simulators may be used to verify interfaces. Anomalies occurring during a test are documented on the appropriate reporting system, and a proposed resolution should be defined before testing continues. Major anomalies, or those that are not easily dispositioned, may require resolution by a collaborative effort of the systems engineer and the design, test, and other organizations. Where appropriate, analyses and models are validated and updated as test data are acquired.

Acceptance verification verifies workmanship, not design. Test levels are set to stress items so that failures arise from defects in parts, materials, and workmanship. As such, test levels are those anticipated during flight/operations with no additional margin.

Deployment Verification

The pre-launch verification stage begins with the arrival of the flight/operations article at the launch site and concludes at liftoff. During this stage, the flight/operations article is processed and integrated with the launch vehicle. The launch vehicle could be the shuttle or some other launch vehicle, or the flight/operations article could be part of the launch vehicle. Verifications performed during this stage ensure that no visible damage to the system has occurred during shipment and that the system continues to function properly.

If system elements are shipped separately and integrated at the launch site, testing of the system and system interfaces is generally required. If the system is integrated into a carrier, the interface to the carrier must also be verified. Other verifications include those that occur following integration into the launch vehicle and those that occur at the launch pad; these are intended to ensure that the system is functioning and in its proper launch configuration. Contingency verifications and procedures are developed for any contingencies that can be foreseen to occur during pre-launch and countdown. These contingency verifications and procedures are critical in that some contingencies may require a return of the launch vehicle or flight/operations article from the launch pad to a processing facility.

Operational and Disposal Verification

Operational verification begins in Phase E and provides the assurance that the system functions properly in a relevant environment. These verifications are performed through system activation and operation, rather than through a verification activity. Systems that are assembled on-orbit must have each interface verified and must function properly during end-to-end testing. Mechanical interfaces that provide fluid and gas flow must be verified to ensure no leakage occurs and that pressures and flow rates are within specification. Environmental systems must be verified.

Disposal verification provides the assurance that the safe deactivation and disposal of all system products and processes has occurred. The disposal stage begins in Phase F at the appropriate time (i.e., either as scheduled, or earlier in the event of premature failure or accident) and concludes when all mission data have been acquired and verifications necessary to establish compliance with disposal requirements are finished.

Both operational and disposal verification activities may also include validation assessments, that is, assessments of the degree to which the system accomplished the desired mission goals/objectives.

5.3.2.3 Verification Procedures

Verification procedures provide step-by-step instructions for performing a given verification activity. This procedure could be a test, demonstration, or any other verification-related activity. The procedure to be used is written and submitted for review and approval at the Test Readiness Review (TRR) for the verification activity. (See Test Readiness Review discussion in Subsection 6.7.2.1.)

Procedures are also used to verify the acceptance of facilities, electrical and mechanical ground support equipment, and special test equipment. The information generally contained in a procedure is as follows, but it may vary according to the activity and test article:

- Nomenclature and identification of the test article or material;
- Identification of test configuration and any differences from flight/operations configuration;
- Identification of objectives and criteria established for the test by the applicable verification specification;
- Characteristics and design criteria to be inspected or tested, including values, with tolerances, for acceptance or rejection;
- Description, in sequence, of steps and operations to be taken;
- Identification of computer software required;
- Identification of measuring, test, and recording equipment to be used, specifying range, accuracy, and type;
- Credentials showing that required computer test programs/support equipment and software have been verified prior to use with flight/operations hardware;
- Any special instructions for operating data recording equipment or other automated test equipment as applicable;
- Layouts, schematics, or diagrams showing identification, location, and interconnection of test equipment, test articles, and measuring points;
- Identification of hazardous situations or operations;
- Precautions and safety instructions to ensure safety of personnel and prevent degradation of test articles and measuring equipment;
- Environmental and/or other conditions to be maintained with tolerances;
- Constraints on inspection or testing;
- Special instructions for nonconformances and anomalous occurrences or results; and
- Specifications for facility, equipment maintenance, housekeeping, quality inspection, and safety and handling requirements before, during, and after the total verification activity.

The written procedure may provide blank spaces in the format for the recording of results and narrative comments so that the completed procedure can serve as part of the verification report. The as-run and certified copy of the procedure is maintained as part of the project's archives.

5.3.2.4 Verification Reports

A verification report should be provided for each analysis and, at a minimum, for each major test activity—such as functional testing, environmental testing, and end-to-end compatibility testing—occurring over long periods of time or separated by other activities. Verification reports may be needed for each individual test activity, such as functional testing, acoustic testing, vibration testing, and thermal vacuum/thermal balance testing. Verification reports should be completed within a few weeks following a test and should provide evidence of compliance with the verification requirements for which it was conducted.

The verification report should include as appropriate:

- Verification objectives and the degree to which they were met;
- Description of verification activity;
- Test configuration and differences from flight/operations configuration;
- Specific result of each test and each procedure, including annotated tests;
- Specific result of each analysis;
- Test performance data tables, graphs, illustrations, and pictures;
- Descriptions of deviations from nominal results, problems/failures, approved anomaly corrective actions, and retest activity;
- Summary of nonconformance/discrepancy reports, including dispositions;
- Conclusions and recommendations relative to success of verification activity;
- Status of support equipment as affected by test;
- Copy of as-run procedure; and
- Authentication of test results and authorization of acceptability.

5.3.2.5 End-to-End System Testing

The objective of end-to-end testing is to demonstrate interface compatibility and desired total functionality among different elements of a system, between systems,

> Note: It is important to understand that, over the lifetime of a system, requirements may change or component obsolescence may make a design solution too difficult to produce from either a cost or technical standpoint. In these instances, it is critical to employ the systems engineering design processes at a lower level to ensure the modified design provides a proper design solution. An evaluation should be made to determine the magnitude of the change required, and the process should be tailored to address the issues appropriately. A modified qualification, verification, and validation process may be required to baseline a new design solution, consistent with the intent previously described for those processes. The acceptance testing will also need to be updated as necessary to verify that the new product has been manufactured and coded in compliance with the revised baselined design.

and within a system as a whole. End-to-end tests performed on the integrated ground and flight system include all elements of the payload, its control, stimulation, communications, and data processing to demonstrate that the entire system is operating in a manner to fulfill all mission requirements and objectives.

End-to-end testing includes executing complete threads or operational scenarios across multiple configuration items, ensuring that all mission and performance requirements are verified. Operational scenarios are used extensively to ensure that the system (or collections of systems) will successfully execute mission requirements. Operational scenarios are a step-by-step description of how the system should operate and interact with its users and its external interfaces (e.g., other systems). Scenarios should be described in a manner that will allow engineers to walk through them and gain an understanding of how all the various parts of the system should function and interact as well as verify that the system will satisfy the user's needs and expectations. Operational scenarios should be described for all operational modes, mission phases (e.g., installation, startup, typical examples of normal and contingency operations, shutdown, and maintenance), and critical sequences of activities for all classes of users identified. Each scenario should include events, actions, stimuli, information, and interactions as appropriate to provide a comprehensive understanding of the operational aspects of the system.

Figure 5.3-3 presents an example of an end-to-end data flow for a scientific satellite mission. Each arrow in the diagram represents one or more data or control flows between two hardware, software, subsystem, or system configuration items. End-to-end testing verifies that the data flows throughout the multisystem environment are correct, that the system provides the required functionality, and that the outputs at the eventual end points correspond to expected results. Since the test environment is as close an approximation as possible to the operational environment, performance requirements testing is also included. This figure is not intended to show the full extent of end-to-end testing. Each system shown would need to be broken down into a further level of granularity for completeness.

End-to-end testing is an integral part of the verification and validation of the total system and is an activity that is employed during selected hardware, software, and system phases throughout the life cycle. In comparison with configuration item testing, end-to-end testing addresses each configuration item only down to the level where it interfaces externally to other configuration items, which can be either hardware, software, or human based. Internal interfaces (e.g., software subroutine calls, analog-to-digital conversion) of a configuration item are not within the scope of end-to-end testing.

How to Perform End-to-End Testing

End-to-end testing is probably the most significant element of any project verification program and the test should be designed to satisfy the edict to "test the way we fly." This means assembling the system in its realistic configuration, subjecting it to a realistic environment and then "flying" it through all of its expected operational modes. For a scientific robotic mission, targets and stimuli should be designed to provide realistic inputs to the scientific instruments. The output signals from the instruments would flow through the satellite data-handling system and then be transmitted to the actual ground station through the satellite communications system. If data are transferred to the ground station through one or more satellite or ground relays (e.g., the Tracking and Data Relay Satellite System (TDRSS)) then those elements must be included as part of the test.

The end-to-end compatibility test encompasses the entire chain of operations that will occur during all mission modes in such a manner as to ensure that the system will fulfill mission requirements. The mission environment should be simulated as realistically as possible, and the instruments should receive stimuli of the kind they will receive during the mission. The Radio Frequency (RF) links, ground station operations, and software functions should be fully exercised. When acceptable simulation

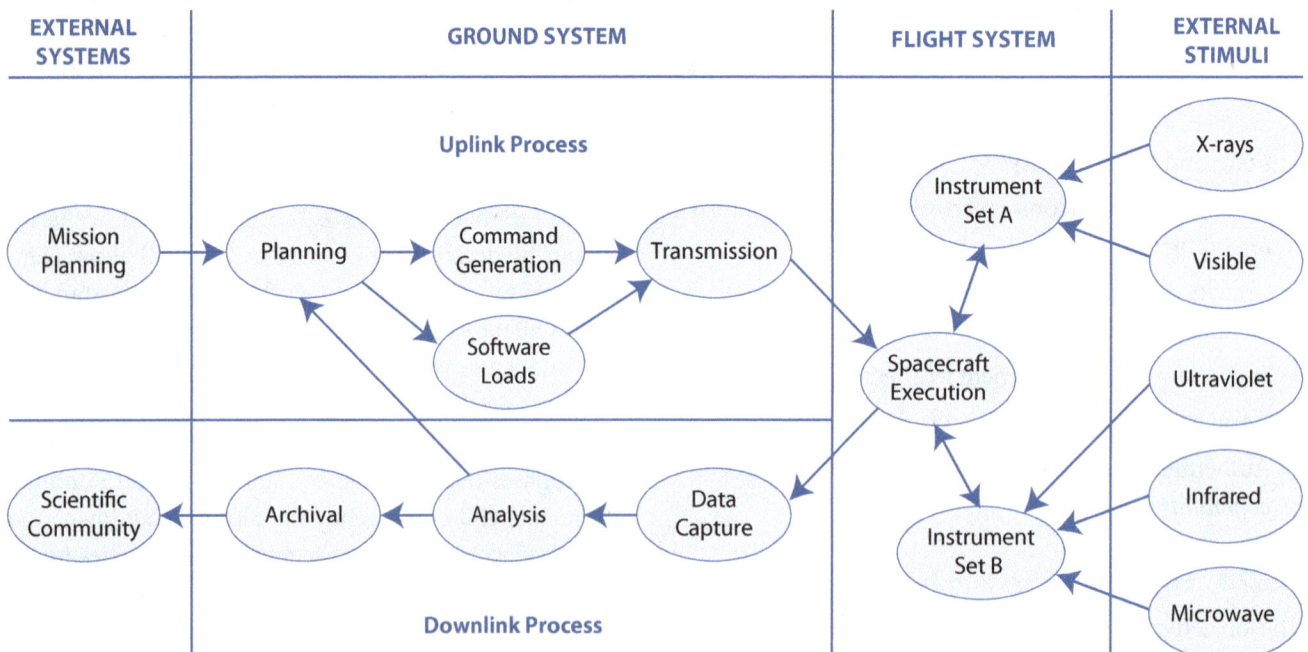

Figure 5.3-3 Example of end-to-end data flow for a scientific satellite mission

facilities are available for portions of the operational systems, they may be used for the test instead of the actual system elements. The specific environments under which the end-to-end test is conducted and the stimuli, payload configuration, RF links, and other system elements to be used must be determined in accordance with the characteristics of the mission.

Although end-to-end testing is probably the most complex test in any system verification program, the same careful preparation is necessary as for any other system-level test. For example, a test lead must be appointed and the test team selected and trained. Adequate time must be allocated for test planning and coordination with the design team. Test procedures and test software must be documented, approved, and placed under configuration control.

Plans, agreements, and facilities must be put in place well in advance of the test to enable end-to-end testing between all components of the system.

Once the tests are run, the test results are documented and any discrepancies carefully recorded and reported. All test data must be maintained under configuration control.

> Note: This is particularly important when missions are developed with international or external partners.

Before completing end-to-end testing, the following activities are completed for each configuration item:

- All requirements, interfaces, states, and state transitions of each configuration item should be tested through the exercise of comprehensive test procedures and test cases to ensure the configuration items are complete and correct.

- A full set of operational range checking tests should be conducted on software variables to ensure that the software performs as expected within its complete range and fails, or warns, appropriately for out-of-range values or conditions.

End-to-end testing activities include the following:

1. Operational scenarios are created that span all of the following items (during nominal, off-nominal, and stressful conditions) that could occur during the mission:

- Mission phase, mode, and state transitions;
- First-time events;
- Operational performance limits;
- Fault protection routines;
- Failure Detection, Isolation, and Recovery (FDIR) logic;
- Safety properties;
- Operational responses to transient or off-nominal sensor signals; and
- Communication uplink and downlink.

2. The operational scenarios are used to test the configuration items, interfaces, and end-to-end performance as early as possible in the configuration items' development life cycle. This typically means simulators or software stubs have to be created to implement a full scenario. It is extremely important to produce a skeleton of the actual system to run full scenarios as soon as possible with both simulated/stubbed-out and actual configuration items.

3. A complete diagram and inventory of all interfaces are documented.

4. Test cases are executed to cover human-human, human-hardware, human-software, hardware-software, software-software, and subsystem-subsystem interfaces and associated inputs, outputs, and modes of operation (including safing modes).

5. It is strongly recommended that during end-to-end testing, an operations staff member who has not previously been involved in the testing activities be designated to exercise the system as it is intended to be used to determine if it will fail.

6. The test environment should approximate/simulate the actual operational conditions when possible. The fidelity of the test environment should be authenticated. Differences between the test and operational environment should be documented in the test or verification plan.

7. When testing of a requirement is not possible, verification is demonstrated by other means (e.g., model checking, analysis, or simulation). If true end-to-end testing cannot be achieved, then the testing must be done piecemeal and patched together by analysis and simulation. An example of this would be a system that is assembled on orbit where the various elements come together for the first time on orbit.

8. When an error in the developed system is identified and fixed, regression testing of the system or compo-

nent is performed to ensure that modifications have not caused unintended effects and that the system or component still complies with previously tested specified requirements.

9. When tests are aborted or a test is known to be flawed (e.g., due to configuration, test environment), the test should be rerun after the identified problem is fixed.

10. The operational scenarios should be used to formulate the final operations plan.

11. Prior to system delivery, as part of the system qualification testing, test cases should be executed to cover all of the plans documented in the operations plan in the order in which they are expected to occur during the mission.

End-to-end test documentation includes the following:

- Inclusion of end-to-end testing plans as a part of the test or verification plan.
- A document, matrix, or database under configuration control that traces the end-to-end system test suite to the results. Data that are typically recorded include the test-case identifier, subsystems/hardware/program sets exercised, list of the requirements being verified, interfaces exercised, date, and outcome of test (i.e., whether the test actual output met the expected output).
- End-to-end test cases and procedures (including inputs and expected outputs).
- A record of end-to-end problems/failures/anomalies.

End-to-end testing can be integrated with other project testing activities; however, the documentation mentioned in this subsection should be readily extractable for review, status, and assessment.

5.3.2.6 Modeling and Simulation

For the Product Verification Process, a model is a physical, mathematical, or logical representation of an end product to be verified. Modeling and Simulation (M&S) can be used to augment and support the Product Verification Process and is an effective tool for performing the verification whether in early life-cycle phases or later. Both the facilities and the model itself are developed using the system design and product realization processes.

The model used, as well as the M&S facility, are enabling products and must use the 17 technical processes (see *NPR 7123.1, NASA Systems Engineering Processes and Require-*

> Note: The development of the physical, mathematical, or logical model includes evaluating whether the model to be used as representative of the system end product was realized according to its design–solution-specified requirements for a model and whether it will be valid for use as a model. In some cases, the model must also be accredited to certify the range of specific uses for which the model can be used. Like any other enabling product, budget and time must be planned for creating and evaluating the model to be used to verify the applicable system end product.

ments) for their development and realization (including acceptance by the operational community) to ensure that the model and simulation adequately represent the operational environment and performance of the modeled end product. Additionally, in some cases certification is required before models and simulations can be used.

M&S assets can come from a variety of sources; for example, contractors, other Government agencies, or laboratories can provide models that address specific system attributes.

5.3.2.7 Hardware-in-the-Loop

Fully functional end products, such as an actual piece of hardware, may be combined with models and simulations that simulate the inputs and outputs of other end products of the system. This is referred to as "Hardware-in-the-Loop" (HWIL) testing. HWIL testing links all elements (subsystems and test facilities) together within a synthetic environment to provide a high-fidelity, real-time operational evaluation for the real system or subsystems. The operator can be intimately involved in the testing, and HWIL resources can be connected to other facilities for distributed test and analysis applications. One of the uses of HWIL testing is to get as close to the actual concept of operation as possible to support verification and validation when the operational environment is difficult or expensive to recreate.

During development, this HWIL verification normally takes place in an integration laboratory or test facility. For example, HWIL could be a complete spacecraft in a special test chamber, with the inputs/outputs being provided as output from models that simulate the system in an operational environment. Real-time computers are used to control the spacecraft and subsystems in projected operational scenarios. Flight dynamics, responding to the commands issued by the guidance and control system hardware/software, are simulated in real-time to determine the trajectory and to calculate system flight conditions. HWIL testing verifies that the end product being evaluated meets the interface requirements, properly transforming inputs to required outputs. HWIL modeling can provide a valuable means of testing physical end products lower in the system structure by providing simulated inputs to the end product or receiving outputs from the end product to evaluate the quality of those outputs. This tool can be used throughout the life cycle of a program or project. The shuttle program uses an HWIL to verify software and hardware updates for the control of the shuttle main engines.

Modeling, simulation, and hardware/human-in-the-loop technology, when appropriately integrated and sequenced with testing, provide a verification method at a reasonable cost. This integrated testing process specifically (1) reduces the cost of life-cycle testing, (2) provides significantly more engineering/performance insights into each system evaluated, and (3) reduces test time and lowers project risk. This process also significantly reduces the number of destructive tests required over the life of the product. The integration of M&S into verification testing provides insights into trends and tendencies of system and subsystem performance that might not otherwise be possible due to hardware limitations.

5.4 Product Validation

The Product Validation Process is the second of the verification and validation processes conducted on a realized end product. While verification proves whether "the system was done right," validation proves whether "the right system was done." In other words, verification provides objective evidence that every "shall" statement was met, whereas validation is performed for the benefit of the customers and users to ensure that the system functions in the expected manner when placed in the intended environment. This is achieved by examining the products of the system at every level of the structure.

Validation confirms that realized end products at any position within the system structure conform to their set of stakeholder expectations captured in the ConOps, and ensures that any anomalies discovered during validation are appropriately resolved prior to product delivery. This section discusses the process activities, types of validation, inputs and outputs, and potential deficiencies.

Distinctions Between Product Verification and Product Validation

From a process perspective, Product Verification and Product Validation may be similar in nature, but the objectives are fundamentally different.

From a customer point of view, the interest is in whether the end product provided will do what they intend within the environment of use. It is essential to confirm that the realized product is in conformance with its specifications and design description documentation because these specifications and documents will establish the configuration baseline of the product, which may have to be modified at a later time. Without a verified baseline and appropriate configuration controls, such later modifications could be costly or cause major performance problems.

When cost-effective and warranted by analysis, various combined tests are used. The expense of validation testing alone can be mitigated by ensuring that each end product in the system structure was correctly realized in accordance with its specified requirements before conducting validation.

5.4.1 Process Description

Figure 5.4-1 provides a typical flow diagram for the Product Validation Process and identifies typical inputs, outputs, and activities to consider in addressing product validation.

5.4.1.1 Inputs

Key inputs to the process are:

- Verified product,
- Validation plan,
- Baselined stakeholder expectations (including ConOps and mission needs and goals), and
- Any enabling products needed to perform the Product Validation Process.

Differences Between Verification and Validation Testing

- **Verification Testing:** Verification testing relates back to the approved requirements set (such as an SRD) and can be performed at different stages in the product life cycle. Verification testing includes: (1) any testing used to assist in the development and maturation of products, product elements, or manufacturing or support processes; and/or (2) any engineering-type test used to verify status of technical progress, to verify that design risks are minimized, to substantiate achievement of contract technical performance, and to certify readiness for initial validation testing. Verification tests use instrumentation and measurements, and are generally accomplished by engineers, technicians, or operator-maintainer test personnel in a controlled environment to facilitate failure analysis.

- **Validation Testing:** Validation relates back to the ConOps document. Validation testing is conducted under realistic conditions (or simulated conditions) on any end product for the purpose of determining the effectiveness and suitability of the product for use in mission operations by typical users; and the evaluation of the results of such tests. Testing is the detailed quantifying method of both verification and validation. However, testing is required to validate final end products to be produced and deployed.

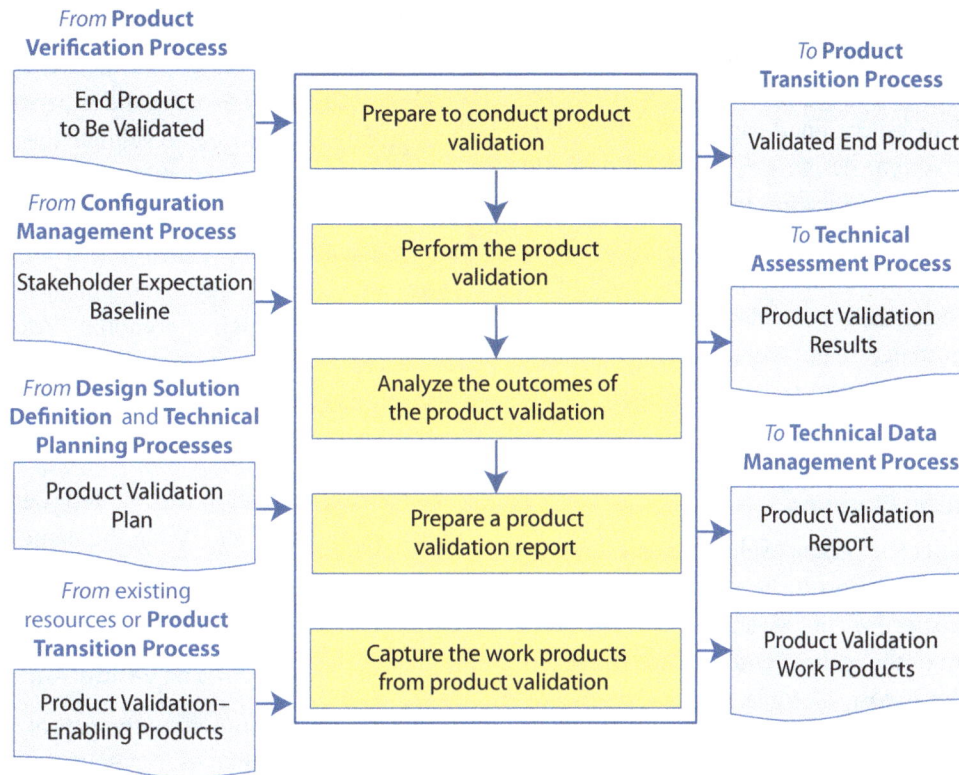

Figure 5.4-1 Product Validation Process

5.4.1.2 Process Activities

The Product Validation Process demonstrates that the realized end product satisfies its stakeholder (customer and other interested party) expectations within the intended operational environments, with validation performed by anticipated operators and/or users. The type of validation is a function of the life-cycle phase and the position of the end product within the system structure.

There are five major steps in the validation process: (1) validation planning (prepare to implement the validation plan), (2) validation preparation (prepare for conducting validation), (3) conduct planned validation (perform validation), (4) analyze validation results, and (5) capture the validation work products.

The objectives of the Product Validation Process are:
- To confirm that
  - ▶ The right product was realized—the one wanted by the customer,
  - ▶ The realized product can be used by intended operators/users, and
  - ▶ The Measures of Effectiveness (MOEs) are satisfied.
- To confirm that the realized product fulfills its intended use when operated in its intended environment:

  - ▶ Validation is performed for each realized (implemented or integrated) product from the lowest end product in a system structure branch up to the top WBS model end product.
  - ▶ Evidence is generated as necessary to confirm that products at each layer of the system structure meet the capability and other operational expectations of the customer/user/operator and other interested parties.

- To ensure that any problems discovered are appropriately resolved prior to delivery of the realized product (if validation is done by the supplier of the product) or prior to integration with other products into a higher level assembled product (if validation is done by the receiver of the product).

Verification and validation events are illustrated as separate processes, but when used, can considerably overlap. When cost effective and warranted by analysis, various combined tests are used. However, while from a process perspective verification and validation are similar in nature, their objectives are fundamentally different.

From a customer's point of view, the interest is in whether the end product provided will supply the needed capabilities within the intended environments of use. The expense of validation testing alone can be mitigated by ensuring that each end product in the system structure was correctly realized in accordance with its specified requirements prior to validation, during verification. It is possible that the system design was not done properly and, even though the verification tests were successful (satisfying the specified requirements), the validation tests would still fail (stakeholder expectations not satisfied). Thus, it is essential that validation of lower products in the system structure be conducted as well as verification so as to catch design failures or deficiencies as early as possible.

Product Validation Planning

Planning to conduct the product validation is a key first step. The type of validation to be used (e.g., analysis, demonstration, inspection, or test) should be established based on the form of the realized end product, the applicable life-cycle phase, cost, schedule, resources available, and location of the system product within the system structure. (See Appendix I for a sample verification and validation plan outline.)

An established set or subset of requirements to be validated should be identified and the validation plan reviewed (an output of the Technical Planning Process, based on design solution outputs) for any specific procedures, constraints, success criteria, or other validation requirements. The conditions and environment under which the product is to be validated should be established and the validation planned based on the relevant life-cycle phase and associated success criteria identified. The Decision Analysis Process should be used to help finalize the planning details.

It is important to review the validation plans with relevant stakeholders and understand the relationship between the context of the validation and the context of use (human involvement). As part of the planning process, validation-enabling products should be identified, and scheduling and/or acquisition initiated.

Procedures should be prepared to conduct validation based on the type (e.g., analysis, inspection, demonstration, or test) planned. These procedures are typically developed during the design phase of the project life cycle and matured as the design is matured. Operational and use-case scenarios are thought through so as to explore all possible validation activities to be performed.

Validation Plan and Methods

The validation plan is one of the work products of the Technical Planning Process and is generated during the Design Solution Process to validate the realized product against the baselined stakeholder expectations. This plan can take many forms. The plan describes the total Test and Evaluation (T&E) planning from development of lower end through higher end products in the system structure and through operational T&E into production and acceptance. It may include the verification and validation plan. (See Appendix I for a sample verification and validation plan outline.)

The types of validation include test, demonstration, inspection, and analysis. While the name of each method

---

Types of Validation

- **Analysis:** The use of mathematical modeling and analytical techniques to predict the suitability of a design to stakeholder expectations based on calculated data or data derived from lower system structure end product validations. It is generally used when a prototype; engineering model; or fabricated, assembled, and integrated product is not available. Analysis includes the use of both modeling and simulation.

- **Demonstration:** The use of a realized end product to show that a set of stakeholder expectations can be achieved. It is generally used for a basic confirmation of performance capability and is differentiated from testing by the lack of detailed data gathering. Validation is done under realistic conditions for any end product within the system structure for the purpose of determining the effectiveness and suitability of the product for use in NASA missions or mission support by typical users and evaluating the results of such tests.

- **Inspection:** The visual examination of a realized end product. It is generally used to validate physical design features or specific manufacturer identification.

- **Test:** The use of a realized end product to obtain detailed data to validate performance or to provide sufficient information to validate performance through further analysis. Testing is the detailed quantifying method of both verification and validation but it is required in order to validate final end products to be produced and deployed.

is the same as the name of the methods for verification, the purpose and intent are quite different.

Validation is conducted by the user/operator or by the developer, as determined by NASA Center directives or the contract with the developers. Systems-level validation (e.g., customer T&E and some other types of validation) may be performed by an acquirer testing organization. For those portions of validation performed by the developer, appropriate agreements must be negotiated to ensure that validation proof-of-documentation is delivered with the realized product.

All realized end products, regardless of the source (buy, make, reuse, assemble and integrate) and the position in the system structure, should be validated to demonstrate/confirm satisfaction of stakeholder expectations. Variations, anomalies, and out-of-compliance conditions, where such have been detected, are documented along with the actions taken to resolve the discrepancies. Validation is typically carried out in the intended operational environment under simulated or actual operational conditions, not under the controlled conditions usually employed for the Product Verification Process.

Validation can be performed recursively throughout the project life cycle and on a wide variety of product forms. For example:

- Simulated (algorithmic models, virtual reality simulator);
- Mockup (plywood, brassboard, breadboard);
- Concept description (paper report);
- Prototype (product with partial functionality);
- Engineering unit (fully functional but may not be same form/fit);
- Design validation test units (form, fit and function may be the same, but they may not have flight parts);
- Qualification unit (identical to flight unit but may be subjected to extreme environments); or
- Flight unit (end product that is flown).

Any of these types of product forms may be in any of these states:

- Produced (built, fabricated, manufactured, or coded);
- Reused (modified internal nondevelopmental products or off-the-shelf product); or
- Assembled and integrated (a composite of lower level products).

> Note: The final, official validation of the end product should be for a controlled unit. Typically, attempting final validation against operational concepts on a prototype is not acceptable: it is usually completed on a qualification, flight, or other more final, controlled unit.

Outcomes of validation planning include the following:

- The validation type that is appropriate to confirm that the realized product or products conform to stakeholder expectations (based on the form of the realized end product) has been identified.
- Validation procedures are defined based on: (1) the needed procedures for each type of validation selected, (2) the purpose and objective of each procedure step, (3) any pre-test and post-test actions, and (4) the criteria for determining the success or failure of the procedure.
- A validation environment (e.g., facilities, equipment, tools, simulations, measuring devices, personnel, and operational conditions) in which the validation procedures will be implemented has been defined.

> Note: In planning for validation, consideration should be given to the extent to which validation testing will be done. In many instances, off-nominal operational scenarios and nominal operational scenarios should be utilized. Off-nominal testing offers insight into a system's total performance characteristics and often assists in identification of design issues and human-machine interface, training, and procedural changes required to meet the mission goals and objectives. Off-nominal testing, as well as nominal testing, should be included when planning for validation.

Product Validation Preparation

To prepare for performing product validation, the appropriate set of expectations against which the validation is to be made should be obtained. Also, the product to be validated (output from implementation, or integration and verification), as well as the validation-enabling products and support resources (requirements identified and acquisition initiated by design solution activities) with which validation will be conducted, should be collected.

The validation environment is then prepared (set up the equipments, sensors, recording devices, etc., that will be involved in the validation conduct) and the validation procedures reviewed to identify and resolve any issues impacting validation.

Outcomes of validation preparation include the following:

- Preparation for doing the planned validation is completed;
- Appropriate set of stakeholder expectations are available and on hand;
- Articles or models to be used for validation with the validation product and enabling products are integrated within the validation environment according to plans and schedules;
- Resources are available according to validation plans and schedules; and
- The validation environment is evaluated for adequacy, completeness, readiness, and integration.

Conduct Planned Product Validation

The act of validating the end product is conducted as spelled out in the validation plans and procedures and conformance established to each specified validation requirement. The responsible engineer should ensure that the procedures were followed and performed as planned, the validation-enabling products were calibrated correctly, and the data were collected and recorded for required validation measures.

When poor validation conduct, design, or conditions cause anomalies, the validation should be replanned as necessary, the environment preparation anomalies corrected, and the validation conducted again with improved or correct procedures and resources. The Decision Analysis Process should be used to make decisions for issues identified that may require alternative choices to be evaluated and a selection made or when needed changes to the validation plans, environment, and/or conduct are required.

Outcomes of conducting validation include the following:

- A validated product is established with supporting confirmation that the appropriate results were collected and evaluated to show completion of validation objectives.
- A determination is made as to whether the fabricated/ manufactured or assembled and integrated products (including software or firmware builds, as applicable) comply with their respective stakeholder expectations.
- A determination is made that the validated product was appropriately integrated with the validation environment and the selected stakeholder expectations set was properly validated.
- A determination is made that the product being validated functions together with interfacing products throughout their performance envelopes.

Analyze Product Validation Results

Once the validation activities have been completed, the results are collected and the data are analyzed to confirm that the end product provided will supply the customer's needed capabilities within the intended environments of use, validation procedures were followed, and enabling products and supporting resources functioned correctly. The data are also analyzed for quality, integrity, correctness, consistency, and validity and any unsuitable products or product attributes are identified and reported.

It is important to compare the actual validation results to the expected results and to conduct any required system design and product realization process activities to resolve deficiencies. The deficiencies, along with recommended corrective actions and resolution results, should be recorded and validation repeated, as required.

Outcomes of analyzing validation results include the following:

- Product deficiencies and/or issues are identified.
- Assurances that appropriate replanning, redefinition of requirements, design, and revalidation have been accomplished for resolution of anomalies, variations, or out-of-compliance conditions (for problems not caused by poor validation conduct).
- Discrepancy and corrective action reports are generated as needed.
- The validation report is completed.

Validation Notes

The types of validation used are dependent on the life-cycle phase; the product's location in the system structure; and cost, schedule, and resources available. Validation of products within a single system model may be conducted together (e.g., an end product with its relevant enabling products, such as operational (control center or a radar with its display), maintenance (required tools work with product), or logistical (launcher or transporter).

Each realized product of system structure should be validated against stakeholder expectations before being integrated into a higher level product.

Reengineering

Based on the results of the Product Validation Process, it could become necessary to reengineer a deficient end product. Care should be taken that correcting a deficiency, or set of deficiencies, does not generate a new issue with a part or performance that had previously operated satisfactorily. Regression testing, a formal process of rerunning previously used acceptance tests primarily used for software, is one method to ensure a change did not affect function or performance that was previously accepted.

Validation Deficiencies

Validation outcomes can be unsatisfactory for several reasons. One reason is poor conduct of the validation (e.g., enabling products and supporting resources missing or not functioning correctly, untrained operators, procedures not followed, equipment not calibrated, or improper validation environmental conditions) and failure to control other variables not involved in validating

a set of stakeholder expectations. A second reason could be a shortfall in the verification process of the end product. This could create the need for:

- Reengineering end products lower in the system structure that make up the end product that was found to be deficient (which failed to satisfy validation requirements) and/or
- Reperforming any needed verification and validation processes.

Other reasons for validation deficiencies (particularly when M&S are involved) may be incorrect and/or inappropriate initial or boundary conditions; poor formulation of the modeled equations or behaviors; the impact of approximations within the modeled equations or behaviors; failure to provide the required geometric and physics fidelities needed for credible simulations for the intended purpose; referent for comparison of poor or unknown uncertainty quantification quality; and/or poor spatial, temporal, and perhaps, statistical resolution of physical phenomena used in M&S.

> Note: Care should be exercised to ensure that the corrective actions identified to remove validation deficiencies do not conflict with the baselined stakeholder expectations without first coordinating such changes with the appropriate stakeholders.

Capture Product Validation Work Products

Validation work products (inputs to the Technical Data Management Process) take many forms and involve many sources of information. The capture and recording of validation-related data is a very important, but often underemphasized, step in the Product Validation Process.

Validation results, deficiencies identified, and corrective actions taken should be captured, as should all relevant results from the application of the Product Validation Process (related decisions, rationale for decisions made, assumptions, and lessons learned).

Outcomes of capturing validation work products include the following:

- Work products and related information generated while doing Product Validation Process activities and tasks are recorded; i.e., type of validation conducted, the form of the end product used for validation, validation procedures used, validation environments, outcomes, decisions, assumptions, corrective actions, lessons learned, etc. (often captured in a matrix or other tool—see Appendix E).

- Deficiencies (e.g., variations and anomalies and out-of-compliance conditions) are identified and documented, including the actions taken to resolve.

- Proof is provided that the realized product is in conformance with the stakeholder expectation set used in the validation.

- Validation report including:
  ▶ Recorded validation results/data;
  ▶ Version of the set of stakeholder expectations used;
  ▶ Version and form of the end product validated;
  ▶ Version or standard for tools and equipment used, together with applicable calibration data;
  ▶ Outcome of each validation including pass or fail declarations; and
  ▶ Discrepancy between expected and actual results.

> Note: For systems where only a single deliverable item is developed, the Product Validation Process normally completes acceptance testing of the system. However, for systems with several production units, it is important to understand that continuing verification and validation is not an appropriate approach to use for the items following the first deliverable. Instead, acceptance testing is the preferred means to ensure that subsequent deliverables comply with the baselined design.

5.4.1.3 Outputs

Key outputs of validation are:

- Validated product,
- Discrepancy reports and identified corrective actions, and
- Validation reports.

Success criteria for this process include: (1) objective evidence of performance and the results of each system-of-interest validation activity are documented, and (2) the validation process should not be considered or designated as complete until all issues and actions are resolved.

5.4.2 Product Validation Guidance

The following is some generic guidance for the Product Validation Process.

5.4.2.1 Modeling and Simulation

As stressed in the verification process material, M&S is also an important validation tool. M&S usage considerations involve the verification, validation, and certification of the models and simulations.

> #### Model Verification and Validation
>
> - **Model Verification:** Degree to which a model accurately meets its specifications. Answers "Is it what I intended?"
> - **Model Validation:** The process of determining the degree to which a model is an accurate representation of the real world from the perspective of the intended uses of the model.
> - **Model Certification:** Certification for use for a specific purpose. Answers, "Should I endorse this model?"

5.4.2.2 Software

Software verification is a software engineering activity that demonstrates the software products meet specified requirements. Methods of software verification include peer reviews/inspections of software engineering products for discovery of defects, software verification of requirements by use of simulations, black box and white box testing techniques, analyses of requirement implementation, and software product demonstrations.

Software validation is a software engineering activity that demonstrates the as-built software product or software product component satisfies its intended use in its intended environment. Methods of software validation include: peer reviews/inspections of software product component behavior in a simulated environment, acceptance testing against mathematical models, analyses, and operational environment demonstrations. The project's approach for software verification and validation is documented in the software development plan. Specific Agency-level requirements for software verification and validation, peer reviews (see Appendix N), testing and reporting are contained in *NPR 7150.2, NASA Software Requirements.*

The rigor and techniques used to verify and validate software depend upon software classifications (which are different from project and payload classifications). A complex project will typically contain multiple systems and subsystems having different software classifications. It is important for the project to classify its software and plan verification and validation approaches that appropriately address the risks associated with each class.

In some instances, NASA management may select a project for additional independent software verification and validation by the NASA Software Independent Verification and Validation (IV&V) Facility in Fairmount, West Virginia. In this case a Memorandum of Understanding (MOU) and separate software IV&V plan will be created and implemented.

5.5 Product Transition

The Product Transition Process is used to transition a verified and validated end product that has been generated by product implementation or product integration to the customer at the next level in the system structure for integration into an end product or, for the top-level end product, transitioned to the intended end user. The form of the product transitioned will be a function of the product-line life-cycle phase success criteria and the location within the system structure of the WBS model in which the end product exits.

Product transition occurs during all phases of the life cycle. During the early phases, the technical team's products are documents, models, studies, and reports. As the project moves through the life cycle, these paper or soft products are transformed through implementation and integration processes into hardware and software solutions to meet the stakeholder expectations. They are repeated with different degrees of rigor throughout the life cycle. The Product Transition Process includes product transitions from one level of the system architecture upward. The Product Transition Process is the last of the product realization processes, and it is a bridge from one level of the system to the next higher level.

The Product Transition Process is the key to bridge from one activity, subsystem, or element to the overall engineered system. As the system development nears completion, the Product Transition Process is again applied for the end product, but with much more rigor since now the transition objective is delivery of the system-level end product to the actual end user. Depending on the kind or category of system developed, this may involve a Center or the Agency and impact thousands of individuals storing, handling, and transporting multiple

end products; preparing user sites; training operators and maintenance personnel; and installing and sustaining, as applicable. Examples are transitioning the external tank, solid rocket boosters, and orbiter to Kennedy Space Center (KSC) for integration and flight.

5.5.1 Process Description

Figure 5.5-1 provides a typical flow diagram for the Product Transition Process and identifies typical inputs, outputs, and activities to consider in addressing product transition.

5.5.1.1 Inputs

Inputs to the Product Transition Process depend primarily on the transition requirements, the product that is being transitioned, the form of the product transition that is taking place, and where the product is transitioning to. Typical inputs are shown in Figure 5.5-1 and described below.

- **The End Product or Products To Be Transitioned (from Product Validation Process):** The product to be transitioned can take several forms. It can be a sub-

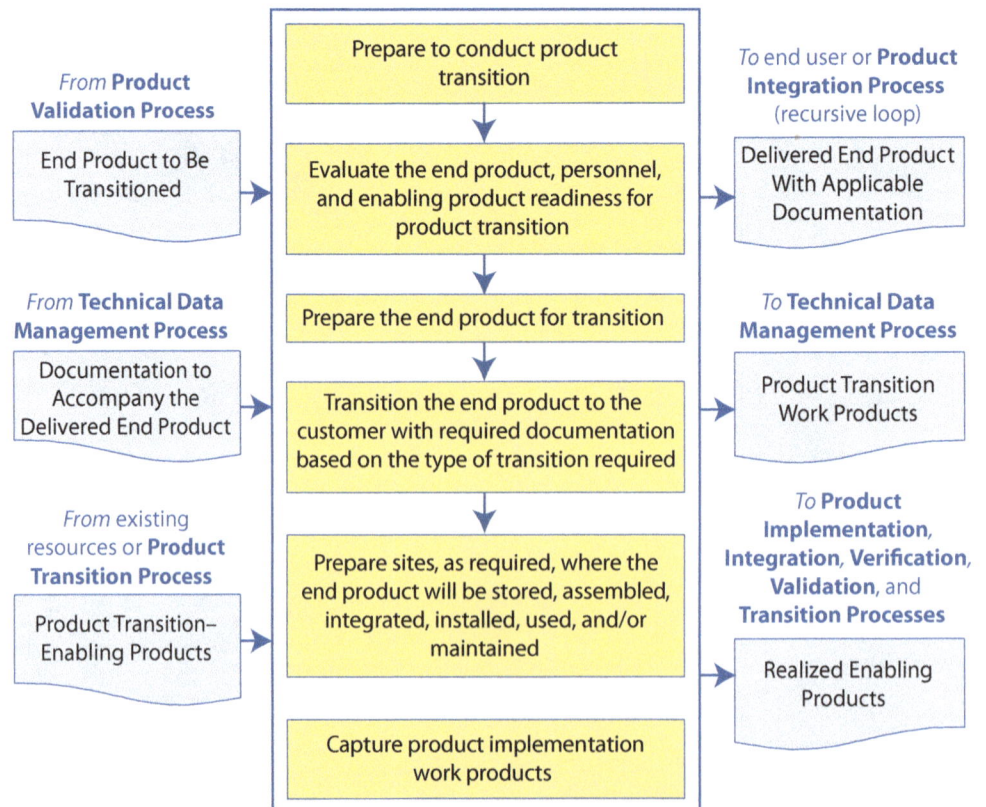

Figure 5.5-1 Product Transition Process

system component, system assembly, or top-level end product. It can be hardware or software. It can be newly built, purchased, or reused. A product can transition from a lower system product to a higher one by being integrated with other transitioned products. This process may be repeated until the final end product is achieved. Each succeeding transition requires unique input considerations when preparing for the validated product for transition to the next level.

Early phase products can take the form of information or data generated from basic or applied research using analytical or physical models and often are in paper or electronic form. In fact, the end product for many NASA research projects or science activities is a report, paper, or even an oral presentation. In a sense, the dissemination of information gathered through NASA research and development is an important form of product transition.

- **Documentation Including Manuals, Procedures, and Processes That Are To Accompany the End Product (from Technical Data Management Process):** The documentation required for the Product Transition Process depends on the specific end product; its current location within the system structure; and the requirements identified in various agreements, plans, or requirements documents. Typically, a product has a unique identification (i.e., serial number) and may have a pedigree (documentation) that specifies its heritage and current state. Pertinent information may be documented through a configuration management system or work order system as well as design drawings and test reports. Documentation often includes proof of verification and validation conformance. A COTS product would typically contain a manufacturer's specification or fact sheet. Documentation may include operations manuals, installation instructions, and other information.

The documentation level of detail is dependent upon where the product is within the product hierarchy and the life cycle. Early in the life cycle, this documentation may be preliminary in nature. Later in the life cycle, the documentation may be detailed design documents, user manuals, drawings, or other work products. Documentation that is gathered during the input process for the transition phase may require editing, assembling, or repackaging to ensure it is in the required condition for acceptance by the customer.

Special consideration must be given to safety, including clearly identifiable tags and markings that identify the use of hazardous materials, special handling instructions, and storage requirements.

- **Product-Transition-Enabling Products, Including Packaging Materials; Containers; Handling Equipment; and Storage, Receiving, and Shipping Facilities (from Existing Resources or Product Transition Process for Enabling Product Realization):** Product-transition-enabling products may be required to facilitate the implementation, integration, evaluation, transition, training, operations, support, and/or retirement of the transition product at its next higher level or for the transition of the final end product. Some or all of the enabling products may be defined in transition-related agreements, system requirements documents, or project plans. In some cases, product-transition-enabling products are developed during the realization of the product itself or may be required to be developed during the transition stage.

As a product is developed, special containers, holders, or other devices may also be developed to aid in the storing and transporting of the product through development and realization. These may be temporary accommodations that do not satisfy all the transition requirements, but allow the product to be initiated into the transition process. In such cases, the temporary accommodations will have to be modified or new accommodations will need to be designed and built or procured to meet specific transportation, handling, storage, and shipping requirements.

Sensitive or hazardous products may require special enabling products such as monitoring equipment, inspection devices, safety devices, and personnel training to ensure adequate safety and environmental requirements are achieved and maintained.

5.5.1.2 Process Activities

Transitioning the product can take one of two forms:

- The delivery of lower system end products to higher ones for integration into another end product or
- The delivery of the final end product to the customer or user that will use it in its operational environment.

In the first case, the end product is one of perhaps several other pieces that will ultimately be integrated together to form the item in the second case for final delivery to the customer. For example, the end product might be one of several circuit cards that will be integrated together to form the final unit that is delivered. Or that unit might also be one of several units that have to be integrated together to form the final product.

The form of the product transitioned is not only a function of the location of that product within the system product hierarchy (i.e., WBS model), but also a function of the life-cycle phase. Early life-cycle phase products may be in the form of paper, electronic files, physical models, or technology demonstration prototypes. Later phase products may be preproduction prototypes (engineering models), the final study report, or the flight units.

Figure 5.5-1 shows what kind of inputs, outputs, and activities are performed during product transition regardless of where in the product hierarchy or life cycle the product is. These activities include preparing to conduct the transition; making sure the end product, all personnel, and any enabling products are ready for transitioning; preparing the site; and performing the transition including capturing and documenting all work products.

How these activities are performed and what form the documentation takes will depend on where the end items are in the product hierarchy (WBS model) and its life-cycle phase.

Prepare to Implement Transition

The first task is to identify which of the two forms of transition is needed: (1) the delivery of lower system end products to higher ones for integration into another end product or (2) the delivery of the final end product to the customer or user that will use the end product in its operational environment. The form of the product being transitioned will affect transition planning and the kind of packaging, handling, storage, and transportation that will be required. The customer and other stakeholder expectations, as well as the specific design solution, may indicate special transition procedures or enabling product needs for packaging, storage, handling, shipping/transporting, site preparation, installation, and/or sustainability. These requirements need to be reviewed during the preparation stage.

Other tasks in preparing to transition a product involve making sure the end product, personnel, and any enabling products are ready for that transition. This includes the availability of the documentation that will be sent with the end product, including proof of verification and validation conformance. The appropriateness of detail for that documentation depends upon where the product is within the product hierarchy and the life cycle. Early in the life cycle, this documentation may be preliminary in nature. Later in the life cycle, the documentation may be detailed design documents, user manuals, drawings, or other work products. Procedures necessary for conducting the transition should be reviewed and approved by this time. This includes all necessary approvals by management, legal, safety, quality, property, or other organizations as identified in the SEMP.

Finally, the availability and skills of personnel needed to conduct the transition as well as the availability of any necessary packaging materials/containers, handling equipment, storage facilities, and shipping/transporter services should also be reviewed. Any special training necessary for the personnel to perform their tasks needs to be performed by this time.

Prepare the Product for Transition

Whether transitioning a product to the next room for integration into the next higher assembly, or for final transportation across the country to the customer, care must be taken to ensure the safe transportation of the product. The requirements for packaging, handling, storage, and transportation should have been identified during system design. Preparing for the packaging for protection, security, and prevention of deterioration is critical for products placed in storage or when it is necessary to transport or ship between and within organizational facilities or between organizations by land, air, and/or water vehicles. Particular emphasis needs to be on protecting surfaces from physical damage, preventing corrosion, eliminating damage to electronic wiring or cabling, shock or stress damage, heat warping or cold fractures, moisture, and other particulate intrusion that could damage moving parts.

The design requirements should have already addressed the ease of handling or transporting the product such as component staking, addition of transportation hooks, crating, etc. The ease and safety of packing and unpacking the product should also have been addressed. Additional measures may also need to be implemented to show accountability and to securely track the product during transportation. In cases where hazardous materials are involved, special labeling or handling needs including transportation routes need to be in place.

Prepare the Site to Receive the Product

For either of the forms of product transition, the receiving site needs to be prepared to receive the product. Here the end product will be stored, assembled, integrated, installed, used, and/or maintained, as appropriate for the life-cycle phase, position of the end product in the system structure, and customer agreement.

A vast number of key complex activities, many of them outside direct control of the technical team, have to be synchronized to ensure smooth transition to the end user. If transition activities are not carefully controlled, there can be impacts on schedule, cost, and safety of the end product.

A site survey may need to be performed to determine the issues and needs. This should address the adequacy of existing facilities to accept, store, and operate the new end product and identify any logistical-support-enabling products and services required but not planned for. Additionally, any modifications to existing facilities must be planned well in advance of fielding; therefore, the site survey should be made during an early phase in the product life cycle. These may include logistical enabling products and services to provide support for end-product use, operations, maintenance, and disposal. Training for users, operators, maintainers, and other support personnel may need to be conducted. National Environmental Policy Act documentation or approvals may need to be obtained prior to the receipt of the end product.

Prior to shipment or after receipt, the end product may need to be stored in suitable storage conditions to protect and secure the product and prevent damage or the deterioration of it. These conditions should have been identified early in the design life cycle.

Transition the Product

The end product is then transitioned (i.e., moved, transported, or shipped) with required documentation to the customer based on the type of transition required, e.g., to the next higher level item in the Product Breakdown Structure (PBS) for product integration or to the end user. Documentation may include operations manuals, installation instructions, and other information.

The end product is finally installed into the next higher assembly or into the customer/user site using the preapproved installation procedures.

Confirm Ready to Support

After installation, whether into the next higher assembly or into the final customer site, functional and acceptance testing of the end product should be conducted. This ensures no damage from the shipping/handling process has occurred and that the product is ready for support. Any final transitional work products should be captured as well as documentation of product acceptance.

5.5.1.3 Outputs

- **Delivered End Product for Integration to Next Level up in System Structure:** This includes the appropriate documentation. The form of the end product and applicable documentation are a function of the life-cycle phase and the placement within the system structure. (The form of the end product could be hardware, software, model, prototype, first article for test, or single operational article or multiple production article.) Documentation includes applicable draft installation, operation, user, maintenance, or training manuals; applicable baseline documents (configuration baseline, specifications, and stakeholder expectations); and test results that reflect completion of verification and validation of the end product.

- **Delivered Operational End Product for End Users:** The appropriate documentation is to be delivered with the delivered end product as well as the operational end product appropriately packaged. Documentation includes applicable final installation, operation, user, maintenance, or training manuals; applicable baseline documents (configuration baseline, specifications, stakeholder expectations); and test results that reflect completion of verification and validation of the end product. If the end user will perform end product validation, sufficient documentation to support end user validation activities is delivered with the end product.

- **Work Products from Transition Activities to Technical Data Management:** Work products could include the transition plan, site surveys, measures, training modules, procedures, decisions, lessons learned, corrective actions, etc.

- **Realized Enabling End Products to Appropriate Life-Cycle Support Organization:** Some of the enabling products that were developed during the various phases could include fabrication or integration specialized machines; tools; jigs; fabrication processes and manuals; integration processes and manuals; specialized inspection, analysis, demonstration, or test equipment; tools; test stands; specialized packaging materials and containers; handling equipment; storage-site environments; shipping or transportation vehicles or equipment; specialized courseware; instructional site environments; and delivery of the training instruction. For the later life-cycle phases, enabling products that are to be delivered may include specialized mission control equipment; data collection equipment; data analysis equipment; operations manuals; specialized maintenance equipment, tools, manuals, and spare parts; specialized recovery equipment; disposal equipment; and readying recovery or disposal site environments.

The process is complete when the following activities have been accomplished:

- The end product is validated against stakeholder expectations unless the validation is to be done by the integrator before integration is accomplished.

- For deliveries to the integration path, the end product is delivered to intended usage sites in a condition suitable for integration with other end products or composites of end products. Procedures, decisions, assumptions, anomalies, corrective actions, lessons learned, etc., resulting from transition for integration are recorded.

- For delivery to the end user path, the end products are installed at the appropriate sites; appropriate acceptance and certification activities are completed; training of users, operators, maintainers, and other necessary personnel is completed; and delivery is closed out with appropriate acceptance documentation.

- Any realized enabling end products are also delivered as appropriate including procedures, decisions, assumptions, anomalies, corrective actions, lessons learned, etc., resulting from transition-enabling products.

5.5.2 Product Transition Guidance

5.5.2.1 Additional Product Transition Input Considerations

It is important to consider all customer, stakeholder, technical, programmatic, and safety requirements when evaluating the input necessary to achieve a successful Product Transition Process. This includes the following:

- **Transportability Requirements:** If applicable, requirements in this section define the required configuration of the system of interest for transport. Further, this section details the external systems (and the interfaces to those systems) required for transport of the system of interest.

- **Environmental Requirements:** Requirements in this section define the environmental conditions in which the system of interest is required to be during transition (including storage and transportation).

- **Maintainability Requirements:** Requirements in this section detail how frequently, by whom, and by what means the system of interest will require maintenance (also any "care and feeding," if required).

- **Safety Requirements:** Requirements in this section define the life-cycle safety requirements for the system of interest and associated equipment, facilities, and personnel.

- **Security Requirements:** This section defines the Information Technology (IT) requirements, Federal and international export and security requirements, and physical security requirements for the system of interest.

- **Programmatic Requirements:** Requirements in this section define cost and schedule requirements.

5.5.2.2 After Product Transition to the End User—What Next?

As mentioned in Chapter 2.0, there is a relationship between the SE engine and the activities performed after the product is transitioned to the end user. As shown in Figure 2.3-8, after the final deployment to the end user, the end product is operated, managed, and maintained through sustaining engineering functions. The technical management processes described in Section 6.0 are used during these activities. If at any time a new capability, upgrade, or enabling product is needed, the developmental processes of the engine are reengaged. When the end product's use is completed, the plans developed early in the life cycle to dispose, retire, or phase out the product are enacted.

6.0 Crosscutting Technical Management

This chapter describes the activities in the technical management processes listed in Figure 2.1-1. The chapter is separated into sections corresponding to steps 10 through 17 listed in Figure 2.1-1. The processes within each step are discussed in terms of the inputs, the activities, and the outputs. Additional guidance is provided using examples that are relevant to NASA projects.

The technical management processes are the bridges between project management and the technical team. In this portion of the engine, eight crosscutting processes provide the integration of the crosscutting functions that allow the design solution to be realized. Even though every technical team member might not be directly involved with these eight processes, they are indirectly affected by these key functions. Every member of the technical team relies on technical planning; management of requirements, interfaces, technical risk, configuration, and technical data; technical assessment; and decision analysis to meet the project's objectives. Without these crosscutting processes, individual members and tasks cannot be integrated into a functioning system that meets the ConOps within cost and schedule. The project management team also uses these crosscutting functions to execute project control on the apportioned tasks.

This effort starts with the technical team conducting extensive planning early in Pre-Phase A. With this early, detailed baseline plan, technical team members will understand the roles and responsibilities of each team member, and the project can establish its program cost and schedule goals and objectives. From this effort, the Systems Engineering Management Plan (SEMP) is developed and baselined. Once a SEMP has been established, it must be synchronized with the project master plans and schedule. In addition, the plans for establishing and executing all technical contracting efforts are identified.

This is a recursive and iterative process. Early in the life cycle, the plans are established and synchronized to run the design and realization processes. As the system matures and progresses through the life cycle, these plans must be updated as necessary to reflect the current environment and resources and to control the project's performance, cost, and schedule. At a minimum, these updates will occur at every Key Decision Point (KDP). However, if there is a significant change in the project, such as new stakeholder expectations, resource adjustments, or other constraints, all plans must be analyzed for the impact of these changes to the baselined project.

The next sections describe each of the eight technical management processes and their associated products for a given NASA mission.

Crosscutting Technical Management Keys

- Thoroughly understand and plan the scope of the technical effort by investing time upfront to develop the technical product breakdown structure, the technical schedule and workflow diagrams, and the technical resource requirements and constraints (funding, budget, facilities, and long-lead items) that will be the technical planning infrastructure.

- Define all interfaces and assign interface authorities and responsibilities to each, both intra- and interorganizational. This includes understanding potential incompatibilities and defining the transition processes.

- Control of the configuration is critical to understanding how changes will impact the system. For example, changes in design and environment could invalidate previous analysis results.

- Conduct milestone reviews to enable a critical and valuable assessment to be performed. These reviews are not to be used to meet contractual or scheduling incentives. These reviews have specific entrance criteria and should be conducted when these are met.

- Understand any biases, assumptions, and constraints that impact the analysis results.

- Place all analysis under configuration control to be able to track the impact of changes and understand when the analysis needs to be reevaluated.

6.1 Technical Planning

The Technical Planning Process, the first of the eight technical management processes contained in the systems engineering engine, establishes a plan for applying and managing each of the common technical processes that will be used to drive the development of system products and associated work products. This process also establishes a plan for identifying and defining the technical effort required to satisfy the project objectives and life-cycle phase success criteria within the cost, schedule, and risk constraints of the project.

6.1.1 Process Description

Figure 6.1-1 provides a typical flow diagram for the Technical Planning Process and identifies typical inputs, outputs, and activities to consider in addressing technical planning.

6.1.1.1 Inputs

Input to the Technical Planning Process comes from both the project management and technical teams as

outputs from the other common technical processes. Initial planning utilizing external inputs from the project to determine the general scope and framework of the technical effort will be based on known technical and programmatic requirements, constraints, policies, and processes. Throughout the project's life cycle, the technical team continually incorporates results into the technical planning strategy and documentation and any internal changes based on decisions and assessments generated by the other processes of the SE engine or from requirements and constraints mandated by the project.

As the project progresses through the life-cycle phases, technical planning for each subsequent phase must be assessed and continually updated. When a project transitions from one life-cycle phase to the next, the technical planning for the upcoming phase must be assessed and updated to reflect the most recent project data.

- **External Inputs from the Project:** The project plan provides the project's top-level technical requirements, the available budget allocated to the project

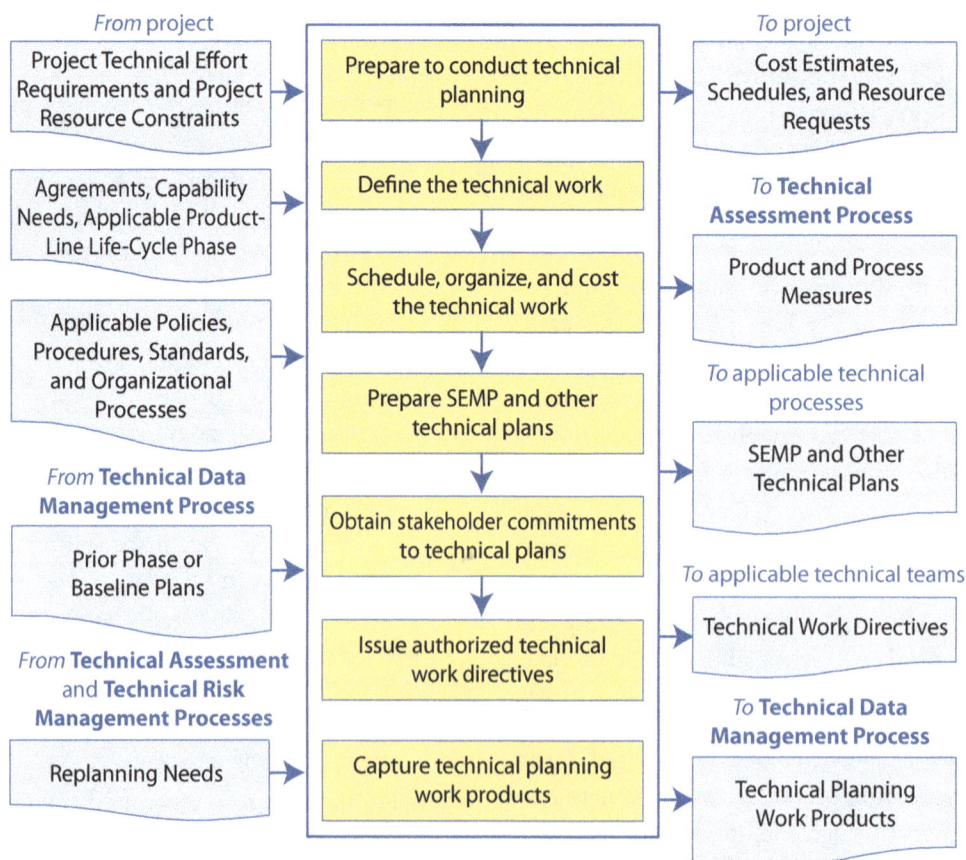

Figure 6.1-1 Technical Planning Process

from the program, and the desired schedule for the project to support overall program needs. Although the budget and schedule allocated to the project will serve as constraints on the project, the technical team will generate a technical cost estimate and schedule based on the actual work required to satisfy the project's technical requirements. Discrepancies between the project's allocated budget and schedule and the technical team's actual cost estimate and schedule must be reconciled continuously throughout the project's life cycle.

The project plan also defines the applicable project life-cycle phases and milestones, as well as any internal and external agreements or capability needs required for successful project execution. The project's life-cycle phases and programmatic milestones will provide the general framework for establishing the technical planning effort and for generating the detailed technical activities and products required to meet the overall project milestones in each of the life-cycle phases.

Finally, the project plan will include all programmatic policies, procedures, standards, and organizational processes that must be adhered to during execution of the technical effort. The technical team must develop a technical approach that ensures the project requirements will be satisfied and that any technical procedures, processes, and standards to be used in developing the intermediate and final products comply with the policies and processes mandated in the project plan.

- **Internal Inputs from Other Common Technical Processes:** The latest technical plans (either baselined or from the previous life-cycle phase) from the Data Management or Configuration Management Processes should be used in updating the technical planning for the upcoming life-cycle phase.

Technical planning updates may be required based on results from technical reviews conducted in the Technical Assessment Process, issues identified during the Technical Risk Management Process, or from decisions made during the Decision Analysis Process.

6.1.1.2 Process Activities

Technical planning as it relates to systems engineering at NASA is intended to identify, define, and plan how the 17 common technical processes in *NPR 7123.1, NASA Systems Engineering Processes and Requirements* will be

applied in each life-cycle phase for all levels of the WBS model (see Subsection 6.1.2.1) within the system structure to meet product-line life-cycle phase success criteria. A key document generated by this process is the SEMP.

The SEMP is a subordinate document to the project plan. While the SEMP defines to all project participants how the project will be technically managed within the constraints established by the project, the project plan defines how the project will be managed to achieve its goals and objectives within defined programmatic constraints. The SEMP also communicates how the systems engineering management techniques will be applied throughout all phases of the project life cycle.

Technical planning should be tightly integrated with the Technical Risk Management Process (see Section 6.4) and the Technical Assessment Process (see Section 6.7) to ensure corrective action for future activities will be incorporated based on current issues identified within the project.

Technical planning, as opposed to program or project planning, addresses the scope of the technical effort required to develop the system products. While the project manager concentrates on managing the overall project life cycle, the technical team, led by the systems engineer, concentrates on managing the technical aspects of the project. The technical team identifies, defines, and develops plans for performing decomposition, definition, integration, verification, and validation of the system while orchestrating and incorporating the appropriate concurrent engineering. Additional planning will include defining and planning for the appropriate technical reviews, audits, assessments, and status reports and determining any specialty engineering and/or design verification requirements.

This section describes how to perform the activities contained in the Technical Planning Process shown in Figure 6.1-1. The initial technical planning at the beginning of the project will establish the technical team members; their roles and responsibilities; and the tools, processes, and resources that will be utilized in executing the technical effort. In addition, the expected activities the technical team will perform and the products it will produce will be identified, defined, and scheduled. Technical planning will continue to evolve as actual data from completed tasks are received and details of near-term and future activities are known.

Technical Planning Preparation

For technical planning to be conducted properly, the processes and procedures to conduct technical planning should be identified, defined, and communicated. As participants are identified, their roles and responsibilities and any training and/or certification activities should be clearly defined and communicated.

Once the processes, people, and roles and responsibilities are in place, a planning strategy may be formulated for the technical effort. A basic technical planning strategy should address the following:

- The level of planning documentation required for the SEMP and all other technical planning documents;
- Identifying and collecting input documentation;
- The sequence of technical work to be conducted, including inputs and outputs;
- The deliverable products from the technical work;
- How to capture the work products of technical activities;
- How technical risks will be identified and managed;
- The tools, methods, and training needed to conduct the technical effort;
- The involvement of stakeholders in each facet of the technical effort;
- How the NASA technical team will be involved with the technical efforts of external contractors;
- The entry and success criteria for milestones, such as technical reviews and life-cycle phases;
- The identification, definition, and control of internal and external interfaces;
- The identification and incorporation of relevant lessons learned into the technical planning;
- The approach for technology development and how the resulting technology will be incorporated into the project;
- The identification and definition of the technical metrics for measuring and tracking progress to the realized product;
- The criteria for make, buy, or reuse decisions and incorporation criteria for Commercial Off-the-Shelf (COTS) software and hardware;
- The plan to identify and mitigate off-nominal performance;
- The "how-tos" for contingency planning and replanning;

- The plan for status assessment and reporting; and
- The approach to decision analysis, including materials needed, required skills, and expectations in terms of accuracy.

By addressing these items and others unique to the project, the technical team will have a basis for understanding and defining the scope of the technical effort, including the deliverable products that the overall technical effort will produce, the schedule and key milestones for the project that the technical team must support, and the resources required by the technical team to perform the work.

A key element in defining the technical planning effort is understanding the amount of work associated with performing the identified activities. Once the scope of the technical effort begins to coalesce, the technical team may begin to define specific planning activities and to estimate the amount of effort and resources required to perform each task. Historically, many projects have underestimated the resources required to perform proper planning activities and have been forced into a position of continuous crisis management in order to keep up with changes in the project.

Define the Technical Work

The technical effort must be thoroughly defined. When performing the technical planning, realistic values for cost, schedule, and labor resources should be used. Whether extrapolated from historical databases or from interactive planning sessions with the project and stakeholders, realistic values must be calculated and provided to the project team. Contingency should be included in any estimate and based on complexity and criticality of the effort. Contingency planning must be conducted. The following are examples of contingency planning:

- Additional, unplanned-for software engineering resources are typically needed during hardware and systems development and testing to aid in troubleshooting errors/anomalies. Frequently, software engineers are called upon to help troubleshoot problems and pinpoint the source of errors in hardware and systems development and testing (e.g., for writing addition test drivers to debug hardware problems). Additional software staff should be planned into the project contingencies to accommodate inevitable component and system debugging and avoid cost and schedule overruns.

- Hardware-in-the-Loop (HWIL) must be accounted for in the technical planning contingencies. HWIL testing is typically accomplished as a debugging exercise where the hardware and software are brought together for the first time in the costly environment of an HWIL. If upfront work is not done to understand the messages and errors arising during this test, additional time in the HWIL facility may result in significant cost and schedule impacts. Impacts may be mitigated through upfront planning, such as making appropriate debugging software available to the technical team prior to the test, etc.

Schedule, Organize, and Cost the Technical Effort

Once the technical team has defined the technical work to be done, efforts can focus on producing a schedule and cost estimate for the technical portion of the project. The technical team must organize the technical tasks according to the project WBS in a logical sequence of events, taking into consideration the major project milestones, phasing of available funding, and timing of availability of supporting resources.

Scheduling

Products described in the WBS are the result of activities that take time to complete. These activities have time precedence relationships among them that may used to create a network schedule explicitly defining the dependencies of each activity on other activities, the availability of resources, and the receipt of receivables from outside sources.

Scheduling is an essential component of planning and managing the activities of a project. The process of creating a network schedule provides a standard method for defining and communicating what needs to be done, how long it will take, and how each element of the project WBS might affect other elements. A complete network schedule may be used to calculate how long it will take to complete a project; which activities determine that duration (i.e., critical path activities); and how much spare time (i.e., float) exists for all the other activities of the project.

"Critical path" is the sequence of dependent tasks that determines the longest duration of time needed to complete the project. These tasks drive the schedule and continually change, so they must be updated. The critical path may encompass only one task or a series of inter-related tasks. It is important to identify the critical path and the resources needed to complete the critical tasks along the path if the project is to be completed on time and within its resources. As the project progresses, the critical path will change as the critical tasks are completed or as other tasks are delayed. This evolving critical path with its identified tasks needs to be carefully monitored during the progression of the project.

Network scheduling systems help managers accurately assess the impact of both technical and resource changes on the cost and schedule of a project. Cost and technical problems often show up first as schedule problems. Understanding the project's schedule is a prerequisite for determining an accurate project budget and for tracking performance and progress. Because network schedules show how each activity affects other activities, they assist in assessing and predicting the consequences of schedule slips or accelerations of an activity on the entire project.

Network Schedule Data and Graphical Formats

Network schedule data consist of:

- Activities and associated tasks;
- Dependencies among activities (e.g., where an activity depends upon another activity for a receivable);
- Products or milestones that occur as a result of one or more activities; and
- Duration of each activity.

A network schedule contains all four of the above data items. When creating a network schedule, creating graphical formats of these data elements may be a useful first step in planning and organizing schedule data.

Workflow Diagrams

A workflow diagram is a graphical display of the first three data items. Two general types of graphical formats are used as shown in Figure 6.1-2. One places activities on arrows, with products and dependencies at the beginning and end of the arrow. This is the typical format of the Program Evaluation and Review Technique (PERT) chart.

The second format, called precedence diagrams, uses boxes to represent activities; dependencies are then shown by arrows. The precedence diagram format allows for simple depiction of the following logical relationships:

- Activity B begins when Activity A begins (start-start).
- Activity B begins only after Activity A ends (finish-start).
- Activity B ends when Activity A ends (finish-finish).

Each of these three activity relationships may be modified by attaching a lag (+ or –) to the relationship, as shown in Figure 6.1-2. It is possible to summarize a number of low-level activities in a precedence diagram with a single activity. One takes the initial low-level activity and attaches a summary activity to it using the start-start relationship described above. The summary activity is then attached to the final low-level activity using the finish-start relationship. The most common relationship used in precedence diagrams is the finish-start one. The activity-on-arrow format can represent the identical time-precedence logic as a precedence diagram by creating artificial events and activities as needed.

Establishing a Network Schedule

Scheduling begins with project-level schedule objectives for delivering the products described in the upper levels of the WBS. To develop network schedules that are consistent with the project's objectives, the following six steps are applied to each element at the lowest available level of the WBS.

Step 1: Identify activities and dependencies needed to complete each WBS element. Enough activities should be identified to show exact schedule dependencies between activities and other WBS elements. This first step is most easily accomplished by:

- Ensuring that the WBS model is extended downward to describe all significant products including documents, reports, and hardware and software items.
- For each product, listing the steps required for its generation and drawing the process as a workflow diagram.
- Indicating the dependencies among the products, and any integration and verification steps within the work package.

Step 2: Identify and negotiate external dependencies. External dependencies are any receivables from outside of, and any deliverables that go outside of, the WBS element. Negotiations should occur to ensure that there is agreement with respect to the content, format, and labeling of products that move across WBS elements so that lower level schedules can be integrated.

Step 3: Estimate durations of all activities. Assumptions behind these estimates (workforce, availability of facilities, etc.) should be written down for future reference.

Step 4: Enter the data for each WBS element into a scheduling program to obtain a network schedule and an estimate of the critical path for that element. It is not unusual at this point for some iteration of steps 1 to 4 to obtain a satisfactory schedule. Reserve is often added to critical-path activities to ensure that schedule commitments can be met within targeted risk levels.

Step 5: Integrate schedules of lower level WBS elements so that all dependencies among elements are correctly included in a project network. It is important to include the impacts of holidays, weekends, etc., by this point. The critical path for the project is discovered at this step in the process.

Step 6: Review the workforce level and funding profile over time and make a final set of adjustments to logic and durations so that workforce levels and funding levels are within project constraints. Adjustments to the logic and the durations of activities may be needed to con-

Activity-on-Arrow Diagram

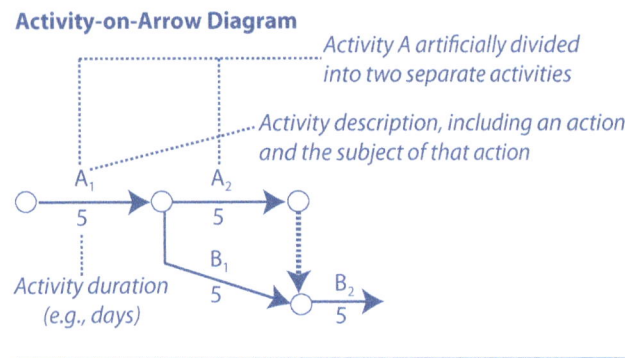

Precedence Diagram

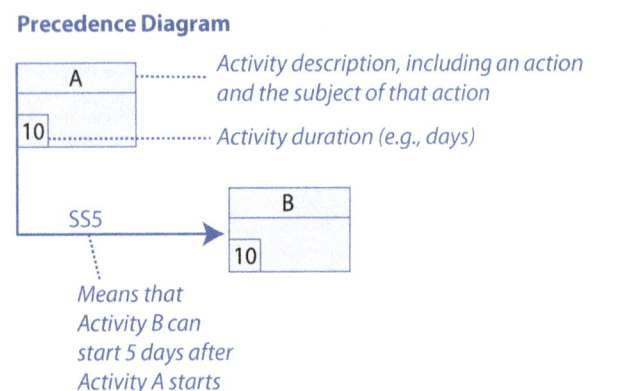

Figure 6.1-2 Activity-on-arrow and precedence diagrams for network schedules

verge to the schedule targets established at the project level. Adjustments may include adding more activities to some WBS elements, deleting redundant activities, increasing the workforce for some activities that are on the critical path, or finding ways to do more activities in parallel, rather than in series.

Again, it is good practice to have some schedule reserve, or float, as part of a risk mitigation strategy. The product of these last steps is a feasible baseline schedule for each WBS element that is consistent with the activities of all other WBS elements. The sum of all of these schedules should be consistent with both the technical scope and the schedule goals of the project. There should be enough float in this integrated master schedule so that schedule and associated cost risk are acceptable to the project and to the project's customer. Even when this is done, time estimates for many WBS elements will have been underestimated or work on some WBS elements will not start as early as had been originally assumed due to late arrival of receivables. Consequently, replanning is almost always needed to meet the project's goals.

Reporting Techniques

Summary data about a schedule is usually described in charts. A Gantt chart is a bar chart that depicts a project schedule using start and finish dates of the appropriate product elements tied to the project WBS of a project. Some Gantt charts also show the dependency (i.e., precedence and critical path) relationships among activities and also current status. A good example of a Gantt chart is shown in Figure 6.1-3. (See box on Gantt chart features.)

Another type of output format is a table that shows the float and recent changes in float of key activities. For example, a project manager may wish to know precisely how much schedule reserve has been consumed by critical path activities, and whether reserves are being consumed or are being preserved in the latest reporting period. This table provides information on the rate of change of schedule reserve.

Resource Leveling

Good scheduling systems provide capabilities to show resource requirements over time and to make adjustments so that the schedule is feasible with respect to resource constraints over time. Resources may include workforce level, funding profiles, important facilities, etc. The objective is to move the start dates of tasks that have float to

---

Gantt Chart Features

The Gantt chart shown in Figure 6.1-3 illustrates the following desirable features:

- A heading that describes the WBS element, identifies the responsible manager, and provides the date of the baseline used and the date that status was reported.
- A milestone section in the main body (lines 1 and 2).
- An activity section in the main body. Activity data shown includes:
  - ▸ WBS elements (lines 3, 5, 8, 12, 16, and 21);
  - ▸ Activities (indented from WBS elements);
  - ▸ Current plan (shown as thick bars);
  - ▸ Baseline plan (same as current plan, or if different, represented by thin bars under the thick bars);
  - ▸ Slack for each activity (dotted horizontal line before the milestone on line 12);
  - ▸ Schedule slips from the baseline (dotted horizontal lines after the current plan bars);
  - ▸ The critical path is shown encompassing lines 18 through 21 and impacting line 24; and
  - ▸ Status line (dotted vertical line from top to bottom of the main body of the chart) at the date the status was reported.
- A legend explaining the symbols in the chart.

This Gantt chart shows only 24 lines, which is a summary of the activities currently being worked for this WBS element. It is appropriate to tailor the amount of detail reported to those items most pertinent at the time of status reporting.

---

points where the resource profile is feasible. If that is not sufficient, then the assumed task durations for resource-intensive activities should be reexamined and, accordingly, the resource levels changed.

Budgeting

Budgeting and resource planning involve the establishment of a reasonable project baseline budget and the capability to analyze changes to that baseline resulting from technical and/or schedule changes. The project's WBS, baseline schedule, and budget should be viewed as mutually dependent, reflecting the technical content, time, and cost of meeting the project's goals and objectives. The budgeting process needs to take into account

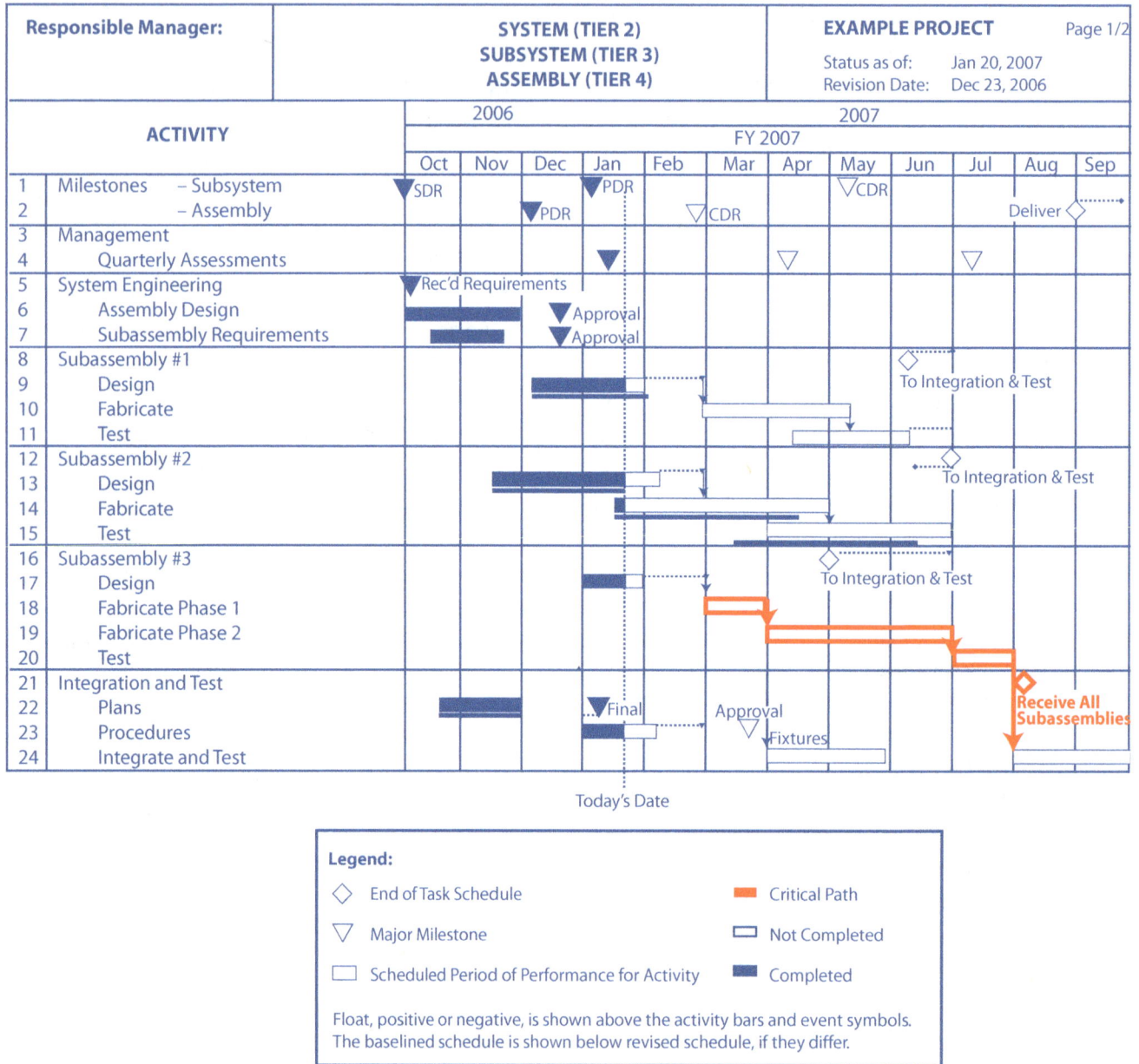

Figure 6.1-3 Gantt chart

whether a fixed cost cap or cost profile exists. When no such cap or profile exists, a baseline budget is developed from the WBS and network schedule. This specifically involves combining the project's workforce and other resource needs with the appropriate workforce rates and other financial and programmatic factors to obtain cost element estimates. These elements of cost include:

- Direct labor costs,
- Overhead costs,
- Other direct costs (travel, data processing, etc.),
- Subcontract costs,
- Material costs,
- General and administrative costs,

- Cost of money (i.e., interest payments, if applicable),
- Fee (if applicable), and
- Contingency.

When there is a cost cap or a fixed cost profile, there are additional logic gates that must be satisfied before completing the budgeting and planning process. A determination needs to be made whether the WBS and network schedule are feasible with respect to mandated cost caps and/or cost profiles. If not, it will be necessary to consider stretching out a project (usually at an increase in the total cost) or descoping the project's goals and objectives, requirements, design, and/or implementation approach.

If a cost cap or fixed cost profile exists, it is important to control costs after they have been baselined. An important aspect of cost control is project cost and schedule status reporting and assessment, methods for which are discussed in Section 6.7. Another is cost and schedule risk planning, such as developing risk avoidance and workaround strategies. At the project level, budgeting and resource planning must ensure that an adequate level of contingency funds is included to deal with unforeseen events.

The maturity of the Life-Cycle Cost Estimate (LCCE) should progress as follows:

- Pre-Phase A: Initial LCCE (70 percent confidence level; however, much uncertainty is expected)
- Phase A: Preliminary commitment to LCCE
- Phase B: Approve LCCE (70 percent confidence level at PDR commitment)
- Phase C, D, and E report variances to LCCE baseline using Earned Value Management (EVM) and LCCE updates

Credibility of the cost estimate is suspect if:

- WBS cost estimates are expressed only in dollars with no other identifiable units, indicating that requirements are not sufficiently defined for processes and resources to be identified.
- The basis of estimates does not contain sufficient detail for independent verification that work scope and estimated cost (and schedule) are reasonable.
- Actual costs vary significantly from the LCCE.
- Work is performed that was not originally planned, causing cost or schedule variance.
- Schedule and cost earned value performance trends readily indicate unfavorable performance.

Prepare the SEMP and Other Technical Plans

The SEMP is the primary, top-level technical management document for the project and is developed early in the Formulation phase and updated throughout the project life cycle. The SEMP is driven by the type of project, the phase in the project life cycle, and the technical development risks and is written specifically for each project or project element. While the specific content of the SEMP is tailored to the project, the recommended content is discussed in Appendix J.

The technical team, working under the overall project plan, develops and updates the SEMP as necessary. The technical team works with the project manager to review the content and obtain concurrence. This allows for thorough discussion and coordination of how the proposed technical activities would impact the programmatic, cost, and schedule aspects of the project. The SEMP provides the specifics of the technical effort and describes what technical processes will be used, how the processes will be applied using appropriate activities, how the project will be organized to accomplish the activities, and the cost and schedule associated with accomplishing the activities.

The physical length of a SEMP is not what is important. This will vary from project to project. The plan needs to be adequate to address the specific technical needs of the project. It is a *living* document that is updated as often as necessary to incorporate new information as it becomes available and as the project develops through Implementation. The SEMP should not duplicate other project documents; however, the SEMP should reference and summarize the content of other technical plans.

The systems engineer and project manager must identify additional required technical plans based on the project scope and type. If plans are not included in the SEMP, they should be referenced and coordinated in the development of the SEMP. Other plans, such as system safety and the probabilistic risk assessment, also need to be planned for and coordinated with the SEMP. If a technical plan is a stand-alone, it should be referenced in the SEMP. Depending on the size and complexity of the project, these may be separate plans or may be included within the SEMP. Once identified, the plans can be developed, training on these plans established, and the plans implemented. Examples of technical plans in addition to the SEMP are listed in Appendix K.

The SEMP must be developed concurrently with the project plan. In developing the SEMP, the technical approach to the project and, hence, the technical aspect of the project life cycle is developed. This determines the project's length and cost. The development of the programmatic and technical management approaches requires that the key project personnel develop an understanding of the work to be performed and the relationships among the various parts of that work. Refer to Subsections 6.1.2.1 and 6.1.1.2 on WBSs and network scheduling, respectively.

The SEMP's development requires contributions from knowledgeable programmatic and technical experts from all areas of the project that can significantly influence the project's outcome. The involvement of recognized experts is needed to establish a SEMP that is credible to the project manager and to secure the full commitment of the project team.

Role of the SEMP

The SEMP is the rule book that describes to all participants how the project will be technically managed. The NASA field center responsible for the project should have a SEMP to describe how it will conduct its technical management, and each contractor should have a SEMP to describe how it will manage in accordance with both its contract and NASA's technical management practices. Each Center that is involved with the project should also have a SEMP for its part of the project, which would interface with the project SEMP of the responsible NASA Center, but this lower tier SEMP specifically will address that Center's technical effort and how it interfaces with the overall project. Since the SEMP is project- and contract-unique, it must be updated for each significant programmatic change, or it will become outmoded and unused, and the project could slide into an uncontrolled state. The lead NASA field center should have its SEMP developed before attempting to prepare an initial cost estimate, since activities that incur cost, such as technical risk reduction, need to be identified and described beforehand. The contractor should have its SEMP developed during the proposal process (prior to costing and pricing) because the SEMP describes the technical content of the project, the potentially costly risk management activities, and the verification and validation techniques to be used, all of which must be included in the preparation of project cost estimates. The SEMPs from the supporting Centers should be developed along with the primary project SEMP. The project SEMP is the senior technical management document for the project: all other technical plans must comply with it. The SEMP should be comprehensive and describe how a fully integrated engineering effort will be managed and conducted.

Obtain Stakeholder Commitments to Technical Plans

To obtain commitments to the technical plans by the stakeholders, the technical team should ensure that the appropriate stakeholders have a method to provide inputs and to review the project planning for implementation of stakeholder interests. During Formulation, the roles of the stakeholders should be defined in the project plan and the SEMP. Review of these plans and

the agreement from the stakeholders of the content of these plans will constitute buy-in from the stakeholders in the technical approach. Later in the project life cycle, stakeholders may be responsible for delivery of products to the project. Initial agreements regarding the responsibilities of the stakeholders are key to ensuring that the project technical team obtains the appropriate deliveries from stakeholders.

The identification of stakeholders is one of the early steps in the systems engineering process. As the project progresses, stakeholder expectations are flowed down through the Logical Decomposition Process, and specific stakeholders are identified for all of the primary and derived requirements. A critical part of the stakeholders' involvement is in the definition of the technical requirements. As requirements and ConOps are developed, the stakeholders will be required to agree to these products. Inadequate stakeholder involvement will lead to inadequate requirements and a resultant product that does not meet the stakeholder expectations. Status on relevant stakeholder involvement should be tracked and corrective action taken if stakeholders are not participating as planned.

Throughout the project life cycle, communication with the stakeholders and commitment from the stakeholders may be accomplished through the use of agreements. Organizations may use an Internal Task Agreement (ITA), a Memorandum of Understanding (MOU), or other similar documentation to establish the relationship between the project and the stakeholder. These agreements also are used to document the customer and provider responsibilities for definition of products to be delivered. These agreements should establish the Measures of Effectiveness (MOEs) or Measures of Performance (MOPs) that will be used to monitor the progress of activities. Reporting requirements and schedule requirements should be established in these agreements. Preparation of these agreements will ensure that the stakeholders' roles and responsibilities support the project goals and that the project has a method to address risks and issues as they are identified.

During development of the project plan and the SEMP, forums are established to facilitate communication and document decisions during the life cycle of the project. These forums include meetings, working groups, decision panels, and control boards. Each of these forums should establish a charter to define the scope and authority of the forum and identify necessary voting or nonvoting participants. Ad hoc members may be identified when the expertise or input of specific stakeholders is needed when specific topics are addressed. Ensure that stakeholders have been identified to support the forum.

Issue Technical Work Directives

The technical team provides technical work directives to Cost Account Managers (CAMs). This enables the CAMs to prepare detailed plans that are mutually consistent and collectively address all of the work to be performed. These plans include the detailed schedules and budgets for cost accounts that are needed for cost management and EVM.

Issuing technical work directives is an essential activity during Phase B of a project, when a detailed planning baseline is required. If this activity is not implemented, then the CAMs are often left with insufficient guidance for detailed planning. The schedules and budgets that are needed for EVM will then be based on assumptions and local interpretations of project-level information. If this is the case, it is highly likely that substantial variances will occur between the baseline plan and the work performed. Providing technical work directives to CAMs produces a more organized technical team. This activity may be repeated when replanning occurs.

This activity is not limited to systems engineering. This is a normal part of project planning wherever there is a need for an accurate planning baseline.

The technical team will provide technical directives to CAMs for every cost account within the SE element of the WBS. These directives may be in any format, but should clearly communicate the following information for each account:

- Technical products expected;
- Documents and technical reporting requirements for each cost account;
- Critical events, and specific products expected from a particular CAM in support of this event (e.g., this cost account is expected to deliver a presentation on specific topics at the PDR);
- References to applicable requirements, policies, and standards;
- Identification of particular tools that should be used;
- Instructions on how the technical team wants to coordinate and review cost account plans before they go to project management; and

- Decisions that have been made on how work is to be performed and who is to perform it.

CAMs receive these technical directives, along with the project planning guidelines, and prepare cost account plans. These plans may be in any format and may have various names at different Centers, but minimally they will include:

- Scope of the cost account, which includes:
  - ▶ Technical products delivered;
  - ▶ Other products developed that will be needed to complete deliverables (e.g., a Configuration Management (CM) system may need development in order to deliver the product of a "managed configuration");
  - ▶ A brief description of the procedures that will be followed to complete work on these products, such as:
    - Product X will be prepared in-house, using the local procedure A, which is commonly used in Organization ABC,
    - Product X will be verified/validated in the following manner…,
    - Product X will be delivered to the project in the following manner…,
    - Product X delivery will include the following reports (e.g., delivery of a CM system to the project would include regular reports on the status of the configuration, etc.),
    - Product Y will be procured in accordance with procurement procedure B.
- A schedule attached to this plan in a format compatible with project guidelines for schedules. This schedule would contain each of the procedures and deliverables mentioned above and provide additional information on the activity steps of each procedure.
- A budget attached to this plan in a system compatible with project guidelines for budgets. This budget would be consistent with the resources needed to accomplish the scheduled activities.
- Any necessary agreements and approvals.

If the project is going to use EVM, then the scope of a cost account needs to further identify a number of "work packages," which are units of work that can be scheduled and given cost estimates. Work packages should be based on completed products to the greatest extent possible, but may also be based on completed procedures (e.g., completion of validation). Each work package will have its own schedule and a budget. The budget for this work package becomes part of the Budgeted Cost of Work Scheduled (BCWS) in the EVM system. When this unit of work is completed, the project's earned value will increase by this amount. There may be future work in this cost account that is not well enough defined to be described as a set of work packages. For example, launch operations will be supported by the technical team, but the details of what will be done often have not been worked out during Phase B. In this case, this future work is called a "planning package," which has a high-level schedule and an overall budget. When this work is understood better, the planning package will be broken up into work packages, so that the EVM system can continue to operate during launch operations.

Cost account plans should be reviewed and approved by the technical team and by the line manager of the cost account manager's home organization. Planning guidelines may identify additional review and approval requirements.

The planning process described above is not limited to systems engineering. This is the expected process for all elements of a flight project. One role that the systems engineer may have in planning is to verify that the scope of work described in cost account plans across the project is consistent with the project WBS dictionary, and that the WBS dictionary is consistent with the architecture of the project.

Capture Technical Planning Work Products

The work products from the Technical Planning process should be managed using either the Technical Data Management Process or the Configuration Management Process as required. Some of the more important products of technical planning (i.e., the WBS, the SEMP, and the schedule, etc.) are kept under configuration control and captured using the CM process. The Technical Data Management Process is used to capture trade studies, cost estimates, technical analyses, reports, and other important documents not under formal configuration control. Work products, such as meeting minutes and correspondence (including e-mail) containing decisions or agreements with stakeholders also should be retained and stored in project files for later reference.

6.1.1.3 Outputs

Typical outputs from technical planning activities are:

- Technical work cost estimates, schedules, and resource needs, e.g., funds, workforce, facilities, and equipment (to project), within the project resources;
- Product and process measures needed to assess progress of the technical effort and the effectiveness of processes (to Technical Assessment Process);
- Technical planning strategy, WBS, SEMP, and other technical plans that support implementation of the technical effort (to all processes; applicable plans to technical processes);
- Technical work directives, e.g., work packages or task orders with work authorization (to applicable technical teams); and
- Technical Planning Process work products needed to provide reports, records, and nondeliverable outcomes of process activities (to Technical Data Management Process).

The resulting technical planning strategy would constitute an outline, or rough draft, of the SEMP. This would serve as a starting part for the overall Technical Planning Process after initial preparation is complete. When preparations for technical planning are complete, the technical team should have a cost estimate and schedule for the technical planning effort. The budget and schedule to support the defined technical planning effort can then be negotiated with the project manager to resolve any discrepancies between what is needed and what is available. The SEMP baseline needs to be completed. Planning for the update of the SEMP based on programmatic changes needs to be developed and implemented. The SEMP needs to be approved by the appropriate level of authority.

This "technical work directives" step produces: (1) planning directives to cost account managers that result in (2) a consistent set of cost account plans. Where EVM is called for, it produces (3) an EVM planning baseline, including a BCWS.

6.1.2 Technical Planning Guidance

6.1.2.1 Work Breakdown Structure

A work breakdown structure is a hierarchical breakdown of the work necessary to complete a project. The WBS should be a product-based, hierarchical division of deliverable items and associated services. As such, it should contain the project's Product Breakdown Structure (PBS) with the specified prime product(s) at the top and the systems, segments, subsystems, etc., at successive lower levels. At the lowest level are products such as hardware items, software items, and information items (documents, databases, etc.) for which there is a cognizant engineer or manager. Branch points in the hierarchy should show how the PBS elements are to be integrated. The WBS is built, in part, from the PBS by adding, at each branch point of the PBS, any necessary service elements, such as management, systems engineering, Integration and Verification (I&V), and integrated logistics support. If several WBS elements require similar equipment or software, then a higher level WBS element might be defined from the system level to perform a block buy or a development activity (e.g., system support equipment). Figure 6.1-4 shows the relationship between a system, a PBS, and a WBS. In summary, the WBS is a combination of the PBS and input from the system level. The system level is incorporated to capture and integrate similarities across WBS elements.

A project WBS should be carried down to the cost account level appropriate to the risks to be managed. The appropriate level of detail for a cost account is determined by management's desire to have visibility into costs, balanced against the cost of planning and reporting. Contractors may have a Contract WBS (CWBS) that is appropriate to their need to control costs. A summary CWBS, consisting of the upper levels of the full CWBS, is usually included in the project WBS to report costs to the contracting organization. WBS elements should be identified by title and by a numbering system that performs the following functions:

- Identifies the level of the WBS element,
- Identifies the higher level element into which the WBS element will be integrated, and
- Shows the cost account number of the element.

A WBS should also have a companion WBS dictionary that contains each element's title, identification number, objective, description, and any dependencies (e.g., receivables) on other WBS elements. This dictionary provides a structured project description that is valuable for orienting project members and other interested parties. It fully describes the products and/or services expected from each WBS element. This subsection provides some techniques for developing a WBS and points out some mistakes to avoid.

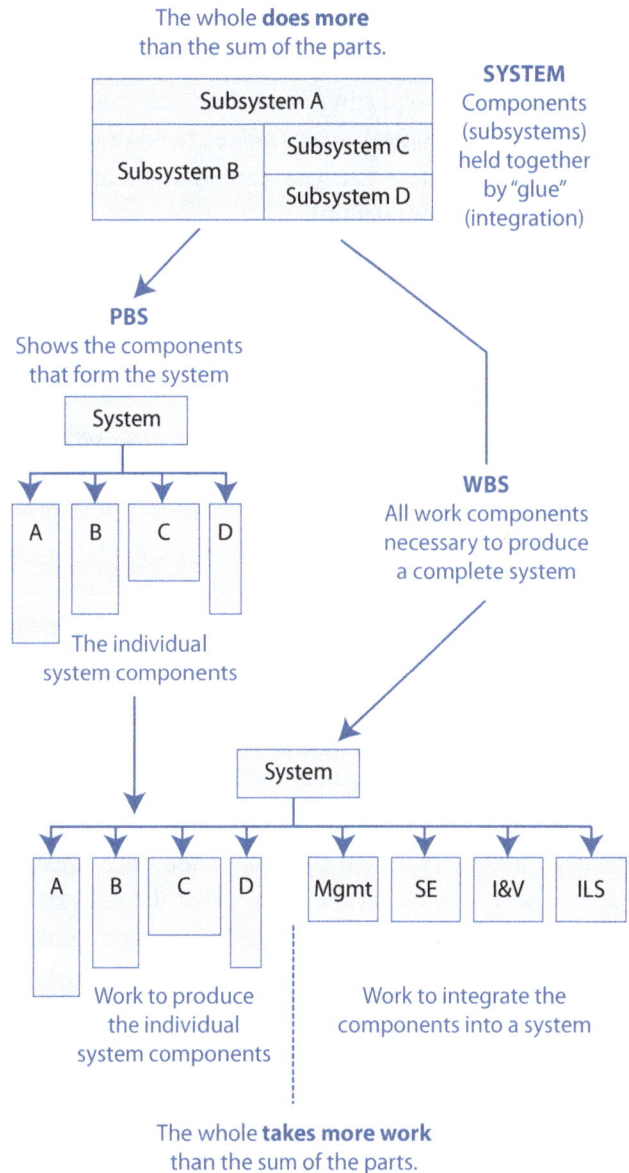

Figure 6.1-4 Relationship between a system, a PBS, and a WBS

Role of the WBS

The technical team should receive planning guidelines from the project office. The technical team should provide the project office with any appropriate tailoring or expansion of the systems engineering WBS element, and have project-level concurrence on the WBS and WBS dictionary before issuing technical work directives.

A product-based WBS is the organizing structure for:

- Project and technical planning and scheduling.
- Cost estimation and budget formulation. (In particular, costs collected in a product-based WBS can be compared to historical data. This is identified as a primary objective by DOD standards for WBSs.)

- Defining the scope of statements of work and specifications for contract efforts.

- Project status reporting, including schedule, cost, workforce, technical performance, and integrated cost/schedule data (such as earned value and estimated cost at completion).

- Plans, such as the SEMP, and other documentation products, such as specifications and drawings.

It provides a logical outline and vocabulary that describes the entire project, and integrates information in a consistent way. If there is a schedule slip in one element of a WBS, an observer can determine which other WBS elements are most likely to be affected. Cost impacts are more accurately estimated. If there is a design change in one element of the WBS, an observer can determine which other WBS elements will most likely be affected, and these elements can be consulted for potential adverse impacts.

Techniques for Developing the WBS

Developing a successful project WBS is likely to require several iterations through the project life cycle since it is not always obvious at the outset what the full extent of the work may be. Prior to developing a preliminary WBS, there should be some development of the system architecture to the point where a preliminary PBS can be created. The PBS and associated WBS can then be developed level by level from the top down. In this approach, a project-level systems engineer finalizes the PBS at the project level and provides a draft PBS for the next lower level. The WBS is then derived by adding appropriate services such as management and systems engineering to that lower level. This process is repeated recursively until a WBS exists down to the desired cost account level. An alternative approach is to define all levels of a complete PBS in one design activity and then develop the complete WBS. When this approach is taken, it is necessary to take great care to develop the PBS so that all products are included and all assembly/I&V branches are correct. The involvement of people who will be responsible for the lower level WBS elements is recommended.

Common Errors in Developing a WBS

There are three common errors found in WBSs.

- **Error 1:** The WBS describes functions, not products. This makes the project manager the only one formally responsible for products.

- **Error 2:** The WBS has branch points that are not consistent with how the WBS elements will be integrated. For instance, in a flight operations system with a distributed architecture, there is typically software associated with hardware items that will be integrated and verified at lower levels of a WBS. It would then be inappropriate to separate hardware and software as if they were separate systems to be integrated at the system level. This would make it difficult to assign accountability for integration and to identify the costs of integrating and testing components of a system.

- **Error 3:** The WBS is inconsistent with the PBS. This makes it possible that the PBS will not be fully implemented and generally complicates the management process.

Some examples of these errors are shown in Figure 6.1-5. Each one prevents the WBS from successfully performing its roles in project planning and organizing. These errors are avoided by using the WBS development techniques described above.

Common to both the project management and systems engineering disciplines is the requirement for organizing and managing a system throughout its life cycle within a systematic and structured framework, reflective of the work to be performed and the associated cost, schedule, technical, and risk data to be accumulated, summarized, and reported. (See NPR 7120.5.)

A key element of this framework is a hierarchical, product-oriented WBS. Derived from both the physical and system architectures, the WBS provides a systematic, logical approach for defining and translating initial mission goals and technical concepts into tangible project goals, system products, and life-cycle support (or enabling) functions.

When appropriately structured and used in conjunction with sound engineering principles, the WBS supplies a common framework for subdividing the total project into clearly defined, product-oriented work components, logically related and sequenced according to hierarchy, schedule, and responsibility assignment.

The composition and level of detail required in the WBS hierarchy is determined by the project management and technical teams based on careful consideration of the project's size and the complexity, constraints, and risk associated with the technical effort. The initial WBS will

Error 1: Functions Without Products

Error 2: Inappropriate Branches

Error 3: Inconsistency With PBS

WBS   PBS

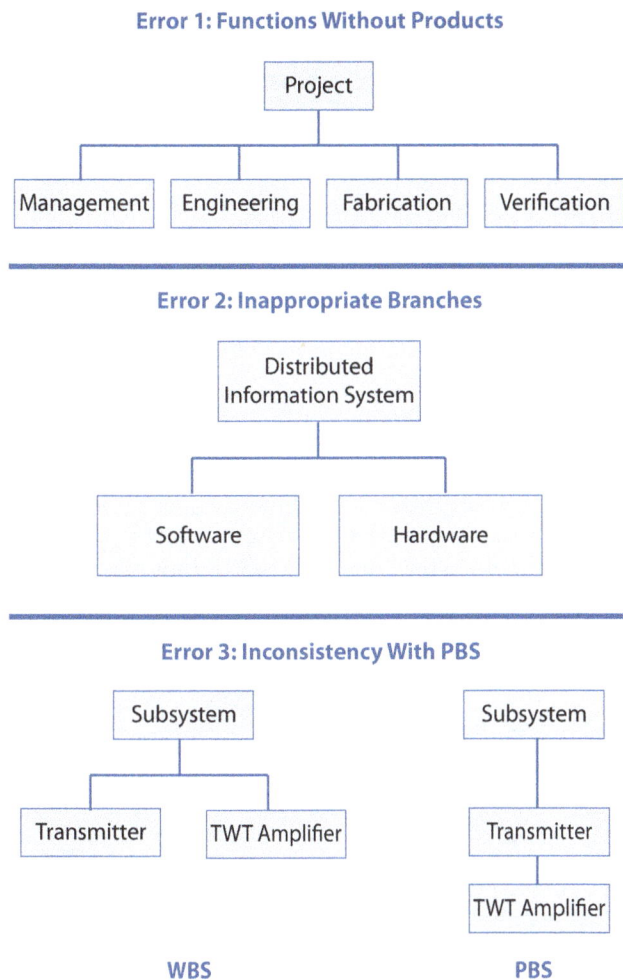

Figure 6.1-5 Examples of WBS development errors

provide a structured framework for conceptualizing and defining the program/project objectives and for translating the initial concepts into the major systems, component products, and services to be developed, produced, and/or obtained. As successive levels of detail are defined, the WBS hierarchy will evolve to reflect a comprehensive, complete view of both the total project effort and each system or end product to be realized throughout the project's life cycle.

Decomposition of the major deliverables into unique, tangible product or service elements should continue to a level representative of how each WBS element will be planned and managed. Whether assigned to in-house or contractor organizations, these lower WBS elements will be subdivided into subordinate tasks and activities and aggregated into the work packages and control accounts utilized to populate the project's cost plans, schedules, and performance metrics.

At a minimum, the WBS should reflect the major system products and services to be developed and/or procured, the enabling (support) products and services, and any high-cost and/or high-risk product elements residing at lower levels in the hierarchy.[1] The baseline WBS configuration will be documented as part of the program plan and utilized to structure the SEMP. The cost estimates and the WBS dictionary are maintained throughout the project's life cycle to reflect the project's current scope.

The preparation and approval of three key program/project documents, the Formulation Authorization Document (FAD), the program commitment agreement, and the program/project plans are significant contributors to early WBS development.

The initial contents of these documents will establish the purpose, scope, objectives, and applicable agreements for the program of interest and will include a list of approved projects, control plans, management approaches, and any commitments and constraints identified.

The technical team selects the appropriate system design processes to be employed in the top-down definition of each product in the system structure. Subdivision of the project and system architecture into smaller, more manageable components will provide logical summary points for assessing the overall project's accomplishments and for measuring cost and schedule performance.

Once the initial mission goals and objectives have evolved into the build-to or final design, the WBS will be refined and updated to reflect the evolving scope and architecture of the project and the bottom-up realization of each product in the system structure.

Throughout the applicable life-cycle phases, the WBS and WBS dictionary will be updated to reflect the project's current scope and to ensure control of high-risk and cost/schedule performance issues.

6.1.2.2 Cost Definition and Modeling

This subsection deals with the role of cost in the systems analysis and engineering process, how to measure it, how to control it, and how to obtain estimates of it. The reason costs and their estimates are of great importance

---

[1]IEEE Standard 1220, Section C.3, "The system products and life cycle enabling products should be jointly engineered and once the enabling products and services are identified, should be treated as systems in the overall system hierarchy."

WBS Hierarchies for Systems

It is important to note that while product-oriented in nature, the standard WBS mandated for NASA space flight projects in NPR 7120.5 approaches WBS development from a project and not a system perspective. The WBS mandated reflects the scope of a major Agency project and, therefore, is structured to include the development, operation, and disposal of more than one major system of interest during the project's normal life cycle.

WBS hierarchies for NASA's space flight projects will include high-level system products, such as payload, spacecraft, and ground systems, and enabling products and services, such as project management, systems engineering, and education. These standard product elements have been established to facilitate alignment with the Agency's accounting, acquisition, and reporting systems.

Unlike the project-view WBS approach described in NPR 7120.5, creation of a technical WBS focuses on the development and realization of both the overall end product and each subproduct included as a lower level element in the overall system structure.

NPR 7123.1, NASA Systems Engineering Processes and Requirements mandates a standard, systematic technical approach to system or end-product development and realization. Utilizing a building-block or product-hierarchy approach, the system architecture is successively defined and decomposed into subsystems (elements performing the operational functions of the system) and associated and interrelated subelements (assemblies, components, parts, and enabling life-cycle products).

The resulting hierarchy or family-product tree depicts the entire system architecture in a PBS. Recognized by Government and industry as a "best practice," utilization of the PBS and its building-block configuration facilitates both the application of NPR 7123.1's 17 common technical processes at all levels of the PBS structure and the definition and realization of successively lower level elements of the system's hierarchy.

Definition and application of the work effort to the PBS structure yields a series of functional subproducts or "children" WBS models. The overall parent or system WBS model is realized through the rollup of successive levels of these product-based, subelement WBS models.

Each WBS model represents one unique unit or functional end product in the overall system configuration and, when related by the PBS into a hierarchy of individual models, represents one functional system end product or "parent" WBS model.

(See *NPR 7120.5, NASA Space Flight Program and Project Management Requirements*.)

in systems engineering goes back to a principal objective of systems engineering: fulfilling the system's goals in the most cost-effective manner. The cost of each alternative should be one of the most important outcome variables in trade studies performed during the systems engineering process.

One role, then, for cost estimates is in helping to choose rationally among alternatives. Another is as a control mechanism during the project life cycle. Cost measures produced for project life-cycle reviews are important in determining whether the system goals and objectives are still deemed valid and achievable, and whether constraints and boundaries are worth maintaining. These measures are also useful in determining whether system goals and objectives have properly flowed down through to the various subsystems.

As system designs and ConOps mature, cost estimates should mature as well. At each review, cost estimates need to be presented and compared to the funds likely to be available to complete the project. The cost estimates presented at early reviews must be given special attention since they usually form the basis for the initial cost commitment for the project. The systems engineer must be able to provide realistic cost estimates to the project manager. In the absence of such estimates, overruns are likely to occur, and the credibility of the entire system development process, both internal and external, is threatened.

Life-Cycle Cost and Other Cost Measures

A number of questions need to be addressed so that costs are properly treated in systems analysis and engineering. These questions include:

- What costs should be counted?
- How should costs occurring at different times be treated?
- What about costs that cannot easily be measured in dollars?

What Costs Should Be Counted

The most comprehensive measure of the cost of an alternative is its life-cycle cost. According to NPR 7120.5, a system's life-cycle cost is, "the total of the direct, indirect, recurring, nonrecurring, and other related expenses incurred, or estimated to be incurred, in the design, development, verification, production, operation, maintenance, support, and disposal of a project. The life-cycle cost of a project or system can also be defined as the total cost of ownership over the project or system's life cycle from Formulation through Implementation. It includes all design, development, deployment, operation and maintenance, and disposal costs."

Costs Occurring Over Time

The life-cycle cost combines costs that typically occur over a period of several years. To facilitate engineering trades and comparison of system costs, these real year costs are deescalated to constant year values. This removes the impact of inflation from all estimates and allows ready comparison of alternative approaches. In those instances where major portfolio architectural trades are being conducted, it may be necessary to perform formal cost benefit analyses or evaluate leasing versus purchase alternatives. In those trades, engineers and cost analysts should follow the guidance provided in Office of Management and Budget (OMB) Circular A-94 on rate of return and net present value calculation in comparing alternatives.

Difficult-to-Measure Costs

In practice, estimating some costs poses special problems. These special problems, which are not unique to NASA systems, usually occur in two areas: (1) when alternatives have differences in the irreducible chances of loss of life, and (2) when externalities are present. Two examples of externalities that impose costs are pollution caused by some launch systems and the creation of orbital debris. Because it is difficult to place a dollar figure on these resource uses, they are generally called "incommensurable costs." The general treatment of these types of costs in trade studies is not to ignore them, but instead to keep track of them along with other costs. If these ele-

ments are part of the trade space, it is generally advisable to apply Circular A-94 approaches to those trades.

Controlling Life-Cycle Costs

The project manager/systems engineer must ensure that the probabilistic life-cycle cost estimate is compatible with NASA's budget and strategic priorities. The current policy is that projects are to submit budgets sufficient to ensure a 70 percent probability of achieving the objectives within the proposed resources. Project managers and systems engineers must establish processes to estimate, assess, monitor, and control the project's life-cycle cost through every phase of the project.

Early decisions in the systems engineering process tend to have the greatest effect on the resultant system life-cycle cost. Typically, by the time the preferred system architecture is selected, between 50 and 70 percent of the system's life-cycle cost has been locked in. By the time a preliminary system design is selected, this figure may be as high as 90 percent. This presents a major dilemma to the systems engineer, who must lead this selection process. Just at the time when decisions are most critical, the state of information about the alternatives is least certain. Uncertainty about costs is a fact of systems engineering, and that uncertainty must be accommodated by complete and careful analysis of the project risks and provision of sufficient margins (cost, technical, and schedule) to ensure success. There are a number of estimating techniques to assist the systems engineer and project manager in providing for uncertainty and unknown requirements. Additional information on these techniques can be found in the *NASA Cost Estimating Handbook*.

This suggests that efforts to acquire better information about the life-cycle cost of each alternative early in the project life cycle (Phases A and B) potentially have very high payoffs. The systems engineer needs to identify the principal life-cycle cost drivers and the risks associated with the system design, manufacturing, and operations. Consequently, it is particularly important with such a system to bring in the specialty engineering disciplines such as reliability, maintainability, supportability, and operations engineering early in the systems engineering process, as they are essential to proper life-cycle cost estimation.

One mechanism for controlling life-cycle cost is to establish a *life-cycle cost management program* as part of the project's management approach. (Life-cycle cost management has sometimes been called "design-to-life-cycle cost.") Such a program establishes life-cycle cost as a design goal, perhaps with subgoals for acquisition costs or operations and support costs. More specifically, the objectives of a life-cycle cost management program are to:

- Identify a common set of ground rules and assumptions for life-cycle cost estimation;

- Manage to a cost baseline and maintain traceability to the technical baseline with documentation for subsequent cost changes;

- Ensure that best-practice methods, tools, and models are used for life-cycle cost analysis;

- Track the estimated life-cycle cost throughout the project life cycle; *and, most important*

- Integrate life-cycle cost considerations into the design and development process via trade studies and formal change request assessments.

Trade studies and formal change request assessments provide the means to balance the effectiveness and life-cycle cost of the system. The complexity of integrating life-cycle cost considerations into the design and development process should not be underestimated, but neither should the benefits, which can be measured in terms of greater cost-effectiveness. The existence of a rich set of potential life-cycle cost trades makes this complexity even greater.

Cost-Estimating Methods

Various cost-estimating methodologies are utilized throughout a program's life cycle. These include parametric, analogous, and engineering (grassroots).

- **Parametric:** Parametric cost models are used in the early stages of project development when there is limited program and technical definition. Such models involve collecting relevant historical data at an aggregated level of detail and relating it to the area to be estimated through the use of mathematical techniques to create cost-estimating relationships. Normally, less detail is required for this approach than for other methods.

- **Analogous:** This is based on most new programs originated or evolved from existing programs or simply representing a new combination of existing components. It uses actual costs of similar existing or past programs and adjusts for complexity, technical, or physical differences to derive the new system esti-

mate. This method would be used when there is insufficient actual cost data to use as a basis for a detailed approach but there is a sufficient amount of program and technical definition.

- **Engineering (Grassroots):** These bottom-up estimates are the result of rolling up the costs estimated by each organization performing work described in the WBS. Properly done, grassroots estimates can be quite accurate, but each time a "what if" question is raised, a new estimate needs to be made. Each change of assumptions voids at least part of the old estimate. Because the process of obtaining grassroots estimates is typically time consuming and labor intensive, the number of such estimates that can be prepared during trade studies is in reality severely limited.

The type of cost estimating method used will depend on the adequacy of program definition, level of detail required, availability of data, and time constraints. For example, during the early stages of a program, a conceptual study considering several options would dictate an estimating method requiring no actual cost data and limited program definition on the systems being estimated. A parametric model would be a sound approach at this point. Once a design is baselined and the program is more adequately defined, an analogy approach becomes appropriate. As detailed actual cost data are accumulated, a grassroots methodology is used.

More information on cost-estimating methods and the development of cost estimates can be found in the *NASA Cost Estimating Handbook*.

Integrating Cost Model Results for a Complete Life-Cycle Cost Estimate

A number of parametric cost models are available for costing NASA systems. A list of the models currently in use may be found in an appendix in the *NASA Cost Estimating Handbook*. Unfortunately, none alone is sufficient to estimate life-cycle cost. Assembling an estimate of life-cycle cost often requires that several different models (along with the other two techniques) be used together. Whether generated by parametric models, analogous, or grassroots methods, the estimated cost of the hardware element must frequently be "wrapped" or have factors applied to estimate the costs associated with management, systems engineering, test, etc., of the systems being estimated. The NASA full-cost factors also must be applied separately.

To integrate the costs being estimated by these different models, the systems engineer should ensure that the inputs to and assumptions of the models are consistent, that all relevant life-cycle cost components are covered, and that the phasing of costs is correct. Estimates from different sources are often expressed in different year *constant* dollars which must be combined. Appropriate inflation factors must be applied to enable construction of a total life-cycle cost estimate in real year dollars. Guidance on the use of inflation rates for new projects and for budget submissions for ongoing projects can be found in the annual NASA strategic guidance.

Cost models frequently produce a cost estimate for the first unit of a hardware item, but where the project requires multiple units a learning curve can be applied to the first unit cost to obtain the required multiple-unit estimate. Learning curves are based on the concept that resources required to produce each additional unit decline as the total number of units produced increases. The learning curve concept is used primarily for uninterrupted manufacturing and assembly tasks, which are highly repetitive and labor intensive. The major premise of learning curves is that each time the product quantity doubles, the resources (labor hours) required to produce the product will reduce by a determined percentage of the prior quantity resource requirements. The two types of learning curve approaches are unit curve and cumulative average curve. The systems engineer can learn more about the calculation and use of learning curves in the *NASA Cost Estimating Handbook*.

Models frequently provide a cost estimate of the total acquisition effort without providing a recommended phasing of costs over the life cycle. The systems engineer can use a set of phasing algorithms based on the typical ramping-up and subsequent ramping-down of acquisition costs for that type of project if a detailed project schedule is not available to form a basis for the phasing of the effort. A normal distribution curve, or beta curve, is one type of function used for spreading parametrically derived cost estimates and for R&D contracts where costs build up slowly during the initial phases and then escalate as the midpoint of the contract approaches. A beta curve is a combination of percent spent against percent time elapsed between two points in time. More about beta curves can be found in an appendix of the *NASA Cost Estimating Handbook*.

Although parametric cost models for space systems are already available, their proper use usually requires a considerable investment in learning how to appropriately utilize the models. For projects outside of the domains of these existing cost models, new cost models may be needed to support trade studies. Efforts to develop these models need to begin early in the project life cycle to ensure their timely application during the systems engineering process. Whether existing models or newly created ones are used, the SEMP and its associated life-cycle cost management plan should identify which (and how) models are to be used during each phase of the project life cycle.

6.1.2.3 Lessons Learned

No section on technical planning guidance would be complete without the effective integration and incorporation of the lessons learned relevant to the project.

Systems Engineering Role in Lessons Learned

Systems engineers are the main users and contributors to lessons learned systems. A lesson learned is knowledge or understanding gained by experience—either a successful test or mission or a mishap or failure. Systems engineers compile lessons learned to serve as historical documents, requirements' rationales, and other supporting data analysis. Systems engineering practitioners collect lessons learned during program and project plans, key decision points, life-cycle phases, systems engineering processes and technical reviews. Systems engineers' responsibilities include knowing how to utilize, manage, create, and store lessons learned and knowledge management best practices.

Utilization of Lessons Learned Best Practice

Lessons learned are important to future programs, projects, and processes because they show hypotheses and conclusive insights from previous projects or processes. Practitioners determine how previous lessons from processes or tasks impact risks to current projects and implement those lessons learned that improve design and/or performance.

To pull in lessons learned at the start of a project or task:

- Search the NASA Lessons Learned Information System (LLIS) database using keywords of interest to the new program or project. The process for recording lessons learned is explained in *NPR 7120.6*,

Lessons Learned Process. In addition, other organizations doing similar work may have publicly available databases with lessons learned. For example, the Chemical Safety Board has a good series of case study reports on mishaps.

- Supporting lessons from each engineering discipline should be reflected in the program and project plans. Even if little information was found, the search for lessons learned can be documented.

- Compile lessons by topic and/or discipline.

- Review and select knowledge gained from particular lessons learned.

- Determine how these lessons learned may represent potential risk to the current program or project.

- Incorporate knowledge gained into the project database for risk management, cost estimate, and any other supporting data analysis.

As an example, a systems engineer working on the concept for an instrument for a spacecraft might search the lessons learned database using the keywords "environment," "mishap," or "configuration management." One of the lessons learned that search would bring up is #1514. The lesson was from Chandra. A rebaseline of the program in 1992 removed two instruments, changed Chandra's orbit from low Earth to high elliptical, and simplified the thermal control concept from the active control required by one of the descoped instruments to passive "cold-biased" surface plus heaters. This change in thermal control concept mandated silver Teflon thermal control surfaces. The event driving the lesson was a severe spacecraft charging and an electrostatic discharge environment. The event necessitated an aggressive electrostatic discharge test and circuit protection effort that cost over $1 million, according to the database. The Teflon thermal control surfaces plus the high elliptical orbit created the electrostatic problem. Design solutions for one environment were inappropriate in another environment. The lesson learned was that any orbit modifications should trigger a complete new iteration of the systems engineering processes starting from requirements definition. Rebaselining a program should take into account change in the natural environment before new design decisions are made. This lesson would be valuable to keep in mind when changes occur to baselines on the program currently being worked on.

Management of Lessons Learned Best Practice

Capturing lessons learned is a function of good management practice and discipline. Too often lessons learned are missed because they should have been developed and managed within, across, or between life-cycle phases. There is a tendency to wait until resolution of a situation to document a lesson learned, but the unfolding of a problem at the beginning is valuable information and hard to recreate later. It is important to document a lesson learned as it unfolds, particularly as resolution may not be reached until a later phase. Since detailed lessons are often hard for the human mind to recover, waiting until a technical review or the end of a project to collect the lessons learned hinders the use of lessons and the evolution of practice. A mechanism for managing and leveraging lessons as they occur, such as monthly lessons learned briefings or some periodic sharing forums, facilitates incorporating lessons into practice and carrying lessons into the next phase.

At the end of each life-cycle phase, practitioners should use systems engineering processes and procedural tasks as control gate cues. All information passed across control gates must be managed in order to successfully enter the next phase, process, or task.

The systems engineering practitioner should make sure all lessons learned in the present phase are concise and conclusive. Conclusive lessons learned contain series of events that formulate abstracts and driving events. Irresolute lessons learned may be rolled into the next phase to await proper supporting evidence. Project managers and the project technical team are to make sure lessons learned are recorded in the Agency database at the end of all life-cycle phases, major systems engineering processes, key decision points, and technical reviews.

6.2 Requirements Management

Requirements management activities apply to the management of all stakeholder expectations, customer requirements, and technical product requirements down to the lowest level product component requirements (hereafter referred to as expectations and requirements). The Requirements Management Process is used to:

- Manage the product requirements identified, baselined, and used in the definition of the WBS model products during system design;

- Provide bidirectional traceability back to the top WBS model requirements; and

- Manage the changes to established requirement baselines over the life cycle of the system products.

6.2.1 Process Description

Figure 6.2-1 provides a typical flow diagram for the Requirements Management Process and identifies typical inputs, outputs, and activities to consider in addressing requirements management.

6.2.1.1 Inputs

There are several fundamental inputs to the Requirements Management Process.

> Note: Requirements can be generated from nonobvious stakeholders and may not directly support the current mission and its objectives, but instead provide an opportunity to gain additional benefits or information that can support the Agency or the Nation. Early in the process, the systems engineer can help identify potential areas where the system can be used to collect unique information that is not directly related to the primary mission. Often outside groups are not aware of the system goals and capabilities until it is almost too late in the process.

- Requirements and stakeholder expectations are identified during the system design processes, primarily from the Stakeholder Expectation Definition Process and the Technical Requirements Definition Process.

- The Requirements Management Process must be prepared to deal with requirement change requests that can be generated at any time during the project life cycle or as a result of reviews and assessments as part of the Technical Assessment Process.

- TPM estimation/evaluation results from the Technical Assessment Process provide an early warning of

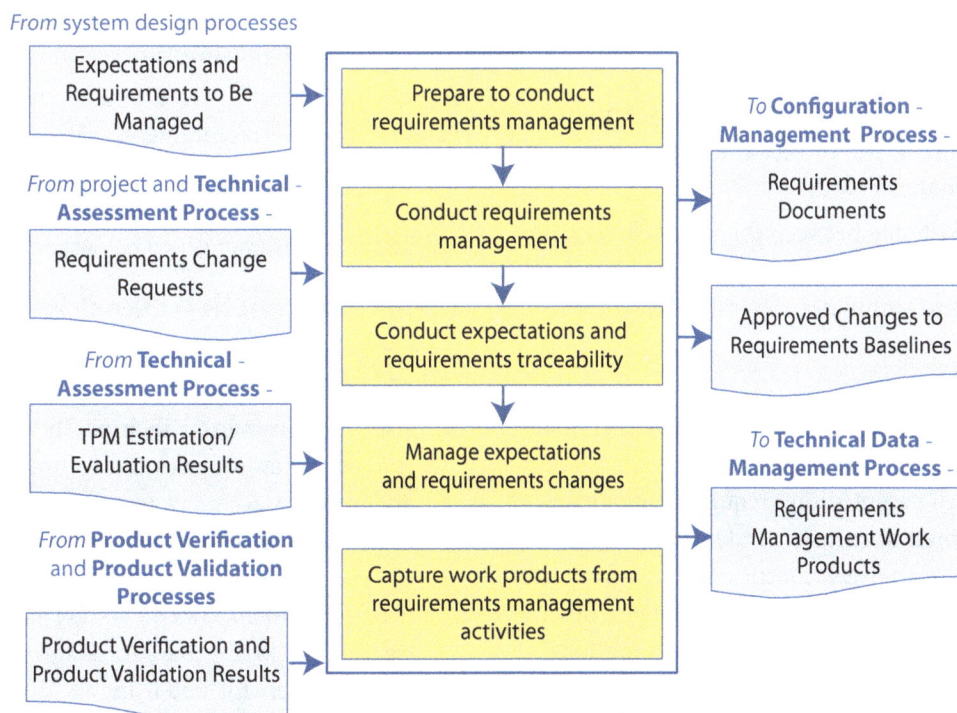

Figure 6.2-1 Requirements Management Process

the adequacy of a design in satisfying selected critical technical parameter requirements. Variances from expected values of product performance may trigger changes to requirements.

- Product verification and product validation results from the Product Verification and Product Validation Processes are mapped into the requirements database with the goal of verifying and validating all requirements.

6.2.1.2 Process Activities

The Requirements Management Process involves managing all changes to expectations and requirements baselines over the life of the product and maintaining bidirectional traceability between stakeholder expectations, customer requirements, technical product requirements, product component requirements, design documents, and test plans and procedures. The successful management of requirements involves several key activities:

- Establish a plan for executing requirements management.
- Receive requirements from the system design processes and organize them in a hierarchical tree structure.
- Establish bidirectional traceability between requirements.
- Validate requirements against the stakeholder expectations, the mission objectives and constraints, the operational objectives, and the mission success criteria.
- Define a verification method for each requirement.
- Baseline requirements.
- Evaluate all change requests to the requirements baseline over the life of the project and make changes if approved by change board.
- Maintain consistency between the requirements, the ConOps, and the architecture/design and initiate corrective actions to eliminate inconsistencies.

Requirements Traceability

As each requirement is documented, its bidirectional traceability should be recorded. Each requirement should be traced back to a parent/source requirement or expectation in a baselined document or identify the requirement as *self-derived* and seek concurrence on it from the next higher level requirements sources. Examples of self-derived requirements are requirements that are locally adopted as good practices or are the result of design decisions made while performing the activities of the Logical Decomposition and Design Solution Processes.

The requirements should be evaluated, independently if possible, to ensure that the requirements trace is correct and that it fully addresses its parent requirements. If it does not, some other requirement(s) must complete fulfillment of the parent requirement and be included in the traceability matrix. In addition, ensure that all top-level parent document requirements have been allocated to the lower level requirements. If there is no parent for a particular requirement and it is not an acceptable self-derived requirement, it should be assumed either that the traceability process is flawed and should be redone or that the requirement is "gold plating" and should be eliminated. Duplication between levels must be resolved. If a requirement is simply repeated at a lower level and it is not an externally imposed constraint, perhaps the requirement does not belong at the higher level. Requirements traceability is usually recorded in a requirements matrix. (See Appendix D.)

Definitions

Traceability: A discernible association between two or more logical entities such as requirements, system elements, verifications, or tasks.

Bidirectional traceability: An association between two or more logical entities that is discernible in either direction (i.e., to and from an entity).

Requirements Validation

An important part of requirements management is the validation of the requirements against the stakeholder expectations, the mission objectives and constraints, the operational objectives, and the mission success criteria. Validating requirements can be broken into three steps:

1. **Are Requirements Written Correctly:** Identify and correct requirements "shall" statement format errors and editorial errors.

2. **Are Requirements Technically Correct:** A few trained reviewers from the technical team identify and remove as many technical errors as possible before having all the relevant stakeholders review the requirements. The reviewers should check that the requirement statements (1) have bidirectional traceability to the baselined stakeholder expectations; (2) were formed using valid assumptions; and (3) are essential to, and consistent with, designing and realizing the appropriate product solution form

that will satisfy the applicable product-line life-cycle phase success criteria.

3. **Do Requirements Satisfy Stakeholders:** All relevant stakeholder groups identify and remove defects.

Requirements validation results are often a deciding factor in whether to proceed with the next process of Logical Decomposition or Design Solution Definition. The project team should be prepared to: (1) demonstrate that the project requirements are complete and understandable; (2) demonstrate that prioritized evaluation criteria are consistent with requirements and the operations and logistics concepts; (3) confirm that requirements and evaluation criteria are consistent with stakeholder needs; (4) demonstrate that operations and architecture concepts support mission needs, goals, objectives, assumptions, guidelines, and constraints; and (5) demonstrate that the process for managing change in requirements is established, documented in the project information repository, and communicated to stakeholders.

Managing Requirement Changes

Throughout Phases A and B, changes in requirements and constraints will occur. It is imperative that all changes be thoroughly evaluated to determine the impacts on the architecture, design, interfaces, ConOps, and higher and lower level requirements. Performing functional and sensitivity analyses will ensure that the requirements are realistic and evenly allocated. Rigorous requirements verification and validation ensure that the requirements can be satisfied and conform to mission objectives. All changes must be subjected to a review and approval cycle to maintain traceability and to ensure that the impacts are fully assessed for all parts of the system.

Once the requirements have been validated and reviewed in the System Requirements Review they are placed under formal configuration control. Thereafter, any changes to the requirements must be approved by the Configuration Control Board (CCB). The systems engineer, project manager, and other key engineers usually participate in the CCB approval processes to assess the impact of the change including cost, performance, programmatic, and safety.

The technical team should also ensure that the approved requirements are communicated in a timely manner to all relevant people. Each project should have already established the mechanism to track and disseminate the latest project information. Further information on Con-

figuration Management (CM) can be found in Section 6.5.

Key Issues for Requirements Management

Requirements Changes

Effective management of requirements changes requires a process that assesses the impact of the proposed changes prior to approval and implementation of the change. This is normally accomplished through the use of the Configuration Management Process. In order for CM to perform this function, a baseline configuration must be documented and tools used to assess impacts to the baseline. Typical tools used to analyze the change impacts are as follows:

- **Performance Margins:** This tool is a list of key performance margins for the system and the current status of the margin. For example, the propellant performance margin will provide the necessary propellant available versus the propellant necessary to complete the mission. Changes should be assessed for their impact to performance margins.

- **CM Topic Evaluators List:** This list is developed by the project office to ensure that the appropriate persons are evaluating the changes and providing impacts to the change. All changes need to be routed to the appropriate individuals to ensure that the change has had all impacts identified. This list will need to be updated periodically.

- **Risk System and Threats List:** The risk system can be used to identify risks to the project and the cost, schedule, and technical aspects of the risk. Changes to the baseline can affect the consequences and likelihood of identified risk or can introduce new risk to the project. A threats list is normally used to identify the costs associated with all the risks for the project. Project reserves are used to mitigate the appropriate risk. Analyses of the reserves available versus the needs identified by the threats list assist in the prioritization for reserve use.

The process for managing requirements changes needs to take into account the distribution of information related to the decisions made during the change process. The Configuration Management Process needs to communicate the requirements change decisions to the affected organizations. During a board meeting to approve a change, actions to update documentation need to be included as part of the change package. These actions should be tracked to ensure that affected documentation is updated in a timely manner.

Feedback to the Requirements Baseline

During development of the system components, it will be necessary to provide feedback to the requirements. This feedback is usually generated during the product design, validation, and verification processes. The feedback to the project will include design implementation issues that impact the interfaces or operations of the system. In many cases, the design may introduce constraints on how the component can be operated, maintained, or stored. This information needs to be communicated to the project team to evaluate the impact to the affected system operation or architecture. Each system component will optimize the component design and operation. It is the systems engineering function to evaluate the impact of this optimization at the component level to the optimization of the entire system.

Requirements Creep

"Requirements creep" is the term used to describe the subtle way that requirements grow imperceptibly during the course of a project. The tendency for the set of requirements is to relentlessly increase in size during the course of development, resulting in a system that is more expensive and complex than originally intended. Often the changes are quite innocent and what appear to be changes to a system are really enhancements in disguise.

However, some of the requirements creep involves truly new requirements that did not exist, and could not have been anticipated, during the Technical Requirements Definition Process. These new requirements are the result of evolution, and if we are to build a relevant system, we cannot ignore them.

There are several techniques for avoiding or at least minimizing requirements creep:

- In the early requirements definition phase, flush out the conscious, unconscious, and undreamt-of requirements that might otherwise not be stated.

- Establish a strict process for assessing requirement changes as part of the Configuration Management Process.

- Establish official channels for submitting change requests. This will determine who has the authority to generate requirement changes and submit them formally to the CCB (e.g., the contractor-designated representative, project technical leads, customer/science team lead, or user).

- Measure the functionality of each requirement change request and assess its impact on the rest of the system. Compare this impact with the consequences of not approving the change. What is the risk if the change is not approved?

- Determine if the proposed change can be accommodated within the fiscal and technical resource budgets. If it cannot be accommodated within the established resource margins, then the change most likely should be denied.

6.2.1.3 Outputs

Typical outputs from the requirements management activities are:

- **Requirements Documents:** Requirements documents are submitted to the Configuration Management Process when the requirements are baselined. The official controlled versions of these documents are generally maintained in electronic format within the requirements management tool that has been selected by the project. In this way they are linked to the requirements matrix with all of its traceable relationships.

- **Approved Changes to the Requirements Baselines:** Approved changes to the requirements baselines are issued as an output of the Requirements Management Process after careful assessment of all the impacts of the requirements change across the entire product or system. A single change can have a far-reaching ripple effect which may result in several requirement changes in a number of documents.

- **Various Requirements Management Work Products:** Requirements management work products are any reports, records, and undeliverable outcomes of the Requirements Management Process. For example, the bidirectional traceability status would be one of the work products that would be used in the verification and validation reports.

6.2.2 Requirements Management Guidance

6.2.2.1 Requirements Management Plan

The technical team should prepare a plan for performing the requirements management activities. This plan is normally part of the SEMP but also can stand alone. The plan should:

- Identify the relevant stakeholders who will be involved in the Requirements Management Process (e.g., those who may be affected by, or may affect, the product as well as the processes).

- Provide a schedule for performing the requirements management procedures and activities.

- Assign responsibility, authority, and adequate resources for performing the requirements management activities, developing the requirements management work products, and providing the requirements management services defined in the activities (e.g., staff, requirements management database tool, etc.).

- Define the level of configuration management/data management control for all requirements management work products.

- Identify the training for those who will be performing the requirements management activities.

6.2.2.2 Requirements Management Tools

For small projects and products, the requirements can usually be managed using a spreadsheet program. However, the larger programs and projects require the use of one of the available requirements management tools. In selecting a tool, it is important to define the project's procedure for specifying how the requirements will be organized in the requirements management database tool and how the tool will be used. It is possible, given modern requirements management tools, to create a requirements management database that can store and sort requirements data in multiple ways according to the particular needs of the technical team. The organization of the database is not a trivial exercise and has consequences on how the requirements data can be viewed for the life of the project. Organize the database so that it has all the views into the requirements information that the technical team is likely to need. Careful consideration should be given to how flowdown of requirements and bidirectional traceability will be represented in the database. Sophisticated requirements management database tools also have the ability to capture numerous requirement attributes in the tools' requirements matrix, including the requirements traceability and allocation links. For each requirement in the requirements matrix, the verification method(s), level, and phase are documented in the verification requirements matrix housed in the requirements management database tool (e.g., the tool associates the attributes of method, level, and phase with each requirement). It is important to make sure that the requirements management database tool is compatible with the verification and validation tools chosen for the project.

6.3 Interface Management

The management and control of interfaces is crucial to successful programs or projects. Interface management is a process to assist in controlling product development when efforts are divided among parties (e.g., Government, contractors, geographically diverse technical teams, etc.) and/or to define and maintain compliance among the products that must interoperate.

6.3.1 Process Description

Figure 6.3-1 provides a typical flow diagram for the Interface Management Process and identifies typical inputs, outputs, and activities to consider in addressing interface management.

6.3.1.1 Inputs

Typical inputs needed to understand and address interface management would include the following:

- **System Description:** This allows the design of the system to be explored and examined to determine where system interfaces exist. Contractor arrangements will also dictate where interfaces are needed.

- **System Boundaries:** Document physical boundaries, components, and/or subsystems, which are all drivers for determining where interfaces exist.

- **Organizational Structure:** Decide which organization will dictate interfaces, particularly when there is the need to jointly agree on shared interface param-

eters of a system. The program and project WBS will also provide interface boundaries.

- **Boards Structure:** The SEMP should provide insight into organizational interface responsibilities and drive out interface locations.

- **Interface Requirements:** The internal and external functional and physical interface requirements developed as part of the Technical Requirements Definition Process for the product(s).

- **Interface Change Requests:** These include changes resulting from program or project agreements or changes on the part of the technical team as part of the Technical Assessment Process.

6.3.1.2 Process Activities

During project Formulation, the ConOps of the product is analyzed to identify both external and internal interfaces. This analysis will establish the origin, destination, stimuli, and special characteristics of the interfaces that need to be documented and maintained. As the system structure and architecture emerges, interfaces will be added and existing interfaces will be changed and must be maintained. Thus, the Interface Management Process has a close relationship to other areas, such as requirements definition and configuration management during this period. Typically, an Interface Working Group (IWG) establishes communication links between those

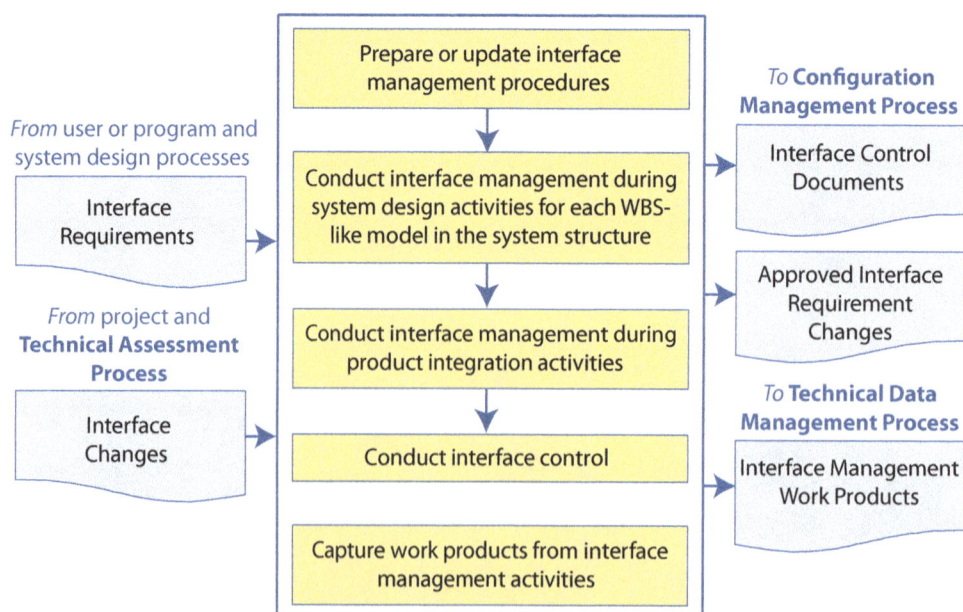

Figure 6.3-1 Interface Management Process

responsible for interfacing systems, end products, enabling products, and subsystems. The IWG has the responsibility to ensure accomplishment of the planning, scheduling, and execution of all interface activities. An IWG is typically a technical team with appropriate technical membership from the interfacing parties (e.g., the project, the contractor, etc.).

During product integration, interface management activities would support the review of integration and assembly procedures to ensure interfaces are properly marked and compatible with specifications and interface control documents. The interface management process has a close relationship to verification and validation. Interface control documentation and approved interface requirement changes are used as inputs to the Product Verification Process and the Product Validation Process, particularly where verification test constraints and interface parameters are needed to set the test objectives and test plans. Interface requirements verification is a critical aspect of the overall system verification.

6.3.1.3 Outputs

Typical outputs needed to capture interface management would include interface control documentation. This is the documentation that identifies and captures the interface information and the approved interface change requests. Types of interface documentation include the Interface Requirements Document (IRD), Interface Control Document/Drawing (ICD), Interface Definition Document (IDD), and Interface Control Plan (ICP). These outputs will then be maintained and approved using the Configuration Management Process and become a part of the overall technical data package for the project.

6.3.2 Interface Management Guidance

6.3.2.1 Interface Requirements Document

An interface requirement defines the functional, performance, electrical, environmental, human, and physical requirements and constraints that exist at a common boundary between two or more functions, system elements, configuration items, or systems. Interface requirements include both logical and physical interfaces. They include, as necessary, physical measurements, definitions of sequences of energy or information transfer, and all other significant interactions between items. For example, communication interfaces involve the movement and transfer of data and information within the system, and between the system and its environment. Proper

evaluation of communications requirements involves definition of both the structural components of communications (e.g., bandwidth, data rate, distribution, etc.) and content requirements (what data/information is being communicated, what is being moved among the system components, and the criticality of this information to system functionality). Interface requirements can be derived from the functional allocation if function inputs and outputs have been defined. For example:

- If function F1 outputs item A to function F2, and
- Function F1 is allocated to component C1, and
- Function F2 is allocated to component C2,
- Then there is an implicit requirement that the interface between components C1 and C2 pass item A, whether item A is a liquid, a solid, or a message containing data.

The IRD is a document that defines all physical, functional, and procedural interface requirements between two or more end items, elements, or components of a system and ensures project hardware and software compatibility. An IRD is composed of physical and functional requirements and constraints imposed on hardware configuration items and/or software configuration items. The purpose of the IRD is to control the interfaces between interrelated components of the system under development, as well as between the system under development and any external systems (either existing or under development) that comprise a total architecture. Interface requirements may be contained in the SRD until the point in the development process where the individual interfaces are determined. IRDs are useful when separate organizations are developing components of the system or when the system must levy requirements on other systems outside program/project control. During both Phase A and Phase B, multiple IRDs are drafted for different levels of interfaces. By SRR, draft IRDs would be complete for system-to-external-system interfaces (e.g., the shuttle to the International Space Station), and segment-to-segment interfaces (e.g., the shuttle to the launch pad). An IRD generic outline is described in Appendix L.

6.3.2.2 Interface Control Document or Interface Control Drawing

An interface control document or drawing details the physical interface between two system elements, including the number and types of connectors, electrical parameters, mechanical properties, and environmental constraints. The ICD identifies the design solution to the interface requirement. ICDs are useful when separate organizations are developing design solutions to be adhered to at a particular interface.

6.3.2.3 Interface Definition Document

An IDD is a unilateral document controlled by the end-item provider, and it basically provides the details of the interface for a design solution that is already established. This document is sometimes referred to as a "one-sided ICD." The user of the IDD is provided connectors, electrical parameters, mechanical properties, environmental constraints, etc., of the existing design. The user must then design the interface of the system to be compatible with the already existing design interface.

6.3.2.4 Interface Control Plan

An ICP should be developed to address the process for controlling identified interfaces and the related interface documentation. Key content for the ICP is the list of interfaces by category and who owns the interface. The ICP should also address the configuration control forum and mechanisms to implement the change process (e.g., Preliminary Interface Revision Notice (PIRN)/Interface Revision Notice (IRN)) for the documents.

Typical Interface Management Checklist

- Use the generic outline provided when developing the IRD. Define a "reserved" placeholder if a paragraph or section is not applicable.

- Ensure that there are two or more specifications that are being used to serve as the parent for the IRD specific requirements.

- Ensure that "shall" statements are used to define specific requirements.

- Each organization must approve and sign the IRD.

- A control process must be established to manage changes to the IRD.

- Corresponding ICDs are developed based upon the requirements in the IRD.

- Confirm connectivity between the interface requirements and the Product Verification and Product Validation Processes.

- Define the SEMP content to address interface management.

- Each major program or project should include an ICP to describe the how and what of interface management products.

6.4 Technical Risk Management

The Technical Risk Management Process is one of the crosscutting technical management processes. Risk is defined as the combination of (1) the probability that a program or project will experience an undesired event and (2) the consequences, impact, or severity of the undesired event, were it to occur. The undesired event might come from technical or programmatic sources (e.g., a cost overrun, schedule slippage, safety mishap, health problem, malicious activities, environmental impact, or failure to achieve a needed scientific or technological objective or success criterion). Both the probability and consequences may have associated uncertainties. Technical risk management is an organized, systematic risk-informed decisionmaking discipline that proactively identifies, analyzes, plans, tracks, controls, communicates, documents, and manages risk to increase the likelihood of achieving project goals. The Technical Risk Management Process focuses on project objectives,

---

Key Concepts in Technical Risk Management

- **Risk:** Risk is a measure of the inability to achieve overall program objectives within defined cost, schedule, and technical constraints and has two components: (1) the probability of failing to achieve a particular outcome and (2) the consequences/impacts of failing to achieve that outcome.

- **Cost Risk:** This is the risk associated with the ability of the program/project to achieve its life-cycle cost objectives and secure appropriate funding. Two risk areas bearing on cost are (1) the risk that the cost estimates and objectives are not accurate and reasonable and (2) the risk that program execution will not meet the cost objectives as a result of a failure to handle cost, schedule, and performance risks.

- **Schedule Risk:** Schedule risks are those associated with the adequacy of the time estimated and allocated for the development, production, implementation, and operation of the system. Two risk areas bearing on schedule risk are (1) the risk that the schedule estimates and objectives are not realistic and reasonable and (2) the risk that program execution will fall short of the schedule objectives as a result of failure to handle cost, schedule, or performance risks.

- **Technical Risk:** This is the risk associated with the evolution of the design and the production of the system of interest affecting the level of performance necessary to meet the stakeholder expectations and technical requirements. The design, test, and production processes (process risk) influence the technical risk and the nature of the product as depicted in the various levels of the PBS (product risk).

- **Programmatic Risk:** This is the risk associated with action or inaction from outside the project, over which the project manager has no control, but which may have significant impact on the project. These impacts may manifest themselves in terms of technical, cost, and/or schedule. This includes such activities as: International Traffic in Arms Requirements (ITAR), import/export control, partner agreements with other domestic or foreign organizations, congressional direction or earmarks, Office of Management and Budget direction, industrial contractor restructuring, external organizational changes, etc.

- **Hazard Versus Risk:** Hazard is distinguished from risk. A hazard represents a potential for harm, while risk includes consideration of not only the potential for harm, but also the scenarios leading to adverse outcomes and the likelihood of these outcomes. In the context of safety, "risk" considers the likelihood of undesired consequences occurring.

- **Probabilistic Risk Assessment (PRA):** PRA is a scenario-based risk assessment technique that quantifies the likelihoods of various possible undesired scenarios and their consequences, as well as the uncertainties in the likelihoods and consequences. Traditionally, design organizations have relied on surrogate criteria such as system redundancy or system-level reliability measures, partly because the difficulties of directly quantifying actual safety impacts, as opposed to simpler surrogates, seemed insurmountable. Depending on the detailed formulation of the objectives hierarchy, PRA can be applied to quantify Technical Performance Measures (TPMs) that are very closely related to fundamental objectives (e.g., Probability of Loss of Crew (P(LOC))). PRA focuses on the development of a comprehensive scenario set, which has immediate application to identify key and candidate contributors to risk. In all but the simplest systems, this requires the use of models to capture the important scenarios, to assess consequences, and to systematically quantify scenario likelihoods. These models include reliability models, system safety models, simulation models, performance models, and logic models.

bringing to bear an analytical basis for risk management decisions and the ensuing management activities, and a framework for dealing with uncertainty.

Strategies for risk management include transferring performance risk, eliminating the risk, reducing the likelihood of undesired events, reducing the negative effects of the risk (i.e., reducing consequence severity), reducing uncertainties if warranted, and accepting some or all of the consequences of a particular risk. Once a strategy is selected, technical risk management ensures its successful implementation through planning and implementation of the risk tracking and controlling activities. Technical risk management focuses on risk that relates to technical performance. However, management of technical risk has an impact on the nontechnical risk by affecting budget, schedule, and other stakeholder expectations. This discussion of technical risk management is applicable to technical and nontechnical risk issues, but the focus of this section is on technical risk issues.

6.4.1 Process Description

Figure 6.4-1 provides a typical flow diagram for the Technical Risk Management Process and identifies typical inputs, activities, and outputs to consider in addressing technical risk management.

6.4.1.1 Inputs

The following are typical inputs to technical risk management:

- **Plans and Policies:** Risk management plan, risk reporting requirements, systems engineering management plan, form of technical data products, and policy input to metrics and thresholds.

- **Technical Inputs:** Technical performance measures, program alternatives to be assessed, technical issues, and current program baseline.

- **Inputs Needed for Risk Analysis of Alternatives:** Design information and relevant experience data.

6.4.1.2 Process Activities

Technical risk management is an iterative process that considers activity requirements, constraints, and priorities to:

- Identify and assess the risks associated with the implementation of technical alternatives;

- Analyze, prioritize, plan, track and control risk and the implementation of the selected alternative;

- Plan, track, and control the risk and the implementation of the selected alternative;

- Implement contingency action plans as triggered;

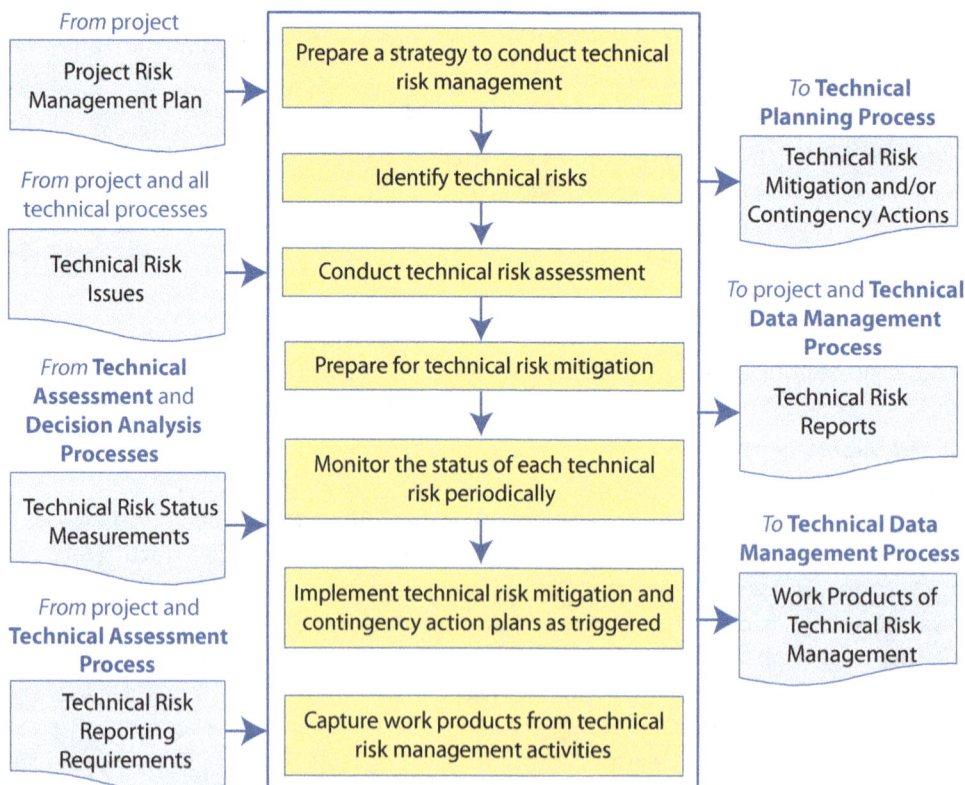

Figure 6.4-1 Technical Risk Management Process

- Communicate, deliberate, and document work products and the risk; and

- Iterate with previous steps in light of new information throughout the life cycle.

6.4.1.3 Outputs

Following are key technical risk outputs from activities:

- **Plans and Policies:** Baseline-specific plan for tracking and controlling risk

- **Technical Outputs:** Technical risk mitigation or contingency actions and tracking results, status findings, and emergent issues

- **Outputs from Risk Analysis of Alternatives:** Identified, analyzed, prioritized, and assigned risk; and risk analysis updates

6.4.2 Technical Risk Management Guidance

A widely used conceptualization of risk is the scenarios, likelihoods, and consequences concept as shown in Figures 6.4-2 and 6.4-3.

The scenarios, along with consequences, likelihoods, and associated uncertainties, make up the complete risk triplet (risk as a set of triplets—scenarios, likelihoods, consequences). The triplet concept applies in principle to all risk types, and includes the information needed for quantifying simpler measures, such as expected consequences. Estimates of expected consequences (probability or frequency multiplied by consequences) alone do not adequately inform technical decisions. Scenario-based analyses provide more of the information that risk-informed decisions need. For example, a rare but severe risk contributor may warrant a response different from that warranted by a frequent, less severe contributor, even though both have the same expected consequences. In all but the simplest systems, this requires the use of detailed models to capture the important scenarios, to assess consequences, and to systematically quantify scenario likelihoods. For additional information on probabilistic risk assessments, refer to *NPR 8705.3, Probabilistic Risk Assessment Procedures Guide for NASA Managers and Practitioners.*

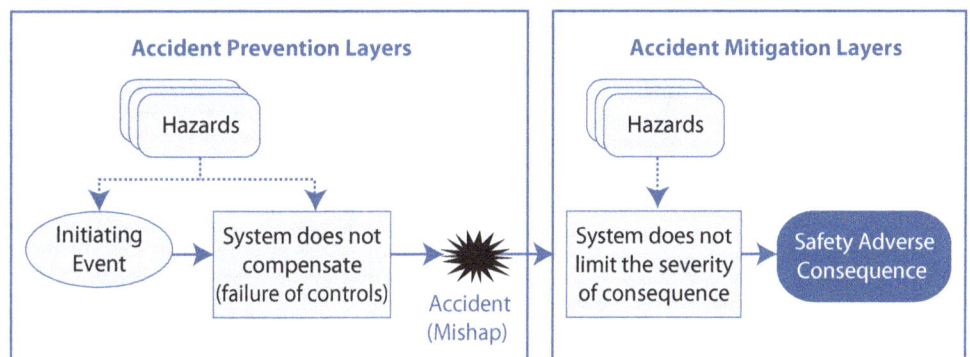

Figure 6.4-2 Scenario-based modeling of hazards

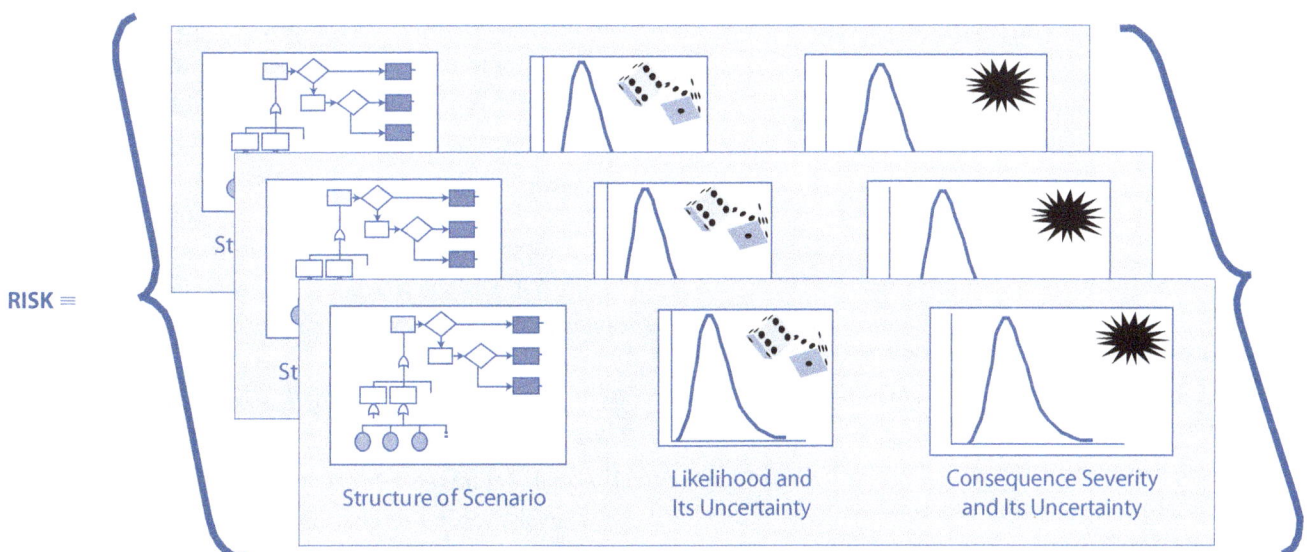

Figure 6.4-3 Risk as a set of triplets

6.4.2.1 Role of Continuous Risk Management in Technical Risk Management

Continuous Risk Management (CRM) is a widely used technique within NASA, initiated at the beginning and continuing throughout the program life cycle to monitor and control risk. It is an iterative and adaptive process, which promotes the successful handling of risk. Each step of the paradigm builds on the previous step, leading to improved designs and processes through the feedback of information generated. Figure 6.4-4 suggests this adaptive feature of CRM.

Figure 6.4-4 Continuous risk management

A brief overview of CRM is provided below for reference:

- **Identify:** Identify program risk by identifying scenarios having adverse consequences (deviations from program intent). CRM addresses risk related to safety, technical performance, cost, schedule, and other risk that is specific to the program.

- **Analyze:** Estimate the likelihood and consequence components of the risk through analysis, including uncertainty in the likelihoods and consequences, and the timeframes in which risk mitigation actions must be taken.

- **Plan:** Plan the track and control actions. Decide what will be tracked, decision thresholds for corrective action, and proposed risk control actions.

- **Track:** Track program observables relating to TPMs (performance data, schedule variances, etc.), measuring how close the program performance is compared to its plan.

- **Control:** Given an emergent risk issue, execute the appropriate control action and verify its effectiveness.

- **Communicate, Deliberate, and Document:** This is an element of each of the previous steps. Focus on un-

derstanding and communicating all risk information throughout each program phase. Document the risk, risk control plans, and closure/acceptance rationale. Deliberate on decisions throughout the CRM process.

6.4.2.2 The Interface Between CRM and Risk-Informed Decision Analysis

Figure 6.4-5 shows the interface between CRM and risk-informed decision analysis. (See Subsection 6.8.2 for more on the Decision Analysis Process.) The following steps are a risk-informed Decision Analysis Process:

1. Formulate the objectives hierarchy and TPMs.

2. Propose and identify decision alternatives. Alternatives from this process are combined with the alternatives identified in the other systems engineering processes, including design solution, verification, and validation as well as production.

3. Perform risk analysis and rank decision alternatives.

4. Evaluate and recommend decision alternative.

5. Track the implementation of the decision.

These steps support good decisions by focusing first on objectives, next on developing decision alternatives with those objectives clearly in mind, and using decision alternatives that have been developed under other systems engineering processes. The later steps of the decision analysis interrelate heavily with the Technical Risk Management Process, as indicated in Figure 6.4-5.

The risk analysis of decision alternatives (third box) not only guides selection of a preferred alternative, it also carries out the "identify" and "analyze" steps of CRM. Selection of a preferred alternative is based in part on an understanding of the risks associated with that alternative. Alternative selection is followed immediately by a planning activity in which key implementation aspects are addressed, namely, risk tracking and control, including risk mitigation if necessary. Also shown conceptually on Figure 6.4-5 is the interface between risk management and other technical and programmatic processes.

Risk Analysis, Performing Trade Studies and Ranking

The goal of this step is to carry out the kinds and amounts of analysis needed to characterize the risk for two purposes: ranking risk alternatives, and performing the "identify" and "analyze" steps of CRM.

Risk-Informed Decision Analysis

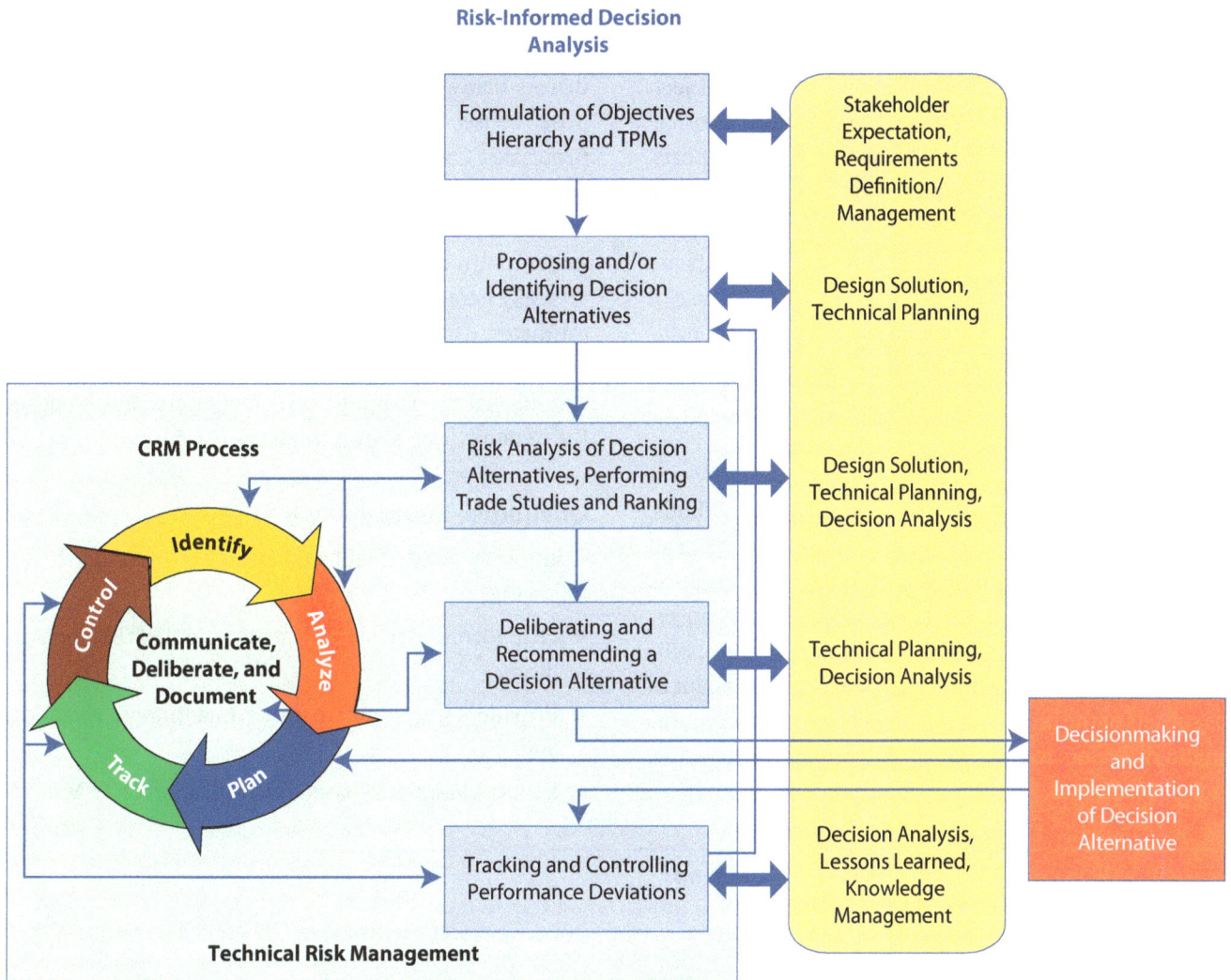

Figure 6.4-5 **The interface between CRM and risk-informed decision analysis**

To support ranking, trade studies may be performed. TPMs that can affect the decision outcome are quantified including uncertainty as appropriate.

To support the "identify" and "analyze" steps of CRM, the risk associated with the preferred alternative is analyzed in detail. Refer to Figure 6.4-6. Risk analysis can take many forms, ranging from qualitative risk identification (essentially scenarios and consequences, without performing detailed quantification of likelihood using techniques such as Failure Mode and Effects Analysis (FMEA) and fault trees), to highly quantitative methods such as PRA. The analysis stops when the technical case is made; if simpler, more qualitative methods suffice, then more detailed methods need not be applied. The process is then identified, planned for, and continuously checked. Selection and application of appropriate methods is discussed as follows.

6.4.2.3 Selection and Application of Appropriate Risk Methods

The nature and context of the problem, and the specific TPMs, determine the methods to be used. In some projects, qualitative methods are adequate for making decisions; in others, these methods are not precise enough to appropriately characterize the magnitude of the problem, or to allocate scarce risk reduction resources. The technical team needs to decide whether risk identification and judgment-based characterization are adequate, or whether the improved quantification of TPMs through more detailed risk analysis is justified. In making that determination, the technical team must balance the cost of risk analysis against the value of the additional information to be gained. The concept of "value of information" is central to making the determination of what analysis is appropriate and to what extent uncertainty needs to be quantified.

A review of the lessons learned files, data, and reports from previous similar projects can produce insights and information for hazard identification on a new project. This includes studies from similar systems and historical documents, such as mishap files and near-miss reports. The key to applying this technique is in recognizing what aspects of the old projects and the current project are analogous, and what data from the old projects are relevant to the current project. In some cases the use of quantitative methods can compensate for limited availability of information because these techniques pull the most value from the information that is available.

Types of Risk

As part of selecting appropriate risk analysis methods, it is useful to categorize types of risk. Broadly, risk can be related to cost, schedule, and technical performance. Many other categories exist, such as safety, organizational, management, acquisition, supportability, political, and programmatic risk, but these can be thought of as subsets of the broad categories. For example, programmatic risk refers to risk that affects cost and/or schedule, but not technical.

In the early stages of a risk analysis, it is typically necessary to screen contributors to risk to determine the drivers that warrant more careful analysis. For this purpose, conservative bounding approaches may be appropriate. Overestimates of risk significance will be corrected when more detailed analysis is performed. However, it can be misleading to allow bounding estimates to drive risk ranking. For this reason, analysis will typically iterate on a problem, beginning with screening estimates, using these to prioritize subsequent analysis, and moving on to a more defensible risk profile based on careful analysis of significant contributors. This is part of the iteration loop shown in Figure 6.4-6.

Qualitative Methods

Commonly used qualitative methods accomplish the following:

- Help identify scenarios that are potential risk contributors,

- Provide some input to more quantitative methods, and

- Support judgment-based quantification of TPMs.

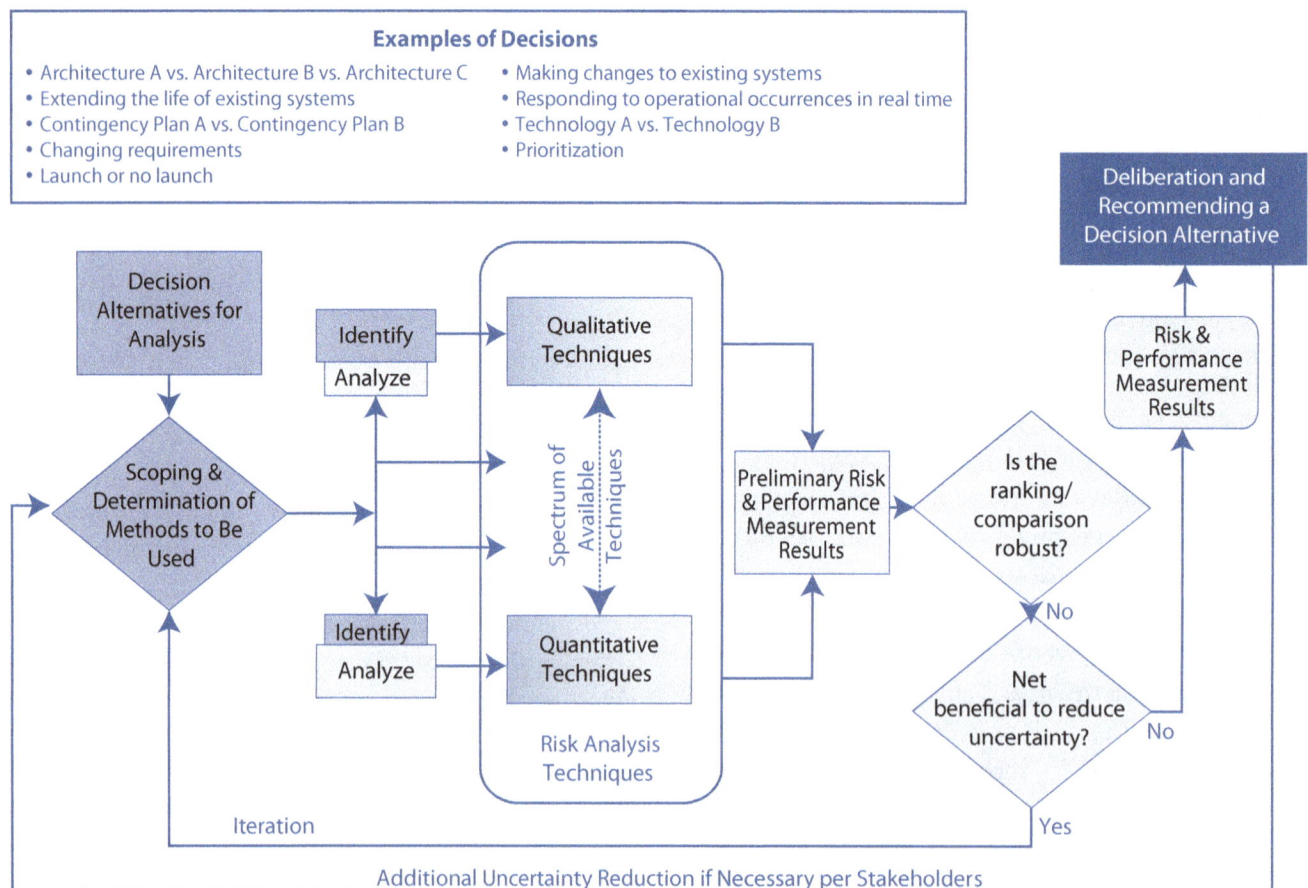

Figure 6.4-6 Risk analysis of decision alternatives

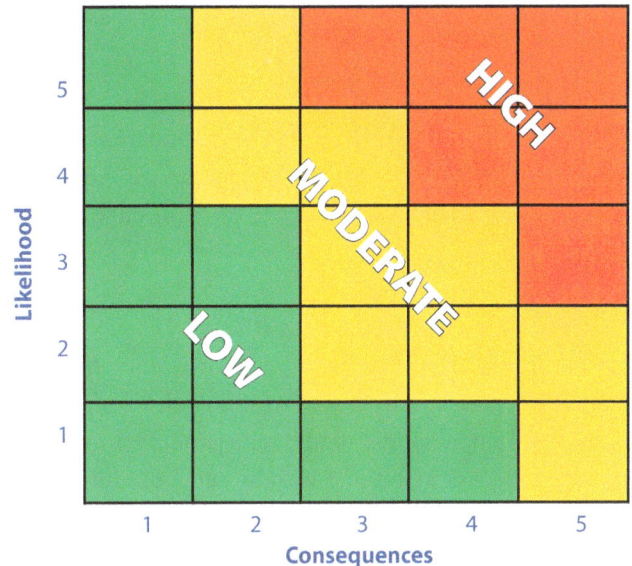

Figure 6.4-7 Risk matrix

Example Sources of Risk

In the "identify" activity, checklists such as this can serve as a reminder to analysts regarding areas in which risks have been identified previously.

- Unrealistic schedule estimates or allocation
- Unrealistic cost estimates or budget allocation
- Inadequate staffing or skills
- Uncertain or inadequate contractor capability
- Uncertain or inadequate vendor capability
- Insufficient production capacity
- Operational hazards
- Issues, hazards, and vulnerabilities that could adversely affect the program's technical effort
- Unprecedented efforts without estimates
- Poorly defined requirements
- No bidirectional traceability of requirements
- Infeasible design
- Inadequate configuration management
- Unavailable technology
- Inadequate test planning
- Inadequate quality assurance
- Requirements prescribing nondevelopmental products too low in the product tree
- Lack of concurrent development of enabling products for deployment, training, production, operations, support, or disposal

These qualitative methods are discussed briefly below.

Risk Matrices

"NxM" (most commonly 5x5) risk matrices provide assistance in managing and communicating risk. (See Figure 6.4-7.) They combine qualitative and semi-quantitative measures of likelihood with similar measures of consequences. The risk matrix is not an assessment tool, but can facilitate risk discussions. Specifically, risk matrices help to:

- Track the status and effects of risk-handling efforts, and
- Communicate risk status information.

When ranking risk, it is important to use a common methodology. Different organizations, and sometimes projects, establish their own format. This can cause con-

fusion and miscommunication. So before using a ranking system, the definitions should be clearly established and communicated via a legend or some other method. For the purposes of this handbook, a definition widely used by NASA, other Government organizations, and industry is provided.

- **Low (Green) Risk:** Has little or no potential for increase in cost, disruption of schedule, or degradation of performance. Actions within the scope of the planned program and normal management attention should result in controlling acceptable risk.

- **Moderate (Yellow) Risk:** May cause some increase in cost, disruption of schedule, or degradation of per-

Limitations of Risk Matrices

- Interaction between risks is not considered. Each risk is mapped onto the matrix individually. (These risks can be related to each item using FMECA or a fault tree.)
- Inability to deal with aggregate risks (i.e., total risk).
- Inability to represent uncertainties. A risk is assumed to exist within one likelihood range and consequence range, both of which are assumed to be known.
- Fixed tradeoff between likelihood and consequence. Using the standardized 5x5 matrix, the significance of different levels of likelihood and consequence are fixed and unresponsive to the context of the program.

formance. Special action and management attention may be required to handle risk.

- **High (Red) Risk:** Likely to cause significant increase in cost, disruption of schedule, or degradation of performance. Significant additional action and high-priority management attention will be required to handle risk.

FMECAs, FMEAs, and Fault Trees

FMEA; Failure Modes, Effects, and Criticality Analysis (FMECA); and fault trees are methodologies designed to identify potential failure modes for a product or process, to assess the risk associated with those failure modes, to rank the issues in terms of importance, and to identify and carry out corrective actions to address the most serious concerns. These methodologies focus on the hardware components as well as processes that make up the system. According to *MIL-STD-1629, Failure Mode and Effects Analysis*, FMECA is an ongoing procedure by which each potential failure in a system is analyzed to determine the results or effects thereof on the system, and to classify each potential failure mode according to its consequence severity. A fault tree evaluates the combinations of failures that can lead to the top event of interest. (See Figure 6.4-8.)

Quantitative and Communication Methods

PRA is a comprehensive, structured, and logical analysis method aimed at identifying and assessing risks in complex technological systems for the purpose of cost-effectively improving their safety and performance.

Risk management involves prevention of (reduction of the frequency of) adverse scenarios (ones with undesirable consequences) and promotion of favorable scenarios. This requires understanding the elements of adverse scenarios so that they can be prevented and the elements of successful scenarios so that they can be promoted.

PRA quantifies risk metrics. "Risk metric" refers to the kind of quantities that might appear in a decision model: such things as the frequency or probability of consequences of a specific magnitude or perhaps expected consequences. Risk metrics of interest for NASA include probability of loss of vehicle for some specific mission type, probability of mission failure, and probability of large capital loss. Figures of merit such as system failure probability can be used as risk metrics, but the phrase "risk metric" ordinarily suggests a higher level, more

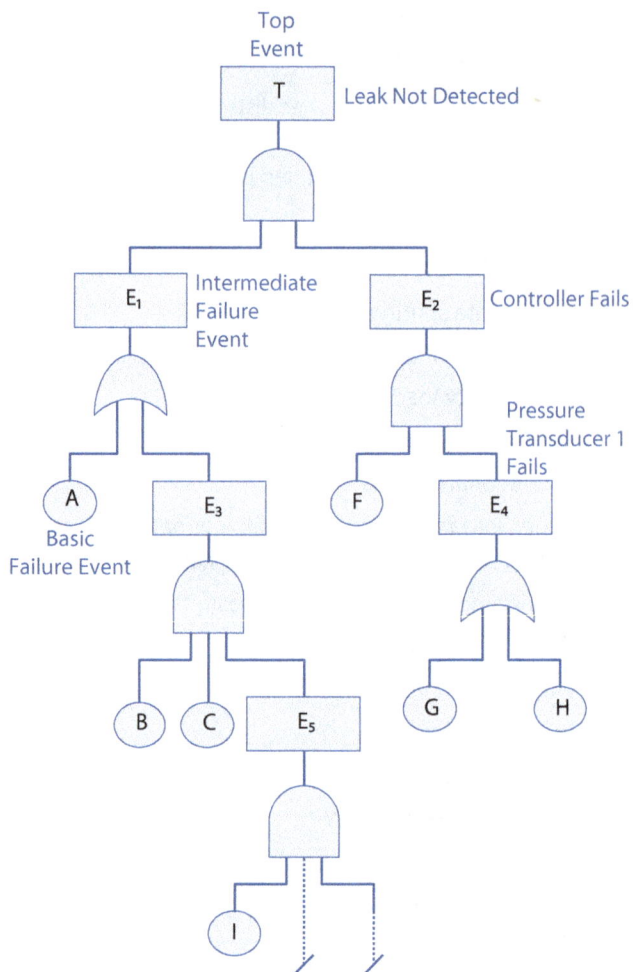

Figure 6.4-8 Example of a fault tree

consequence-oriented figure of merit. The resources needed for PRA are justified by the importance of the consequences modeled or until the cost in time and resources of further analysis is no longer justified by the expected benefits.

The NASA safety and risk directives determine the scope and the level of rigor of the risk assessments. *NPR 8715.3, NASA General Safety Program Requirements* assigns the project a priority ranking based on its consequence category and other criteria. *NPR 8705.5, Probabilistic Risk As-*

sessment (PRA) Procedures for NASA Programs and Projects then determines the scope and the level of rigor and details for the assessment based on the priority ranking and the level of design maturity.

Quantification

TPMs are quantified for each alternative and used to quantify an overall performance index or an overall measure of effectiveness for each alternative. These results are then used for ranking alternatives.

Bounding approaches are often used for initial screening of possible risk contributors. However, realistic assessments must ultimately be performed on the risk drivers. Bounding approaches are inappropriate for ranking alternatives because they bias each TPM in which they are applied, and are very difficult to do at a quantitatively consistent level from one analysis to the next.

Because different tools employ different simplifications and approximations, it is difficult to compare analysis results in a consistent manner if they are based on different tools or done by different analysts. These sources of inconsistency need to be considered when the work is planned and when the results are applied. Vetting risk and TPM results with these factors in mind is one benefit of deliberation (discussed below).

Consideration of Uncertainty Reduction Measures

In some cases, the preliminary ranking of alternatives will not be robust. A "robust" ranking is one that is not sensitive to small changes in model parameters or as-

sumptions. As an example, suppose that differences in TPMs of different decision alternatives are sufficiently small that variations of key parameters within the stated uncertainty bounds could change the ranking. This could arise in a range of decision situations, including architecture decisions and risk management decisions for a given architecture. In the latter case, the alternatives result in different risk mitigation approaches.

In such cases, it may be worthwhile to invest in work to reduce uncertainties. Quantification of the "value of information" can help the decisionmaker determine whether uncertainty reduction is an efficient use of resources.

Deliberation and Recommendation of Decision Alternative

Deliberation

Deliberation is recommended in order to make use of collective wisdom to promote selection of an alternative for actual implementation, or perhaps, in the case of complex and high-stakes decisions, to recommend a final round of trade studies or uncertainty reduction efforts, as suggested by the analysis arrow in Figure 6.4-9.

Capturing the Preferred Alternative and the Basis for Its Selection

Depending on the level at which this methodology is being exercised (project level, subtask level, etc.), the technical team chooses an alternative, basing the choice on deliberation to the extent appropriate. The decision itself is made by appropriate authority inside of the systems engineering processes. The purpose of calling out

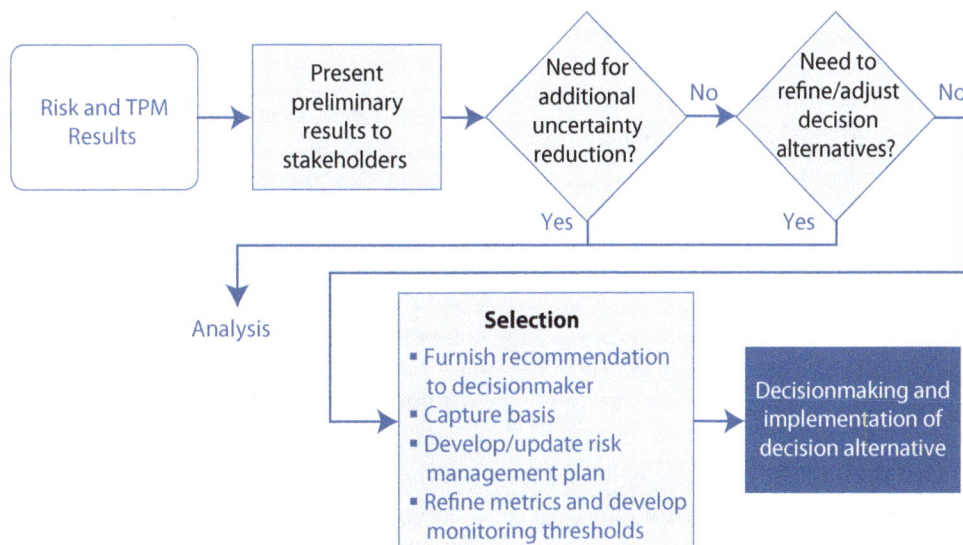

Figure 6.4-9 Deliberation

this step is to emphasize that key information about the alternative needs to be captured and that this key information includes the perceived potential program vulnerabilities that are input to the "planning" activity within CRM. By definition, the selection of the alternative is based at least in part on the prospective achievement of certain values of the TPMs. For purposes of monitoring and implementation, these TPM values help to define success, and are key inputs to the determination of monitoring thresholds.

Planning Technical Risk Management of the Selected Alternative

At this point, a single alternative has been chosen. During analysis, the risk of each alternative will have been evaluated for purposes of TPM quantification, but detailed risk management plans will not have been drawn up. At this stage, detailed planning for technical risk management of the selected alternative takes place and a formal risk management plan is drafted. In the planning phase:

- Provisional decisions are made on risk control actions (eliminate, mitigate, research, watch, or accept);
- Observables are determined for use in measurement of program performance;
- Thresholds are determined for the observables such that nonexceedance of the thresholds indicates satisfactory program performance;
- Protocols are determined that guide how often observables are to be measured, what to do when a threshold is exceeded, how often to update the analyses, decision authority, etc.; and
- Responsibility for the risk tracking is assigned.

General categories of risk control actions from *NPR 8000.4, Risk Management Procedural Requirements* are summarized here. Each identified and analyzed risk can be managed in one of five ways:

- Eliminate the risk,
- Mitigate the risk,
- Research the risk,
- Watch the risk, or
- Accept the risk.

Steps should be taken to eliminate or mitigate the risk if it is well understood and the benefits realized are commensurate with the cost. Benefits are determined using the TPMs from the program's objectives hierarchy. The consequences of mitigation alternatives need to be analyzed to ensure that they do not introduce unwarranted new contributions to risk.

If mitigation is not justified, other activities are considered. Suppose that there is substantial uncertainty regarding the risk. For example, there may be uncertainty in the probability of a scenario or in the consequences. This creates uncertainty in the benefits of mitigation, such that a mitigation decision cannot be made with confidence. In this case, research may be warranted to reduce uncertainty and more clearly indicate an appropriate choice for the control method. Research is only an interim measure, eventually leading either to risk mitigation or to acceptance.

If neither risk mitigation nor research is justified and the consequence associated with the risk is small, then it may need to be accepted. The risk acceptance process considers the likelihood and the severity of consequences. NPR 8000.4 delineates the program level with authority to accept risk and requires accepted risk to be reviewed periodically (minimum of every 6 months) to ensure that conditions and assumptions have not changed requiring the risk acceptance to be reevaluated. These reviews should take the form of quantitative and qualitative analyses, as appropriate.

The remaining cases are those in which neither risk mitigation nor research are justified, and the consequence associated with the risk is large. If there is large uncertainty in the risk, then it may need to be watched. This allows the uncertainty to reduce naturally as the program progresses and knowledge accumulates, without a research program targeting that risk. As with research, watching is an interim measure, eventually leading either to risk mitigation or to acceptance, along guidelines previously cited.

Effective Planning

The balance of this subsection is aimed primarily at ensuring that the implementation plan for risk monitoring is net beneficial.

A good plan has a high probability of detecting significant deviations from program intent in a timely fashion, without overburdening the program. In order to accomplish this, a portfolio of observables and thresholds needs to be identified. Selective plan implementation then checks for deviations of actual TPM values from planned TPM values, and does so in a way that adds net value by not overburdening the project with reporting requirements. Elements of the plan include financial and progress reporting requirements, which are somewhat predetermined, and additional program-specific observables, audits, and program reviews.

The selection of observables and thresholds should have the following properties:

- Measurable parameters (direct measurement of the parameter or of related parameters that can be used to calculate the parameter) exist to monitor system performance against clearly defined, objective thresholds;

- The monitoring program is set up so that, when a threshold is exceeded, it provides timely indication of performance issues; and

- The program burden associated with the activity is the minimum needed to satisfy the above.

For example, probability of loss of a specific mission cannot be directly measured, but depends on many quantities that can be measured up to a point, such as lower level reliability and availability metrics.

Monitoring protocols are established to clarify requirements, assign responsibility, and establish intervals for monitoring. The results of monitoring are collected and analyzed, and responses are triggered if performance thresholds are exceeded. These protocols also determine when the analyses must be updated. For example, technical risk management decisions should be reassessed with analysis if the goals of the program change. Due to the long lead time required for the high-technology products required by NASA programs, program requirements often change before the program completes its life cycle. These changes may include technical requirements, budget or schedule, risk tolerance, etc.

Tracking and Controlling Performance Deviations

As shown in Figure 6.4-10, tracking is the process by which parameters are observed, compiled, and reported according to the risk management plan. Risk mitigation/control is triggered when a performance threshold is exceeded, when risk that was assumed to be insignificant is found to be significant, or when risk that was not addressed during the analyses is discovered. Control may also be required if there are significant changes to the

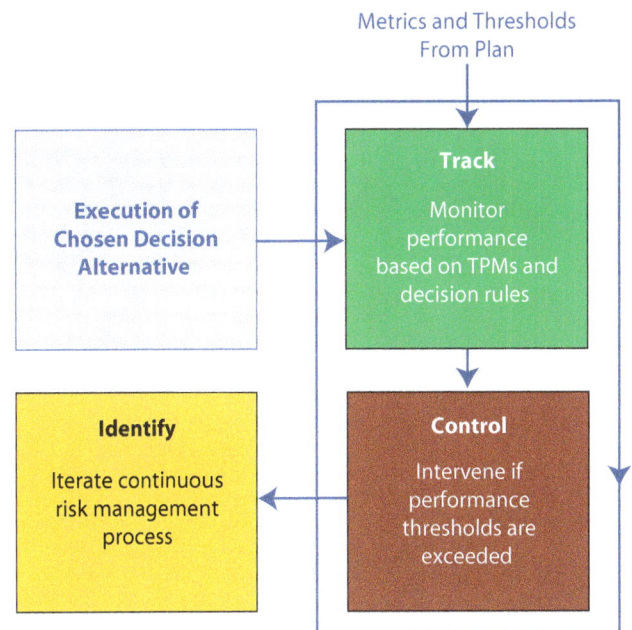

Figure 6.4-10 Performance monitoring and control of deviations

program. The need to invoke risk control measures in light of program changes is determined in the risk management plan. Alternatives are proposed and analyzed, and a preferred alternative is chosen based on the performance of the alternatives with respect to the TPM, sensitivity and uncertainty analyses, and deliberation by the stakeholders. The new preferred alternative is then subjected to planning, tracking, and control.

During the planning phase, control alternatives were proactively conceived before required. Once a threshold is triggered, a risk control action (as described in Subsection 6.4.2.3) is required. At this point, there may be considerably more information available to the decisionmaker than existed when the control alternatives were proposed. Therefore, new alternatives or modifications of existing alternatives should be considered in addition to the existing alternatives by iterating this technical risk management process.

Figure 6.4-11 shows an example of tracking and controlling performance by tracking TPM margins against predetermined thresholds. At a point in time corresponding with the vertical break, the TPM's margin is less than the required margin. At this point, the alternative was changed, such that the margin and margin requirement increased.

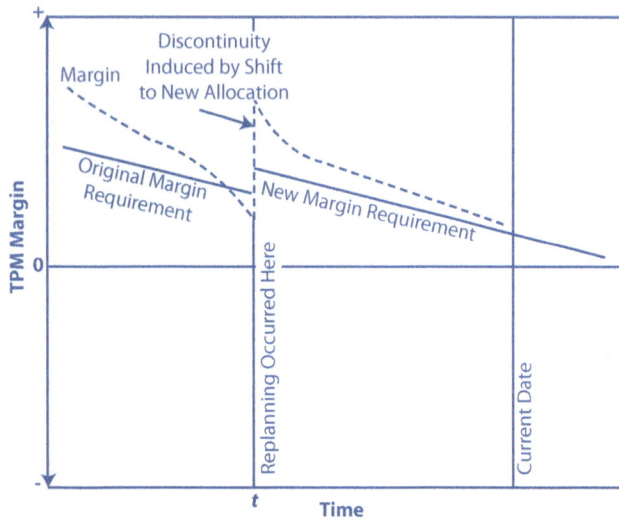

Figure 6.4-11 Margin management method

Technical risk management is not exited until the program terminates, although the level of activity varies according to the current position of the activity in the life cycle. The main outputs are the technical risk reports, including risk associated with proposed alternatives, risk control alternatives, and decision support data. Risk control alternatives are fed back to technical planning as more information is learned about the alternatives' risk. This continues until the risk management plan is established. This learning process also produces alternatives, issues, or problems and supporting data that are fed back into the project. Once a project baseline is chosen, technical risk management focuses on measuring the deviation of project risk from this baseline, and driving decision support requests based on these measurements.

6.5 Configuration Management

Configuration Management is a management discipline applied over the product's life cycle to provide visibility into and to control changes to performance and functional and physical characteristics. CM ensures that the configuration of a product is known and reflected in product information, that any product change is beneficial and is effected without adverse consequences, and that changes are managed.

CM reduces technical risks by ensuring correct product configurations, distinguishes among product versions, ensures consistency between the product and information about the product, and avoids the embarrassment of stakeholder dissatisfaction and complaint. NASA adopts the CM principles as defined by ANSI/EIA 649, NASA methods of implementation as defined by NASA CM professionals, and as approved by NASA management.

When applied to the design, fabrication/assembly, system/subsystem testing, integration, operational and sustaining activities of complex technology items, CM represents the "backbone" of the enterprise structure. It instills discipline and keeps the product attributes and documentation consistent. CM enables all stakeholders in the technical effort, at any given time in the life of a product, to use identical data for development activities and decision-making. CM principles are applied to keep the documentation consistent with the approved engineering, and to ensure that the product conforms to the functional and physical requirements of the approved design.

6.5.1 Process Description

Figure 6.5-1 provides a typical flow diagram for the Configuration Management Process and identifies typical inputs, outputs, and activities to consider in addressing CM.

6.5.1.1 Inputs

The required inputs for this process are:

- CM plan,
- Work products to be controlled, and
- Proposed baseline changes.

6.5.1.2 Process Activities

There are five elements of CM (see Figure 6.5-2):

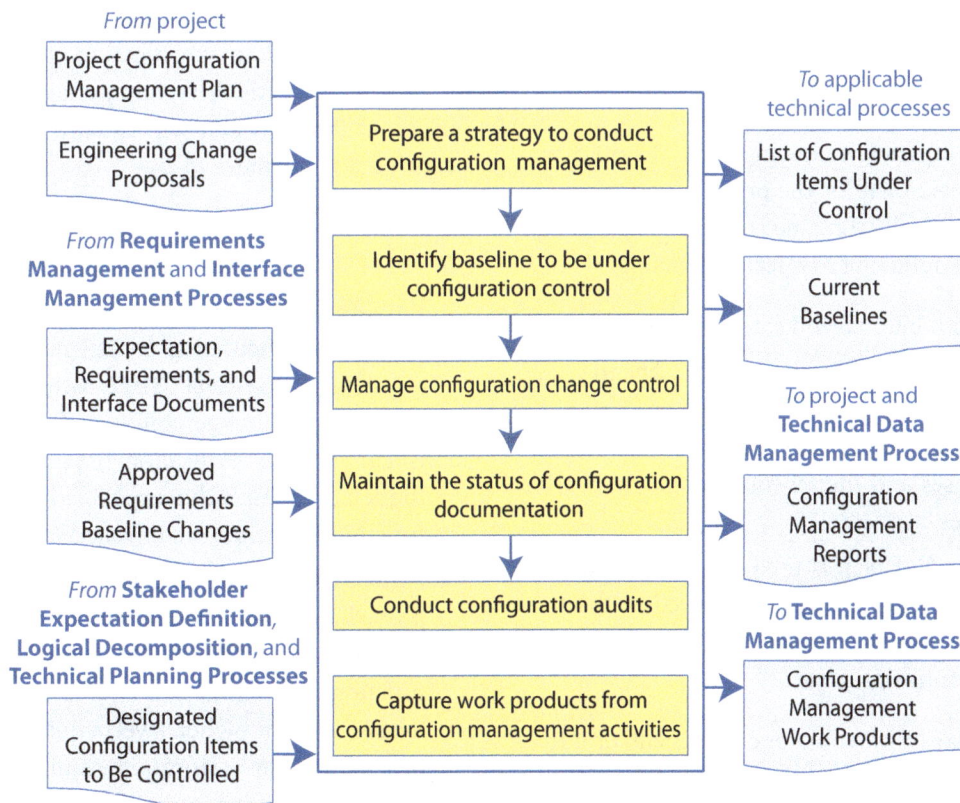

Figure 6.5-1 CM Process

- Configuration planning and management
- Configuration identification,
- Configuration change management,
- Configuration Status Accounting (CSA), and
- Configuration verification.

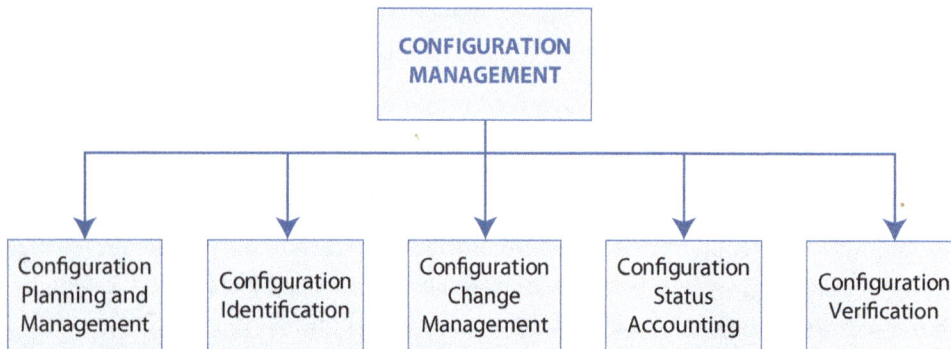

Figure 6.5-2 Five elements of configuration management

CM Planning and Management

CM planning starts at a program's or project's inception. The CM office must carefully weigh the value of prioritizing resources into CM tools or into CM surveillance of the contractors. Reviews by the Center Configuration Management Organization (CMO) are warranted and will cost resources and time, but the correction of systemic CM problems before they erupt into losing configuration control are always preferable to explaining why incorrect or misidentified parts are causing major problems in the program/project.

One of the key inputs to preparing for CM implementation is a strategic plan for the project's complete CM process. This is typically contained in a CM plan. See Appendix M for an outline of a typical CM plan.

This plan has both internal and external uses:

- **Internal:** It is used within the project office to guide, monitor, and measure the overall CM process. It describes both the CM activities planned for future acquisition phases and the schedule for implementing those activities.

- **External:** The CM plan is used to communicate the CM process to the contractors involved in the program. It establishes consistent CM processes and working relationships.

The CM plan may be a stand-alone document, or it may be combined with other program planning documents. It should describe the criteria for each technical baseline creation, technical approvals, and audits.

Configuration Identification

Configuration identification is the systematic process of selecting, organizing, and stating the product attributes. Identification requires unique identifiers for a product and its configuration documentation. The CM activity associated with identification includes selecting the Configuration Items (CIs), determining CIs' associated configuration documentation, determining the appropriate change control authority, issuing unique identifiers for both CIs and CI documentation, releasing configuration documentation, and establishing configuration baselines.

NASA has four baselines, each of which defines a distinct phase in the evolution of a product design. The baseline identifies an agreed-to description of attributes of a CI at a point in time and provides a known configuration to which changes are addressed. Baselines are established by agreeing to (and documenting) the stated definition of a CI's attributes. The approved "current" baseline defines the basis of the subsequent change. The system specification is typically finalized following the SRR. The functional baseline is established at the SDR and will usually transfer to NASA's control at that time.

The four baselines (see Figure 6.5-3) normally controlled by the program, project, or Center are the following:

- **Functional Baseline:** The functional baseline is the approved configuration documentation that describes a system's or top-level CI's performance requirements (functional, interoperability, and interface characteristics) and the verification required to demonstrate the achievement of those specified characteristics. The functional baseline is controlled by NASA.

- **Allocated Baseline:** The allocated baseline is the approved performance-oriented configuration documentation for a CI to be developed that describes the functional and interface characteristics that are allocated from a higher level requirements document or a CI and the verification required to demonstrate achievement of those specified characteristics. The allocated baseline extends the top-level performance

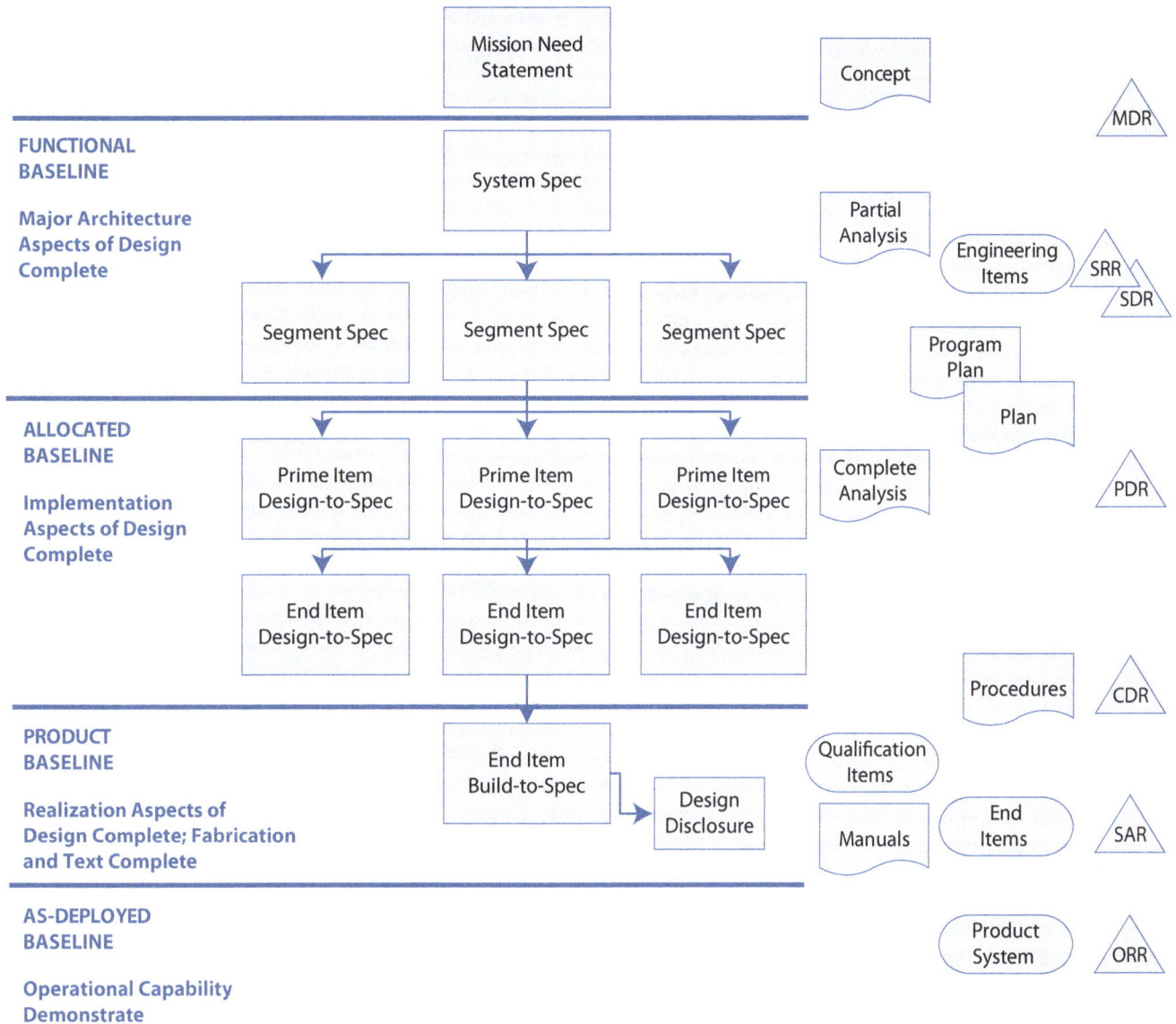

Figure 6.5-3 Evolution of technical baseline

requirements of the functional baseline to sufficient detail for initiating manufacturing or coding of a CI. The allocated baseline is usually controlled by the design organization until all design requirements have been verified. The allocated baseline is typically established at the successful completion of the PDR. Prior to CDR, NASA normally reviews design output for conformance to design requirements through incremental deliveries of engineering data. NASA control of the allocated baseline occurs through review of the engineering deliveries as data items.

- **Product Baseline:** The product baseline is the approved technical documentation that describes the configuration of a CI during the production, fielding/ deployment, and operational support phases of its life

cycle. The established product baseline is controlled as described in the configuration management plan that was developed during Phase A. The product baseline is typically established at the completion of the CDR. The product baseline describes:

- ▶ Detailed physical or form, fit, and function characteristics of a CI;
- ▶ The selected functional characteristics designated for production acceptance testing; and
- ▶ The production acceptance test requirements.

- **As-Deployed Baseline:** The as-deployed baseline occurs at the ORR. At this point, the design is considered to be functional and ready for flight. All changes will have been incorporated into the documentation.

Configuration Change Management

Configuration change management is a process to manage approved designs and the implementation of approved changes. Configuration change management is achieved via the systematic proposal, justification, and evaluation of proposed changes, followed by incorporation of approved changes and verification of implementation. Implementing configuration change management in a given program requires unique knowledge of the program objectives and requirements. The first step establishes a robust and well-disciplined internal NASA Configuration Control Board (CCB) system, which is chaired by someone with program change authority. CCB members represent the stakeholders with authority to commit the team they represent. The second step creates configuration change management surveillance of the contractor's activity. The CM office advises the NASA program or project manager to achieve a balanced configuration change management implementation that suits the unique program/project situation. See Figure 6.5-4 for an example of a typical configuration change management control process.

Types of Configuration Change Management Changes

- **Engineering Change:** An engineering change is an iteration in the baseline (draft or established). Changes can be major or minor. They may or may not include a specification change. Changes affecting an external interface must be coordinated and approved by all stakeholders affected.

  ► A "major" change is a change to the baseline configuration documentation that has significant impact (i.e., requires retrofit of delivered products or affects the baseline specification, cost, safety, compatibility with interfacing products, or operator, or maintenance training).

  ► A "minor" change corrects or modifies configuration documentation or processes without impact to the interchangeability of products or system elements in the system structure.

- **Waiver:** A waiver is a documented agreement intentionally releasing a program or project from meeting a requirement. (Some Centers use deviations prior to Implementation and waivers during Implementation.) Authorized waivers do not constitute a change to a baseline.

Configuration Status Accounting

Configuration Status Accounting (CSA) is the recording and reporting of configuration data necessary to manage CIs effectively. An effective CSA system provides timely and accurate configuration information such as:

- Complete current and historical configuration documentation and unique identifiers.
- Status of proposed changes, deviations, and waivers from initiation to implementation.
- Status and final disposition of identified discrepancies and actions identified during each configuration audit.

Some useful purposes of the CSA data include:

- An aid for proposed change evaluations, change decisions, investigations of design problems, warranties, and shelf-life calculations.
- Historical traceability.
- Software trouble reporting.
- Performance measurement data.

The following are critical functions or attributes to consider if designing or purchasing software to assist with the task of managing configuration.

- Ability to share data real time with internal and external stakeholders securely;
- Version control and comparison (track history of an object or product);
- Secure user checkout and check in;
- Tracking capabilities for gathering metrics (i.e., time, date, who, time in phases, etc.);
- Web based;
- Notification capability via e-mail;
- Integration with other databases or legacy systems;
- Compatible with required support contractors and/or suppliers (i.e., can accept data from a third party as required);
- Integration with drafting and modeling programs as required;
- Provide neutral format viewer for users;
- License agreement allows for multiple users within an agreed-to number;
- Workflow and life-cycle management;
- Limited customization;
- Migration support for software upgrades;
- User friendly;
- Consideration for users with limited access;

ORIGINATOR	CONFIGURATION MANAGEMENT ORGANIZATION	EVALUATORS	RESPONSIBLE OFFICE	CONFIGURATION CONTROL BOARD	ACTIONEES

1.
Prepares change request and sends to configuration management organization

2.
Checks format and content

Assigns number and enters into CSA

Determines evaluators (with originator)

Schedules CCB date if required

Prepares CCB agenda

3.
Evaluate change package

4.
Consolidates evaluations

Formulates recommended disposition

Presents to CCB

5.
Chair dispositions change request

Chair assigns action items if required

8.
Complete assigned actions

6.
Finalizes CCB directive

Updates CSA

Creates CCB minutes

Tracks action items

7.
Chair signs final directive

9.
Updates document, hardware, or software

10.
Performs final check on documents

11.
Releases document per directive

Stores and distributes as required

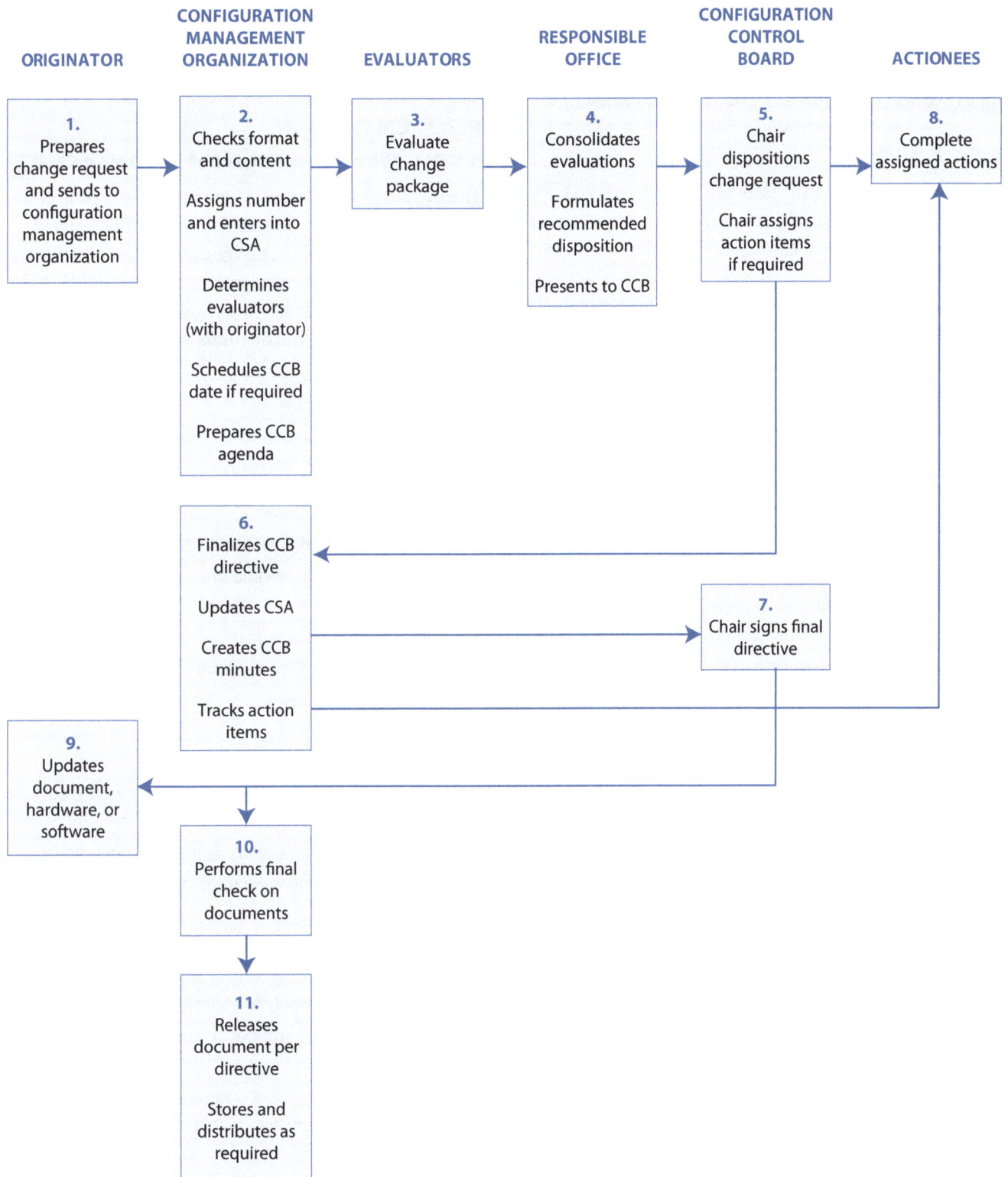

Figure 6.5-4 Typical change control process

- Ability to attach standard format files from desktop
- Workflow capability (i.e., route a CI as required based on a specific set of criteria); and
- Capable of acting as the one and only source for re-leased information.

Configuration Verification

Configuration verification is accomplished by inspecting documents, products, and records; reviewing proce-dures, processes, and systems of operations to verify that the product has achieved its required performance requirements and functional attributes; and verifying that the product's design is documented. This is sometimes divided into functional and physical configuration audits. (See Section 6.7 for more on technical reviews.)

6.5.1.3 Outputs

NPR 7120.5 defines a project's life cycle in progressive phases. Beginning with Pre-Phase A, these steps in turn are grouped under the headings of Formulation and Implementation. Approval is required to transition between these phases. Key Decision Points (KDPs) define transitions between the phases. CM plays an important role in

determining whether a KDP has been met. Major outputs of CM are procedures, approved baseline changes, configuration status, and audit reports.

6.5.2 CM Guidance

6.5.2.1 What Is the Impact of Not Doing CM?

The impact of not doing CM may result in a project being plagued by confusion, inaccuracies, low productivity, and unmanageable configuration data. During the Columbia accident investigation, the Columbia Accident Investigation Board found inconsistencies related to the hardware and the documentation with "unincor-

---

Warning Signs/Red Flags (How Do You Know When You're in Trouble?)

General warning signs of an improper implementation of CM include the following:

- Failure of program to define the "top-level" technical requirement ("We don't need a spec").
- Failure of program to recognize the baseline activities that precede and follow design reviews.
- Program office reduces the time to evaluate changes to one that is impossible for engineering, SMA, or other CCB members to meet.
- Program office declares "there will be no dissent in the record" for CCB documentation.
- Contract is awarded without CM requirements concurred with by CMO supporting the program office.
- Redlines used inappropriately on production floor to keep track of changes to design.
- Material Review Board does not know the difference between critical, major, and minor nonconformances and the appropriate classification of waivers.
- Drawings are not of high quality and do not contain appropriate notes to identify critical engineering items for configuration control or appropriate tolerancing.
- Vendors do not understand the implication of submitting waivers to safety requirements as defined in engineering.
- Subcontractors/vendors change engineering design without approval of integrating contractor, do not know how to coordinate and write an engineering change request, etc.
- Manufacturing tooling engineering does not keep up with engineering changes that affect tooling concepts. Manufacturing tools lose configuration control and acceptability for production.
- Verification data cannot be traced to released part number and specification that apply to verification task.
- Operational manuals and repair instructions cannot be traced to latest released part number and repair drawings that apply to repair/modification task.
- Maintenance and ground support tools and equipment cannot be traced to latest released part number and specification that applies to equipment.
- Parts and items cannot be identified due to improper identification markings.
- Digital closeout photography cannot be correlated to the latest release engineering.
- NASA is unable to verify the latest released engineering through access to the contractor's CM Web site.
- Tools required per installation procedures do not match the fasteners and nuts and bolts used in the design of CIs.
- CIs do not fit into their packing crates and containers due to losing configuration control in the design of the shipping and packing containers.
- Supporting procurement/fabrication change procedures do not adequately involve approval by originating engineering organization.

---

porated documentation changes" that led to failure. No CM issues were cited as a cause of the accident. The usual impact of not implementing CM can be described as "losing configuration control." Within NASA, this has resulted in program delays and engineering issues, especially in fast prototyping developments (X-37 Program) where schedule has priority over recording what is being done to the hardware. If CM is implemented properly, discrepancies identified during functional and physical configuration audits will be addressed. The following impacts are possible and have occurred in the past:

- Mission failure and loss of property and life due to improperly configured or installed hardware or software,
- Mission failure to gather mission data due to improperly configured or installed hardware or software,
- Significant mission delay incurring additional cost due to improperly configured or installed hardware or software, and
- Significant mission costs or delay due to improperly certified parts or subsystems due to fraudulent verification data.

If CM is not implemented properly, problems may occur in manufacturing, quality, receiving, procurement, etc. The user will also experience problems if ILS data are not maintained. Using a shared software system that can route and track tasks provides the team with the resources necessary for a successful project.

6.5.2.2 When Is It Acceptable to Use Redline Drawings?

"Redline" refers to the control process of marking up drawings and documents during design, fabrication, production, and testing that are found to contain errors or inaccuracies. Work stoppages could occur if the documents were corrected through the formal change process.

All redlines require the approval of the responsible hardware manager and quality assurance manager at a minimum. These managers will determine whether redlines are to be incorporated into the plan or procedure.

The important point is that each project must have a controlled procedure for redlines that specifies redline procedures and approvals.

Redlines Were identified as One of the Major Causes of the NOAA N-Prime Mishap

Excerpts from the *NOAA N-Prime Mishap Investigation Final Report*:

"Several elements contributed to the NOAA N-PRIME incident, the most significant of which were the lack of proper TOC [Turn Over Cart] verification, including the lack of proper PA [Product Assurance] witness, the change in schedule and its effect on the crew makeup, the failure of the crew to recognize missing bolts while performing the interface surface wipe down, the failure to notify in a timely fashion or at all the Safety, PA, and Government representatives, and the improper use of procedure redlines leading to a difficult-to-follow sequence of events. The interplay of the several elements allowed a situation to exist where the extensively experienced crew was not focusing on the activity at hand. There were missed opportunities that could have averted this mishap.

"In addition, the operations team was utilizing a heavily redlined procedure that required considerable 'jumping' from step to step, and had not been previously practiced. The poorly written procedure and novel redlines were preconditions to the decision errors made by the RTE [Responsible Test Engineer].

"The I&T [Integration and Test] supervisors allowed routine poor test documentation and routine misuse of procedure redlines.

"Key processes that were found to be inadequate include those that regulate operational tempo, operations planning, procedure development, use of redlines, and GSE [Ground Support Equipment] configurations. For instance, the operation during which the mishap occurred was conducted using extensively redlined procedures. The procedures were essentially new at the time of the operation—that is, they had never been used in that particular instantiation in any prior operation. The rewritten procedure had been approved through the appropriate channels even though such an extensive use of redlines was unprecedented. Such approval had been given without hazard or safety analyses having been performed."

6.6 Technical Data Management

The Technical Data Management Process is used to plan for, acquire, access, manage, protect, and use data of a technical nature to support the total life cycle of a system. Data Management (DM) includes the development, deployment, operations and support, eventual retirement, and retention of appropriate technical, to include mission and science, data beyond system retirement as required by *NPR 1441.1, NASA Records Retention Schedules.*

DM is illustrated in Figure 6.6-1. Key aspects of DM for systems engineering include:

- Application of policies and procedures for data identification and control,
- Timely and economical acquisition of technical data,
- Assurance of the adequacy of data and its protection,
- Facilitating access to and distribution of the data to the point of use,
- Analysis of data use,
- Evaluation of data for future value to other programs/projects, and
- Process access to information written in legacy software.

6.6.1 Process Description

Figure 6.6-1 provides a typical flow diagram for the Technical Data Management Process and identifies typical inputs, outputs, and activities to consider in addressing technical data management.

6.6.1.1 Inputs

Inputs include technical data, regardless of the form or method of recording and whether the data are generated by the contractor or Government during the life cycle of the system being developed. Major inputs to the Technical Data Management Process include:

- Program DM plan,
- Data products to be managed, and
- Data requests.

6.6.1.2 Process Activities

Each Center is responsible for policies and procedures for technical DM. NPR 7120.5 and NPR 7123.1 define the need to manage data, but leave specifics to the individual Centers. However, NPR 7120.5 does require that DM planning be provided as either a section in the program/project plan or as a separate document. The program or project manager is responsible for ensuring that the data required are captured and stored, data integrity is maintained, and data are disseminated as required.

Other NASA policies address the acquisition and storage of data and not just the technical data used in the life cycle of a system.

Role of Data Management Plan

The recommended procedure is that the DM plan be a separate plan apart from the program/project plan. DM issues are usually of sufficient magnitude to justify a separate plan. The lack of specificity in Agency policy and procedures provides further justification for more detailed DM planning. The plan should cover the following major DM topics:

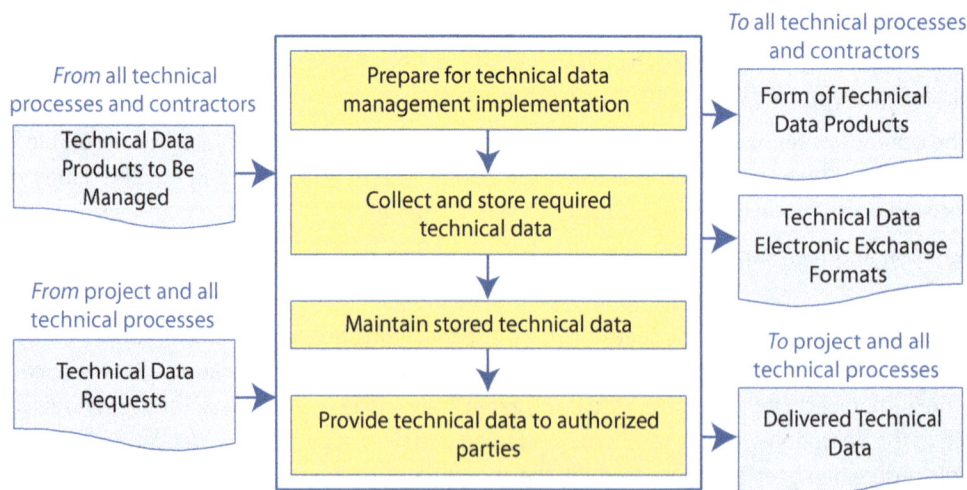

Figure 6.6-1 Technical Data Management Process

- Identification/definition of data requirements for all aspects of the product life cycle.

- Control procedures—receipt, modification, review, and approval.

- Guidance on how to access/search for data for users.

- Data exchange formats that promote data reuse and help to ensure that data can be used consistently throughout the system, family of systems, or system of systems.

- Data rights and distribution limitations such as export-control Sensitive But Unclassified (SBU).

- Storage and maintenance of data, including master lists where documents and records are maintained and managed.

Technical Data Management Key Considerations

Subsequent activities collect, store, and maintain technical data and provide it to authorized parties as required. Some considerations that impact these activities for implementing Technical Data Management include:

- Requirements relating to the flow/delivery of data to or from a contractor should be specified in the technical data management plan and included in the Request for Proposal (RFP) and contractor agreement.

- NASA should not impose changes on existing contractor data management systems unless the program technical data management requirements, including data exchange requirements, cannot otherwise be met.

- Responsibility for data inputs into the technical data management system lies solely with the originator or generator of the data.

- The availability/access of technical data will lie with the author, originator, or generator of the data in conjunction with the manager of the technical data management system.

- The established availability/access description and list should be baselined and placed under configuration control.

- For new programs, a digital generation and delivery medium is desired. Existing programs must weigh the cost/benefit trades of digitizing hard copy data.

General Data Management Roles

The Technical Data Management Process provides the basis for applying the policies and procedures to identify and control data requirements; to responsively and economically acquire, access, and distribute data; and to analyze data use.

Adherence to DM principles/rules enables the sharing, integration, and management of data for performing technical efforts by Government and industry, and ensures that information generated from managed technical data satisfies requests or meets expectations.

The Technical Data Management Process has a leading role in capturing and organizing technical data and providing information for the following uses:

- Identifying, gathering, storing and maintaining the work products generated by other systems engineering technical and technical management processes as well as the assumptions made in arriving at those work products;

- Enabling collaboration and life-cycle use of system product data;

- Capturing and organizing technical effort inputs, as well as current, intermediate, and final outputs;

- Data correlation and traceability among requirements, designs, solutions, decisions, and rationales;

- Documenting engineering decisions, including procedures, methods, results, and analyses;

- Facilitating technology insertion for affordability improvements during reprocurement and post-production support; and

- Supporting other technical management and technical processes, as needed.

Data Identification/Definition

Each program/project determines data needs during the life cycle. Data types may be defined in standard documents. Center and Agency directives sometimes specify content of documents and are appropriately used for in-house data preparation. The standard description is modified to suit program/project-specific needs, and appropriate language is included in SOWs to implement actions resulting from the data evaluation. "Data suppliers" may be a contractor, academia, or the Government. Procurement of data from an outside supplier is a formal procurement action that requires a procurement document; in-house requirements may be handled in a less formal method. Below are the different types of data that might be utilized within a program/project:

- **Data**
  - ▶ "Data" is defined in general as "recorded information regardless of the form or method of recording." However, the terms "data" and "information" are frequently used interchangeably. To be more precise, data generally must be processed in some manner to generate useful, actionable information.
  - ▶ "Data," as used in SE DM, includes technical data; computer software documentation; and representation of facts, numbers, or data of any nature that can be communicated, stored, and processed to form information required by a contract or agreement to be delivered to, or accessed by, the Government.
  - ▶ Data include that associated with system development, modeling and simulation used in development or test, test and evaluation, installation, parts, spares, repairs, usage data required for product sustainability, and source and/or supplier data.
  - ▶ Data specifically not included in Technical Data Management would be data relating to general NASA workforce operations information, communications information (except where related to a specific requirement), financial transactions, personnel data, transactional data, and other data of a purely business nature.
- **Data Call:** Solicitation from Government stakeholders (specifically Integrated Product Team (IPT) leads and functional managers) identifies and justifies their data requirements from a proposed contracted procurement. Since data provided by contractors have a cost to the Government, a data call (or an equivalent activity) is a common control mechanism used to ensure that the requested data are truly needed. If approved by the data call, a description of each data item needed is then developed and placed on contract.
- **Information:** Information is generally considered as processed data. The form of the processed data is dependent on the documentation, report, review formats, or templates that are applicable.
- **Technical Data Package:** A technical data package is a technical description of an item adequate for supporting an acquisition strategy, production, engineering, and logistics support. The package defines the required design configuration and procedures to ensure adequacy of item performance. It consists of all applicable items such as drawings, associated lists, specifications, standards, performance requirements, quality assurance provisions, and packaging details.

- **Technical Data Management System:** The strategies, plans, procedures, tools, people, data formats, data exchange rules, databases, and other entities and descriptions required to manage the technical data of a program.

Inappropriate Uses of Technical Data

Examples of inappropriate uses of technical data include:

- Unauthorized disclosure of classified data or data otherwise provided in confidence;
- Faulty interpretation based on incomplete, out-of-context, or otherwise misleading data; and
- Use of data for parts or maintenance procurement for which at least Government purpose rights have not been obtained.

Ways to help prevent inappropriate use of technical data include the following:

- Educate stakeholders on appropriate data use and
- Control access to data.

Initial Data Management System Structure

When setting up a DM system, it is not necessary to acquire (that is, to purchase and take delivery of) all technical data generated on a project. Some data may be stored in other locations with accessibility provided on a need-to-know basis. Data should be purchased only when such access is not sufficient, timely, or secure enough to provide for responsive life-cycle planning and system maintenance. Data calls are a common control mechanism to help address this need.

Data Management Planning

- Prepare a technical data management strategy. This strategy can document how the program data management plan will be implemented by the technical effort or, in the absence of such a program-level plan, be used as the basis for preparing a detailed technical data management plan, including:
  - ▶ Items of data that will be managed according to program or organizational policy, agreements, or legislation;
  - ▶ The data content and format;

▸ A framework for data flow within the program and to/from contractors including the language(s) to be employed in technical effort information exchanges;

▸ Technical data management responsibilities and authorities regarding the origin, generation, capture, archiving, security, privacy, and disposal of data products;

▸ Establishing the rights, obligations, and commitments regarding the retention of, transmission of, and access to data items; and

▸ Relevant data storage, transformation, transmission, and presentation standards and conventions to be used according to program or organizational policy, agreements, or legislative constraints.

● Obtain strategy/plan commitment from relevant stakeholders.

● Prepare procedures for implementing the technical data management strategy for the technical effort and/or for implementing the activities of the technical data management plan.

● Establish a technical database(s) to use for technical data maintenance and storage or work with the program staff to arrange use of the program database(s) for managing technical data.

● Establish data collection tools, as appropriate to the technical data management scope and available resources. (See Section 7.3.)

● Establish electronic data exchange interfaces in accordance with international standards/agreements and applicable NASA standards.

● Train appropriate stakeholders and other technical personnel in the established technical data management strategy/plan, procedures, and data collection tools, as applicable.

● Expected outcomes:

▸ A strategy and/or plan for implementing technical data management;

▸ Established procedures for performing planned Technical Data Management activities;

▸ Master list of managed data and its classification by category and use;

▸ Data collection tools established and available; and

▸ Qualified technical personnel capable of conducting established technical data management procedures and using available data collection tools.

Key Considerations for Planning Data Management and for Tool Selection

● All data entered into the technical data management system or delivered to a requester from the databases of the system should have traceability to the author, originator, or generator of the data.

● All technical data entered into the technical data management system should carry objective evidence of current status (for approval, for agreement, for information, etc.), version/control number, and date.

● The technical data management approach should be covered as part of the program's SEMP.

● Technical data expected to be used for reprocurement of parts, maintenance services, etc., might need to be reviewed by the Center's legal counsel.

Careful consideration should be taken when planning the data access and storage of data that will be generated from a project or program. If a system or tool is needed, many times the CM tool can be used with less formality. If a separate tool is required to manage the data, refer to the section below for some best practices when evaluating a data management tool. Priority must be placed on being able to access the data and ease of inputting the data. Second priority should be the consideration of the value of the specific data to current project/program, future programs/projects, NASA's overall efficiency, and uniqueness to NASA's engineering knowledge.

The following are critical functions or attributes to consider if designing or purchasing software to assist with the task of managing data:

● Ability to share data with internal and external stakeholders securely;

● Version control and comparison, to track history of an object or product;

● Secure user updating;

● Access control down to the file level;

● Web based;

● Ability to link data to CM system or elements;

● Compatible with required support contractors and/or suppliers, i.e., can accept data from a third party as required;

● Integrate with drafting and modeling programs as required;

● Provide neutral format viewer for users;

● License agreement allows for multiuser seats;

- Workflow and life-cycle management is a suggested option;
- Limited customization;
- Migration support between software version upgrades;
- User friendly;
- Straightforward search capabilities; and
- Ability to attach standard format files from desktop.

Value of Data

Storage of engineering data needs to be planned at the beginning of a program or project. Some of the data types will fall under the control of *NPR 1441.1, Records Retention Schedules*; those that do not will have to be addressed. It is best to evaluate all data that will be produced and decide how long it is of value to the program or project or to NASA engineering as a whole. There are four basic questions to ask when evaluating data's value:

- Do the data describe the product/system that is being developed or built?
- Are the data required to accurately produce the product/system being developed or built?
- Do the data offer insight for similar future programs or projects?
- Do the data hold key information that needs to be maintained in NASA's knowledge base for future engineers to use or kept as a learning example?

Technical Data Capture Tasks

Table 6.6-1 defines the tasks required to capture technical data.

Protection for Data Deliverables

All data deliverables should include distribution statements and procedures to protect all data that contain critical technology information, as well as to ensure that limited distribution data, intellectual property data, or proprietary data are properly handled during systems engineering activities. This injunction applies whether the data are hard copy or digital.

As part of overall asset protection planning, NASA has established special procedures for the protection of Critical Program Information (CPI). CPI may include components; engineering, design, or manufacturing processes; technologies; system capabilities and vulner-

---

> **Data Collection Checklist**
>
> - Have the frequency of collection and the points in the technical and technical management processes when data inputs will be available been determined?
> - Has the timeline that is required to move data from the point of origin to storage repositories or stakeholders been established?
> - Who is responsible for the input of the data?
> - Who is responsible for data storage, retrieval, and security?
> - Have necessary supporting tools been developed or acquired?

---

abilities; and any other information that gives a system its distinctive operational capability.

CPI protection should be a key consideration for the Technical Data Management effort and is part of the asset protection planning process, as shown in Appendix Q.

6.6.1.3 Outputs

Outputs include timely, secure availability of needed data in various representations to those authorized to receive it. Major outputs from the Technical Data Management Process include (refer to Figure 6.6-1):

- Technical data management procedures,
- Data representation forms,
- Data exchange formats, and
- Requested data/information delivered.

6.6.2 Technical Data Management Guidance

6.6.2.1 Data Security and ITAR

NASA generates an enormous amount of information, much of which is unclassified/nonsensitive in nature with few restrictions on its use and dissemination. NASA also generates and maintains Classified National Security Information (CNSI) under a variety of Agency programs, projects, and through partnerships and collaboration with other Federal agencies, academia, and private enterprises. SBU markings requires the author, distributor, and receiver to keep control of the sensitive

Table 6.6-1 Technical Data Tasks

Description	Tasks	Expected Outcomes
Technical data capture	Collect and store inputs and technical effort outcomes from the technical and technical management processes, including: ● results from technical assessments; ● descriptions of methods, tools, and metrics used; ● recommendations, decisions, assumptions, and impacts of technical efforts and decisions; ● lessons learned; ● deviations from plan; ● anomalies and out-of-tolerances relative to requirements; and ● other data for tracking requirements Perform data integrity checks on collected data to ensure compliance with content and format as well as technical data check to ensure there are no errors in specifying or recording the data. Report integrity check anomalies or variances to the authors or generators of the data for correction. Prioritize, review, and update data collection and storage procedures as part of regularly scheduled maintenance.	Sharable data needed to perform and control the technical and technical management processes is collected and stored. Stored data inventory.
Technical data maintenance	Implement technical management roles and responsibilities with technical data products received. Manage database(s) to ensure that collected data have proper quality and integrity; and are properly retained, secure, and available to those with access authority. Periodically review technical data management activities to ensure consistency and identify anomalies and variances. Review stored data to ensure completeness, integrity, validity, availability, accuracy, currency, and traceability. Perform technical data maintenance, as required. Identify and document significant issues, their impacts, and changes made to technical data to correct issues and mitigate impacts. Maintain, control, and prevent the stored data from being used inappropriately. Store data in a manner that enables easy and speedy retrieval. Maintain stored data in a manner that protects the technical data against foreseeable hazards, e.g., fire, flood, earthquake, etc.	Records of technical data maintenance. Technical effort data, including captured work products, contractor-delivered documents and acquirer-provided documents, are controlled and maintained. Status of data stored is maintained, to include: version description, timeline, and security classification.

(continued)

document and data or pass the control to an established control process. Public release is prohibited, and a document/data marked as such must be transmitted by secure means. Secure means are encrypted e-mail, secure fax, or person-to-person tracking. WebEx is a nonsecure environment. Standard e-mail is not permitted to transmit SBU documents and data. A secure way to send SBU information via e-mail is using the Public Key Infrastructure (PKI) to transmit the file(s). PKI is a system that manages keys to lock and unlock computer data. The basic purpose of PKI is to enable you to share your data keys with other people in a secure manner. PKI provides desktop security, as well as security for desktop and network applications, including electronic and Internet commerce.

Data items such as detailed design data (models, drawings, presentations, etc.), limited rights data, source selection data, bid and proposal information, financial data, emergency contingency plans, and restricted computer software are all examples of SBU data. Items that

Table 6.6-1 Technical Data Tasks (continued)

Description	Tasks	Expected Outcomes
Technical data/ information distribution	Maintain an information library or reference index to provide technical data availability and access instructions. Receive and evaluate requests to determine data requirements and delivery instructions. Process special requests for technical effort data or information according to established procedures for handling such requests. Ensure that required and requested data are appropriately distributed to satisfy the needs of the acquirer and requesters in accordance with the agreement, program directives, and technical data management plans and procedures. Ensure that electronic access rules are followed before database access is allowed or any requested data are electronically released/transferred to the requester. Provide proof of correctness, reliability, and security of technical data provided to internal and external recipients.	Access information (e.g., available data, access means, security procedures, time period for availability, and personnel cleared for access) is readily available. Technical data are provided to authorized requesters in the appropriate format, with the appropriate content, and by a secure mode of delivery, as applicable.
Data management system maintenance	Implement safeguards to ensure protection of the technical database and of en route technical data from unauthorized access or intrusion. Establish proof of coherence of the overall technical data set to facilitate effective and efficient use. Maintain, as applicable, backups of each technical database. Evaluate the technical data management system to identify collection and storage performance issues and problems; satisfaction of data users; risks associated with delayed or corrupted data, unauthorized access, or survivability of information from hazards such as fire, flood, earthquake, etc. Review systematically the technical data management system, including the database capacity, to determine its appropriateness for successive phases of the Defense Acquisition Framework. Recommend improvements for discovered risks and problems: ● Handle risks identified as part of technical risk management. ● Control recommended changes through established program change management activities.	Current technical data management system. Technical data are appropriately and regularly backed up to prevent data loss.

are deemed SBU must be clearly marked in accordance with *NPR 1600.1, NASA Security Program Procedural Requirements*. Data or items that cannot be directly marked, such as computer models and analyses, must have an attached copy of NASA Form 1686 that indicates the entire package is SBU data. Documents are required to have a NASA Form 1686 as a cover sheet. SBU documents and data should be safeguarded. Some examples of ways to safeguard SBU data are: access is limited on a need-to-know basis, items are copy controlled, items are attended while being used, items are properly marked (document header, footer, and NASA Form 1686), items are stored in locked containers or offices and secure servers, transmitted by secure means, and destroyed by approved

methods (shredding, etc.). For more information on SBU data, see NPR 1600.1.

International Traffic in Arms Regulations (ITAR) implement the Arms Export Control Act, and contain the United States Munitions List (USML). The USML lists articles, services, and related technical data that are designated as "defense articles" and "defense services," pursuant to Sections 38 and 47(7) of the Arms Export Control Act. The ITAR is administered by the U.S. Department of State. "Technical data" as defined in the ITAR does not include information concerning general scientific, mathematical, or engineering principles commonly taught in schools, colleges, and universities or information in the public domain (as that term is defined in 22 CFR 120.11). It also does not include basic marketing information on function and purpose or general system descriptions. For purposes of the ITAR, the following definitions apply:

- **"Defense Article" (22 CFR 120.6):** A defense article is any item or technical data on the USML. The term includes technical data recorded or stored in any physical form, models, mockups, or other items that reveal technical data directly relating to items designated in the USML. Examples of defense articles included on the USML are (1) launch vehicles, including their specifically designed or modified components, parts, accessories, attachments, and associated equipment; (2) remote sensing satellite systems, including ground control stations for telemetry, tracking, and control of such satellites, as well as passive ground stations if such stations employ any cryptographic items controlled on the USML or employ any uplink command capability; and (3) all components, parts, accessories, attachments, and associated equipment (including ground support equipment) that is specifically designed, modified, or configured for such systems. (See 22 CFR 121.1 for the complete listing.)

- **"Technical Data" (22 CFR 120.10):** Technical data are information required for the design, development, production, manufacture, assembly, operation, repair, testing, maintenance, or modification of defense articles. This includes information in the form of blueprints, drawings, photographs, plans, instructions, and documentation.

- **Classified Information Relating to Defense Articles and Defense Services:** Classified information is covered by an invention secrecy order (35 U.S.C. 181 et seq.; 35 CFR Part 5).

- **Software Directly Related to Defense Articles:** Controlled software includes, but is not limited to, system functional design, logic flow, algorithms, application programs, operating systems, and support software for design, implementation, test, operations, diagnosis, and repair related to defense articles.

6.7 Technical Assessment

Technical assessment is the crosscutting process used to help monitor technical progress of a program/project through Periodic Technical Reviews (PTRs). It also provides status information to support assessing system design, product realization, and technical management decisions.

6.7.1 Process Description

Figure 6.7-1 provides a typical flow diagram for the Technical Assessment Process and identifies typical inputs, outputs, and activities to consider in addressing technical assessment.

6.7.1.1 Inputs

Typical inputs needed for the Technical Assessment Process would include the following:

- **Technical Plans**: These are the planning documents that will outline the technical reviews/assessment process as well as identify the technical product/process measures that will be tracked and assessed to determine technical progress. Examples of these plans will be the SEMP, review plans, and EVM plan.

- **Technical Measures**: These are the identified technical measures that will be tracked to determine technical progress. These measures are also referred to as MOEs, MOPs, and TPMs.

- **Reporting Requirements**: These are the requirements on the methodology in which the status of the technical measures will be reported in regard to risk, cost, schedule, etc. The methodology and tools used for reporting the status will be established on a project-by-project basis.

6.7.1.2 Process Activities

As outlined in Figure 6.7-1, the technical plans (e.g., SEMP, review plans) provide the initial inputs into the Technical Assessment Process. These documents will outline the technical reviews/assessment approach as well as identify the technical measures that will be tracked and assessed to determine technical progress. An important part of the technical planning is determining what is needed in time, resources, and performance to complete a system that meets desired goals and objectives. Project

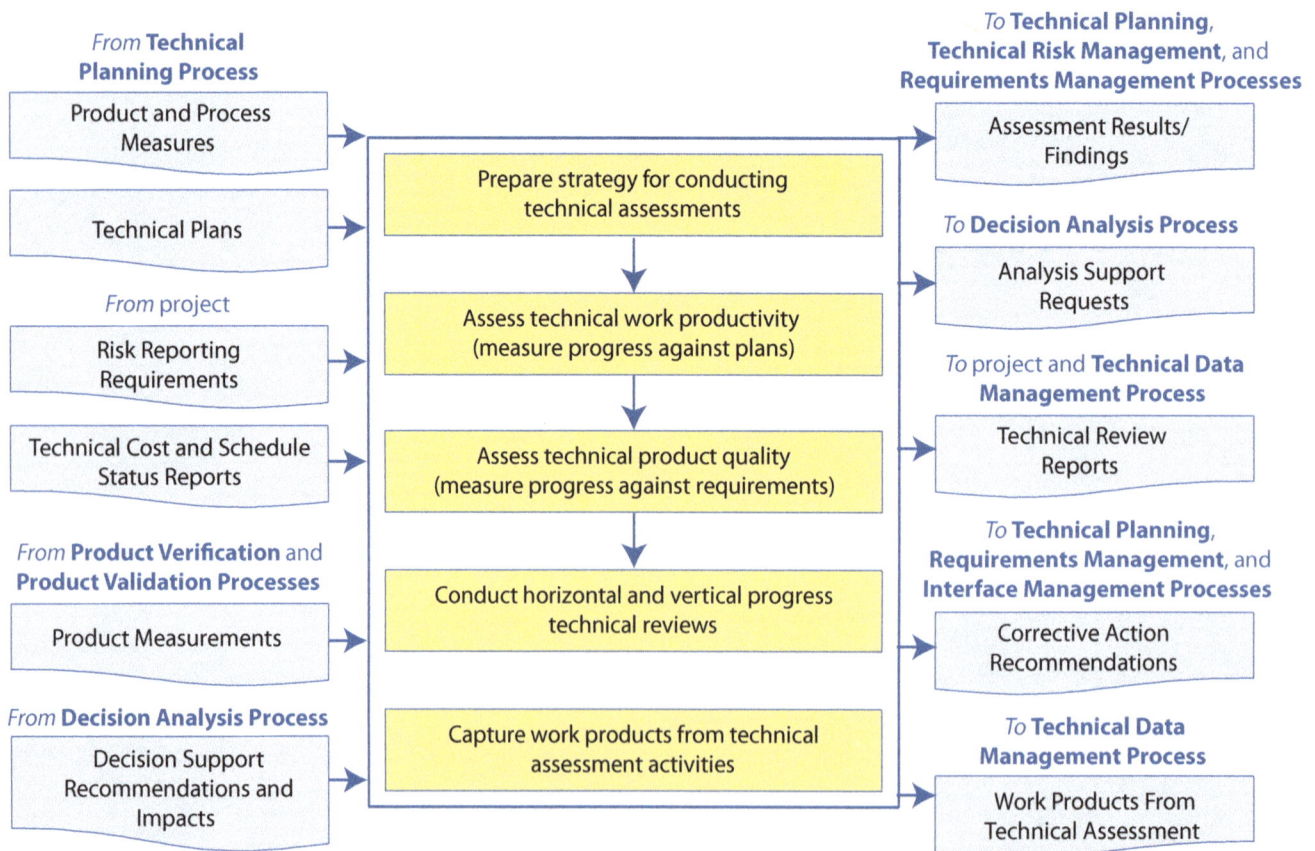

Figure 6.7-1 Technical Assessment Process

managers need visibility into the progress of those plans in order to exercise proper management control. Typical activities in determining progress against the identified technical measures will include status reporting and assessing the data. Status reporting will identify where the project stands in regard to a particular technical measure. Assessing will analytically convert the output of the status reporting into a more useful form from which trends can be determined and variances from expected results can be understood. Results of the assessment activity will then feed into the Decision Analysis Process (see Section 6.8) where potential corrective action is necessary.

These activities together form the feedback loop depicted in Figure 6.7-2.

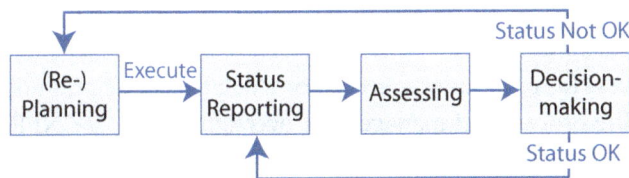

Figure 6.7-2 Planning and status reporting feedback loop

This loop takes place on a continual basis throughout the project life cycle. This loop is applicable at each level of the project hierarchy. Planning data, status reporting data, and assessments flow up the hierarchy with appropriate aggregation at each level; decisions cause actions to be taken down the hierarchy. Managers at each level determine (consistent with policies established at the next higher level of the project hierarchy) how often, and in what form, reporting data and assessments should be made. In establishing these status reporting and assessment requirements, some principles of good practice are:

- Use an agreed-upon set of well-defined technical measures. (See Subsection 6.7.2.2.)

- Report these technical measures in a consistent format at all project levels.

- Maintain historical data for both trend identification and cross-project analyses.

- Encourage a logical process of rolling up technical measures (e.g., use the WBS for project progress status).

- Support assessments with quantitative risk measures.

- Summarize the condition of the project by using color-coded (red, yellow, and green) alert zones for all technical measures.

Regular, periodic (e.g., monthly) tracking of the technical measures is recommended, although some measures should be tracked more often when there is rapid change or cause for concern. Key reviews, such as PDRs and CDRs, are points at which technical measures and their trends should be carefully scrutinized for early warning signs of potential problems. Should there be indications that existing trends, if allowed to continue, will yield an unfavorable outcome, corrective action should begin as soon as practical. Subsection 6.7.2.2 provides additional information on status reporting and assessment techniques for costs and schedules (including EVM), technical performance, and systems engineering process metrics.

The measures are predominantly assessed during the program and project technical reviews. Typical activities performed for technical reviews include (1) identifying, planning, and conducting phase-to-phase technical reviews; (2) establishing each review's purpose, objective, and entry and success criteria; (3) establishing the makeup of the review team; and (4) identifying and resolving action items resulting from the review. Subsection 6.7.2.1 summarizes the types of technical reviews typically conducted on a program/project and the role of these reviews in supporting management decision processes. It also identifies some general principles for holding reviews, but leaves explicit direction for executing a review to the program/project team to define.

The process of executing technical assessment has close relationships to other areas, such as risk management, decision analysis, and technical planning. These areas may provide input into the Technical Assessment Process or be the benefactor of outputs from the process.

6.7.1.3 Outputs

Typical outputs of the Technical Assessment Process would include the following:

- **Assessment Results, Findings, and Recommendations:** This is the collective data on the established measures from which trends can be determined and variances from expected results can be understood. Results will then feed into the Decision Analysis Process where potential corrective action is necessary.

- **Technical Review Reports/Minutes:** This is the collective information coming out of each review that captures the results, recommendations, and actions in regard to meeting the review's success criteria.

6.7.2 Technical Assessment Guidance

6.7.2.1 Reviews, Audits, and Key Decision Points

To gain a general understanding of the various technical reviews called out in Agency policy (e.g., NPR 7120.5 and NPR 7123.1), we need to examine the intent of the policy within each of the above-mentioned documents. These reviews inform the decision authority. NPR 7120.5's primary focus is to inform the decision authority as to the readiness of a program/project to proceed into the next phase of the life cycle. This is done for each milestone review and is tied to a KDP throughout the life cycle. For KDP/milestone reviews, external independent reviewers known as Standing Review Board (SRB) members evaluate the program/project and, in the end, report their findings to the decision authority. For a program or project to prepare for the SRB, the technical team must conduct their own internal peer review process. This process typically includes both informal and formal peer reviews at the subsystem and system level. This handbook attempts to provide sufficient insight and guidance into both policy documents so that practitioners can understand how they are to be successfully integrated; however, the main focus in this handbook will be on the internal review process.

The intent and policy for reviews, audits, and KDPs should be developed during Phase A and defined in the program/project plan. The specific implementation of these activities should be consistent with the types of reviews and audits described in this section, and with the NASA program and project life-cycle charts (see Figures 3.0-1 and 3.0-2). However, the timing of reviews, audits, and KDPs should accommodate the need of each specific project.

Purpose and Definition

The purpose of a review is to furnish the forum and process to provide NASA management and their contractors assurance that the most satisfactory approach, plan, or design has been selected; that a configuration item has been produced to meet the specified requirements; or that a configuration item is ready. Reviews help to develop a better understanding among task or project participants, open communication channels, alert participants and management to problems, and open avenues for solutions. Reviews are intended to add value to the project and enhance project quality and the likelihood of success. This is aided by inviting outside experts to confirm the viability of the presented approach, concept, or baseline or to recommend alternatives. Reviews may be program life-cycle reviews, project life-cycle reviews, or internal reviews.

The purpose of an audit is to provide NASA management and its contractors a thorough examination of adherence to program/project policies, plans, requirements, and specifications. Audits are the systematic examination of tangible evidence to determine adequacy, validity, and effectiveness of the activity or documentation under review. An audit may examine documentation of policies and procedures, as well as verify adherence to them.

The purpose of a KDP is to provide a scheduled event at which the decision authority determines the readiness of a program/project to progress to the next phase of the life cycle (e.g., B to C, C to D, etc.) or to the next KDP. KDPs are part of NASA's oversight and approval process for programs/projects. For a detailed description of the process and management oversight teams, see NPR 7120.5. Essentially, KDPs serve as gates through which programs and projects must pass. Within each phase, a KDP is preceded by one or more reviews, including the governing Program Management Council (PMC) review. Allowances are made within a phase for the differences between human and robotic space flight programs and projects, but phases always end with the KDP. The potential outcomes at a KDP include:

- Approval for continuation to the next KDP.
- Approval for continuation to the next KDP, pending resolution of actions.
- Disapproval for continuation to the next KDP. In such cases, follow-up actions may include a request for more information and/or a delta independent review; a request for a Termination Review (described below) for the program or the project (Phases B, C, D, and E only); direction to continue in the current phase; or redirection of the program/project.

The decision authority reviews materials submitted by the governing PMC, SRB, Program Manager (PM), project manager, and Center Management Council (CMC) in addition to agreements and program/project documentation to support the decision process. The decision authority makes decisions by considering a number of factors, including continued relevance to Agency strategic needs, goals, and objectives; continued cost affordability with respect to the Agency's resources; the viability and the readiness to proceed to the next phase; and remaining program or project risk (cost, schedule, technical, safety). Appeals against the final decision of the decision authority go to the next higher decision authority.

Project Termination

It should be noted that project termination, while usually disappointing to project personnel, may be a proper reaction to changes in external conditions or to an improved understanding of the system's projected cost-effectiveness.

Termination Review

A termination review is initiated by the decision authority to secure a recommendation as to whether to continue or terminate a program or project. Failing to stay within the parameters or levels specified in controlling documents will result in consideration of a termination review.

At the termination review, the program and the project teams present status, including any material requested by the decision authority. Appropriate support organizations are represented (e.g., procurement, external affairs, legislative affairs, public affairs) as needed. The decision and basis of the decision are fully documented and reviewed with the NASA Associate Administrator prior to final implementation.

General Principles for Reviews

Several factors can affect the implementation plan for any given review, such as design complexity, schedule, cost, visibility, NASA Center practices, the review itself, etc. As such, there is no set standard for conducting a review across the Agency; however, there are key elements, or principles, that should be included in a review plan. These include definition of review scope, objectives, success criteria (consistent with NPR 7123.1), and process. Definition of the review process should include identification of schedule, including duration of the face-to-face meeting (and draft agenda), definition of roles and responsibilities of participants, identification of presentation material and data package contents, and a copy of the form to be used for Review Item Disposition (RID)/Request For Action (RFA)/Comment. The review process for screening and processing discrepancies/requests/comments should also be included in the plan. The review plan must be agreed to by the technical team lead, project manager, and for SRB-type reviews, the SRB chair prior to the review.

It is recommended that all reviews consist of oral presentations of the applicable project requirements and the approaches, plans, or designs that satisfy those requirements. These presentations are normally provided by the cognizant design engineers or their immediate supervisor. It is also recommended that, in addition to the SRB, the review audience include key stakeholders, such as the science community, program executive, etc. This ensures that the project obtains buy-in from the personnel who have control over the project as well as those who benefit from a successful mission. It is also very beneficial to have project personnel in attendance that are not directly associated with the design being reviewed (e.g., EPS attending a thermal discussion). This gives the project an additional opportunity to utilize cross-discipline expertise to identify design shortfalls or recommend improvements. Of course, the audience should also include nonproject specialists from safety, quality and mission assurance, reliability, verification, and testing.

Program Technical Life-Cycle Reviews

Within NASA there are various types of programs:

- Single-project programs (e.g., James Webb Space Telescope Program) tend to have long development and/or operational lifetimes, represent a large investment of Agency resources in one program/project, and have contributions to that program/project from multiple organizations or agencies.

- Uncoupled programs (e.g., Discovery Program, Explorer) are implemented under a broad scientific theme and/or a common program implementation concept, such as providing frequent flight opportunities for cost-capped projects selected through AOs or NASA research announcements. Each such project is independent of the other projects within the program.

- Loosely coupled programs (e.g., Mars Exploration Program or Lunar Precursor and Robotic Program) address specific scientific or exploration objectives through multiple space flight projects of varied scope. While each individual project has an assigned set of mission objectives, architectural and technological synergies and strategies that benefit the program as a whole are explored during the Formulation process. For instance, all orbiters designed for more than one year in Mars orbit are required to carry a communication system to support present and future landers.

- Tightly coupled programs (e.g., Constellation Program) have multiple projects that execute portions of a mission or missions. No single project is capable of implementing a complete mission. Typically, multiple NASA Centers contribute to the program. Individual projects may be managed at different Centers. The program may also include other Agency or international partner contributions.

Regardless of the type, all programs are required to undergo the two technical reviews listed in Table 6.7-1. The main difference lies between uncoupled/loosely coupled programs that tend to conduct "status-type" reviews on their projects after KDP I and single-project/tightly coupled programs that tend to follow the project technical life-cycle review process post KDP I.

Table 6.7-1 Program Technical Reviews

Review	Purpose
Program/System Requirements Review	The P/SRR examines the functional and performance requirements defined for the program (and its constituent projects) and ensures that the requirements and the selected concept will satisfy the program and higher level requirements. It is an internal review. Rough order of magnitude budgets and schedules are presented.
Program/System Definition Review	The P/SDR examines the proposed program architecture and the flowdown to the functional elements of the system.

After KDP I, single-project/tightly coupled programs are responsible for conducting the system-level reviews. These reviews bring the projects together and help ensure the flowdown of requirements and that the overall system/subsystem design solution satisfies the program requirements. The program/program reviews also help resolve interface/integration issues between projects. For the sake of this handbook, single-project programs and tightly coupled programs will follow the project life-cycle review process defined after this table. Best practices and lessons learned drive programs to conduct their "concept and requirements-type" reviews prior to project concept and requirements reviews and "program design and acceptance-type" reviews after project design and acceptance reviews.

Project Technical Life-Cycle Reviews

The phrase "project life cycle/project milestone reviews" has, over the years, come to mean different things to various Centers. Some equate it to mean the project's controlled formal review using RIDS and pre-boards/boards, while others use it to mean the activity tied to RFAs and SRB/KDP process. This document will use the latter process to define the term. Project life-cycle reviews are mandatory reviews convened by the decision authority, which summarize the results of internal technical processes (peer reviews) throughout the project life cycle to NASA management and/or an independent review team, such as an SRB (see NPR 7120.5). These reviews are used to assess the progress and health of a project by providing NASA management assurance that the most satisfactory approach, plan, or design has been selected, that a configuration item has been produced to meet the specified requirements, or that a configuration item is ready for launch/operation. Some examples of life-cycle reviews include System Requirements Review, Preliminary Design Review, Critical Design Review, and Acceptance Review.

Specified life-cycle reviews are followed by a KDP in which the decision authority for the project determines, based on results and recommendations from the life-cycle review teams, whether or not the project can proceed to the next life-cycle phase.

Standing Review Boards

The SRB's role is advisory to the program/project and the convening authorities, and does not have authority over any program/project content. Its review provides expert assessment of the technical and programmatic approach, risk posture, and progress against the program/project baseline. When appropriate, it may offer recommendations to improve performance and/or reduce risk.

Internal Reviews

During the course of a project or task, it is necessary to conduct internal reviews that present technical approaches, trade studies, analyses, and problem areas to a peer group for evaluation and comment. The timing, participants, and content of these reviews is normally defined by the project manager or the manager of the performing organization with support from the technical team. In preparation for the life-cycle reviews a project will initiate an internal review process as defined in the project plan. These reviews are not just meetings to share ideas and resolve issues, but are internal reviews that allow the project to establish baseline requirements, plans, or design through the review of technical approaches, trade studies, and analyses.

Internal peer reviews provide an excellent means for controlling the technical progress of the project. They should also be used to ensure that all interested parties are involved in the development early on and throughout the process. Thus, representatives from areas such as manufacturing and quality assurance should attend the internal reviews as active participants. It is also a good practice to include representatives from other Centers and outside organizations providing support or developing systems or subsystems that may interface to your system/subsystem. They can then, for example, ensure that the design is producible and integratable and that quality is managed through the project life cycle.

Since internal peer reviews will be at a much greater level of detail than the life-cycle reviews, the team may utilize internal and external experts to help develop and assess approaches and concepts at the internal reviews. Some organizations form a red team to provide an internal, independent, peer review to identify deficiencies and offer recommendations. Projects often refer to their internal reviews as "tabletop" reviews or "interim" design reviews. Whatever the name, the purpose is the same: to ensure the readiness of the baseline for successful project life-cycle review.

It should be noted that due to the importance of these reviews each review should have well-defined entrance and success criteria established prior to the review.

Required Technical Reviews

This subsection describes the purpose, timing, objectives, success criteria, and results of the NPR 7123.1 required technical reviews in the NASA program and project life cycles. This information is intended to provide guidance to program/project managers and systems engineers, and to illustrate the progressive maturation of review activities and systems engineering products. For Flight Systems and Ground Support (FS&GS) projects, the NASA life-cycle phases of Formulation and Implementation divide into seven project phases. The checklists provided below aid in the preparation of specific review entry and success criteria, but do not take their place. To minimize extra work, review material should be keyed to program/project documentation.

Program/System Requirements Review

The P/SRR is used to ensure that the program requirements are properly formulated and correlated with the Agency and mission directorate strategic objectives.

Table 6.7-2 P/SRR Entrance and Success Criteria

Program/System Requirements Review	
Entrance Criteria	**Success Criteria**

1. An FAD has been approved.
2. Program requirements have been defined that support mission directorate requirements on the program.
3. Major program risks and corresponding mitigation strategies have been identified.
4. The high-level program requirements have been documented to include:
  a. performance,
  b. safety, and
  c. programmatic requirements.
5. An approach for verifying compliance with program requirements has been defined.
6. Procedures for controlling changes to program requirements have been defined and approved.
7. Traceability of program requirements to individual projects is documented in accordance with Agency needs, goals, and objectives, as described in the NASA Strategic Plan.
8. Top program/project risks with significant technical, safety, cost, and schedule impacts are identified.
	1. With respect to mission and science requirements, defined high-level program requirements are determined to be complete and are approved.
	2. Defined interfaces with other programs are approved.
	3. The program requirements are determined to provide a cost-effective program.
	4. The program requirements are adequately levied on either the single-program project or the multiple projects of the program.
	5. The plans for controlling program requirement changes have been approved.
	6. The approach for verifying compliance with program requirements has been approved.
	7. The mitigation strategies for handling identified major risks have been approved.

Program/System Definition Review

The P/SDR applies to all NASA space flight programs to ensure the readiness of these programs to enter an approved Program Commitment Agreement (PCA). The approved PCA permits programs to transition from the program Formulation phase to the program Implementation phase. A Program Approval Review (PAR) is conducted as part of the P/SDR to provide Agency management with an independent assessment of the readiness of the program to proceed into implementation.

The P/SDR examines the proposed program architecture and the flowdown to the functional elements of the system. The proposed program's objectives and the concept for meeting those objectives are evaluated. Key technologies and other risks are identified and assessed. The baseline program plan, budgets, and schedules are presented. The technical team provides the technical content to support the P/SDR. The P/SDR examines the proposed program architecture and the flowdown to the functional elements of the system.

Table 6.7-3 P/SDR Entrance and Success Criteria

Program/System Definition Review	
Entrance Criteria	**Success Criteria**

1. A P/SRR has been satisfactorily completed.
2. A program plan has been prepared that includes the following:
  a. how the program will be managed;
  b. a list of specific projects;
  c. the high-level program requirements (including risk criteria);
  d. performance, safety, and programmatic requirements correlated to Agency and directorate strategic objectives;
  e. description of the systems to be developed (hardware and software), legacy systems, system interfaces, and facilities; and
  f. identification of major constraints affecting system development (e.g., cost, launch window, required launch vehicle, mission planetary environment, engine design, international partners, and technology drivers).
3. Program-level SEMP that includes project technical approaches and management plans to implement the allocated program requirements including constituent launch, flight, and ground systems; and operations and logistics concepts.
4. Independent cost analyses (ICAs) and independent cost estimates (ICEs).
5. Management plan for resources other than budget.
6. Documentation for obtaining the PCA that includes the following:
  a. the feasibility of the program mission solution with a cost estimate within acceptable cost range,
  b. project plans adequate for project formulation initiation,
  c. identified and prioritized program concept evaluation criteria to be used in project evaluations,
  d. estimates of required annual funding levels,
  e. credible program cost and schedule allocation estimates to projects,
  f. acceptable risk and mitigation strategies (supported by a technical risk assessment),
  g. organizational structures and defined work assignments,
  h. defined program acquisition strategies,
  i. interfaces to other programs and partners,
  j. a draft plan for program implementation, and
  k. a defined program management system.
7. A draft program control plan that includes:
  a. how the program plans to control program requirements, technical design, schedule, and cost to achieve its high-level requirements;
  b. how the requirements, technical design, schedule, and cost of the program will be controlled;
  c. how the program will utilize its technical, schedule, and cost reserves to control the baseline;
  d. how the program plans to report technical, schedule, and cost status to the MDAA, including frequency and the level of detail; and
  e. how the program will address technical waivers and how dissenting opinions will be handled.
8. For each project, a top-level description has been documented.
	1. An approved program plan and management approach.
	2. Approved SEMP and technical approach.
	3. Estimated costs are adequate.
	4. Documentation for obtaining the PCA is approved.
	5. An approved draft program control plan.
	6. Agreement that the program is aligned with Agency needs, goals, and objectives.
	7. The technical approach is adequate.
	8. The schedule is adequate and consistent with cost, risk, and mission goals.
	9. Resources other than budget are adequate and available.

Mission Concept Review

The MCR will affirm the mission need and examine the proposed mission's objectives and the concept for meeting those objectives. It is an internal review that usually occurs at the cognizant organization for system development. The MCR should be completed prior to entering the concept development phase (Phase A).

Objectives

The objectives of the review are to:

- Ensure a thorough review of the products supporting the review.

- Ensure the products meet the entrance criteria and success criteria.

- Ensure issues raised during the review are appropriately documented and a plan for resolution is prepared.

Results of Review

A successful MCR supports the determination that the proposed mission meets the customer need, and has sufficient quality and merit to support a field Center management decision to propose further study to the cognizant NASA program associate administrator as a candidate Phase A effort.

Table 6.7-4 MCR Entrance and Success Criteria

Mission Concept Review	
Entrance Criteria	**Success Criteria**
1. Mission goals and objectives. 2. Analysis of alternative concepts to show at least one is feasible. 3. ConOps. 4. Preliminary mission descope options. 5. Preliminary risk assessment including technologies and associated risk management/mitigation strategies and options. 6. Conceptual test and evaluation strategy. 7. Preliminary technical plans to achieve next phase. 8. Defined MOEs and MOPs. 9. Conceptual life-cycle support strategies (logistics, manufacturing, operation, etc.).	1. Mission objectives are clearly defined and stated and are unambiguous and internally consistent. 2. The preliminary set of requirements satisfactorily provides a system that will meet the mission objectives. 3. The mission is feasible. A solution has been identified that is technically feasible. A rough cost estimate is within an acceptable cost range. 4. The concept evaluation criteria to be used in candidate systems evaluation have been identified and prioritized. 5. The need for the mission has been clearly identified. 6. The cost and schedule estimates are credible. 7. An updated technical search was done to identify existing assets or products that could satisfy the mission or parts of the mission. 8. Technical planning is sufficient to proceed to the next phase. 9. Risk and mitigation strategies have been identified and are acceptable based on technical assessments.

System Requirements Review

The SRR examines the functional and performance requirements defined for the system and the preliminary program or project plan and ensures that the requirements and selected concept will satisfy the mission. The SRR is conducted during the concept development phase (Phase A) and before conducting the SDR or MDR.

Objectives

The objectives of the review are to:

- Ensure a thorough review of the products supporting the review.

- Ensure the products meet the entrance criteria and success criteria.

- Ensure issues raised during the review are appropriately documented and a plan for resolution is prepared.

Results of Review

Successful completion of the SRR freezes program/project requirements and leads to a formal decision by the cognizant program associate administrator to proceed with proposal request preparations for project implementation.

Table 6.7-5 SRR Entrance and Success Criteria

System Requirements Review	
Entrance Criteria	**Success Criteria**

1. Successful completion of the MCR and responses made to all MCR RFAs and RIDs.
2. A preliminary SRR agenda, success criteria, and charge to the board have been agreed to by the technical team, project manager, and review chair prior to the SRR.
3. The following technical products for hardware and software system elements are available to the cognizant participants prior to the review:
  a. system requirements document;
  b. system software functionality description;
  c. updated ConOps;
  d. updated mission requirements, if applicable;
  e. baselined SEMP;
  f. risk management plan;
  g. preliminary system requirements allocation to the next lower level system;
  h. updated cost estimate;
  i. technology development maturity assessment plan;
  j. updated risk assessment and mitigations (including PRA, as applicable);
  k. logistics documentation (e.g., preliminary maintenance plan);
  l. preliminary human rating plan, if applicable;
  m. software development plan;
  n. system SMA plan;
  o. CM plan;
  p. initial document tree;
  q. verification and validation approach;
  r. preliminary system safety analysis; and
  s. other specialty disciplines, as required.
	1. The project utilizes a sound process for the allocation and control of requirements throughout all levels, and a plan has been defined to complete the definition activity within schedule constraints.
	2. Requirements definition is complete with respect to top-level mission and science requirements, and interfaces with external entities and between major internal elements have been defined.
	3. Requirements allocation and flowdown of key driving requirements have been defined down to subsystems.
	4. Preliminary approaches have been determined for how requirements will be verified and validated down to the subsystem level.
	5. Major risks have been identified and technically assessed, and viable mitigation strategies have been defined.

Mission Definition Review (Robotic Missions Only)

The MDR examines the proposed requirements, the mission architecture, and the flowdown to all functional elements of the mission to ensure that the overall concept is complete, feasible, and consistent with available resources.

MDR is conducted during the concept development phase (Phase A) following completion of the concept studies phase (Pre-Phase A) and before the preliminary design phase (Phase B).

Objectives

The objectives of the review are to:

- Ensure a thorough review of the products supporting the review.
- Ensure the products meet the entrance criteria and success criteria.
- Ensure issues raised during the review are appropriately documented and a plan for resolution is prepared.

Results of Review

A successful MDR supports the decision to further develop the system architecture/design and any technology needed to accomplish the mission. The results reinforce the mission's merit and provide a basis for the system acquisition strategy.

Table 6.7-6 MDR Entrance and Success Criteria

Mission Definition Review	
Entrance Criteria	**Success Criteria**

1. Successful completion of the SRR and responses made to all SRR RFAs and RIDs.
2. A preliminary MDR agenda, success criteria, and charge to the board have been agreed to by the technical team, project manager, and review chair prior to the MDR.
3. The following technical products for hardware and software system elements are available to the cognizant participants prior to the review:
  a. system architecture;
  b. updated system requirements document, if applicable;
  c. system software functionality description;
  d. updated ConOps, if applicable;
  e. updated mission requirements, if applicable;
  f. updated SEMP, if applicable;
  g. updated risk management plan, if applicable;
  h. technology development maturity assessment plan;
  i. preferred system solution definition including major trades and options;
  j. updated risk assessment and mitigations (including PRA, as applicable);
  k. updated cost and schedule data;
  l. logistics documentation (e.g., preliminary maintenance plan);
  m. software development plan;
  n. system SMA plan;
  o. CM plan;
  p. updated initial document tree, if applicable;
  q. preliminary system safety analysis; and
  r. other specialty disciplines as required.
	1. The resulting overall concept is reasonable, feasible, complete, responsive to the mission requirements, and is consistent with system requirements and available resources (cost, schedule, mass, and power).
	2. System and subsystem design approaches and operational concepts exist and are consistent with the requirements set.
	3. The requirements, design approaches, and conceptual design will fulfill the mission needs within the estimated costs.
	4. Major risks have been identified and technically assessed, and viable mitigation strategies have been defined.

System Definition Review (Human Space Flight Missions Only)

The SDR examines the proposed system architecture/design and the flowdown to all functional elements of the system. SDR is conducted at the end of the concept development phase (Phase A) and before the preliminary design phase (Phase B) begins.

Objectives

The objectives of the review are to:

- Ensure a thorough review of the products supporting the review.

- Ensure the products meet the entrance criteria and success criteria.

- Ensure issues raised during the review are appropriately documented and a plan for resolution is prepared.

Results of Review

As a result of successful completion of the SDR, the system and its operation are well enough understood to warrant design and acquisition of the end items. Approved specifications for the system, its segments, and preliminary specifications for the design of appropriate functional elements may be released. A configuration management plan is established to control design and requirement changes. Plans to control and integrate the expanded technical process are in place.

Table 6.7-7 SDR Entrance and Success Criteria

System Definition Review	
Entrance Criteria	**Success Criteria**

1. Successful completion of the SRR and responses made to all SRR RFAs and RIDs.
2. A preliminary SDR agenda, success criteria, and charge to the board have been agreed to by the technical team, project manager, and review chair prior to the SDR.
3. SDR technical products listed below for both hardware and software system elements have been made available to the cognizant participants prior to the review:
  a. system architecture;
  b. preferred system solution definition including major trades and options;
  c. updated baselined documentation, as required;
  d. preliminary functional baseline (with supporting tradeoff analyses and data);
  e. preliminary system software functional requirements;
  f. SEMP changes, if any;
  g. updated risk management plan;
  h. updated risk assessment and mitigations (including PRA, as applicable);
  i. updated technology development maturity assessment plan;
  j. updated cost and schedule data;
  k. updated logistics documentation;
  l. based on system complexity, updated human rating plan;
  m. software test plan;
  n. software requirements document(s);
  o. interface requirements documents (including software);
  p. technical resource utilization estimates and margins;
  q. updated SMA plan; and
  r. updated preliminary safety analysis.
	1. Systems requirements, including mission success criteria and any sponsor-imposed constraints, are defined and form the basis for the proposed conceptual design.
	2. All technical requirements are allocated, and the flowdown to subsystems is adequate. The requirements, design approaches, and conceptual design will fulfill the mission needs consistent with the available resources (cost, schedule, mass, and power).
	3. The requirements process is sound and can reasonably be expected to continue to identify and flow detailed requirements in a manner timely for development.
	4. The technical approach is credible and responsive to the identified requirements.
	5. Technical plans have been updated, as necessary.
	6. The tradeoffs are completed, and those planned for Phase B adequately address the option space.
	7. Significant development, mission, and safety risks are identified and technically assessed, and a risk process and resources exist to manage the risks.
	8. Adequate planning exists for the development of any enabling new technology.
	9. The ConOps is consistent with proposed design concept(s) and is in alignment with the mission requirements.

Preliminary Design Review

The PDR demonstrates that the preliminary design meets all system requirements with acceptable risk and within the cost and schedule constraints and establishes the basis for proceeding with detailed design. It will show that the correct design options have been selected, interfaces have been identified, approximately 10 percent of engineering drawings have been created, and verification methods have been described. PDR occurs near the completion of the preliminary design phase (Phase B) as the last review in the Formulation phase.

Objectives

The objectives of the review are to:

- Ensure a thorough review of the products supporting the review.

- Ensure the products meet the entrance criteria and success criteria.

- Ensure issues raised during the review are appropriately documented and a plan for resolution is prepared.

Results of Review

As a result of successful completion of the PDR, the design-to baseline is approved. A successful review result also authorizes the project to proceed into implementation and toward final design.

Table 6.7-8 PDR Entrance and Success Criteria

Preliminary Design Review	
Entrance Criteria	**Success Criteria**

1. Successful completion of the SDR or MDR and responses made to all SDR or MDR RFAs and RIDs, or a timely closure plan exists for those remaining open.
2. A preliminary PDR agenda, success criteria, and charge to the board have been agreed to by the technical team, project manager, and review chair prior to the PDR.
3. PDR technical products listed below for both hardware and software system elements have been made available to the cognizant participants prior to the review:
  a. Updated baselined documentation, as required.
  b. Preliminary subsystem design specifications for each configuration item (hardware and software), with supporting tradeoff analyses and data, as required. The preliminary software design specification should include a completed definition of the software architecture and a preliminary database design description as applicable.
  c. Updated technology development maturity assessment plan.
  d. Updated risk assessment and mitigation.
  e. Updated cost and schedule data.
  f. Updated logistics documentation, as required.
  g. Applicable technical plans (e.g., technical performance measurement plan, contamination control plan, parts management plan, environments control plan, EMI/EMC control plan, payload-to-carrier integration plan, producibility/manufacturability program plan, reliability program plan, quality assurance plan).
  h. Applicable standards.
  i. Safety analyses and plans.
  j. Engineering drawing tree.
  k. Interface control documents.
  l. Verification and validation plan.
  m. Plans to respond to regulatory (e.g., National Environmental Policy Act) requirements, as required.
  n. Disposal plan.
  o. Technical resource utilization estimates and margins.
  p. System-level safety analysis.
  q. Preliminary LLIL.
	1. The top-level requirements—including mission success criteria, TPMs, and any sponsor-imposed constraints—are agreed upon, finalized, stated clearly, and consistent with the preliminary design.
	2. The flowdown of verifiable requirements is complete and proper or, if not, an adequate plan exists for timely resolution of open items. Requirements are traceable to mission goals and objectives.
	3. The preliminary design is expected to meet the requirements at an acceptable level of risk.
	4. Definition of the technical interfaces is consistent with the overall technical maturity and provides an acceptable level of risk.
	5. Adequate technical interfaces are consistent with the overall technical maturity and provide an acceptable level of risk.
	6. Adequate technical margins exist with respect to TPMs.
	7. Any required new technology has been developed to an adequate state of readiness, or backup options exist and are supported to make them a viable alternative.
	8. The project risks are understood and have been credibly assessed, and plans, a process, and resources exist to effectively manage them.
	9. SMA (e.g., safety, reliability, maintainability, quality, and EEE parts) has been adequately addressed in preliminary designs and any applicable SMA products (e.g., PRA, system safety analysis, and failure modes and effects analysis) have been approved.
	10. The operational concept is technically sound, includes (where appropriate) human factors, and includes the flowdown of requirements for its execution.

Critical Design Review

The purpose of the CDR is to demonstrate that the maturity of the design is appropriate to support proceeding with full scale fabrication, assembly, integration, and test, and that the technical effort is on track to complete the flight and ground system development and mission operations to meet mission performance requirements within the identified cost and schedule constraints. Approximately 90 percent of engineering drawings are approved and released for fabrication. CDR occurs during the final design phase (Phase C).

Objectives

The objectives of the review are to:

- Ensure a thorough review of the products supporting the review.

- Ensure the products meet the entrance criteria and success criteria.

- Ensure issues raised during the review are appropriately documented and a plan for resolution is prepared.

Results of Review

As a result of successful completion of the CDR, the build-to baseline, production, and verification plans are approved. A successful review result also authorizes coding of deliverable software (according to the build-to baseline and coding standards presented in the review), and system qualification testing and integration. All open issues should be resolved with closure actions and schedules.

Table 6.7-9 CDR Entrance and Success Criteria

Critical Design Review	
Entrance Criteria	**Success Criteria**

1. Successful completion of the PDR and responses made to all PDR RFAs and RIDs, or a timely closure plan exists for those remaining open.
 	
2. A preliminary CDR agenda, success criteria, and charge to the board have been agreed to by the technical team, project manager, and review chair prior to the CDR.
 	
3. CDR technical work products listed below for both hardware and software system elements have been made available to the cognizant participants prior to the review:
 	
  a. updated baselined documents, as required;
  b. product build-to specifications for each hardware and software configuration item, along with supporting tradeoff analyses and data;
  c. fabrication, assembly, integration, and test plans and procedures;
  d. technical data package (e.g., integrated schematics, spares provisioning list, interface control documents, engineering analyses, and specifications);
  e. operational limits and constraints;
  f. technical resource utilization estimates and margins;
  g. acceptance criteria;
  h. command and telemetry list;
  i. verification plan (including requirements and specifications);
  j. validation plan;
  k. launch site operations plan;
  l. checkout and activation plan;
  m. disposal plan (including decommissioning or termination);
  n. updated technology development maturity assessment plan;
  o. updated risk assessment and mitigation;
  p. update reliability analyses and assessments;
  q. updated cost and schedule data;
  r. updated logistics documentation;
  s. software design document(s) (including interface design documents);
  t. updated LLIL;
  u. subsystem-level and preliminary operations safety analyses;
  v. system and subsystem certification plans and requirements (as needed); and
  w. system safety analysis with associated verifications.
	1. The detailed design is expected to meet the requirements with adequate margins at an acceptable level of risk.
	
	2. Interface control documents are appropriately matured to proceed with fabrication, assembly, integration, and test, and plans are in place to manage any open items.
	
	3. High confidence exists in the product baseline, and adequate documentation exists or will exist in a timely manner to allow proceeding with fabrication, assembly, integration, and test.
	
	4. The product verification and product validation requirements and plans are complete.
	
	5. The testing approach is comprehensive, and the planning for system assembly, integration, test, and launch site and mission operations is sufficient to progress into the next phase.
	
	6. Adequate technical and programmatic margins and resources exist to complete the development within budget, schedule, and risk constraints.
	
	7. Risks to mission success are understood and credibly assessed, and plans and resources exist to effectively manage them.
	
	8. SMA (e.g., safety, reliability, maintainability, quality, and EEE parts) have been adequately addressed in system and operational designs, and any applicable SMA plan products (e.g., PRA, system safety analysis, and failure modes and effects analysis) have been approved.

Production Readiness Review

A PRR is held for FS&GS projects developing or acquiring multiple or similar systems greater than three or as determined by the project. The PRR determines the readiness of the system developers to efficiently produce the required number of systems. It ensures that the production plans; fabrication, assembly, and integration-enabling products; and personnel are in place and ready to begin production. PRR occurs during the final design phase (Phase C).

Objectives

The objectives of the review are to:

- Ensure a thorough review of the products supporting the review.
- Ensure the products meet the entrance criteria and success criteria.
- Ensure issues raised during the review are appropriately documented and a plan for resolution is prepared.

Results of Review

As a result of successful completion of the PRR, the final production build-to baseline, production, and verification plans are approved. Approved drawings are released and authorized for production. A successful review result also authorizes coding of deliverable software (according to the build-to baseline and coding standards presented in the review), and system qualification testing and integration. All open issues should be resolved with closure actions and schedules.

Table 6.7-10 PRR Entrance and Success Criteria

Production Readiness Review	
Entrance Criteria	**Success Criteria**
1. The significant production engineering problems encountered during development are resolved. 2. The design documentation is adequate to support production. 3. The production plans and preparation are adequate to begin fabrication. 4. The production-enabling products and adequate resources are available, have been allocated, and are ready to support end product production.	1. The design is appropriately certified. 2. The system requirements are fully met in the final production configuration. 3. Adequate measures are in place to support production. 4. Design-for-manufacturing considerations ensure ease and efficiency of production and assembly. 5. Risks have been identified, credibly assessed, and characterized; and mitigation efforts have been defined. 6. The bill of materials has been reviewed and critical parts identified. 7. Delivery schedules have been verified. 8. Alternative sources for resources have been identified, as appropriate. 9. Adequate spares have been planned and budgeted. 10. Required facilities and tools are sufficient for end product production. 11. Specified special tools and test equipment are available in proper quantities. 12. Production and support staff are qualified. 13. Drawings are certified. 14. Production engineering and planning are sufficiently mature for cost-effective production. 15. Production processes and methods are consistent with quality requirements and compliant with occupational safety, environmental, and energy conservation regulations. 16. Qualified suppliers are available for materials that are to be procured.

System Integration Review

An SIR ensures that the system is ready to be integrated. Segments, components, and subsystems are available and ready to be integrated into the system. Integration facilities, support personnel, and integration plans and procedures are ready for integration. SIR is conducted at the end of the final design phase (Phase C) and before the systems assembly, integration, and test phase (Phase D) begins.

Objectives

The objectives of the review are to:

- Ensure a thorough review of the products supporting the review.

- Ensure the products meet the entrance criteria and success criteria.

- Ensure issues raised during the review are appropriately documented and a plan for resolution is prepared.

Results of Review

As a result of successful completion of the SIR, the final as-built baseline and verification plans are approved. Approved drawings are released and authorized to support integration. All open issues should be resolved with closure actions and schedules. The subsystems/systems integration procedures, ground support equipment, facilities, logistical needs, and support personnel are planned for and are ready to support integration.

Table 6.7-11 SIR Entrance and Success Criteria

System Integration Review	
Entrance Criteria	**Success Criteria**
1. Integration plans and procedures have been completed and approved. 2. Segments and/or components are available for integration. 3. Mechanical and electrical interfaces have been verified against the interface control documentation. 4. All applicable functional, unit-level, subsystem, and qualification testing has been conducted successfully. 5. Integration facilities, including clean rooms, ground support equipment, handling fixtures, overhead cranes, and electrical test equipment, are ready and available. 6. Support personnel have been adequately trained. 7. Handling and safety requirements have been documented. 8. All known system discrepancies have been identified and disposed in accordance with an agreed-upon plan. 9. All previous design review success criteria and key issues have been satisfied in accordance with an agreed-upon plan. 10. The quality control organization is ready to support the integration effort.	1. Adequate integration plans and procedures are completed and approved for the system to be integrated. 2. Previous component, subsystem, and system test results form a satisfactory basis for proceeding to integration. 3. Risk level is identified and accepted by program/project leadership, as required. 4. The integration procedures and workflow have been clearly defined and documented. 5. The review of the integration plans, as well as the procedures, environment, and the configuration of the items to be integrated, provides a reasonable expectation that the integration will proceed successfully. 6. Integration personnel have received appropriate training in the integration and safety procedures.

Test Readiness Review

A TRR ensures that the test article (hardware/software), test facility, support personnel, and test procedures are ready for testing and data acquisition, reduction, and control. A TRR is held prior to commencement of verification or validation testing.

Objectives

The objectives of the review are to:

- Ensure a thorough review of the products supporting the review.

- Ensure the products meet the entrance criteria and success criteria.

- Ensure issues raised during the review are appropriately documented and a plan for resolution is prepared.

Results of Review

A successful TRR signifies that test and safety engineers have certified that preparations are complete, and that the project manager has authorized formal test initiation.

Table 6.7-12 TRR Entrance and Success Criteria

Test Readiness Review	
Entrance Criteria	**Success Criteria**
1. The objectives of the testing have been clearly defined and documented and all of the test plans, procedures, environment, and the configuration of the test item(s) support those objectives.	1. Adequate test plans are completed and approved for the system under test.
2. Configuration of the system under test has been defined and agreed to. All interfaces have been placed under configuration management or have been defined in accordance with an agreed-to plan, and a version description document has been made available to TRR participants prior to the review.	2. Adequate identification and coordination of required test resources are completed.
	3. Previous component, subsystem, and system test results form a satisfactory basis for proceeding into planned tests.
3. All applicable functional, unit-level, subsystem, system, and qualification testing has been conducted successfully.	4. Risk level is identified and accepted by program/competency leadership as required.
4. All TRR-specific materials such as test plans, test cases, and procedures have been made available to all participants prior to conducting the review.	5. Plans to capture any lessons learned from the test program are documented.
5. All known system discrepancies have been identified and disposed in accordance with an agreed-upon plan.	6. The objectives of the testing have been clearly defined and documented, and the review of all the test plans, as well as the procedures, environment, and the configuration of the test item, provide a reasonable expectation that the objectives will be met.
6. All previous design review success criteria and key issues have been satisfied in accordance with an agreed-upon plan.	
7. All required test resources people (including a designated test director), facilities, test articles, test instrumentation, and other enabling products have been identified and are available to support required tests.	7. The test cases have been reviewed and analyzed for expected results, and the results are consistent with the test plans and objectives.
8. Roles and responsibilities of all test participants are defined and agreed to.	8. Test personnel have received appropriate training in test operation and safety procedures.
9. Test contingency planning has been accomplished, and all personnel have been trained.	

System Acceptance Review

The SAR verifies the completeness of the specific end products in relation to their expected maturity level and assesses compliance to stakeholder expectations. The SAR examines the system, its end products and documentation, and test data and analyses that support verification. It also ensures that the system has sufficient technical maturity to authorize its shipment to the designated operational facility or launch site.

Objectives

The objectives of the review are to:

- Ensure a thorough review of the products supporting the review.

- Ensure the products meet the entrance criteria and success criteria.

- Ensure issues raised during the review are appropriately documented and a plan for resolution is prepared.

Results of Review

As a result of successful completion of the SAR, the system is accepted by the buyer, and authorization is given to ship the hardware to the launch site or operational facility, and to install software and hardware for operational use.

Table 6.7-13 SAR Entrance and Success Criteria

System Acceptance Review	
Entrance Criteria	**Success Criteria**

1. A preliminary agenda has been coordinated (nominally) prior to the SAR.
2. The following SAR technical products have been made available to the cognizant participants prior to the review:
  a. results of the SARs conducted at the major suppliers;
  b. transition to production and/or manufacturing plan;
  c. product verification results;
  d. product validation results;
  e. documentation that the delivered system complies with the established acceptance criteria;
  f. documentation that the system will perform properly in the expected operational environment;
  g. technical data package updated to include all test results;
  h. certification package;
  i. updated risk assessment and mitigation;
  j. successfully completed previous milestone reviews; and
  k. remaining liens or unclosed actions and plans for closure.
	1. Required tests and analyses are complete and indicate that the system will perform properly in the expected operational environment.
	2. Risks are known and manageable.
	3. System meets the established acceptance criteria.
	4. Required safe shipping, handling, checkout, and operational plans and procedures are complete and ready for use.
	5. Technical data package is complete and reflects the delivered system.
	6. All applicable lessons learned for organizational improvement and system operations are captured.

Operational Readiness Review

The ORR examines the actual system characteristics and the procedures used in the system or end product's operation and ensures that all system and support (flight and ground) hardware, software, personnel, procedures, and user documentation accurately reflect the deployed state of the system.

Objectives

The objectives of the review are to:

- Ensure a thorough review of the products supporting the review.
- Ensure the products meet the entrance criteria and success criteria.
- Ensure issues raised during the review are appropriately documented and a plan for resolution is prepared.

Results of Review

As a result of successful ORR completion, the system is ready to assume normal operations.

Table 6.7-14 ORR Entrance and Success Criteria

Operational Readiness Review	
Entrance Criteria	**Success Criteria**
1. All validation testing has been completed. 2. Test failures and anomalies from validation testing have been resolved and the results incorporated into all supporting and enabling operational products. 3. All operational supporting and enabling products (e.g., facilities, equipment, documents, updated databases) that are necessary for the nominal and contingency operations have been tested and delivered/installed at the site(s) necessary to support operations. 4. Operations handbook has been approved. 5. Training has been provided to the users and operators on the correct operational procedures for the system. 6. Operational contingency planning has been accomplished, and all personnel have been trained.	1. The system, including any enabling products, is determined to be ready to be placed in an operational status. 2. All applicable lessons learned for organizational improvement and systems operations have been captured. 3. All waivers and anomalies have been closed. 4. Systems hardware, software, personnel, and procedures are in place to support operations.

Flight Readiness Review

The FRR examines tests, demonstrations, analyses, and audits that determine the system's readiness for a safe and successful flight or launch and for subsequent flight operations. It also ensures that all flight and ground hardware, software, personnel, and procedures are operationally ready.

Objectives

The objectives of the review are to:

- Ensure a thorough review of the products supporting the review.

- Ensure the products meet the entrance criteria and success criteria.

- Ensure issues raised during the review are appropriately documented and a plan for resolution is prepared.

Results of Review

As a result of successful FRR completion, technical and procedural maturity exists for system launch and flight authorization and in some cases initiation of system operations.

Table 6.7-15 FRR Entrance and Success Criteria

Flight Readiness Review	
Entrance Criteria	**Success Criteria**
1. Receive certification that flight operations can safely proceed with acceptable risk.	1. The flight vehicle is ready for flight.
2. The system and support elements have been confirmed as properly configured and ready for flight.	2. The hardware is deemed acceptably safe for flight (i.e., meeting the established acceptable risk criteria or documented as being accepted by the PM and DGA).
3. Interfaces are compatible and function as expected.	3. Flight and ground software elements are ready to support flight and flight operations.
4. The system state supports a launch Go decision based on Go or No-Go criteria.	4. Interfaces are checked out and found to be functional.
5. Flight failures and anomalies from previously completed flights and reviews have been resolved and the results incorporated into all supporting and enabling operational products.	5. Open items and waivers have been examined and found to be acceptable.
6. The system has been configured for flight.	6. The flight and recovery environmental factors are within constraints.
	7. All open safety and mission risk items have been addressed.

Post-Launch Assessment Review

A PLAR is a post-deployment evaluation of the readiness of the spacecraft systems to proceed with full, routine operations. The review evaluates the status, performance, and capabilities of the project evident from the flight operations experience since launch. This can also mean assessing readiness to transfer responsibility from the development organization to the operations organization. The review also evaluates the status of the project plans and the capability to conduct the mission with emphasis on near-term operations and mission-critical events. This review is typically held after the early flight operations and initial checkout.

The objectives of the review are to:

- Ensure a thorough review of the products supporting the review.
- Ensure the products meet the entrance criteria and success criteria.
- Ensure issues raised during the review are appropriately documented and a plan for resolution is prepared.

Table 6.7-16 PLAR Entrance and Success Criteria

Post Launch Assessment Review	
Entrance Criteria	**Success Criteria**
1. The launch and early operations performance, including (when appropriate) the early propulsive maneuver results, are available. 2. The observed spacecraft and science instrument performance, including instrument calibration plans and status, are available. 3. The launch vehicle performance assessment and mission implications, including launch sequence assessment and launch operations experience with lessons learned, are completed. 4. The mission operations and ground data system experience, including tracking and data acquisition support and spacecraft telemetry data analysis, are available. 5. The mission operations organization, including status of staffing, facilities, tools, and mission software (e.g., spacecraft analysis, and sequencing), is available. 6. In-flight anomalies and the responsive actions taken, including any autonomous fault protection actions taken by the spacecraft, or any unexplained spacecraft telemetry, including alarms, are documented. 7. The need for significant changes to procedures, interface agreements, software, and staffing has been documented. 8. Documentation is updated, including any updates originating from the early operations experience. 9. Future development/test plans are developed.	1. The observed spacecraft and science payload performance agrees with prediction, or if not, it is adequately understood so that future behavior can be predicted with confidence. 2. All anomalies have been adequately documented, and their impact on operations assessed. Further, anomalies impacting spacecraft health and safety or critical flight operations have been properly disposed. 3. The mission operations capabilities, including staffing and plans, are adequate to accommodate the actual flight performance. 4. Liens, if any, on operations, identified as part of the ORR, have been satisfactorily disposed.

Critical Event Readiness Review

A CERR confirms the project's readiness to execute the mission's critical activities during flight operation.

The objectives of the review are to:

- Ensure a thorough review of the products supporting the review.
- Ensure the products meet the entrance criteria and success criteria.
- Ensure issues raised during the review are appropriately documented and a plan for resolution is prepared.

Table 6.7-17 CERR Entrance and Success Criteria

Critical Event Readiness Review	
Entrance Criteria	**Success Criteria**
1. Mission overview and context for the critical event(s).	1. The critical activity design complies with requirements.
2. Activity requirements and constraints.	2. The preparation for the critical activity, including the verification and validation, is thorough.
3. Critical activity sequence design description including key tradeoffs and rationale for selected approach.	3. The project (including all the systems, supporting services, and documentation) is ready to support the activity.
4. Fault protection strategy.	
5. Critical activity operations plan including planned uplinks and criticality.	4. The requirements for the successful execution of the critical event(s) are complete and understood and have been flowed down to the appropriate levels for implementation.
6. Sequence verification (testing, walk-throughs, peer review) and critical activity validation.	
7. Operations team training plan and readiness report.	
8. Risk areas and mitigations.	
9. Spacecraft readiness report.	
10. Open items and plans.	

Post-Flight Assessment Review

The PFAR evaluates the activities from the flight after recovery. The review identifies all anomalies that occurred during the flight and mission and determines the actions necessary to mitigate or resolve the anomalies for future flights.

The objectives of the review are to:

- Ensure a thorough review of the products supporting the review.
- Ensure the products meet the entrance criteria and success criteria.
- Ensure issues raised during the review are appropriately documented and a plan for resolution is prepared.

Table 6.7-18 PFAR Entrance and Success Criteria

Post Flight Assessment Review	
Entrance Criteria	**Success Criteria**
1. All anomalies that occurred during the mission, as well as during preflight testing, countdown, and ascent, identified.	1. Formal final report documenting flight performance and recommendations for future missions.
2. Report on overall post-recovery condition.	2. All anomalies have been adequately documented and disposed.
3. Report any evidence of ascent debris.	
4. All photo and video documentation available.	
5. Retention plans for scrapped hardware completed.	3. The impact of anomalies on future flight operations has been assessed.
6. Post-flight assessment team operating plan completed.	
7. Disassembly activities planned and scheduled.	4. Plans for retaining assessment documentation and imaging have been made.
8. Processes and controls to coordinate in-flight anomaly troubleshooting and post-flight data preservation developed.	
9. Problem reports, corrective action requests, post-flight anomaly records, and final post-flight documentation completed.	5. Reports and other documentation have been added to a database for performance comparison and trending.
10. All post-flight hardware and flight data evaluation reports completed.	

Decommissioning Review

The DR confirms the decision to terminate or decommission the system and assesses the readiness of the system for the safe decommissioning and disposal of system assets. The DR is normally held near the end of routine mission operations upon accomplishment of planned mission objectives. It may be advanced if some unplanned event gives rise to a need to prematurely terminate the mission, or delayed if operational life is extended to permit additional investigations.

Objectives
The objectives of the review are to:

- Ensure a thorough review of the products supporting the review.
- Ensure the products meet the entrance criteria and success criteria.
- Ensure issues raised during the review are appropriately documented and a plan for resolution is prepared.

Results of Review
A successful DR completion ensures that the decommissioning and disposal of system items and processes are appropriate and effective.

Table 6.7-19 DR Entrance and Success Criteria

Decommissioning Review	
Entrance Criteria	**Success Criteria**

1. Requirements associated with decommissioning and disposal are defined.
 	
 2. Plans are in place for decommissioning, disposal, and any other removal from service activities.
 	
 3. Resources are in place to support decommissioning and disposal activities, plans for disposition of project assets, and archival of essential mission and project data.
 	
 4. Safety, environmental, and any other constraints are described.
 	
 5. Current system capabilities are described.
 	
 6. For off-nominal operations, all contributing events, conditions, and changes to the originally expected baseline are described.
	1. The reasons for decommissioning disposal are documented.
	
	 2. The decommissioning and disposal plan is complete, approved by appropriate management, and compliant with applicable Agency safety, environmental, and health regulations. Operations plans for all potential scenarios, including contingencies, are complete and approved. All required support systems are available.
	
	 3. All personnel have been properly trained for the nominal and contingency procedures.
	
	 4. Safety, health, and environmental hazards have been identified. Controls have been verified.
	
	 5. Risks associated with the disposal have been identified and adequately mitigated. Residual risks have been accepted by the required management.
	
	 6. If hardware is to be recovered from orbit:
	   a. Return site activity plans have been defined and approved.
	   b. Required facilities are available and meet requirements, including those for contamination control, if needed.
	   c. Transportation plans are defined and approved. Shipping containers and handling equipment, as well as contamination and environmental control and monitoring devices, are available.
	
	 7. Plans for disposition of mission-owned assets (i.e., hardware, software, facilities) have been defined and approved.
	
	 8. Plans for archival and subsequent analysis of mission data have been defined and approved. Arrangements have been finalized for the execution of such plans. Plans for the capture and dissemination of appropriate lessons learned during the project life cycle have been defined and approved. Adequate resources (schedule, budget, and staffing) have been identified and are available to successfully complete all decommissioning, disposal, and disposition activities.

Other Technical Reviews

These typical technical reviews are some that have been conducted on previous programs and projects but are not required as part of the NPR 7123.1 systems engineering process.

Design Certification Review

Purpose

The Design Certification Review (DCR) ensures that the qualification verifications demonstrate design compliance with functional and performance requirements.

Timing

The DCR follows the system CDR, and after qualification tests and all modifications needed to implement qualification-caused corrective actions have been completed.

Objectives

The objectives of the review are to:

- Confirm that the verification results met functional and performance requirements, and that test plans and procedures were executed correctly in the specified environments.

- Certify that traceability between test article and production article is correct, including name, identification number, and current listing of all waivers.

- Identify any incremental tests required or conducted due to design or requirements changes made since test initiation, and resolve issues regarding their results.

Criteria for Successful Completion

The following items comprise a checklist to aid in determining the readiness of DCR product preparation:

- Are the pedigrees of the test articles directly traceable to the production units?

- Is the verification plan used for this article current and approved?

- Do the test procedures and environments used comply with those specified in the plan?

- Are there any changes in the test article configuration or design resulting from the as-run tests? Do they require design or specification changes and/or retests?

- Have design and specification documents been audited?

- Do the verification results satisfy functional and performance requirements?

- Do the verification, design, and specification documentation correlate?

Results of Review

As a result of a successful DCR, the end item design is approved for production. All open issues should be resolved with closure actions and schedules.

Functional and Physical Configuration Audits

Configuration audits confirm that the configured product is accurate and complete. The two types of configuration audits are the Functional Configuration Audit (FCA) and the Physical Configuration Audit (PCA). The FCA examines the functional characteristics of the configured product and verifies that the product has met, via test results, the requirements specified in its functional baseline documentation approved at the PDR and CDR. FCAs will be conducted on both hardware or software configured products and will precede the PCA of the configured product. The PCA (also known as a configuration inspection) examines the physical configuration of the configured product and verifies that the product corresponds to the build-to (or code-to) product baseline documentation previously approved at the CDR. PCAs will be conducted on both hardware and software configured products.

Technical Peer Reviews

Peer reviews provide the technical insight essential to ensure product and process quality. Peer reviews are focused, in-depth technical reviews that support the evolving design and development of a product, including critical documentation or data packages. They are often, but not always, held as supporting reviews for technical reviews such as PDR and CDR. A purpose of the peer review is to add value and reduce risk through expert knowledge infusion, confirmation of approach, identification of defects, and specific suggestions for product improvements.

The results of the engineering peer reviews comprise a key element of the review process. The results and issues that surface during these reviews are documented and reported out at the appropriate next higher element level.

The peer reviewers should be selected from outside the project, but they should have a similar technical background, and they should be selected for their skill and experience. Peer reviewers should be concerned with only the technical integrity and quality of the product. Peer reviews should be kept simple and informal. They should concentrate on a review of the documentation and minimize the viewgraph presentations. A roundtable format rather than a stand-up presentation is preferred. The peer reviews should give the full technical picture of items being reviewed.

Table 6.7-20 Functional and Physical Configuration Audits

Representative Audit Data List	
FCA	**PCA**
• Design specifications • Design drawings and parts list • Engineering change proposals/engineering change requests • Deviation/waiver approval requests incorporated and pending • Specification and drawing tree • Fracture control plan • Structural dynamics, analyses, loads, and models documentation • Materials usage agreements/materials identification usage list • Verification and validation requirements, plans, procedures, and reports • Software requirements and development documents • Listing of accomplished tests and test results • CDR completion documentation including RIDs/RFAs and disposition reports • Analysis reports • ALERT (Acute Launch Emergency Restraint Tip) tracking log • Hazard analysis/risk assessment	• Final version of all specifications • Product drawings and parts list • Configuration accounting and status reports • Final version of all software and software documents • Copy of all FCA findings for each product • List of approved and outstanding engineering change proposals, engineering change requests, and deviation/waiver approval requests • Indentured parts list • As-run test procedures • Drawing and specification tree • Manufacturing and inspection "build" records • Inspection records • As-built discrepancy reports • Product log books • As-built configuration list

Technical depth should be established at a level that allows the review team to gain insight into the technical risks. Rules need to be established to ensure consistency in the peer review process. At the conclusion of the review, a report on the findings, recommendations, and actions must be distributed to the technical team.

For those projects where systems engineering is done out-of-house, peer reviews must be part of the contract.

Additional guidance on establishing and conducting peer reviews can be found in Appendix N.

6.7.2.2 Status Reporting and Assessment

This subsection provides additional information on status reporting and assessment techniques for costs and schedules (including EVM), technical performance, and systems engineering process metrics.

Cost and Schedule Control Measures

Status reporting and assessment on costs and schedules provides the project manager and systems engineer visibility into how well the project is tracking against its planned cost and schedule targets. From a management point of view, achieving these targets is on a par with meeting the technical performance requirements of the system. It is useful to think of cost and schedule status reporting and assessment as measuring the performance of the "system that produces the system."

NPR 7120.5 provides specific requirements for the application of EVM to support cost and schedule management. EVM is applicable to both in-house and contracted efforts. The level of EVM system implementation will depend on the dollar value and risk of a project or contract. The standard for EVM systems is ANSI-EIA-748. The project manager/systems engineer will use the guidelines to establish the program and project EVM implementation plan.

Assessment Methods

Performance measurement data are used to assess project cost, schedule, and technical performance and their impacts on the completion cost and schedule of the project. In program control terminology, a difference between actual performance and planned costs or schedule status is called a "variance." Variances must be controlled at the control account level, which is typically at the subsystem WBS level. The person responsible for this activity is fre-

quently called the Control Account Manager (CAM). The CAM develops work and product plans, schedules, and time-phased resource plans. The technical subsystem manager/leads often takes on this role as part of their subsystem management responsibilities.

Figure 6.7-3 illustrates two types of variances, cost and schedule, and some related concepts. A product-oriented WBS divides the project work into discrete tasks and products. Associated with each task and product (at any level in the WBS) is a schedule and a budgeted (i.e., planned) cost. The Budgeted Cost for Work Scheduled ($BCWS_t$) for any set of WBS elements is the sum of the budgeted cost of all work on tasks and products in those elements scheduled to be completed by time t. The Budgeted Cost for Work Performed ($BCWP_t$), also called Earned Value (EV_t), is the sum of the budgeted cost for tasks and products that have actually been produced at time t in the schedule for those WBS elements. The difference, $BCWP_t$ and $BCWS_t$, is called the schedule variance at time t. A negative value indicates that the work is behind schedule.

The Actual Cost of Work Performed ($ACWP_t$) represents the funds that have been expended up to time t on those WBS elements. The difference between the budgeted and actual costs, $BCWP_t - ACWP_t$, is called the cost variance at time t. A negative value here indicates a cost overrun.

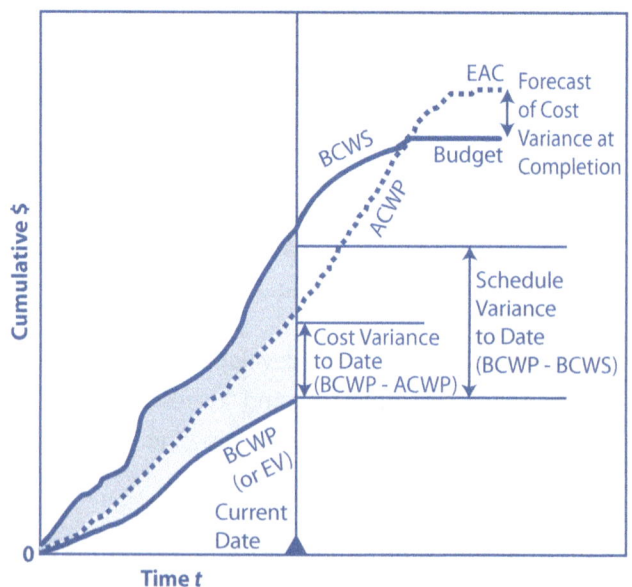

Figure 6.7-3 Cost and schedule variances

When either schedule variance or cost variance exceeds preestablished control-account-level thresholds that represent significant departures from the baseline plan, the conditions must be analyzed to identify why the variance exists. Once the cause is understood, the CAM can make an informed forecast of the time and resources needed to complete the control account. When corrective actions are feasible (can stay within the BCWS), the plan for implementing them must be included in the analysis. Sometimes no corrective action is feasible; overruns or schedule slips may be unavoidable. One must keep in mind that the earlier a technical problem is identified as a result of schedule or cost variances, the more likely the project team can minimize the impact on completion.

Variances may indicate that the cost Estimate at Completion (EAC$_t$) of the project is likely to be different from the Budget at Completion (BAC). The difference between the BAC and the EAC is the Variance at Completion (VAC). A negative VAC is generally unfavorable, while a positive is usually favorable. These variances may also point toward a change in the scheduled completion date of the project. These types of variances enable a program analyst to estimate the EAC at any point in the project life cycle. (See box on analyzing EAC.) These analytically derived estimates should be used only as a "sanity check" against the estimates prepared in the variance analysis process.

If the cost and schedule baselines and the technical scope of the work are not adequately defined and fully integrated, then it is very difficult (or impossible) to estimate the current cost EAC of the project.

Other efficiency factors can be calculated using the performance measurement data. The Schedule Performance Index (SPI) is a measure of work accomplishment in dollars. The SPI is calculated by dividing work accomplished in dollars or BCWP by the dollar value of the work scheduled or BCWS. Just like any other ratio, a value less than one is a sign of a behind-schedule condition, equal to one indicates an on-schedule status, and greater than one denotes that work is ahead of schedule. The Cost Performance Index (CPI) is a measure of cost efficiency and is calculated as the ratio of the earned value or BCWP for a segment of work compared to the cost to complete that same segment of work or ACWP. A CPI will show how much work is being accomplished for every dollar spent on the project. A CPI of less than one reveals negative cost efficiency, equal to one is right on cost, and greater

---

Analyzing the Estimate at Completion

An EAC can be estimated at any point in the project and should be reviewed at least on a monthly basis. The EAC requires a detailed review by the CAM. A statistical estimate can be used as a cross-check of the CAM's estimate and to develop a range to bound the estimate. The appropriate formula used to calculate the statistical EAC depends upon the reasons associated with any variances that may exist. If a variance exists due to a one-time event, such as an accident, then EAC = ACWP + (BAC − BCWP). The CPI and SPI should also be considered in developing the EAC.

If there is a growing number of liens, action items, or significant problems that will increase the difficulty of future work, the EAC might grow at a greater rate than estimated by the above equation. Such factors could be addressed using risk management methods described in the Section 6.4.

---

than one is positive. Note that traditional measures compare planned cost to actual cost; however, this comparison is never made using earned value data. Comparing planned to actual costs is an indicator only of spending and not of overall project performance.

Technical Measures—MOEs, MOPs, and TPMs

Measures of Effectiveness

MOEs are the "operational" measures of success that are closely related to the achievement of mission or operational objectives in the intended operational environment. MOEs are intended to focus on how well mission or operational objectives are achieved, not on how they are achieved, i.e., MOEs should be independent of any particular solution. As such, MOEs are the standards against which the "goodness" of each proposed solution may be assessed in trade studies and decision analyses. Measuring or calculating MOEs not only makes it possible to compare alternative solutions quantitatively, but sensitivities to key assumptions regarding operational environments and to any underlying MOPs can also be investigated. (See MOP discussion below.)

In the systems engineering process, MOEs are used to:

● Define high-level operational requirements from the customer/stakeholder viewpoint.

● Compare and rank alternative solutions in trade studies.

- Investigate the relative sensitivity of the projected mission or operational success to key operational assumptions and performance parameters.

- Determine that the mission or operational success quantitative objectives remain achievable as system development proceeds. (See TPM discussion below.)

Measures of Performance

MOPs are the measures that characterize physical or functional attributes relating to the system, e.g., engine I_{sp}, max thrust, mass, and payload-to-orbit. These attributes are generally measured under specified test conditions or operational environments. MOPs are attributes deemed important in achieving mission or operational success, but do not measure it directly. Usually multiple MOPs contribute to an MOE. MOPs often become system performance requirements that, when met by a design solution, result in achieving a critical threshold for the system MOEs.

The distinction between MOEs and MOPs is that they are formulated from different viewpoints. An MOE refers to the effectiveness of a solution from the mission or operational success criteria expressed by the user/customer/stakeholder. An MOE represents a stakeholder expectation that is critical to the success of the system, and failure to attain a critical value for it will cause the stakeholder to judge the system a failure. An MOP is a measure of actual performance of a (supplier's) particular design solution, which taken alone may only be indirectly related to the customer/stakeholder's concerns.

Technical Performance Measures

TPMs are critical or key mission success or performance parameters that are monitored during implementation by comparing the current actual achievement of the parameters with the values that were anticipated for the current time and projected for future dates. They are used to confirm progress and identify deficiencies that might jeopardize meeting a system requirement or put the project at cost or schedule risk. When a TPM value falls outside the expected range around the anticipated value, it signals a need for evaluation and corrective action.

In the systems engineering process, TPMs are used to:

- Forecast values to be achieved by critical parameters at major milestones or key events during implementation.

- Identify differences between the actual and planned values for those parameters.

- Provide projected values for those parameters in order to assess the implications for system effectiveness.

- Provide early warning for emerging risks requiring management attention (when negative margins exist).

- Provide early identification of potential opportunities to make design trades that reduce risk or cost, or increase system effectiveness (when positive margins exist).

- Support assessment of proposed design changes.

Selecting TPMs

TPMs are typically selected from the defined set of MOEs and MOPs. Understanding that TPM tracking requires allocation of resources, care should be exercised in selecting a small set of succinct TPMs that accurately reflect key parameters or risk factors, are readily measurable, and that can be affected by altering design decisions. In general, TPMs can be generic (attributes that are meaningful to each PBS element, like mass or reliability) or unique (attributes that are meaningful only to specific PBS elements). The relationship of MOEs, MOPs, and TPMs can be found in Figure 6.7-4. The systems engineer needs to decide which generic and unique TPMs are worth tracking at each level of the PBS. (See box for examples of TPMs.) At lower levels of the PBS, TPMs

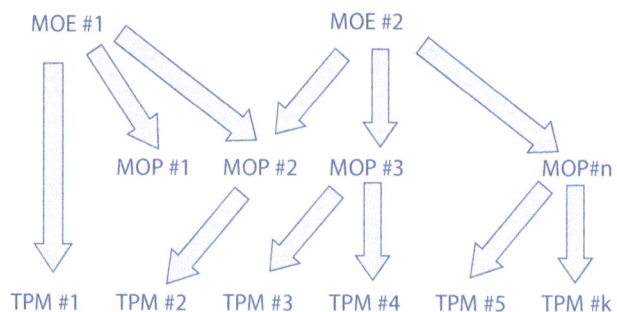

MOEs	Derived from stakeholder expectation statements; deemed critical to mission or operational success of the system
MOPs	Broad physical and performance parameters; means of ensuring meeting the associated MOEs
TPMs	Critical mission success or performance attributes; measurable; progress profile established, controlled, and monitored

Figure 6.7-4 Relationships of MOEs, MOPs, and TPMs

worth tracking can be identified through the functional and performance requirements levied on each individual system, subsystem, etc.

As TPMs are intended to provide an early warning of the adequacy of a design in satisfying selected critical technical parameter requirements, the systems engineer should select TPMs that fall within well-defined (quan-

---

Examples of Technical Performance Measures

TPMs from MOEs

- Mission performance (e.g., total science data volume returned)
- Safety (e.g., probability of loss of crew, probability of loss of mission)
- Achieved availability (e.g., (system uptime)/(system uptime + system downtime))

TPMs from MOPs

- Thrust versus predicted/specified
- I_{sp} versus predicted/specified
- End of Mission (EOM) dry mass
- Injected mass (includes EOM dry mass, baseline mission plus reserve propellant, other consumables and upper stage adaptor mass)
- Propellant margins at EOM
- Other consumables margins at EOM
- Electrical power margins over mission life
- Control system stability margins
- EMI/EMC susceptibility margins
- Onboard data processing memory demand
- Onboard data processing throughput time
- Onboard data bus capacity
- Total pointing error
- Total vehicle mass at launch
- Payload mass (at nominal altitude or orbit)
- Reliability
- Mean time before refurbishment required
- Total crew maintenance time required
- System turnaround time
- Fault detection capability
- Percentage of system designed for on-orbit crew access

---

titative) limits for reasons of system effectiveness or mission feasibility. Usually these limits represent either a firm upper or lower bound constraint. A typical example of such a TPM for a spacecraft is its injected mass, which must not exceed the capability of the selected launch vehicle. Tracking injected mass as a high-level TPM is meant to ensure that this does not happen. A high-level TPM like injected mass must often be "budgeted" and allocated to multiple system elements. Tracking and reporting should be required at these lower levels to gain visibility into the sources of any variances.

In summary, for a TPM to be a valuable status and assessment tool, certain criteria must be met:

- Be a significant descriptor of the system (e.g., weight, range, capacity, response time, safety parameter) that will be monitored at key events (e.g., reviews, audits, planned tests).
- Can be measured (either by test, inspection, demonstration, or analysis).
- Is such that reasonable projected progress profiles can be established (e.g., from historical data or based on test planning).

TPM Assessment and Reporting Methods

Status reporting and assessment of the system's TPMs complement cost and schedule control. There are a number of assessment and reporting methods that have been used on NASA projects, including the planned profile method and the margin management method.

A detailed example of the planned profile method for the Chandra Project weight TPM is illustrated in Figure 6.7-5. This figure depicts the subsystem contributions, various constraints, project limits, and management reserves from project SRR to launch.

A detailed example of the margin management method for the Sojourner mass TPM is illustrated in Figure 6.7-6 This figure depicts the margin requirements (horizontal straight lines) and actual mass margins from project SRR to launch.

Relationship of TPM Assessment Program to the SEMP

The SEMP is the usual document for describing the project's TPM assessment program. This description should include a master list of those TPMs to be tracked, and the measurement and assessment methods to be employed. If analytical methods and models are used to measure

Chandra Project: Weight Changes

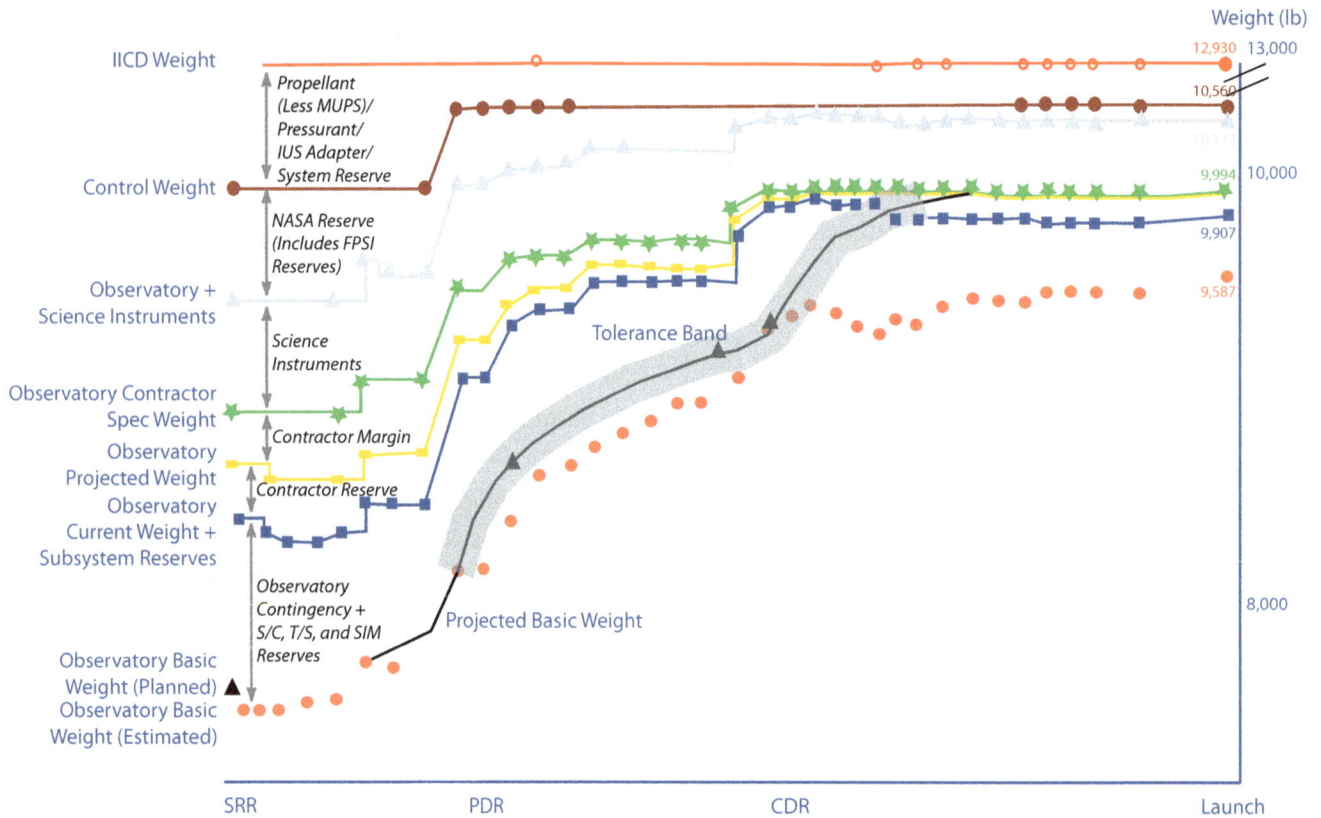

Figure 6.7-5 **Use of the planned profile method for the weight TPM with rebaseline in Chandra Project**

certain high-level TPMs, then these need to be identi-fied. The reporting frequency and timing of assessments should be specified as well. In determining these, the sys-tems engineer must balance the project's needs for accu-rate, timely, and effective TPM tracking against the cost of the TPM tracking program.

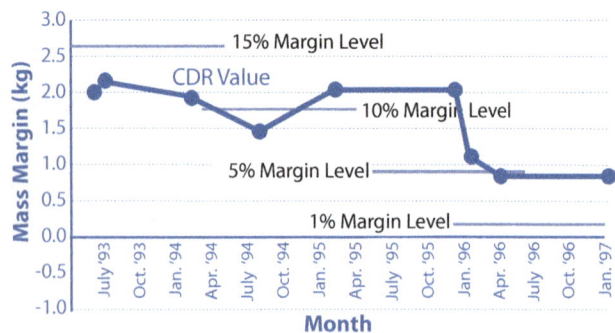

Figure 6.7-6 **Use of the margin management method for the mass TPM in Sojourner**

Note: Current Margin Description: Microrover System (Rover + Lander-Mounted Rover Equipment (LMRE)) Allocation = 16.0 kg; Microrover System (Rover + LMRE) Current Best Estimate = 15.2 kg; Microrover System (Rover + LMRE) Margin = 0.8 kg (5.0%).

The TPM assessment program plan, which may be a part of the SEMP or a stand-alone document for large pro-grams/projects, should specify each TPM's allocation, time-phased planned profile or margin requirement, and alert zones, as appropriate to the selected assessment method.

A formal TPM assessment program should be fully planned and baselined with the SEMP. Tracking TPMs should begin as soon as practical in Phase B. Data to support the full set of selected TPMs may, however, not be available until later in the project life cycle. As the project life cycle proceeds through Phases C and D, the measurement of TPMs should become increasingly more accurate with the availability of more actual data about the system.

For the WBS model in the system structure, typically the following activities are performed:

● Analyze stakeholder expectation statements to estab-lish a set of MOEs by which overall system or product effectiveness will be judged and customer satisfaction will be determined.

- Define MOPs for each identified MOE.
- Define appropriate TPMs and document the TPM assessment program in the SEMP.

Systems Engineering Process Metrics

Status reporting and assessment of systems engineering process metrics provide additional visibility into the performance of the "system that produces the system." As such, these metrics supplement the cost and schedule control measures discussed in this subsection.

Systems engineering process metrics try to quantify the effectiveness and productivity of the systems engineering process and organization. Within a single project, tracking these metrics allows the systems engineer to better understand the health and progress of that project. Across projects (and over time), the tracking of systems engineering process metrics allows for better estimation of the cost and time of performing systems engineering functions. It also allows the systems engineering organization to demonstrate its commitment to continuous improvement.

Selecting Systems Engineering Process Metrics

Generally, systems engineering process metrics fall into three categories—those that measure the progress of the systems engineering effort, those that measure the quality of that process, and those that measure its productivity. Different levels of systems engineering management are generally interested in different metrics. For example, a project manager or lead systems engineer may focus on metrics dealing with systems engineering staffing, project risk management progress, and major trade study progress. A subsystem systems engineer may focus on subsystem requirements and interface definition progress and verification procedures progress. It is useful for each systems engineer to focus on just a few process metrics. Which metrics should be tracked depends on the systems engineer's role in the total systems engineering effort. The systems engineering process metrics worth tracking also change as the project moves through its life cycle.

Collecting and maintaining data on the systems engineering process is not without cost. Status reporting and assessment of systems engineering process metrics divert time and effort from the activity itself. The systems engineer must balance the value of each systems engineering

process metric against its collection cost. The value of these metrics arises from the insights they provide into the activities that cannot be obtained from cost and schedule control measures alone. Over time, these metrics can also be a source of hard productivity data, which are invaluable in demonstrating the potential returns from investment in systems engineering tools and training.

Examples and Assessment Methods

Table 6.7-21 lists some systems engineering process metrics to be considered. This list is not intended to be exhaustive. Because some of these metrics allow for different interpretations, each NASA Center needs to define them in a common-sense way that fits its own processes. For example, each field Center needs to determine what is meant by a "completed" versus an "approved" requirement, or whether these terms are even relevant. As part of this definition, it is important to recognize that not all requirements, for example, need be lumped together. It may be more useful to track the same metric separately for each of several different types of requirements.

Quality-related metrics should serve to indicate when a part of the systems engineering process is overloaded and/or breaking down. These metrics can be defined and tracked in several different ways. For example, requirements volatility can be quantified as the number of newly identified requirements, or as the number of changes to already approved requirements. As another example, Engineering Change Request (ECR) processing could be tracked by comparing cumulative ECRs opened versus cumulative ECRs closed, or by plotting the age profile of open ECRs, or by examining the number of ECRs opened last month versus the total number open. The systems engineer should apply his or her own judgment in picking the status reporting and assessment method.

Productivity-related metrics provide an indication of systems engineering output per unit of input. Although more sophisticated measures of input exist, the most common is the number of systems engineering hours dedicated to a particular function or activity. Because not all systems engineering hours cost the same, an appropriate weighing scheme should be developed to ensure comparability of hours across systems engineering personnel.

Schedule-related metrics can be depicted in a table or graph of planned quantities versus actuals, for example, comparing planned number of verification closure notices against actual. This metric should not be confused

Table 6.7-21 Systems Engineering Process Metrics

Function	Metric	Category
Requirements development and management	Requirements identified versus completed versus approved	S
	Requirements volatility	Q
	Trade studies planned versus completed	S
	Requirements approved per systems engineering hour	P
	Tracking of TBAs, TBDs, and TBRs (to be announced, determined, or resolved) resolved versus remaining	S
Design and development	Specifications planned versus completed	S
	Processing of engineering change proposals (ECPs)/engineering change requests (ECRs)	Q
	Engineering drawings planned versus released	S
Verification and validation	Verification and validation plans identified versus approved	S
	Verification and validation procedures planned versus completed	S
	Functional requirements approved versus verified	S
	Verification and validation plans approved per systems engineering hour	P
	Processing of problem/failure reports	Q
Reviews	Processing of RIDs	Q
	Processing of action items	Q

S = progress or schedule related; Q = quality related; P = productivity related

with EVM described in this subsection. EVM is focused on integrated cost and schedule at the desired level, whereas this metric focuses on an individual process or product within a subsystem, system, or project itself.

The combination of quality, productivity, and schedule metrics can provide trends that are generally more important than isolated snapshots. The most useful kind of assessment method allows comparisons of the trend on a current project with that for a successfully completed project of the same type. The latter provides a benchmark against which the systems engineer can judge his or her own efforts.

6.8 Decision Analysis

The purpose of this section is to provide a description of the Decision Analysis Process, including alternative tools and methodologies. Decision analysis offers individuals and organizations a methodology for making decisions; it also offers techniques for modeling decision problems mathematically and finding optimal decisions numerically. Decision models have the capacity for accepting and quantifying human subjective inputs: judgments of experts and preferences of decisionmakers. Implementation of models can take the form of simple paper-and-pencil procedures or sophisticated computer programs known as decision aids or decision systems. The methodology is broad and must always be adapted to the issue under consideration. The problem is structured by identifying alternatives, one of which must be decided upon; possible events, one of which occurs thereafter; and outcomes, each of which results from a combination of decision and event. Decisions are made throughout a program/project life cycle and often are made through a hierarchy of panels, boards, and teams with increasing complementary authority, wherein each progressively more detailed decision is affected by the assumptions made at the lower level. Not all decisions need a formal process, but it is important to establish a process for those decisions that do require a formal process. Important decisions as well as supporting information (e.g., assumptions made), tools, and models must be completely documented so that new information can be incorporated and assessed and past decisions can be researched in context. The Decision Analysis Process accommodates this iterative environment and occurs throughout the project life cycle.

An important aspect of the Decision Analysis Process is to consider and understand at what time it is appropriate or required for a decision to be made or not made. When considering a decision, it is important to ask questions such as: Why is a decision required at this time? For how long can a decision be delayed? What is the impact of delaying a decision? Is all of the necessary information available to make a decision? Are there other key drivers or dependent factors and criteria that must be in place before a decision can be made?

The outputs from this process support the decisionmaker's difficult task of deciding among competing alternatives without complete knowledge; therefore, it is critical to understand and document the assumptions and limitation of any tool or methodology and integrate them with other factors when deciding among viable options.

Early in the project life cycle, high-level decisions are made regarding which technology could be used, such as solid or liquid rockets for propulsion. Operational scenarios, probabilities, and consequences are determined and the design decision made without specifying the component-level detail of each design alternative. Once high-level design decisions are made, nested systems engineering processes occur at progressively more detailed design levels flowed down through the entire system. Each progressively more detailed decision is affected by the assumptions made at the previous levels. For example, the solid rocket design is constrained by the operational assumptions made during the decision process that selected that design. This is an iterative process among elements of the system. Also early in the life cycle, the technical team should determine the types of data and information products required to support the Decision Analysis Process during the later stages of the project. The technical team should then design, develop, or acquire the models, simulations, and other tools that will supply the required information to decisionmakers. In this section, application of different levels and kinds of analysis are discussed at different stages of the project life cycle.

6.8.1 Process Description

The Decision Analysis Process is used to help evaluate technical issues, alternatives, and their uncertainties to support decisionmaking. A typical process flow diagram is provided in Figure 6.8-1, including inputs, activities, and outputs.

Typical processes that use decision analysis are:

- Determining how to allocate limited resources (e.g., budget, mass, power) among competing subsystem interests to favor the overall outcome of the project;
- Select and test evaluation methods and tools against sample data;
- Configuration management processes for major change requests or problem reports;
- Design processes for making major design decisions and selecting design approaches;

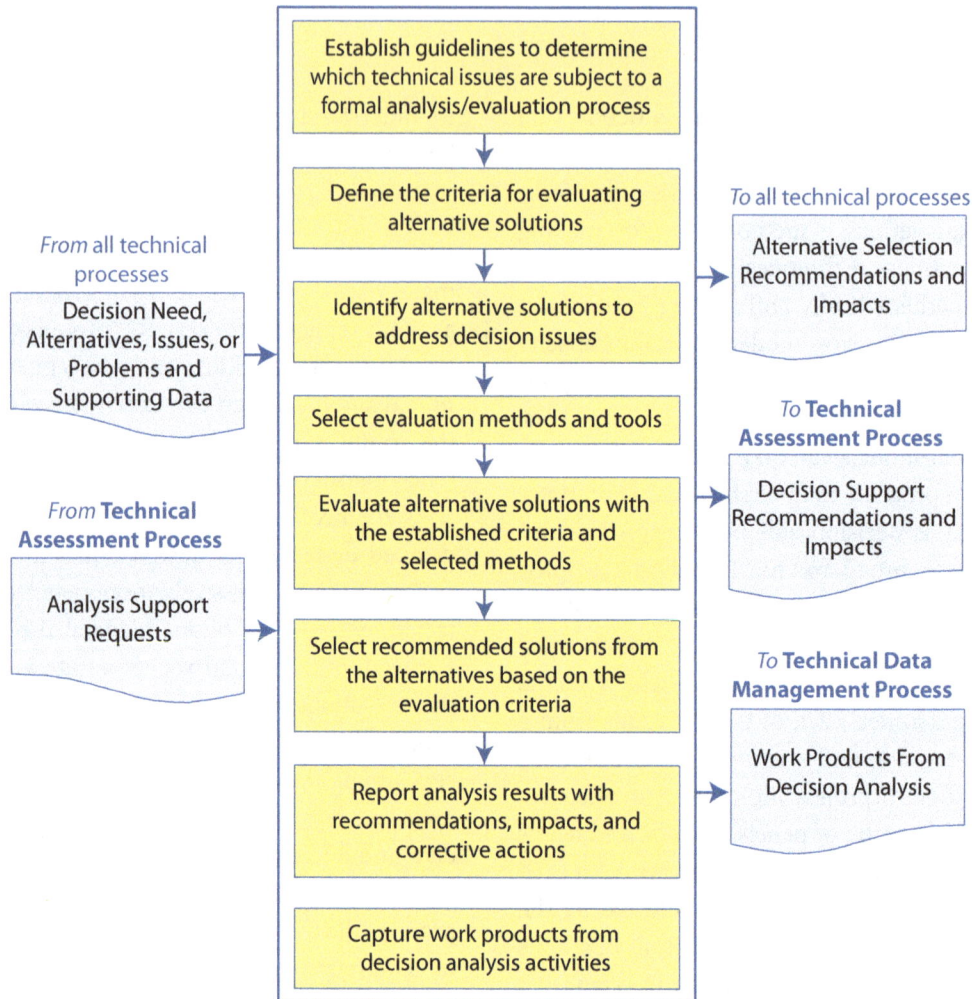

Figure 6.8-1 Decision Analysis Process

- Key decision point reviews or technical review decisions (e.g., PDR, CDR) as defined in NPR 7120.5 and NPR 7123.1;

- Go or No-Go decisions (e.g., FRR):
  - ► Go—authorization to proceed or
  - ► No-Go—repeat some specific aspects of development or conduct further research.

- Project management of major issues, schedule delays, or budget increases;

- Procurement of major items;

- Technology decisions;

- Risk management of major risks (e.g., red or yellow);

- SMA decisions; and

- Miscellaneous decisions (e.g., whether to intervene in the project to address an emergent performance issue).

Decision analysis can also be used in emergency situations. Under such conditions, process steps, procedures,

Note: Studies often deal in new territory, so it is important to test whether there are sufficient data, needed quality, resonance with decision authority, etc., before diving in, especially for large or very complex decision trade spaces.

and meetings may be combined, and the decision analysis documentation may be completed at the end of the process (i.e., after the decision is made). However, a decision matrix should be completed and used during the decision. Decision analysis documentation must be archived as soon as possible following the emergency situation.

6.8.1.1 Inputs

Formal decision analysis has the potential to consume significant resources and time. Typically, its application to a specific decision is warranted only when some of the following conditions are met:

- **High Stakes:** High stakes are involved in the decision, such as significant cost, safety, or mission success criteria.

- **Complexity:** The actual ramifications of alternatives are difficult to understand without detailed analysis.

- **Uncertainty:** Uncertainty in key inputs creates substantial uncertainty in the ranking of alternatives and points to risks that may need to be managed.

- **Multiple Attributes:** Greater numbers of attributes cause a greater need for formal analysis.

- **Diversity of Stakeholders:** Extra attention is warranted to clarify objectives and formulate TPMs when the set of stakeholders reflects a diversity of values, preferences, and perspectives.

Satisfaction of all of these conditions is not a requirement for initiating decision analysis. The point is, rather, that the need for decision analysis increases as a function of the above conditions. When the Decision Analysis Process is triggered, the following are inputs:

- Decision need, identified alternatives, issues, or problems and supporting data (from all technical management processes).

- Analysis support requests (from Technical Assessment Process).

- High-level objectives and constraints (from the program/project).

6.8.1.2 Process Activities

For the Decision Analysis Process, the following activities typically are performed.

Establish Guidelines to Determine Which Technical Issues Are Subject to a Formal Analysis/ Evaluation Process

This step includes determining:

- When to use a formal decisionmaking procedure,

- What needs to be documented,

- Who will be the decisionmakers and their responsibilities and decision authorities, and

- How decisions will be handled that do not require a formal evaluation procedure.

Decisions are based on facts, qualitative and quantitative data, engineering judgment, and open communications to facilitate the flow of information throughout the hierarchy of forums where technical analyses and evalua-

tions are presented and assessed and where decisions are made. The extent of technical analysis and evaluation required should be commensurate with the consequences of the issue requiring a decision. The work required to conduct a formal evaluation is not insignificant and applicability must be based on the nature of the problem to be resolved. Guidelines for use can be determined by the magnitude of the possible consequences of the decision to be made.

For example, the consequence table from a risk scorecard can be used to assign numerical values for applicability according to impacts to mission success, flight safety, cost, and schedule. Actual numerical thresholds for use would then be set by a decision authority. Sample values could be as shown in Table 6.8-1.

Table 6.8-1 Consequence Table

Numerical Value	Consequence	Applicability
Consequence = 5, 4	High	Mandatory
Consequence = 3	Moderate	Optional
Consequence = 1, 2	Low	Not required

Define the Criteria for Evaluating Alternative Solutions

This step includes identifying:

- The types of criteria to consider, such as customer expectations and requirements, technology limitations, environmental impact, safety, risks, total ownership and life-cycle costs, and schedule impact;

- The acceptable range and scale of the criteria; and

- The rank of each criterion by its importance.

Decision criteria are requirements for individually assessing options and alternatives being considered. Typical decision criteria include cost, schedule, risk, safety, mission success, and supportability. However, considerations should include technical criteria specific to the decision being made. Criteria should be objective and measurable. Criteria should also permit distinguishing among options or alternatives. Some criteria may not be meaningful to a decision; however, they should be documented as having been considered. Identify criteria that are mandatory (i.e., "must have") versus the other criteria (i.e., "nice to have"). If mandatory criteria are not met, that option should be disregarded. For complex decisions, criteria can be grouped into categories or objectives. (See the analytical hierarchy process in Subsection 6.8.2.6.)

Ranking or prioritizing the criteria is probably the hardest part of completing a decision matrix. Not all criteria have the same importance, and ranking is typically accomplished by assigning weights to each. To avoid "gaming" the decision matrix (i.e., changing decision outcomes by playing with criteria weights), it is best to agree upon weights before the decision matrix is completed. Weights should only be changed with consensus from all decision stakeholders.

For example, ranking can be done using a simple approach like percentages. Have all the weights for each criterion add up to 100. Assign percents based on how important the criterion is. (The higher the percentage, the more important, such as a single criterion worth 50 percent.) The weights need to be divided by 100 to calculate percents. Using this approach, the option with the highest percentage is typically the recommended option. Ranking can also be done using sophisticated decision tools. For example, pair-wise comparison is a decision technique that calculates the weights using paired comparisons among criteria and options. Other methods include:

- Formulation of objectives hierarchy and TPMs;
- Analytical hierarchy process, which addresses criteria and paired comparisons; and
- Risk-informed decision analysis process with weighting of TPMs.

Identify Alternative Solutions to Address Decision Issues

This step includes considering alternatives in addition to those that may be provided with the issue.

Almost every decision will have options to choose from. Brainstorm decision options, and document option summary names for the available options. For complex decisions, it is also a best practice to perform a literature search to identify options. Reduce the decision options to a reasonable set (e.g., seven plus or minus two). Some options will obviously be bad options. Document the fact that these options were considered. The use of mandatory criteria also can help reduce the number of options. A few decisions might only have one option. It is a best practice to document a decision matrix even for one option if it is a major decision. (Sometimes doing nothing or not making a decision is an option.)

Select Evaluation Methods and Tools

Select evaluation methods and tools/techniques based on the purpose for analyzing a decision and on the availability of the information used to support the method and/or tool.

Typical evaluation methods include: simulations; weighted tradeoff matrices; engineering, manufacturing, cost, and technical opportunity of trade studies; surveys; extrapolations based on field experience and prototypes; user review and comment; and testing.

Tools and techniques to be used should be selected based on the purpose for analyzing a decision and on the availability of the information used to support the method and/or tool.

Additional evaluation methods include:

- Decision matrix (see Figure 6.8-2);
- Decision analysis process support, evaluation methods, and tools;
- Risk-informed decision analysis process; and
- Trade studies and decision alternatives.

Evaluate Alternative Solutions with the Established Criteria and Selected Methods

Regardless of the methods or tools used, results must include:

- Evaluation of assumptions related to evaluation criteria and of the evidence that supports the assumptions, and
- Evaluation of whether uncertainty in the values for alternative solutions affects the evaluation.

Alternatives can be compared to evaluation criteria via the use of a decision matrix as shown in Figure 6.8-2. Evaluation criteria typically are in the rows on the left side of the matrix. Alternatives are typically the column headings on the top of the matrix (and to the right top). Criteria weights are typically assigned to each criterion. In the example shown, there are also mandatory criteria. If mandatory criteria are not met, the option is scored at 0 percent.

When decision criteria have different measurement bases (e.g., numbers, money, weight, dates), normalization can be used to establish a common base for mathematical operations. The process of "normalization" is making a scale so that all different kinds of criteria can

Decision Matrix Example for Battery			ENTER SCORES ↘	Extend Old Battery Life	Buy New Batteries	Collect Experiment Data With Alternative Experiment	Cancelled Experiment
CRITERIA	Mandatory (Y=1/N=0)?	Weight	SCALE				
Mission Success (Get Experiment Data)	1	30	3 = Most Supportive 1 = Least Supportive	2	3	3	0
Cost per Option	0	10	3 = Least Expensive 1 = Most Expensive	1	2	3	1
Risk (Overall Option Risk)	0	15	3 = Least Risk 1 = Most Risk	2	1	2	3
Schedule	0	10	3 = Shortest Schedule 1 = Longest Schedule	3	2	1	3
Safety	1	15	3 = Most Safe 1 = Least Safe	2	1	2	3
Uninterrupted Data Collection	0	20	3 = Most Supportive 1 = Least Supportive	3	1	2	1
WEIGHTED TOTALS in %		100%	3	73%	60%	77%	0%
			SCALE 1-3				

Figure 6.8-2 Example of a decision matrix

be compared or added together. This can be done informally (e.g., low, medium, high), on a scale (e.g., 1-3-9), or more formally with a tool. No matter how normalization is done, the most important thing to remember is to have operational definitions of the scale. An operational definition is a repeatable, measurable number. For example, "high" could mean "a probability of 67 percent and above." "Low" could mean "a probability of 33 percent and below." For complex decisions, decision tools usually provide an automated way for normalization. Be sure to question and understand the operational definitions for the weights and scales of the tool.

Select Recommended Solutions from the Alternatives Based on the Evaluation Criteria

This step includes documenting the information, including assumptions and limitations of the evaluation methods

Note: Completing the decision matrix can be thought of as a default evaluation method. Completing the decision matrix is iterative. Each cell for each criterion and each option needs to be completed by the team. Use evaluation methods as needed to complete the entire decision matrix.

used, that justifies the recommendations made and gives the impacts of taking the recommended course of action.

The highest score (e.g., percentage, total score) is typically the option that is recommended to management. If a different option is recommended, an explanation must be provided as to why the lower score is preferred. Usually, if a lower score is recommended, the "risks" or "disadvantages" were too great for the highest score. Sometimes the benefits and advantages of a lower or close score outweigh the highest score. Ideally, all risks/benefits and advantages/disadvantages would show up in the decision matrix as criteria, but this is not always possible. Sometimes if there is a lower score being recommended, the weighting or scores given may not be accurate.

Report the Analysis and Evaluation Results and Findings with Recommendations, Impacts, and Corrective Actions

Typically a technical team of subject matter experts makes a recommendation to a NASA decisionmaker (e.g., a NASA board, forum, or panel). It is highly recommended that the team produce a white paper to document all major recommendations to serve as a backup to any presentation materials used. A presentation can also be used, but a paper in conjunction with a decision matrix is preferred (especially for complex decisions). Decisions are typically captured in meeting minutes, but can be captured in the white paper.

Capture Work Products from Decision Analysis Activities

This step includes capturing:

- Decision analysis guidelines generated and strategy and procedures used;
- Analysis/evaluation approach, criteria, and methods and tools used;
- Analysis/evaluation results, assumptions made in arriving at recommendations, uncertainties, and sensitivities of the recommended actions or corrective actions; and
- Lessons learned and recommendations for improving future decision analyses.

Typical information captured in a decision report is shown in Table 6.8-2.

6.8.1.3 Outputs

Decision analysis continues throughout the life cycle. The products from decision analysis include:

- Alternative selection recommendations and impacts (to all technical management processes);
- Decision support recommendations and impacts (to Technical Assessment Process);

Table 6.8-2 Typical Information to Capture in a Decision Report

#	Section	Section Description
1	Executive Summary	Provide a short half-page executive summary of the report: • Recommendation (short summary—1 sentence) • Problem/issue requiring a decision (short summary—1 sentence)
2	Problem/Issue Description	Describe the problem/issue that requires a decision. Provide background, history, the decisionmaker(s) (e.g., board, panel, forum, council), and decision recommendation team, etc.
3	Decision Matrix Setup Rationale	Provide the rationale for setting up the decision matrix: • Criteria selected • Options selected • Weights selected • Evaluation methods selected Provide a copy of the setup decision matrix.
4	Decision Matrix Scoring Rationale	Provide the rationale for the scoring of the decision matrix. Provide the results of populating the scores of the matrix using the evaluation methods selected.
5	Final Decision Matrix	Cut and paste the final spreadsheet into the document. Also include any important snapshots of the decision matrix.
6	Risk/Benefits	For the final options being considered, document the risks and benefits of each option.
7	Recommendation and/or Final Decision	Describe the recommendation that is being made to the decisionmaker(s) and the rationale for why the option was selected. Can also document the final decision in this section.
8	Dissent	If applicable, document any dissent with the recommendation. Document how dissent was addressed (e.g., decision matrix, risk, etc.).
9	References	Provide any references.
A	Appendices	Provide the results of the literature search, including lessons learned, previous related decisions, and previous related dissent. Also document any detailed data analysis and risk analysis used for the decision. Can also document any decision metrics.

- Work products of decision analysis activities (to Technical Data Management Process); and

- Technical risk status measurements (to Technical Risk Management Process).

- TPMs, Performance Indexes (PIs) for alternatives, the program- or project-specific objectives hierarchy, and the decisionmakers' preferences (to all technical management processes).

6.8.2 Decision Analysis Guidance

The purpose of this subsection is to provide guidance, methods, and tools to support the Decision Analysis Process at NASA.

6.8.2.1 Systems Analysis, Simulation, and Performance

Systems analysis can be better understood in the context of the system's overall life cycle. Systems analysis within the context of the life cycle is responsive to the needs of the stakeholder at every phase of the life cycle, from pre-Phase A through Phase B to realizing the final product and beyond (See Figure 6.8-3.)

Systems analysis of a product must support the transformation from a need into a realized, definitive product; be able to support compatibility with all physical and functional requirements; and support the operational

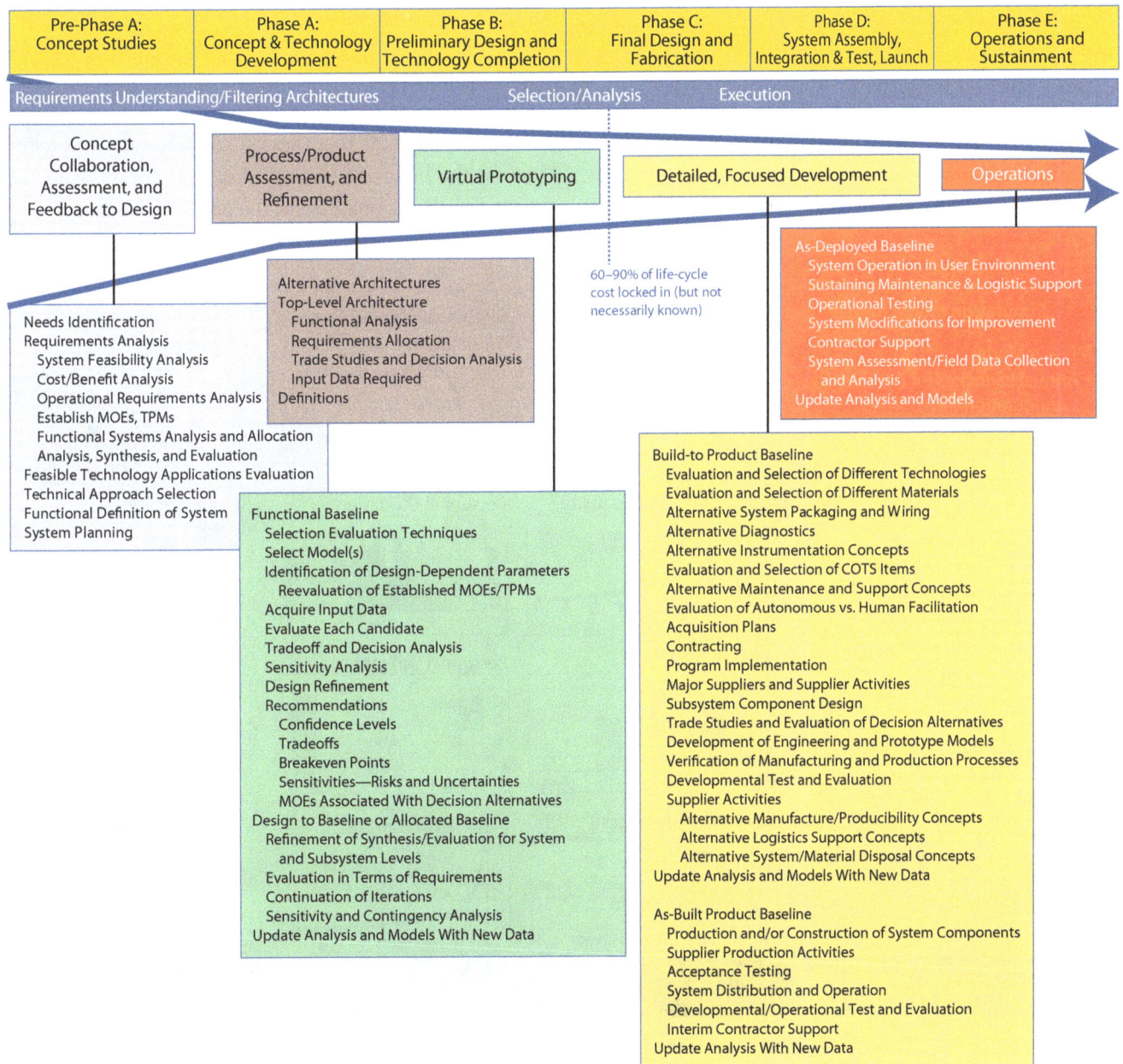

Figure 6.8-3 Systems analysis across the life cycle

scenarios in terms of reliability, maintainability, supportability, serviceability, and disposability, while maintaining performance and affordability.

Systems analysis support is provided from cradle to grave of the system. This covers the product design, verification, manufacturing, operations and support, and disposal. Viewed in this manner, life-cycle engineering is the basis for concurrent engineering.

Systems analysis should support concurrent engineering. Appropriate systems analysis can be conducted early in the life cycle to support planning and development. The intent here is to support seamless systems analysis optimally planned across the entire life cycle. For example, systems engineering early in the life cycle can support optimal performance of the deployment, operations, and disposal facets of the system.

Historically, this has not been the case. Systems analysis would focus only on the life cycle that the project occupied at that time. The systems analyses for the later phases were treated serially, in chronological order. This resulted in major design modifications that were very costly in the later life-cycle phases. Resources can be used more efficiently if the requirements across the life cycle are considered concurrently, providing results for decisionmaking about the system.

Figure 6.8-3 shows a life-cycle chart that indicates how the various general types of systems analyses fit across the phases of the life cycle. The requirements for analysis begin with a broader scope and more types of analysis required in the early phases of the life cycle and funnel or narrow in scope and analysis requirements as decisions are made and project requirements become clearer as the project proceeds through its life cycle. Figure 6.8-4 presents a specific spaceport example and shows how specific operational analysis inputs can provide analysis result outputs pertinent to the operations portion of the life cycle. Note that these simulations are conducted across the life cycle and updated periodically with the new data that is obtained as the project evolves.

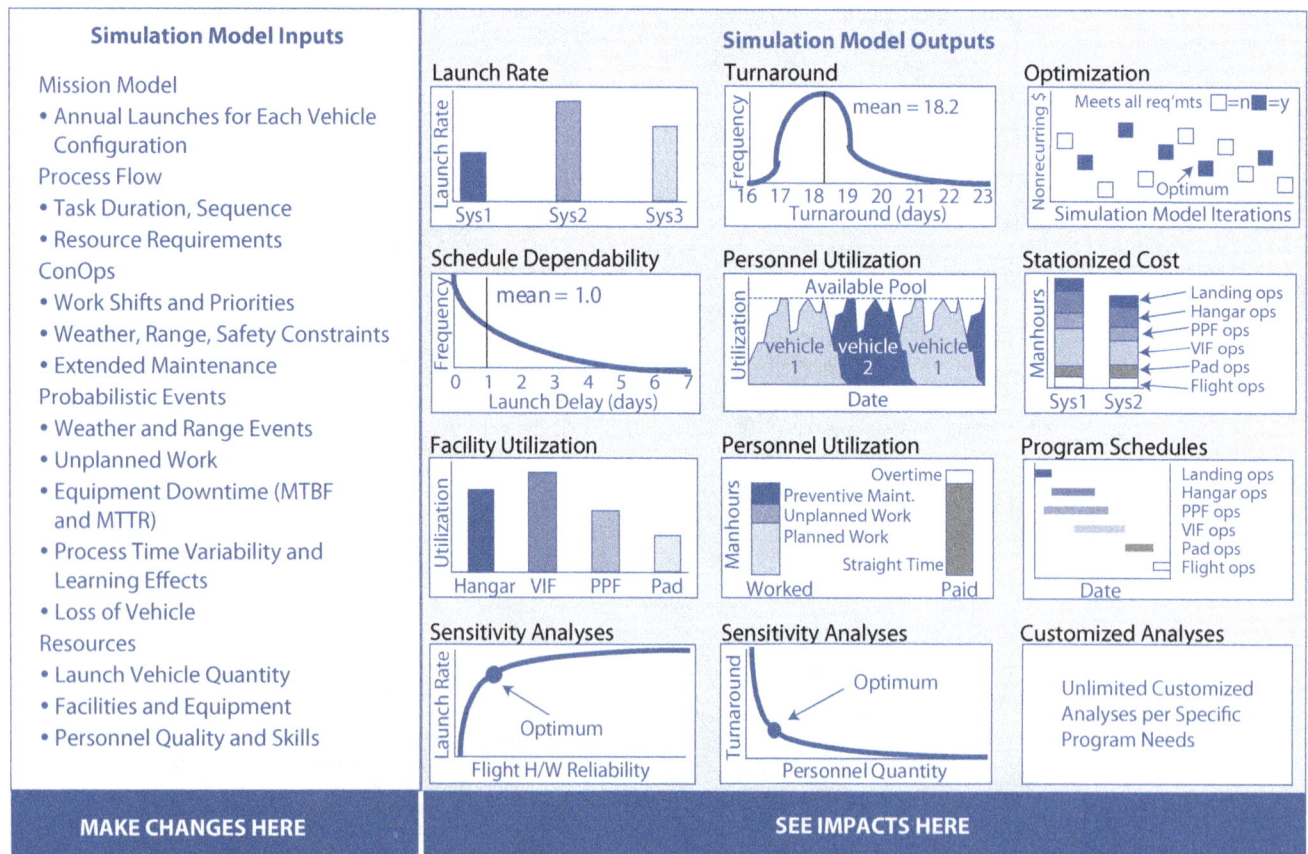

Figure 6.8-4 Simulation model analysis techniques

From: Lockheed Martin presentation to KSC, November 2003, Kevin Brughelli, Lockheed Martin Space Systems Company; Debbie Carstens, Florida Institute of Technology; and Tim Barth, KSC.

During the early life-cycle phases, inputs should include a plan for collecting the quantitative and qualitative data necessary to manage contracts and improve processes and products as the project evolves. This plan should indicate the type of data necessary to determine the cause of problems, nonconformances, and anomalies, and propose corrective action to prevent recurrence. This closed-loop plan involving identification, resolution, and recurrence control systems is critical to producing actual reliability that approaches predicted reliability. It should indicate the information technology infrastructure and database capabilities to provide data sorting, data mining, data analysis, and precursor management. Management of problems, nonconformances, and anomalies should begin with data collection, should be a major part of technical assessment, and should provide critical information for decision analysis.

6.8.2.2 Trade Studies

The trade study process is a critical part of systems engineering. Trade studies help to define the emerging system at each level of resolution. One key message of this subsection is that to be effective, trade studies require the participation of people with many skills and a unity of effort to move toward an optimum system design.

Figure 6.8-5 shows the trade study process in simplest terms, beginning with the step of defining the system's goals and objectives, and identifying the constraints it must meet. In the early phases of the project life cycle, the goals, objectives, and constraints are usually stated in general operational terms. In later phases of the project life cycle, when the architecture and, perhaps, some aspects of the design have already been decided, the goals and objectives may be stated as performance requirements that a segment or subsystem must meet.

At each level of system resolution, the systems engineer needs to understand the full implications of the goals, objectives, and constraints to formulate an appropriate system solution. This step is accomplished by performing a functional analysis. "Functional analysis" is the process of identifying, describing, and relating the functions a system must perform to fulfill its goals and objectives and is described in detail in Section 4.4.

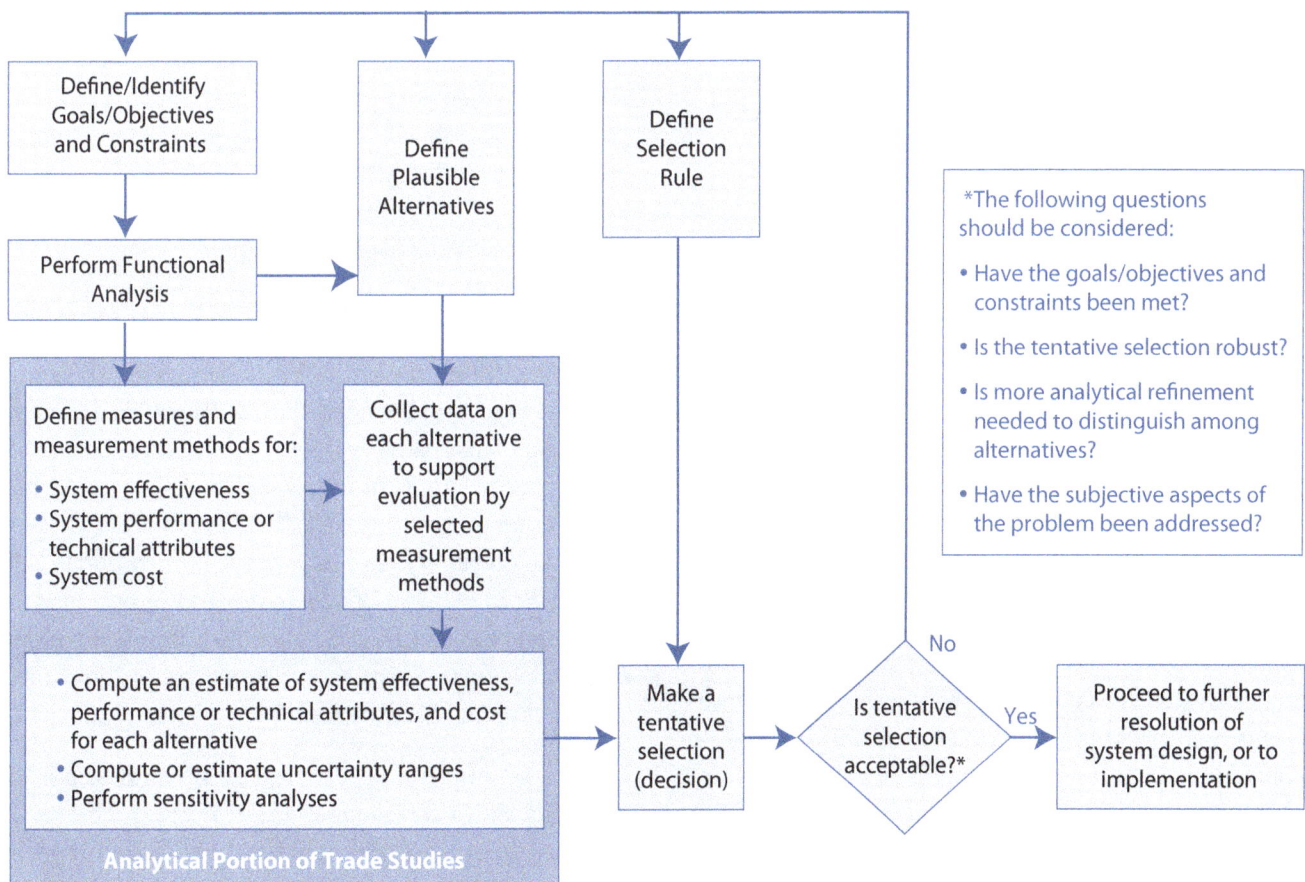

Figure 6.8-5 Trade study process

Closely related to defining the goals and objectives and performing a functional analysis is the step of defining the measures and measurement methods for system effectiveness (when this is practical), system performance or technical attributes, and system cost. (These variables are collectively called outcome variables, in keeping with the discussion in Section 2.3. Some systems engineering books refer to these variables as decision criteria, but this term should not be confused with "selection rule," described below. Sections 2.5 and 6.1 discuss the concepts of system cost and effectiveness in greater detail.) Defining measures and measurement methods begins the analytical portion of the trade study process, since it suggests the involvement of those familiar with quantitative methods.

For each measure, it is important to address how that quantitative measure will be computed—that is, which measurement method is to be used. One reason for doing this is that this step then explicitly identifies those variables that are important in meeting the system's goals and objectives.

Evaluating the likely outcomes of various alternatives in terms of system effectiveness, the underlying performance or technical attributes, and cost before actual fabrication and/or programming usually requires the use of a mathematical model or series of models of the system. So a second reason for specifying the measurement methods is to identify necessary models.

Sometimes these models are already available from previous projects of a similar nature; other times, they need to be developed. In the latter case, defining the measurement methods should trigger the necessary system modeling activities. Since the development of new models can take a considerable amount of time and effort, early identification is needed to ensure they will be ready for formal use in trade studies. Defining the selection rule is the step of explicitly determining how the outcome variables will be used to make a (tentative) selection of the preferred alternative. As an example, a selection rule may be to choose the alternative with the highest estimated system effectiveness that costs less than x dollars (with some given probability), meets safety requirements, and possibly meets other political or schedule constraints. Defining the selection rule is essentially deciding how the selection is to be made. This step is independent from the actual measurement of system effectiveness, system performance or technical attributes, and system cost.

Many different selection rules are possible. The selection rule in a particular trade study may depend on the context in which the trade study is being conducted—in particular, what level of system design resolution is being addressed. At each level of the system design, the selection rule generally should be chosen only after some guidance from the next higher level. The selection rule for trade studies at lower levels of the system design should be in consonance with the higher level selection rule.

Defining plausible alternatives is the step of creating some alternatives that can potentially achieve the goals and objectives of the system. This step depends on understanding (to an appropriately detailed level) the system's functional requirements and operational concept. Running an alternative through an operational timeline or reference mission is a useful way of determining whether it can plausibly fulfill these requirements. (Sometimes it is necessary to create separate behavioral models to determine how the system reacts when a certain stimulus or control is applied, or a certain environment is encountered. This provides insights into whether it can plausibly fulfill time-critical and safety requirements.) Defining plausible alternatives also requires an understanding of the technologies available, or potentially available, at the time the system is needed. Each plausible alternative should be documented qualitatively in a description sheet. The format of the description sheet should, at a minimum, clarify the allocation of required system functions to that alternative's lower level architectural or design components (e.g., subsystems).

One way to represent the trade study alternatives under consideration is by a trade tree.

During Phase A trade studies, the trade tree should contain a number of alternative high-level system architectures to avoid a premature focus on a single one. As the systems engineering process proceeds, branches of the trade tree containing unattractive alternatives will be "pruned," and greater detail in terms of system design will be added to those branches that merit further attention. The process of pruning unattractive early alternatives is sometimes known as doing "killer trades." (See trade tree box.)

Given a set of plausible alternatives, the next step is to collect data on each to support the evaluation of the measures by the selected measurement methods. If models are to be used to calculate some of these measures, then obtaining the model inputs provides some impetus and

An Example of a Trade Tree for a Mars Rover

The figure below shows part of a trade tree for a robotic Mars rover system, whose goal is to find a suitable manned landing site. Each layer represents some aspect of the system that needs to be treated in a trade study to determine the best alternative. Some alternatives have been eliminated a priori because of technical feasibility, launch vehicle constraints, etc. The total number of alternatives is given by the number of end points of the tree. Even with just a few layers, the number of alternatives can increase quickly. (This tree has already been pruned to eliminate low-autonomy, large rovers.) As the systems engineering process proceeds, branches of the tree with unfavorable trade study outcomes are discarded. The remaining branches are further developed by identifying more detailed trade studies that need to be made. A whole family of (implicit) alternatives can be represented in a trade tree by the continuous variable. In this example, rover speed or range might be so represented. By treating a variable this way, mathematical optimization techniques can be applied. Note that a trade tree is, in essence, a decision tree without chance nodes.

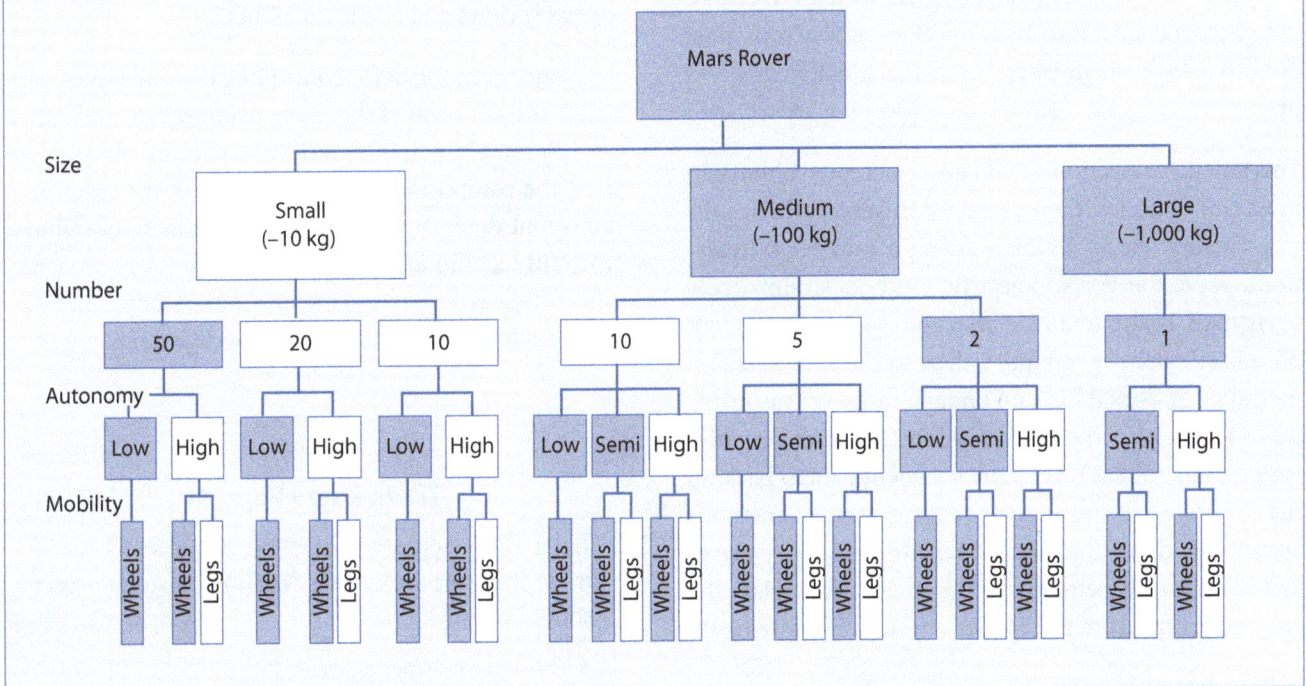

direction to the data collection activity. By providing data, engineers in such disciplines as reliability, maintainability, producibility, integrated logistics, software, testing, operations, and costing have an important supporting role in trade studies. The data collection activity, however, should be orchestrated by the systems engineer. The results of this step should be a quantitative description of each alternative to accompany the qualitative.

Test results on each alternative can be especially useful. Early in the systems engineering process, performance and technical attributes are generally uncertain and must be estimated. Data from breadboard and brassboard testbeds can provide additional confidence that the range of values used as model inputs is correct. Such confidence is also enhanced by drawing on data collected on related, previously developed systems.

The next step in the trade study process is to quantify the outcome variables by computing estimates of system effectiveness, its underlying system performance or technical attributes, and system cost. If the needed data have been collected and the measurement methods (for example, models) are in place, then this step is, in theory, mechanical. In practice, considerable skill is often needed to get meaningful results.

In an ideal world, all input values would be precisely known and models would perfectly predict outcome variables. This not being the case, the systems engineer should supplement point estimates of the outcome variables for each alternative with computed or estimated uncertainty ranges. For each uncertain key input, a range of values should be estimated. Using this range of input values, the sensitivity of the outcome variables can be gauged and their uncertainty ranges calculated. The systems engineer may be able to obtain meaningful probability distributions for the outcome variables using Monte Carlo simulation, but when this is not feasible, the systems engineer must be content with only ranges and sensitivities. See the risk-informed decision analysis process in Subsection 6.8.2.8 for more information on uncertainty.

This essentially completes the analytical portion of the trade study process. The next steps can be described as the judgmental portion. Combining the selection rule with the results of the analytical activity should enable the systems engineer to array the alternatives from most preferred to least, in essence making a tentative selection.

This tentative selection should not be accepted blindly. In most trade studies, there is a need to subject the results to a "reality check" by considering a number of questions. Have the goals, objectives, and constraints truly been met? Is the tentative selection heavily dependent on a particular set of input values to the measurement methods, or does it hold up under a range of reasonable input values? (In the latter case, the tentative selection is said to be robust.) Are there sufficient data to back up the tentative selection? Are the measurement methods sufficiently discriminating to be sure that the tentative selection is really better than other alternatives? Have the subjective aspects of the problem been fully addressed?

If the answers support the tentative selection, then the systems engineer can have greater confidence in a recommendation to proceed to a further resolution of the system design, or to the implementation of that design. The estimates of system effectiveness, its underlying performance or technical attributes, and system cost generated during the trade study process serve as inputs to that further resolution. The analytical portion of the trade study process often provides the means to quantify the performance or technical (and cost) attributes that the system's lower levels must meet. These can be formalized as performance requirements.

If the reality check is not met, the trade study process returns to one or more earlier steps. This iteration may result in a change in the goals, objectives, and constraints; a new alternative; or a change in the selection rule, based on the new information generated during the trade study. The reality check may lead instead to a decision to first

improve the measures and measurement methods (e.g., models) used in evaluating the alternatives, and then to repeat the analytical portion of the trade study process.

Controlling the Trade Study Process

There are a number of mechanisms for controlling the trade study process. The most important one is the SEMP. The SEMP specifies the major trade studies that are to be performed during each phase of the project life cycle. It should also spell out the general contents of trade study reports, which form part of the decision support packages (i.e., documentation submitted in conjunction with formal reviews and change requests).

A second mechanism for controlling the trade study process is the selection of the study team leaders and members. Because doing trade studies is part art and part science, the composition and experience of the team is an important determinant of a study's ultimate usefulness. A useful technique to avoid premature focus on a specific technical design is to include in the study team individuals with differing technology backgrounds.

Trade Study Reports

Trade study reports should be prepared for each trade study. At a minimum, each trade study report should identify:

- The system under analysis
- System goals and objectives (or requirements, as appropriate to the level of resolution), and constraints
- The measures and measurement methods (models) used
- All data sources used
- The alternatives chosen for analysis
- The computational results, including uncertainty ranges and sensitivity analyses performed
- The selection rule used
- The recommended alternative.

Trade study reports should be maintained as part of the system archives so as to ensure traceability of decisions made through the systems engineering process. Using a generally consistent format for these reports also makes it easier to review and assimilate them into the formal change control process.

Another mechanism is limiting the number of alternatives that are to be carried through the study. This number is usually determined by the time and resources available to do the study because the work required in defining additional alternatives and obtaining the necessary data on them can be considerable. However, focusing on too few or too similar alternatives defeats the purpose of the trade study process.

A fourth mechanism for controlling the trade study process can be exercised through the use (and misuse) of models. Lastly, the choice of the selection rule exerts a considerable influence on the results of the trade study process. See Appendix O for different examples of how trade studies are used throughout the life cycle.

6.8.2.3 Cost-Benefit Analysis

A cost-benefit analysis is performed to determine the advantage of one alternative over another in terms of equivalent cost or benefits. The analysis relies on the addition of positive factors and the subtraction of negative factors to determine a net result. Cost-benefit analysis maximizes net benefits (benefits minus costs). A cost-benefit analysis finds, quantifies, and adds all the positive factors. These are the benefits. Then it identifies, quantifies, and subtracts all the negatives, the costs. The difference between the two indicates whether the planned action is a preferred alternative. The real trick to doing a cost-benefit analysis well is making sure to include all the costs and all the benefits and properly quantify them. A similar approach, used when a cost cap is imposed externally, is to maximize effectiveness for a given level of cost. Cost-effectiveness is a systematic quantitative method for comparing the costs of alternative means of achieving the same equivalent benefit for a specific objective. A project is cost-effective if, on the basis of life-cycle cost analysis of competing alternatives, it is determined to have the lowest costs expressed in present value terms for a given amount of benefits.

Cost-effectiveness analysis is appropriate whenever it is impractical to consider the dollar value of the benefits provided by the alternatives. This is the scenario whenever each alternative has the same life-cycle benefits expressed in monetary terms, or each alternative has the same life-cycle effects, but dollar values cannot be assigned to their benefits. After determining the scope of the project on the basis of mission and other requirements, and having identified, quantified, and valued

the costs and benefits of the alternatives, the next step is to identify the least-cost or most cost-effective alternative to achieve the purpose of the project. A comparative analysis of the alternative options or designs is often required. This is illustrated in Figure 4.4-3. In cases in which alternatives can be defined that deliver the same benefits, it is possible to estimate the equivalent rate between each alternative for comparison. Least-cost analysis aims at identifying the least-cost project option for meeting the technical requirements. Least-cost analysis involves comparing the costs of the various technically feasible options and selecting the one with the lowest costs. Project options must be alternative ways of achieving the mission objectives. If differences in results or quality exist, a normalization procedure must be applied that takes the benefits of one option relative to another as a cost to the option that does not meet all of the mission objectives to ensure an equitable comparison. Procedures for the calculation and interpretation of the discounting factors should be made explicit, with the least-cost project being identified by comparing the total life-cycle costs of the project alternatives and calculating the equalizing factors for the difference in costs. The project with the highest equalizing factors for all comparisons is the least-cost alternative.

Cost-effectiveness analysis also deals with alternative means of achieving mission requirements. However, the results may be estimated only indirectly. For example, different types of systems may be under consideration to obtain science data. The effectiveness of each alternative may be measured through obtaining science data through different methods. An example of a cost-effectiveness analysis requires the increase in science data to be divided by the costs for each alternative. The most cost-effective method is the one that raises science data by a given amount for the least cost. If this method is chosen and applied to all similar alternatives, the same increase in science data can be obtained for the lowest cost. Note, however, that the most cost-effective method is not necessarily the most effective method of meeting mission objectives. Another method may be the most effective, but also cost a lot more, so it is not the most cost-effective. The cost-effectiveness ratios—the cost per unit increase in science data for each method—can be compared to see how much more it would cost to implement the most effective method. Which method is chosen for implementation then depends jointly on the desired mission objectives and the extra cost involved in implementing the most effective method.

There will be circumstances where project alternatives have more than one outcome. To assess the cost-effectiveness of the different alternatives, it is necessary to devise a testing system where the results for the different factors can be added together. It also is necessary to decide on weights for adding the different elements together, reflecting their importance in relation to the objectives of the project. Such a cost-effectiveness analysis is called weighted cost-effectiveness analysis. It introduces a subjective element, the weights, into the comparison of project alternatives, both to find the most cost-effective alternative and to identify the extra cost of implementing the most effective alternative.

6.8.2.4 Influence Diagrams

An influence diagram (also called a decision network) is a compact graphical and mathematical representation of a decision state. (See Figure 6.8-6.) Influence diagrams were first developed in the mid-1970s within the decision analysis community as an intuitive approach that is easy to understand. They are now adopted widely and are becoming an alternative to decision trees, which typically suffer from exponential growth in number of branches with each variable modeled. An influence diagram is directly applicable in team decision analysis

since it allows incomplete sharing of information among team members to be modeled and solved explicitly. Its elements are:

- Decision nodes, indicating the decision inputs, and the items directly influenced by the decision outcome;
- Chance nodes, indicating factors that impact the chance outcome, and items influenced by the chance outcome;
- Value nodes, indicating factors that affect the value, and items influenced by the value; and
- Arrows, indicating the relationships among the elements.

An influence diagram does not depict a strictly sequential process. Rather, it illustrates the decision process at a particular point, showing all of the elements important to the decision. The influence diagram for a particular model is not unique. The strength of influence diagrams is their ability to display the structure of a decision problem in a clear, compact form, useful both for communication and to help the analyst think clearly during problem formulation. An influence diagram can be transformed into a decision tree for quantification.

6.8.2.5 Decision Trees

Like the influence diagram, a decision tree portrays a decision model, but a decision tree is drawn from a point of view different from that of the influence diagram. The decision tree exhaustively works out the expected consequences of all decision alternatives by discretizing all "chance" nodes, and, based on this discretization, calculating and appropriately weighting all possible consequences of all alternatives. The preferred alternative is then identified by summing the appropriate outcome variables (MOE or expected utility) from the path end states.

A decision tree grows horizontally from left to right, with the trunk at the left. Typically, the possible alternatives initially available to the decisionmaker stem from the trunk at the left. Moving across the tree, the decisionmaker encounters branch points corresponding to probabilistic outcomes and perhaps additional decision nodes. Thus, the tree branches as it is read from left to right. At the far right side of the decision tree, a vector of TPM scores is listed for each terminal branch, representing each combination of decision outcome and chance outcome. From the TPM scores, and the chosen selection rule, a preferred alternative is determined.

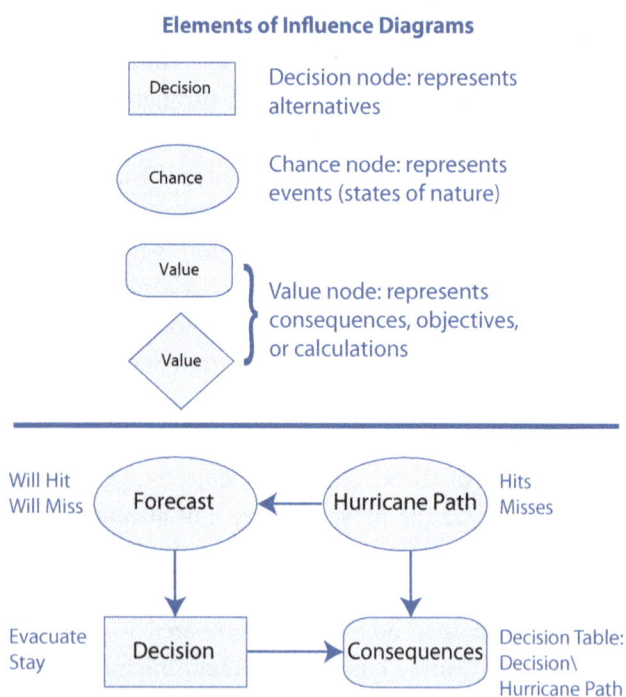

Elements of Influence Diagrams

Decision node: represents alternatives

Chance node: represents events (states of nature)

Value node: represents consequences, objectives, or calculations

Figure 6.8-6 Influence diagrams

In even moderately complicated problems, decision trees can quickly become difficult to understand. Figure 6.8-7 shows a sample of a decision tree. This figure only shows a simplified illustration. A complete decision tree with additional branches would be expanded to the appropriate level of detail as required by the analysis. A commonly employed strategy is to start with an equivalent influence diagram. This often aids in helping to understand the principal issues involved. Some software packages make it easy to develop an influence diagram and then, based on the influence diagram, automatically furnish a decision tree. The decision tree can be edited if this is desired. Calculations are typically based on the decision tree itself.

6.8.2.6 Multi-Criteria Decision Analysis

Multi-Criteria Decision Analysis (MCDA) is a method aimed at supporting decisionmakers who are faced with making numerous and conflicting evaluations. These techniques aim at highlighting the conflicts in alternatives and deriving a way to come to a compromise in a transparent process. For example, NASA may apply MCDA to help assess whether selection of one set of software tools for every NASA application is cost effective. MCDA involves a certain element of subjectiveness; the bias and position of the team implementing MCDA play a significant part in the accuracy and fairness of decisions. One of the MCDA methods is the Analytic Hierarchy Process (AHP).

The Analytic Hierarchy Process

AHP was first developed and applied by Thomas Saaty. AHP is a multi-attribute methodology that provides a proven, effective means to deal with complex decision-making and can assist with identifying and weighting selection criteria, analyzing the data collected for the criteria, and expediting the decisionmaking process. Many different problems can be investigated with the mathematical techniques of this approach. AHP helps capture both subjective and objective evaluation measures, providing a useful mechanism for checking the consistency of the evaluation measures and alternatives suggested by the team, and thus reducing bias in decisionmaking. AHP is supported by pair-wise comparison techniques, and it can support the entire decision process. AHP is normally done in six steps:

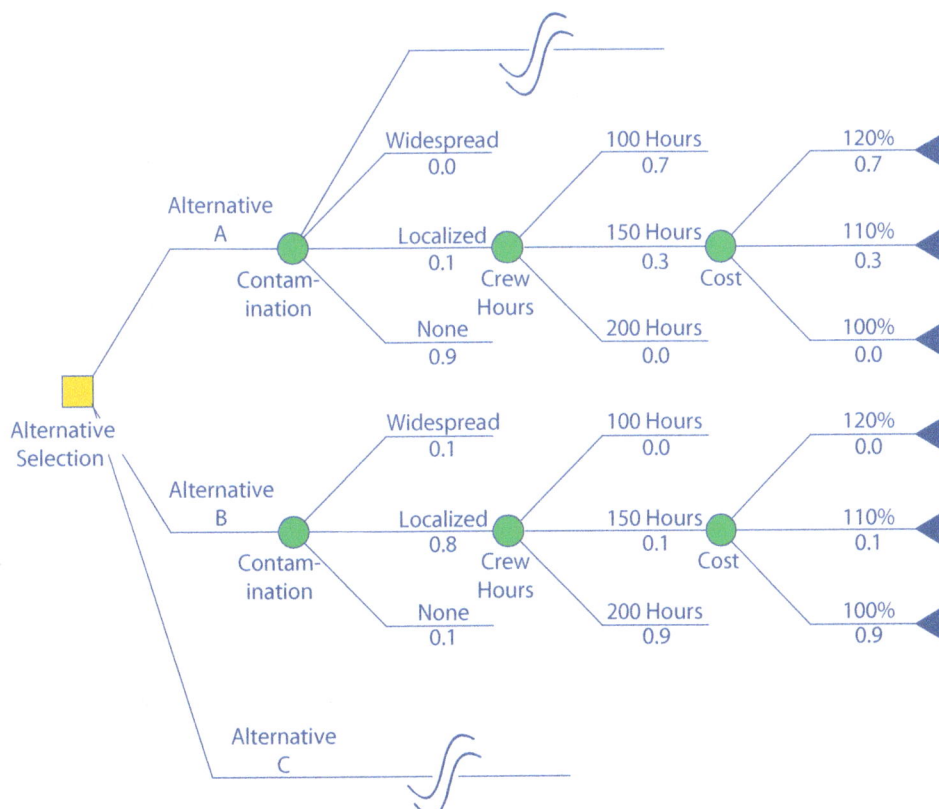

Figure 6.8-7 Decision tree

1. Describe in summary form the alternatives under consideration.

2. Develop a set of high-level objectives.

3. Decompose the high-level objective from general to specific to produce an objectives hierarchy.

4. Determine the relative importance of the evaluation objectives and attributes by assigning weights arrived at by engaging experts through a structured process such as interviews or questionnaires.

5. Have each expert make pair-wise comparisons of the performance of each decision alternative with respect to a TPM. Repeat this for each TPM. Combine the results of these subjective evaluations mathematically using a process or, commonly, an available software tool that ranks the alternatives.

6. Iterate the interviews/questionnaires and AHP evaluation process until a consensus ranking of the alternatives is achieved.

If AHP is used only to produce the TPM weights to be used in a PI or MOE calculation, then only the first four steps listed above are applicable.

With AHP, consensus may be achieved quickly or several feedback rounds may be required. The feedback consists of reporting the computed ranking, for each evaluator and for the group, for each option, along with the reasons for differences in rankings, and identified areas of divergence. Experts may choose to change their judgments on TPM weights. At this point, divergent preferences can be targeted for more detailed study. AHP assumes the existence of an underlying preference vector with magnitudes and directions that are revealed through the pairwise comparisons. This is a powerful assumption, which may at best hold only for the participating experts. The ranking of the alternatives is the result of the experts' judgments and is not necessarily a reproducible result. For further information on AHP, see references by Saaty, *The Analytic Hierarchy Process*.

Flexibility and Extensibility Attributes

In some decision situations, the selection of a particular decision alternative will have implications for the long term that are very difficult to model in the present. In such cases, it is useful to structure the problem as a series of linked decisions, with some decisions to be made in the near future and others to be made later, perhaps on the basis of information to be obtained in the meantime.

There is value in delaying some decisions to the future, when additional information will be available. Some technology choices might foreclose certain opportunities that would be preserved by other choices.

In these cases, it is desirable to consider attributes such as "flexibility" and "extensibility." Flexibility refers to the ability to support more than one current application. Extensibility refers to the ability to be extended to other applications. For example, in choosing an architecture to support lunar exploration, one might consider extensibility to Mars missions. A technology choice that imposes a hard limit on mass that can be boosted into a particular orbit has less flexibility than a choice that is more easily adaptable to boost more. Explicitly adding extensibility and flexibility as attributes to be weighted and evaluated allows these issues to be addressed systematically. In such applications, extensibility and flexibility are being used as *surrogates* for certain future performance attributes.

6.8.2.7 Utility Analysis

"Utility" is a measure of the relative value gained from an alternative. Given this measure, the team looks at increasing or decreasing utility, and thereby explain alternative decisions in terms of attempts to increase their utility. The theoretical unit of measurement for utility is the util.

The utility function maps the range of the TPM into the range of associated utilities, capturing the decisionmaker's preferences and risk attitude. It is possible to imagine simply mapping the indicated range of values linearly onto the interval [0,1] on the utility axis, but in general, this would not capture the decisionmaker's preferences. The decisionmaker's attitude toward risk causes the curve to be convex (risk prone), concave (risk averse), or even some of each.

The utility function directly reflects the decisionmaker's attitude toward risk. When ranking alternatives on the basis of utility, a risk-averse decisionmaker will rank an alternative with highly uncertain performance below an alternative having the same expected performance but less uncertainty. The opposite outcome would result for a risk-prone decisionmaker. When the individual TPM utility functions have been assessed, it is important to check the result for consistency with the decisionmaker's actual preferences (e.g., is it true that intermediate values of TPM_1 and TPM_2 are preferred to a high value of TPM_1 and a low value of TPM_2).

An example of a utility function for the TPM "volume" is shown in Figure 6.8-8. This measure was developed in the context of design of sensors for a space mission. Volume was a precious commodity. The implication of the graph is that low volume is good, large volume is bad, and the decisionmaker would prefer a design alternative with a very well-determined volume of a few thousand cc's to an alternative with the same expected volume but large uncertainty.

Utility

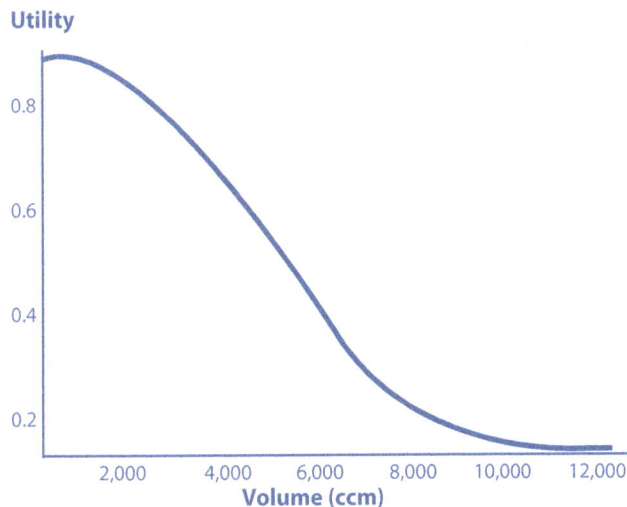

Figure 6.8-8 Utility function for a "volume" performance measure

Value functions can take the place of utility functions when a formal treatment of risk attitude is unnecessary. They appear very similar to utility functions, but have one important difference. Value functions do not consider the risk attitude of the decisionmaker. They do not reflect how the decisionmaker compares certain outcomes to uncertain outcomes.

The assessment of a TPM's value function is relatively straightforward. The "best" end of the TPM's range is assigned a value of 1. The "worst" is assigned a value of 0. The decisionmaker makes direct assessments of the value of intermediate points to establish the preference structure in the space of possible TPM values. The utility function can be treated as a value function, but the value function is not necessarily a utility function.

One way to rank alternatives is to use a Multi-Attribute, Utility Theory (MAUT) approach. With this approach, the "expected utility" of each alternative is quantified, and alternatives are ranked based on their expected utilities.

Sometimes the expected utility is referred to as a PI. An important benefit of applying this method is that it is the best way to deal with significant uncertainties when the decisionmaker is not risk neutral. Probabilistic methods are used to treat uncertainties. A downside of applying this method is the need to quantify the decisionmaker's risk attitudes. Top-level system architecture decisions are natural examples of appropriate applications of MAUT.

6.8.2.8 Risk-Informed Decision Analysis Process Example

Introduction

A decision matrix works for many decisions, but the decision matrix may not scale up to very complex decisions or risky decisions. For some decisions, a tool is needed to handle the complexity. The following subsection describes a detailed Decision Analysis Process that can be used to support a risk-informed decision.

In practice, decisions are made in many different ways. Simple approaches may be useful, but it is important to recognize their limitations and upgrade to better analysis when this is warranted. Some decisionmakers, when faced with uncertainty in an important quantity, determine a best estimate for that quantity, and then reason as if the best estimate were correct. This might be called the "take-your-best-shot" approach. Unfortunately, when the stakes are high, and uncertainty is significant, this best-shot approach may lead to poor decisions.

The following steps are a risk-informed decision analysis process:

1. Formulation of the objectives hierarchy, TPMs.
2. Proposing and identifying decision alternatives. Alternatives from this process are combined with the alternatives identified in the other systems engineering processes including the Design Solution Definition Process, but also including verification and validation as well as production.
3. Risk analysis of decision alternatives and ranking of alternatives.
4. Deliberation and recommendation of decision alternatives.
5. Followup tracking of the implementation of the decision.

These steps support good decisions by focusing first on objectives, next on developing decision alternatives with those objectives clearly in mind and/or using decision alternatives that have been developed under other systems engineering processes. The later steps of the Decision Analysis Process interrelate heavily with the Technical Risk Management Process, as indicated in Figure 6.8-9. These steps include risk analysis of the decision alternatives, deliberation informed by risk analysis results, and recommendation of a decision alternative to the decisionmaker. Implementation of the decision is also important.

Objectives Hierarchy/TPMs

As shown in Figure 6.8-9, risk-informed decision analysis starts with formulation of the objectives hierarchy. Using this hierarchy, TPMs are formulated to quantify performance of a decision with respect to the program objectives. The TPMs should have the following characteristics:

- They can support ranking of major decision alternatives.
- They are sufficiently detailed to be used directly in the risk management process.
- They are preferentially independent. This means that they contribute in distinct ways to the program goal. This property helps to ensure that alternatives are ranked appropriately.

An example of an objectives hierarchy is shown in Figure 6.8-10. Details will vary from program to program, but a construct like Figure 6.8-10 is behind the program-specific objectives hierarchy.

The TPMs in this figure are meant to be generically important for many missions. Depending on the mission, these TPMs are further subdivided to the point where they can be objectively measured. Not all TPMs can be measured directly. For example, safety-related TPMs are

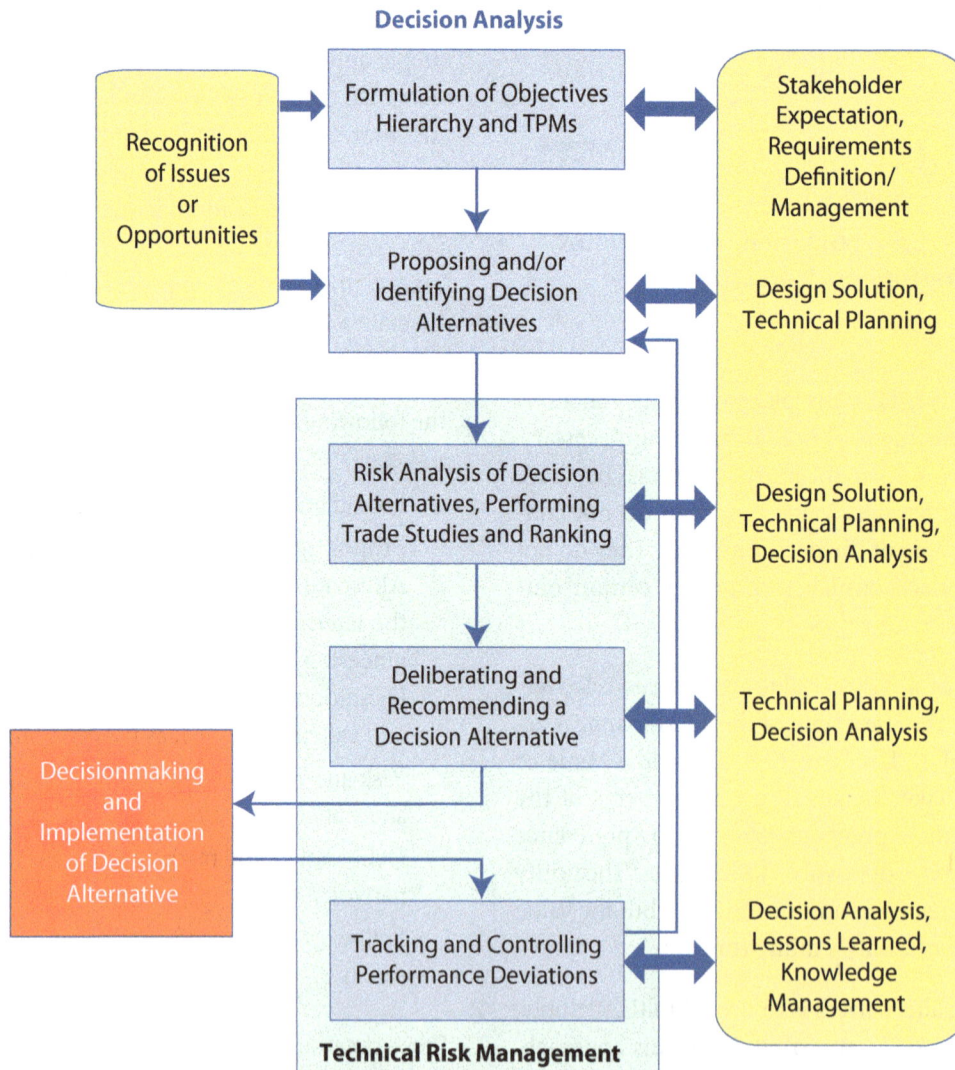

Figure 6.8-9 Risk-informed Decision Analysis Process

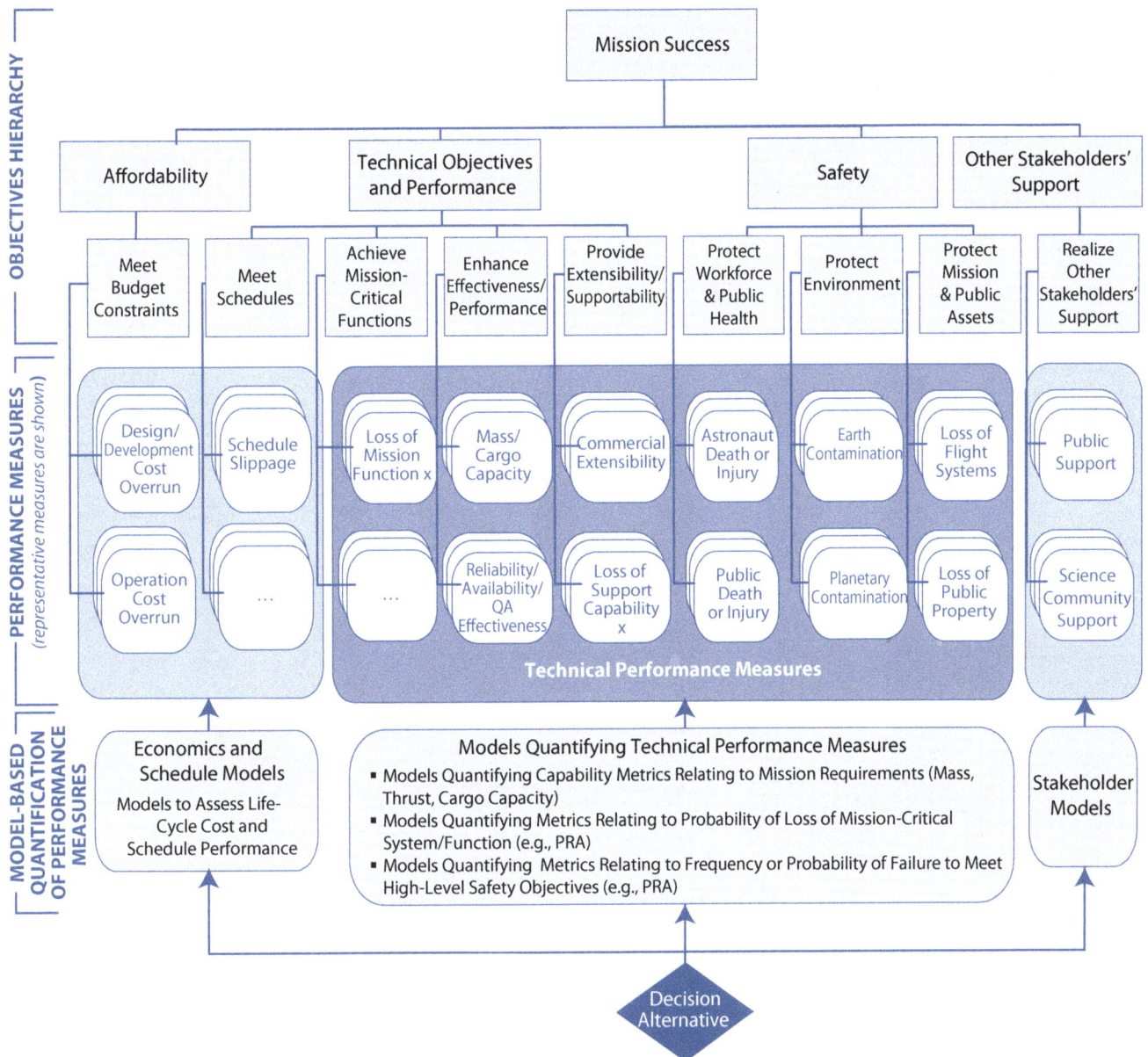

Figure 6.8-10 Example of an objectives hierarchy

defined in terms of the probability of a consequence type of a specific magnitude (e.g., probability of any general public deaths or injuries) or the expected magnitude of a consequence type (e.g., the number of public deaths or injuries). Probability of Loss of Mission and Prob-

ability of Loss of Crew (P(LOM) and P(LOC)) are two particularly important safety-related TPMs for manned space missions. Because an actuarial basis does not suffice for prediction of these probabilities, modeling will be needed to quantify them.

7.0 Special Topics

The topics below are of special interest for enhancing the performance of the systems engineering process or constitute special considerations in the performance of systems engineering. The first section elucidates the process of how the systems engineering principles need to be applied to contracting and contractors that implement NASA processes and create NASA products.

Applying lessons learned enhances the efficiency of the present with the wisdom of the past. Protecting the environment and the Nation's space assets are important considerations in the design and development of requirements and designs. Integrated design can enhance the efficiency and effectiveness of the design process.

7.1 Engineering with Contracts

7.1.1 Introduction, Purpose, and Scope

Historically, most successful NASA projects have depended on effectively blending project management, systems engineering, and technical expertise among NASA, contractors, and third parties. Underlying these successes are a variety of agreements (e.g., contract, memorandum of understanding, grant, cooperative agreement) between NASA organizations or between NASA and other Government agencies, Government organizations, companies, universities, research laboratories, and so on. To simplify the discussions, the term "contract" is used to encompass these agreements.

This section focuses on the engineering activities pertinent to awarding a contract, managing contract performance, and completing a contract. However, interfaces to the procurement process will be covered, since the engineering technical team plays a key role in development and evaluation of contract documentation.

Contractors and third parties perform activities that supplement (or substitute for) the NASA project technical team accomplishment of the common technical process activities and requirements. Since contractors might be involved in any part of the systems engineering life cycle, the NASA project technical team needs to know how to prepare for, perform, and complete surveillance of technical activities that are allocated to contractors.

7.1.2 Acquisition Strategy

Creating an acquisition strategy for a project is a collaborative effort among several NASA HQ offices that leads to approval for project execution. The program and project offices characterize the acquisition strategy in sufficient detail to identify the contracts needed to execute the strategy. Awarding contracts at the project level occurs in the context of the overall program acquisition strategy.

While this section pertains to projects where the decision has been made to have a contractor implement a portion of the project, it is important to remember that the choice between "making" a product in-house by NASA or "buying" it from a contractor is one of the most crucial decisions in systems development. (See Section 5.1.) Questions that should be considered in the "make/buy" decision include the following:

- Is the desired system a development item or more off the shelf?
- What is the relevant experience of NASA versus potential contractors?
- What are the relative importance of risk, cost, schedule, and performance?
- Is there a desire to maintain an "in-house" capability?

As soon as it is clear that a contract will be needed to obtain a system or service, the responsible project manager should contact the local procurement office. The contracting officer will assign a contract specialist to navigate the numerous regulatory requirements that affect NASA procurements and guide the development of contract documentation needed to award a contract. The contract specialist engages the local legal office as needed.

7.1.2.1 Develop an Acquisition Strategy

The project manager, assisted by the assigned procurement and legal offices, first develops a project acquisition strategy or verifies the one provided. The acquisition strategy provides a business and technical management outline for planning, directing, and managing a project and obtaining products and services via contract.

In some cases, it may be appropriate to probe outside sources in order to gather sufficient information to formulate an acquisition strategy. This can be done by issuing a Request for Information (RFI) to industry and other parties that may have interest in potential future contracts. An RFI is a way to obtain information about technology maturity, technical challenges, capabilities, price and delivery considerations, and other market information that can influence strategy decisions.

The acquisition strategy includes:

- Objectives of the acquisition—capabilities to be provided, major milestones;
- Acquisition approach—single step or evolutionary (incremental), single or multiple suppliers/contracts, competition or sole source, funding source(s), phases, system integration, Commercial-Off-the-Shelf (COTS) products;
- Business considerations—constraints (e.g., funding, schedule), availability of assets and technologies, applicability of commercial items versus internal technical product development;
- Risk management of acquired products or services—major risks and risk sharing with the supplier;
- Contract types—performance-based or level of effort, fixed-price or cost reimbursable;
- Contract elements—incentives, performance parameters, rationale for decisions on contract type; and
- Product support strategy—oversight of delivered system, maintenance, and improvements.

The technical team gathers data to facilitate the decision-making process regarding the above items. The technical team knows about issues with the acquisition approach, determining availability of assets and technologies, applicability of commercial items, issues with system in-

tegration, and details of product support. Similarly, the technical team provides corporate knowledge to identify and evaluate risks of acquiring the desired product, especially regarding the proposed contract type and particular contract elements.

7.1.2.2 Acquisition Life Cycle

Contract activities are part of the broader acquisition life cycle, which comprises the phases solicitation, source selection, contract monitoring, and acceptance. (See Figure 7.1-1.) The acquisition life cycle overlaps and interfaces with the systems engineering processes in the project life cycle. Acquisition planning focuses on technical planning when a particular contract (or purchase) is required. (See Section 6.1.) In the figure below, requirements development corresponds to the Technical Requirements Definition Process in the systems engineering engine. (See Figure 2.1-1.) The next four phases—solicitation, source selection, contract monitoring, and acceptance—are the phases of the contract activities. Transition to operations and maintenance represents activities performed to transition acquired products to the organization(s) responsible for operating and maintaining them (which could be contractor(s)). Acquisition management refers to project management activities that are performed throughout the acquisition life cycle by the acquiring organization.

7.1.2.3 NASA Responsibility for Systems Engineering

The technical team is responsible for systems engineering throughout the acquisition life cycle. The technical team contributes heavily to systems engineering decisions and results, whatever the acquisition strategy, for any combination of suppliers, contractors, and subcontractors. The technical team is responsible for systems engineering whether the acquisition strategy calls for the technical team, a prime contractor, or some combination of the two to perform system integration and testing of products from multiple sources.

This subsection provides specific guidance on how to assign responsibility when translating the technical processes onto a contract. Generally, the Technical Planning,

Acquisition Planning	Requirements Development	Solicitation	Source Selection	Contract Monitoring	Acceptance	Transition to Operations & Maintenance

Figure 7.1-1 Acquisition life cycle

Interface Management, Technical Risk Management, Configuration Management, Technical Data Management, Technical Assessment, and Decision Analysis processes should be implemented throughout the project by both the NASA team and the contractor. Stakeholder Expectations Definition, Technical Requirements Definition, Logical Decomposition, Design Solution Definition, Product Implementation and Integration, Product Verification and Validation, Product Transition, and Requirements Management Processes are implemented by NASA or the contractor depending upon the level of the product decomposition.

Table 7.1-1 provides guidance on how to implement the 17 technical processes from NPR 7123.1. The first two columns have the number of the technical process and the requirement statement of responsibility. The next column provides general guidance on how to distinguish who has responsibility for implementing the process. The last column provides a specific example of the application of how to implement the process for a particular project. The particular scenario is a science mission where a contractor is building the spacecraft, NASA assigns Government-Furnished Property (GFP) instruments to the contractor, and NASA operates the mission.

7.1.3 Prior to Contract Award

7.1.3.1 Acquisition Planning

Based on the acquisition strategy, the technical team needs to plan acquisitions and document the plan in developing the SEMP. The SEMP covers the technical team's involvement in the periods before contract award, during contract performance, and upon contract completion. Included in acquisition planning are solicitation preparation, source selection activities, contract phase-in, monitoring contractor performance, acceptance of deliverables, completing the contract, and transition beyond the contract. The SEMP focuses on interface activities with the contractor, including NASA technical team involvement with and monitoring of contracted work.

Often overlooked in project staffing estimates is the amount of time that technical team members are involved in contracting-related activities. Depending on the type of procurement, a technical team member involved in source selection could be consumed nearly full time for 6 to 12 months. After contract award, technical monitoring consumes 30 to 50 percent, peaking at full

time when critical milestones or key deliverables arrive. Keep in mind that for most contractor activities, NASA staff performs supplementary activities.

The technical team is intimately involved in developing technical documentation for the acquisition package. The acquisition package consists of the solicitation (e.g., Request for Proposals (RFPs) and supporting documents. The solicitation contains all the documentation that is advertised to prospective contractors (or offerors). The key technical sections of the solicitation are the SOW (or performance work statement), technical specifications, and contract data requirements list. Other sections of the solicitation include proposal instructions and evaluation criteria. Documents that support the solicitation include a procurement schedule, source evaluation plan, Government cost estimate, and purchase request. Input from the technical team will be needed for some of the supporting documents.

It is the responsibility of the contract specialist, with input from the technical team, to ensure that the appropriate clauses are included in the solicitation. The contract specialist is familiar with requirements in the Federal Acquisition Regulation (FAR) and the NASA FAR Supplement (NFS) that will be included in the solicita-

---

Solicitations

The release of a solicitation to interested parties is the formal indication of a future contract. A solicitation conveys sufficient details of a Government need (along with terms, conditions, and instructions) to allow prospective contractors (or offerors) to respond with a proposal. Depending on the magnitude and complexity of the work, a draft solicitation may be issued. After proposals are received, a source evaluation board (or committee) evaluates technical and business proposals per its source evaluation plan and recommends a contractor selection to the contracting officer. The source evaluation board, led by a technical expert, includes other technical experts and a contracting specialist. The source selection process is completed when the contracting officer signs the contract.

The most common NASA solicitation types are RFP and Announcement of Opportunity (AO). Visit the online NASA Procurement Library for a full range of details regarding procurements and source selection.

Table 7.1-1 Applying the Technical Processes on Contract

#	NPR 7123.1 Process	General Guidance on Who Implements the Process	Application to a Science Mission
1	The Center Directors or designees establish and maintain a process to include activities, requirements, guidelines, and documentation for the definition of stakeholder expectations for the applicable WBS model.	If stakeholders are at the contractor, then the contractor should have responsibility and vice versa.	Stakeholders for the mission/project are within NASA; stakeholders for the spacecraft power subsystem are mostly at the contractor.
2	The Center Directors or designees establish and maintain a process to include activities, requirements, guidelines, and documentation for definition of the technical requirements from the set of agreed-upon stakeholder expectations for the applicable WBS model.	Assignment of responsibility follows the stakeholders, e.g., if stakeholders are at the contractor, then requirements are developed by the contractor and vice versa.	NASA develops the high-level requirements, and the contractor develops the requirements for the power subsystem.
3	The Center Directors or designees establish and maintain a process to include activities, requirements, guidelines, and documentation for logical decomposition of the validated technical requirements of the applicable WBS.	Follows the requirements, e.g., if requirements are developed at the contractor, then the decomposition of those requirements is implemented by the contractor and vice versa.	NASA performs the decomposition of the high-level requirements, and the contractor performs the decomposition of the power subsystem requirements.
4	The Center Directors or designees establish and maintain a process to include activities, requirements, guidelines, and documentation for designing product solution definitions within the applicable WBS model that satisfy the derived technical requirements.	Follows the requirements, e.g., if requirements are developed at the contractor, then the design of the product solution is implemented by the contractor and vice versa.	NASA designs the mission/project, and the contractor designs the power subsystem.
5	The Center Directors or designees establish and maintain a process to include activities, requirements, guidelines, and documentation for implementation of a design solution definition by making, buying, or reusing an end product of the applicable WBS model.	Follows the design, e.g., if the design is developed at the contractor, then the implementation of the design is performed by the contractor and vice versa.	NASA implements (and retains responsibility for) the design for the mission/project, and the contractor does the same for the power subsystem.
6	The Center Directors or designees establish and maintain a process to include activities, requirements, guidelines, and documentation for the integration of lower level products into an end product of the applicable WBS model in accordance with its design solution definition.	Follows the design, e.g., if the design is developed at the contractor, then the integration of the design elements is performed by the contractor and vice versa.	NASA integrates the design for the mission/project, and the contractor does the same for the power subsystem.
7	The Center Directors or designees establish and maintain a process to include activities, requirements, guidelines, and documentation for verification of end products generated by the Product Implementation Process or Product Integration Process against their design solution definitions.	Follows the product integration, e.g., if the product integration is implemented at the contractor, then the verification of the product is performed by the contractor and vice versa.	NASA verifies the mission/project, and the contractor does the same for the power subsystem.

(continued)

Table 7.1-1 Applying the Technical Processes on Contract (continued)

#	NPR 7123.1 Process	General Guidance on Who Implements the Process	Application to a Science Mission
8	The Center Directors or designees establish and maintain a process to include activities, requirements, guidelines, and documentation for validation of end products generated by the Product Implementation Process or Product Integration Process against their stakeholder expectations.	Follows the product integration, e.g., if the product integration is implemented at the contractor, then the validation of the product is performed by the contractor and vice versa.	NASA validates the mission/project, and the contractor does the same for the power subsystem.
9	The Center Directors or designees establish and maintain a process to include activities, requirements, guidelines, and documentation for transitioning end products to the next-higher-level WBS model customer or user.	Follows the product verification and validation, e.g., if the product verification and validation is implemented at the contractor, then the transition of the product is performed by the contractor and vice versa.	NASA transitions the mission/project to operations, and the contractor transitions the power subsystem to the spacecraft level.
10	The Center Directors or designees establish and maintain a process to include activities, requirements, guidelines, and documentation for planning the technical effort.	Assuming both NASA and the contractor have technical work to perform, then both NASA and the contractor need to plan their respective technical efforts.	NASA would plan the technical effort associated with the GFP instruments and the launch and operations of the spacecraft, and the contractor would plan the technical effort associated with the design, build, verification and validation, and delivery and operations of the power subsystem.
11	The Center Directors or designees establish and maintain a process to include activities, requirements, guidelines, and documentation for **management of requirements defined and baselined during the application of the system design processes**.	Follows process #2.	
12	The Center Directors or designees establish and maintain a process to include activities, requirements, guidelines, and documentation for **management of the interfaces defined and generated during the application of the system design processes**.	Interfaces should be managed one level above the elements being interfaced.	The interface from the spacecraft to the project ground system would be managed by NASA, while the power subsystem to attitude control subsystem interface would be managed by the contractor.
13	The Center Directors or designees establish and maintain a process to include activities, requirements, guidelines, and documentation for **management of the technical risk identified during the technical effort**. *NPR 8000.4, Risk Management Procedural Requirements* is to be used as a source document for defining this process; and *NPR 8705.5, Probabilistic Risk Assessment (PRA) Procedures for NASA Programs and Projects* provides one means of identifying and assessing technical risk.	Technical risk management is a process that needs to be implemented by both NASA and the contractor. All elements of the project need to identify their risks and participate in the project risk management process. Deciding which risks to mitigate, when, at what cost is generally a function of NASA project management.	NASA project management should create a project approach to risk management that includes participation from the contractor. Risks identified throughout the project down to the power subsystem level and below should be identified and reported to NASA for possible mitigation.

(continued)

Table 7.1-1 Applying the Technical Processes on Contract (continued)

#	NPR 7123.1 Process	General Guidance on Who Implements the Process	Application to a Science Mission
14	The Center Directors or designees establish and maintain a process to include activities, requirements, guidelines, and documentation for **CM**.	Like risk management, CM is a process that should be implemented throughout the project by both the NASA and contractor teams.	NASA project management should create a project approach to CM that includes participation from the contractor. The contractor's internal CM process will have to be integrated with the NASA approach. CM needs to be implemented throughout the project down to the power subsystem level and below.
15	The Center Directors or designees establish and maintain a process to include activities, requirements, guidelines, and documentation for **management of the technical data generated and used in the technical effort**.	Like risk management and CM, technical data management is a process that should be implemented throughout the project by both the NASA and contractor teams.	NASA project management should create a project approach to technical data management that includes participation from the contractor. The contractor's internal technical data process will have to be integrated with the NASA approach. Management of technical data needs to be implemented throughout the project down to the power subsystem level and below.
16	The Center Directors or designees establish and maintain a process to include activities, requirements, guidelines, and documentation for **making assessments of the progress of planned technical effort and progress toward requirements satisfaction**.	Assessing progress is a process that should be implemented throughout the project by both the NASA and contractor teams.	NASA project management should create a project approach to assessing progress that includes participation from the contractor. Typically this would be the project review plan. The contractor's internal review process will have to be integrated with the NASA approach. Technical reviews need to be implemented throughout the project down to the power subsystem level and below.
17	The Center Directors or designees establish and maintain a process to include activities, requirements, guidelines, and documentation for **making technical decisions**.	Clearly technical decisions are made throughout the project both by NASA and contractor personnel. Certain types of decisions or decisions on certain topics may best be made by either NASA or the contractor depending upon the Center's processes and the type of project.	For this example, decisions affecting high-level requirements or mission success would be made by NASA and those at the lower level, e.g., the power subsystem that did not affect mission success would be made by the contractor.

tion as clauses in full text form or as clauses incorporated by reference. Many of these clauses relate to public laws, contract administration, and financial management. Newer clauses address information technology security, data rights, intellectual property, new technology reporting, and similar items. The contract specialist stays abreast of updates to the FAR and NFS. As the SOW and other parts of the solicitation mature, it is important for the contract specialist and technical team to work closely to avoid duplication of similar requirements.

7.1.3.2 Develop the Statement of Work

Effective surveillance of a contractor begins with the development of the SOW. The technical team establishes the SOW requirements for the product to be developed. The SOW contains process, performance, and management requirements the contractor must fulfill during product development.

As depicted in Figure 7.1-2, developing the SOW requires the technical team to analyze the work, performance, and data needs to be accomplished by the contractor. The process is iterative and supports the development of other documentation needed for the contracting effort. The principal steps in the figure are discussed further in Table 7.1-2.

After a few iterations, baseline the SOW requirements and place them under configuration management. (See Section 6.5.)

Use the SOW checklist, which is in Appendix P, to help ensure that the SOW is complete, consistent, correct, unambiguous, and verifiable. Below are some key items to require in the SOW:

- Technical and management deliverables having the highest risk potential (e.g., the SEMP, development and transition plans); requirements and architecture specifications; test plans, procedures and reports; metrics reports; delivery, installation, and maintenance documentation.

- Contractual or scheduling incentives in a contract should not be tied to the technical milestone reviews. These milestone reviews (for example, SRR, PDR, CDR, etc.) enable a critical and valuable technical assessment to be performed. These reviews have specific entrance criteria that should not be waived. The reviews should be conducted when these criteria are met, rather than being driven by a particular schedule.

- Timely electronic access to data, work products, and interim deliverables to assess contractor progress on final deliverables.

- Provision(s) to flow down requirements to subcontractors and other team members.

- Content and format requirements of deliverables in the contract data requirements list. These requirements are specified in a data requirements document

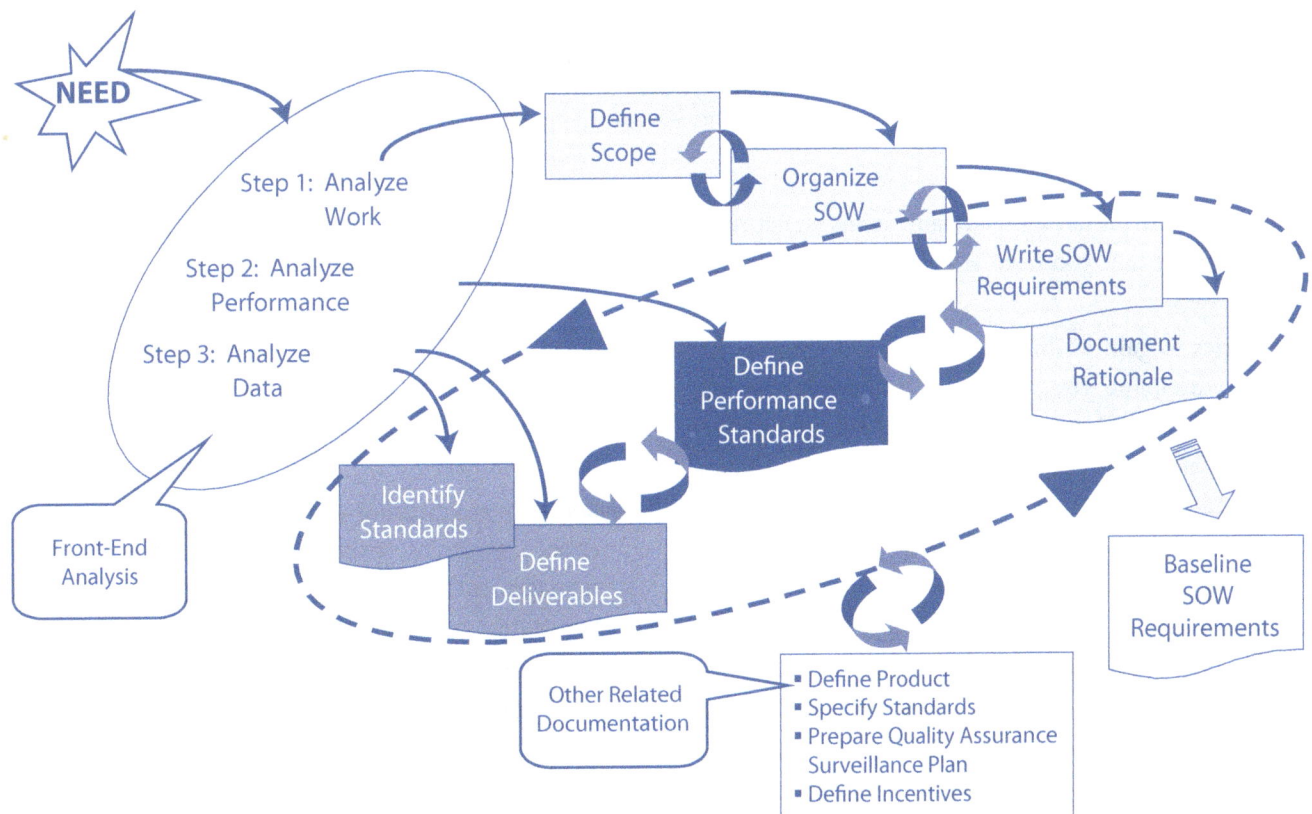

Figure 7.1-2 Contract requirements development process

Table 7.1-2 Steps in the Requirements Development Process

Step	Task	Detail
Step 1: Analyze the Work	Define scope	Document in the SOW that part of the project's scope that will be contracted. Give sufficient background information to orient offerors.
	Organize SOW	Organize the work by products and associated activities (i.e., product WBS).
	Write SOW requirements	Include activities necessary to: ● Develop products defined in the requirements specification; and ● Support, manage, and oversee development of the products. Write SOW requirements in the form "the Contractor shall." Write product requirements in the form "the system shall."
	Document rationale	Document separately from the SOW the reason(s) for including requirements that may be unique, unusual, controversial, political, etc. The rationale is not part of the solicitation.
Step 2: Analyze Performance	Define performance standards	Define what constitutes acceptable performance by the contractor. Common metrics for use in performance standards include cost and schedule. For guidance on metrics to assess the contractor's performance and to assess adherence to product requirements on delivered products, refer to *System and Software Metrics for Performance-Based Contracting*.
Step 3: Analyze Data	Identify standards	Identify standards (e.g., EIA, IEEE, ISO) that apply to deliverable work products including plans, reports, specifications, drawings, etc. Consensus standards and codes (e.g., National Electrical Code, National Fire Protection Association, American Society of Mechanical Engineers) that apply to product development and workmanship are included in specifications.
	Define deliverables	Ensure each deliverable data item (e.g., technical data—requirements specifications, design documents; management data—plans, metrics reports) has a corresponding SOW requirement for its preparation. Ensure each product has a corresponding SOW requirement for its delivery.

or data item description, usually as an attachment. Remember that you need to be able to edit data deliverables.

● Metrics to gain visibility into technical progress for each discipline (e.g., hardware, software, thermal, optics, electrical, mechanical). For guidance on metrics to assess the contractor's performance and to assess adherence to product requirements on delivered products, refer to *System and Software Metrics for Performance-Based Contracting*.

● Quality incentives (defect, error count, etc.) to reduce risk of poor quality deliverables. Be careful because incentives can affect contractor behavior. For example, if you reward early detection and correction of software defects, the contractor may expend effort correcting minor defects and saving major defects for later.

● A continuous management program to include a periodically updated risk list, joint risk reviews, and vendor risk approach.

● Surveillance activities (e.g., status meetings, reviews, audits, site visits) to monitor progress and production, especially access to subcontractors and other team members.

● Specialty engineering (e.g., reliability, quality assurance, cryogenics, pyrotechnics, biomedical, waste management) that is needed to fulfill standards and verification requirements.

● Provisions to assign responsibilities between NASA and contractor according to verification, validation, or similar plans that are not available prior to award.

● Provisions to cause a contractor to disclose changing a critical process. If a process is critical to human safety, require the contractor to obtain approval from the contracting officer before a different process is implemented.

> Note: If you neglect to require something in the SOW, it can be costly to add it later.

The contractors must supply a SEMP that specifies their systems engineering approach for requirements development, technical solution definition, design realization, product evaluation, product transition, and technical planning, control, assessment, and decision analysis. It is best to request a preliminary SEMP in the solicitation. The source evaluation board can use the SEMP to evaluate the offeror's understanding of the requirements, as well as the offeror's capability and capacity to deliver the system. After contract award, the technical team can eliminate any gaps between the project's SEMP and the contractor's SEMP that could affect smooth execution of the integrated set of common technical processes.

Often a technical team has experience developing technical requirements, but little or no experience developing SOW requirements. If you give the contractor a complex set of technical requirements, but neglect to include sufficient performance measures and reporting requirements, you will have difficulty monitoring progress and determining product and process quality. Understanding performance measures and reporting requirements will enable you to ask for the appropriate data or reports that you intend to use.

Traditionally, NASA contracts require contractors to satisfy requirements in NASA policy directives, NASA procedural requirements, NASA standards, and similar documents. These documents are almost never written in language that can be used directly in a contract. Too often, these documents contain requirements that do not apply to contracts. So, before the technical team boldly goes where so many have gone before, it is a smart idea to understand what the requirements mean and if they apply to contracts. The requirements that apply to contracts need to be written in a way that is suitable for contracts.

7.1.3.3 Task Order Contracts

Sometimes, the technical team can obtain engineering products and services through an existing task order contract. The technical team develops a task order SOW and interacts with the contracting officer's technical representative to issue a task order. Preparing the task order SOW is simplified because the contract already establishes baseline requirements for execution. First-time users need to understand the scope of the contract and the degree to which delivery and reporting requirements, performance metrics, incentives, and so forth are already covered. Task contracts offer quick access (days or weeks instead of months) to engineering services for studies, analyses, design, development, and testing and to support services for configuration management, quality assurance, maintenance, and operations. Once a task order is issued, the technical team performs engineering activities associated with managing contract performance and completing a contract (discussed later) as they apply to the task order.

7.1.3.4 Surveillance Plan

The surveillance plan defines the monitoring of the contractor effort and is developed at the same time as the SOW. The technical team works with mission assurance personnel, generally from the local Safety and Mission Assurance (SMA) organization, to prepare the surveillance plan for the contracted effort. Sometimes mission assurance is performed by technical experts on the project. In either case, mission assurance personnel should be engaged from the start of the project. Prior to contract award, the surveillance plan is written at a general level to cover the Government's approach to perceived programmatic risk. After contract award, the surveillance plan describes in detail inspection, testing, and other quality-related surveillance activities that will be performed to ensure the integrity of contract deliverables, given the current perspective on programmatic risks.

Recommended items to include in the surveillance plan follow:

- Review key deliverables within the first 30 days to ensure adequate startup of activities.
- Conduct contractor/subcontractor site visits to monitor production or assess progress.
- Evaluate effectiveness of the contractor's systems engineering processes.

Drafting the surveillance plan when the SOW is developed promotes the inclusion of key requirements in the SOW that enable activities in the surveillance plan. For example, in order for the technical team to conduct site visits to monitor production of a subcontractor, then the SOW must include a requirement that permits site visits, combined with a requirement for the contractor to flow down requirements that directly affect subcontractors.

7.1.3.5 Writing Proposal Instructions and Evaluation Criteria

Once the technical team has written the SOW, the Government cost estimate, and the preliminary surveillance plan and updated the SEMP, the solicitation can be developed. Authors of the solicitation must understand the information that will be needed to evaluate the proposals and write instructions to obtain specifically needed information. In a typical source selection, the source selection board evaluates the offerors' understanding of the requirements, management approach, and cost and their relevant experience and past performance. This information is required in the business and technical proposals. (This section discusses only the technical proposal.) The solicitation also gives the evaluation criteria that the source evaluation board will use. This section corresponds one-for-one to the items requested in the proposal instructions section.

State instructions clearly and correctly. The goal is to obtain enough information to have common grounds for evaluation. The challenge becomes how much information to give the offerors. If you are too prescriptive, the proposals may look too similar. Be careful not to level the playing field too much, otherwise discriminating among offerors will be difficult. Because the technical merits of a proposal compete with nontechnical items of similar importance (e.g., cost), the technical team must wisely choose discriminators to facilitate the source selection.

Source Evaluation Board

One or more members of the technical team serve as members of the source evaluation board. They participate in the evaluation of proposals following applicable NASA and Center source selection procedures. Because source selection is so important, the procurement office works closely with the source evaluation board to ensure that the source selection process is properly executed. The source evaluation board develops a source evaluation plan that describes the evaluation factors, and the method of evaluating the offerors' responses. Unlike decisions made by systems engineers early in a product life cycle, source selection decisions must be carefully managed in accordance with regulations governing the fairness of the selection process.

The source evaluation board evaluates nontechnical (business) and technical items. Items may be evaluated by themselves, or in the context of other technical or nontechnical items. Table 7.1-3 shows technical items to request from offerors and the evaluation criteria with which they correlate.

Evaluation Considerations

The following are important to consider when evaluating proposals:

- Give adequate weight to evaluating the capability of disciplines that could cause mission failure (e.g., hardware, software, thermal, optics, electrical, mechanical).

- Conduct a preaward site visit of production/test facilities that are critical to mission success.

- Distinguish between "pretenders" (good proposal writers) and "contenders" (good performing organizations). Pay special attention to how process descriptions match relevant experience and past performance. While good proposals can indicate good future performance, lesser quality proposals usually predict lesser quality future work products and deliverables.

- Assess the contractor's SEMP and other items submitted with the proposal based on evaluation criteria that include quality characteristics (e.g., complete, unambiguous, consistent, verifiable, and traceable).

The cost estimate that the technical team performs as part of the Technical Planning Process supports evaluation of the offerors' cost proposals, helping the source evaluation board determine the realism of the offerors' technical proposals. (See Section 6.1.) The source evaluation board can determine "whether the estimated proposed cost elements are realistic for the work to be performed; reflect a clear understanding of the requirements; and are consistent with the unique methods of performance and materials described in the offeror's technical proposal."[1]

7.1.3.6 Selection of COTS Products

When COTS products are given as part of the technical solution in a proposal, it is imperative that the selection of a particular product be evaluated and documented by applying the Decision Analysis Process. Bypassing this task or neglecting to document the evaluation suf-

---

[1] FAR 15.404-1(d) (1).

Table 7.1-3 Proposal Evaluation Criteria

Item	Criteria
Preliminary contractor SEMP.	How well the plan can be implemented given the resources, processes, and controls stated. Look at completeness (how well it covers all SOW requirements), internal consistency, and consistency with other proposal items. The SEMP should cover all resources and disciplines needed to meet product requirements, etc.
Process descriptions, including subcontractor's (or team member's) processes.	Effectiveness of processes and compatibility of contractor and sub-contractor processes (e.g., responsibilities, decisionmaking, problem resolution, reporting).
Artifacts (documents) of relevant work completed. Such documentation depicts the probable quality of work products an offeror will provide on your contract. Artifacts provide evidence (or lack) of systems engineering process capability.	Completeness of artifacts, consistency among artifacts on a given project, consistency of artifacts across projects, conformance to standards.
Engineering methods and tools.	Effectiveness of the methods and tools.
Process and product metrics.	How well the offeror measures performance of its processes and quality of its products.
Subcontract management plan (may be part of contractor SEMP).	Effectiveness of subcontract monitoring and control and integration/separation of risk management and CM.
Phase-in plan (may be part of contractor SEMP).	How well the plan can be implemented given the existing workload of resources.

ficiently could lead to a situation where NASA cannot support its position in the event of a vendor protest.

7.1.3.7 Acquisition-Unique Risks

Table 7.1-4 identifies a few risks that are unique to acquisition along with ways to manage them from an engineering perspective. Bear in mind, legal and procurement aspects of these risks are generally covered in contract clauses.

There may also be other acquisition risks not listed in Table 7.1-4. All acquisition risks should be identified and handled the same as other project risks using the Continuous Risk Management (CRM) process. A project can also choose to separate out acquisition risks as a risk-list subset and handle them using the risk-based acquisition management process if so desired.

When the technical team completes the activities prior to contract award, they will have an updated SEMP, the Government cost estimate, an SOW, and a preliminary surveillance plan. Once the contract is awarded, the technical team begins technical oversight.

7.1.4 During Contract Performance

7.1.4.1 Performing Technical Surveillance

Surveillance of a contractor's activities and/or documentation is performed to demonstrate fiscal responsibility, ensure crew safety and mission success, and determine award fees for extraordinary (or penalty fees for substandard) contract execution. Prior to or outside of a contract award, a less formal agreement may be made for the Government to be provided with information for a trade study or engineering evaluation. Upon contract award, it may become necessary to monitor the contractor's adherence to contractual requirements more formally. (For a greater understanding of surveillance requirements, see *NPR 8735.2, Management of Government Quality Assurance Functions for NASA Contracts.*)

Under the authority of the contracting officer, the technical team performs technical surveillance as established in the NASA SEMP. The technical team assesses technical work productivity, evaluates product quality, and conducts technical reviews of the contractor. (Refer to the Technical Assessment Process.) Some of the key activities are discussed below.

Table 7.1-4 Risks in Acquisition

Risk	Mitigation
Supplier goes bankrupt prior to delivery	The source selection process is the strongest weapon. Select a supplier with a proven track record, solid financial position, and stable workforce. As a last resort, the Government may take possession of any materials, equipment, and facilities on the work site necessary for completing the work in-house or via another contract.
Supplier acquired by another supplier with different policies	Determine differences between policies before and after the acquisition. If there is a critical difference, then consult with the procurement and legal offices. Meet with the supplier and determine if the original policy will be honored at no additional cost. If the supplier balks, then follow the advice from legal.
Deliverables include software to be developed	Include an experienced software manager on the technical team. Monitor the contractor's adherence to software development processes. Discuss software progress, issues, and quality at technical interchange meetings.
Deliverables include COTS products (especially software)	Understand the quality of the product: • Look at test results. When test results show a lot of rework to correct defects, then users will probably find more defects. • Examine problem reports. These show whether or not users are finding defects after release. • Evaluate user documentation. • Look at product support.
Products depend on results from models or simulations	Establish the credibility and uncertainty of results. Determine depth and breadth of practices used in verification and validation of the model or simulation. Understand the quality of software upon which the model or simulation is built. For more information, refer to *NASA-STD-(I)-7009, Standard for Models and Simulations.*
Budget changes prior to delivery of all products (and contract was written without interim deliverables)	Options include: • Remove deliverables or services from the contract scope in order to obtain key products. • Relax the schedule in exchange for reduced cost. • Accept deliverables "as is." To avoid this situation, include electronic access to data, work products, and interim deliverables to assess contractor progress on final deliverables in the SOW.
Contractor is a specialty supplier with no experience in a particular engineering discipline; for example, the contractor produces cryogenic systems that use alarm monitoring software from another supplier, but the contractor does not have software expertise	Mitigate risks of COTS product deliverables as discussed earlier. If the contract is for delivery of a modified COTS product or custom product, then include provisions in the SOW to cover the following: • Supplier support (beyond product warranty) that includes subsupplier support • Version upgrade/replacement plans • Surveillance of subsupplier If the product is inexpensive, simply purchasing spares may be more cost effective than adding surveillance requirements.

- **Develop NASA-Contractor Technical Relationship:** At the contract kick-off meeting, set expectations for technical excellence throughout the execution of the contract. Highlight the requirements in the contract SOW that are the most important. Discuss the quality of work and products to be delivered against the technical requirements. Mutually agree on the format of the technical reviews and how to resolve misunderstandings, oversights, and errors.

- **Conduct Technical Interchange Meetings:** Start early in the contract period and meet periodically with the contractor (and subcontractors) to confirm that the contractor has a correct and complete understanding of the requirements and operational concepts. Establish day-to-day NASA-contractor technical communications.

- **Control and Manage Requirements:** Almost inevitably, new or evolving requirements will affect a project. When changes become necessary, the technical team needs to control and manage changes and additions to requirements proposed by either NASA or the contractor. (See Section 6.2.) Communicate changes to any project participants that the changes will affect. Any changes in requirements that affect contract cost, schedule, or performance must be conveyed to the contractor through a formal contract change. Consult the contracting officer's technical representative.

- **Evaluate Systems Engineering Processes:** Evaluate the effectiveness of defined systems engineering processes. Conduct audits and reviews of the processes. Identify process deficiencies and offer assistance with process improvement.

- **Evaluate Work Products:** Evaluate interim plans, reports, specifications, drawings, processes, procedures, and similar artifacts that are created during the systems engineering effort.

- **Monitor Contractor Performance Against Key Metrics:** Monitoring contractor performance extends beyond programmatic metrics to process and product metrics. (See Section 6.7 on technical performance measures.) These metrics depend on acceptable product quality. For example, "50 percent of design drawings completed" is misleading if most of them have defects (e.g., incorrect, incomplete, inconsistent). The amount of work to correct the drawings affects cost and schedule. It is useful to examine reports that show the amount of contractor time invested in product inspection and review.

- **Conduct Technical Reviews:** Assess contractor progress and performance against requirements through technical reviews. (See Section 6.7.)

- **Verify and Validate Products:** Verify and validate the functionality and performance of products before delivery and prior to integration with other system products. To ensure that a product is ready for system integration or to enable further system development, perform verification and validation as early as practical. (See Sections 5.3 and 5.4.)

7.1.4.2 Evaluating Work Products

Work products and deliverables share common attributes that can be used to assess quality. Additionally, relationships among work products and deliverables can be used to assess quality. Some key attributes that help determine quality of work products are listed below:

- Satisfies content and format requirements,
- Understandable,
- Complete,
- Consistent (internally and externally) including terminology (an item is called the same thing throughout the documents, and
- Traceable.

Table 7.1-5 shows some typical work products from the contractor and key attributes with respect to other documents that can be used as evaluation criteria.

7.1.4.3 Issues with Contract-Subcontract Arrangements

In the ideal world, a contractor manages its subcontractors, each subcontract contains all the right requirements, and resources are adequate. In the real world, the technical team deals with contractors and subcontractors that are motivated by profit, (sub)contracts with missing or faulty requirements, and resources that are consumed more quickly than expected. These and other factors cause or influence two key issues in subcontracting:

- Limited or no oversight of subcontractors and
- Limited access to or inability to obtain subcontractor data.

These issues are exacerbated when they apply to second- (or lower) tier subcontractors. Table 7.1-6 looks at these issues more closely along with potential resolutions.

Scenarios other than those above are possible. Resolutions might include reducing contract scope or deliverables in lieu of cost increases or sharing information technology in order to obtain data. Even with the adequate flowdown requirements in (sub)contracts, legal wrangling may be necessary to entice contractors to satisfy the conditions of their (sub)contracts.

Activities during contract performance will generate an updated surveillance plan, minutes documenting meetings, change requests, and contract change orders. Processes will be assessed, deliverables and work products evaluated, and results reviewed.

Table 7.1-5 Typical Work Product Documents

Work Product	Evaluation Criteria
SEMP	Describes activities and products required in the SOW. The SEMP is not complete unless it describes (or references) how each activity and product in the SOW will be accomplished.
Software management/development plan	Consistent with the SEMP and related project plans. Describes how each software-related activity and product in the SOW will be accomplished. Development approach is feasible.
System design	Covers the technical requirements and operational concepts. System can be implemented.
Software design	Covers the technical requirements and operational concepts. Consistent with hardware design. System can be implemented.
Installation plans	Covers all user site installation activities required in the SOW. Presents a sound approach. Shows consistency with the SEMP and related project plans.
Test plans	Covers qualification requirements in the SOW. Covers technical requirements. Approach is feasible.
Test procedures	Test cases are traceable to technical requirements.
Transition plans	Describes all transition activities required in the SOW. Shows consistency with the SEMP and related project plans.
User documentation	Sufficiently and accurately describes installation, operation, or maintenance (depending on the document) for the target audience.
Drawings and documents (general)	Comply with content and format requirements specified in the SOW.

7.1.5 Contract Completion

The contract comes to completion with the delivery of the contracted products, services, or systems and their enabling products or systems. Along with the product, as-built documentation must be delivered and operational instructions including user manuals.

7.1.5.1 Acceptance of Final Deliverables

Throughout the contract period, the technical team reviews and accepts various work products and interim deliverables identified in the contract data requirements list and schedule of deliverables. The technical team also participates in milestone reviews to finalize acceptance of deliverables. At the end of the contract, the technical team ensures that each technical deliverable is received and that its respective acceptance criteria are satisfied.

The technical team records the acceptance of deliverables against the contract data requirements list and the schedule of deliverables. These documents serve as an inventory of items and services to be accepted. Although rejections and omissions are infrequent, the technical team needs to take action in such a case. Good data management and configuration management practices facilitate the effort.

Acceptance criteria include:

- Product verification and validation completed successfully. The technical team performs or oversees verification and validation of products, integration of products into systems, and system verification and validation.

- Technical data package is current (as-built) and complete.

Table 7.1-6 Contract-Subcontract Issues

Issue	Resolution
Oversight of subcontractor is limited because requirement(s) missing from contract	The technical team gives the SOW requirement(s) to the contracting officer who adds the requirement(s) to the contract and negotiates the change order, including additional costs to NASA. The contractor then adds the requirement(s) to the subcontract and negotiates the change order with the subcontractor. If the technical team explicitly wants to perform oversight, then the SOW should indicate what the contractor, its subcontractors, and team members are required to do and provide.
Oversight of subcontractor is limited because requirement(s) not flowed down from contractor to subcontractor	It is the contractor's responsibility to satisfy the requirements of the contract. If the contract includes provisions to flow down requirements to subcontractors, then the technical team can request the contracting officer to direct the contractor to execute the provisions. The contractor may need to add requirements and negotiate cost changes with the subcontractor. If NASA has a cost-plus contract, then expect the contractor to bill NASA for any additional costs incurred. If NASA has a fixed-price contract, then the contractor will absorb the additional costs or renegotiate cost changes with NASA. If the contract does not explicitly include requirements flowdown provisions, the contractor is responsible for performing oversight.
Oversight of second-tier subcontractor is limited because requirement(s) not flowed down from subcontractor to second-tier subcontractor	This is similar to the previous case, but more complicated. Assume that the contractor flowed down requirements to its subcontractor, but the subcontractor did not flow down requirements to the second-tier subcontractor. If the subcontract includes provisions to flow down requirements to lower tier subcontractors, then the technical team can request the contracting officer to direct the contractor to ensure that subcontractors execute the flowdown provisions to their subcontractors. If the subcontract does not explicitly include requirements flowdown provisions, the subcontractor is responsible for performing oversight of lower tier subcontractors.
Access to subcontractor data is limited or not provided because providing the data is not required in the contract	The technical team gives the SOW requirement(s) to the contracting officer who adds the requirement(s) to the contract and negotiates the change order, including additional costs to NASA. The contractor then adds the requirement(s) to the subcontract and negotiates the change order with the subcontractor. If the technical team explicitly wants direct access to subcontractor data, then the SOW should indicate what the contractor, its subcontractors, and team members are required to do and provide.
Access to subcontractor data is limited or not provided because providing the data is not required in the subcontract	It is the contractor's responsibility to obtain data (and data rights) necessary to satisfy the conditions of its contract, including data from subcontractors. If the technical team needs direct access to subcontractor data, then follow the previous case to add flowdown provisions to the contract so that the contractor will add requirements to the subcontract.

- Transfer of certifications, spare parts, warranties, etc., is complete.

- Transfer of software products, licenses, data rights, intellectual property rights, etc., is complete.

- Technical documentation required in contract clauses is complete (e.g., new technology reports).

It is important for NASA personnel and facilities to be ready to receive final deliverables. Key items to have prepared include:

- A plan for support and to transition products to operations;

- Training of personnel;

- Configuration management system in place; and

- Allocation of responsibilities for troubleshooting, repair, and maintenance.

7.1.5.2 Transition Management

Before the contract was awarded, a product support strategy was developed as part of the acquisition strategy. The product support strategy outlines preliminary notions regarding integration, operations, maintenance, improvements, decommissioning, and disposal. Later, after the contract is awarded, a high-level transition plan that expands the product support strategy is recorded in the SEMP. Details of product/system transition are subsequently documented in one or more transition plans. Elements of transition planning are discussed in Section 5.5.

Transition plans must clearly indicate responsibility for each action (NASA or contractor). Also, the contract SOW must have included a requirement that the contractor will execute responsibilities assigned in the transition plan (usually on a cost-reimbursable basis).

Frequently, NASA (or NASA jointly with a prime contractor) is the system integrator on a project. In this situation, multiple contractors (or subcontractors) will execute their respective transition plans. NASA is responsible for developing and managing a system integration plan that incorporates inputs from each transition plan. The provisions that were written in the SOW months or years earlier accommodate the transfer of products and systems from the contractors to NASA.

7.1.5.3 Transition to Operations and Support

The successful transition of systems to operations and support, which includes maintenance and improvements, depends on clear transition criteria that the stakeholders agree on. The technical team participates in the transition, providing continuity for the customer, especially when a follow-on contract is involved. When the existing contract is used, the technical team conducts a formal transition meeting with the contractor. Alternatively, the transition may involve the same contractor under a different contract arrangement (e.g., modified or new contract). Or the transition may involve a different contractor than the developer, using a different contract arrangement.

The key benefits of using the existing contract are that the relevant stakeholders are familiar with the contractor and that the contractor knows the products and systems involved. Ensure that the contractor and other key stakeholders understand the service provisions (requirements) of the contract. This meeting may lead to contract modifications in order to amend or remove service requirements that have been affected by contract changes over the years.

Seeking to retain the development contractor under a different contract can be beneficial. Although it takes time and resources to compete the contract, it permits

NASA to evaluate the contractor and other offerors against operations and support requirements only. The incumbent contractor has personnel with development knowledge of the products and systems, while service providers specialize in optimizing cost and availability of services. In the end, the incumbent may be retained under a contract that focuses on current needs (not several years ago), or else a motivated service provider will work hard to understand how to operate and maintain the systems. If a follow-on contract will be used, consult the local procurement office and exercise the steps that were used to obtain the development contract. Assume that the amount of calendar time to award a follow-on contract will be comparable to the time to award the development contract. Also consider that the incumbent may be less motivated upon losing the competition.

Some items to consider for follow-on contracts during the development of SOW requirements include:

- Staff qualifications;
- Operation schedules, shifts, and staffing levels;
- Maintenance profile (e.g., preventive, predictive, run-to-fail);
- Maintenance and improvement opportunities (e.g., schedule, turnaround time);
- Historical data for similar efforts; and
- Performance-based work.

The transition to operations and support represents a shift from the delivery of products to the delivery of services.

Service contracts focus on the contractor's performance of activities, rather than development of tangible products. Consequently, performance standards reflect customer satisfaction and service efficiency, such as:

- Customer satisfaction ratings;
- Efficiency of service;
- Response time to a customer request;
- Availability (e.g., of system, Web site, facility);
- Time to perform maintenance action;
- Planned versus actual staffing levels;
- Planned versus actual cost;
- Effort and cost per individual service action; and
- Percent decrease in effort and cost per individual service action.

For more examples of standards to assess the contractor's performance, refer to *System and Software Metrics for Performance-Based Contracting*.

7.1.5.4 Decommissioning and Disposal

Contracts offer a means to achieve the safe and efficient decommissioning and disposal of systems and products that require specialized support systems, facilities, and trained personnel, especially when hazardous materials are involved. Consider these needs during development of the acquisition strategy and solidify them before the final design phase. Determine how many contracts will be needed across the product's life cycle.

Some items to consider for decommissioning and disposal during the development of SOW requirements:

- Handling and disposal of waste generated during the fabrication and assembly of the product.
- Reuse and recycling of materials to minimize the disposal and transformation of materials.
- Handling and disposal of materials used in the product's operations.
- End-of-life decommissioning and disposal of the product.
- Cost and schedule to decommission and dispose of the product, waste, and unwanted materials.
- Metrics to measure decommissioning and disposal of the product.

- Metrics to assess the contractor's performance. (Refer to *System and Software Metrics for Performance-Based Contracting*.)

For guidelines regarding disposal, refer to the *Systems Engineering Handbook: A "What To" Guide for all SE Practitioners*.

7.1.5.5 Final Evaluation of Contractor Performance

In preparation for closing out a contract, the technical team gives input to the procurement office regarding the contractor's final performance evaluation. Although the technical team has performed periodic contractor performance evaluations, the final evaluation offers a means to document good and bad performance that continued throughout the contract. Since the evaluation is retained in a database, it can be used as relevant experience and past performance input during a future source selection process.

This phase of oversight is complete with the closeout or modification of the existing contract, award of the follow-on contract, and an operational system. Oversight continues with follow-contract activities.

7.2 Integrated Design Facilities

7.2.1 Introduction

Concurrent Engineering (CE) and integrated design is a systematic approach to integrated product development that emphasizes response to stakeholder expectations and embodies team values of cooperation, trust, and sharing. The objective of CE is to reduce the product development cycle time through a better integration of activities and processes. Parallelism is the prime concept in reducing design lead time and concurrent engineering becomes the central focus. Large intervals of parallel work on different parts of the design are synchronized by comparatively brief exchanges between teams to produce consensus and decisions.[1] CE has become a widely accepted concept and is regarded as an excellent alternative approach to a sequential engineering process.

This section addresses the specific application of CE and integrated design practiced at NASA in Capability for Accelerated Concurrent Engineering (CACE) environments. CACE is comprised of four essential components: people, process, tools, and facility. The CACE environment typically involves the collocation of an in-place leadership team and core multidisciplinary engineering team working with a stakeholder team using well defined processes in a dedicated collaborative, concurrent engineering facility with specialized tools. The engineering and collaboration tools are connected by the facility's integrated infrastructure. The teams work synchronously for a short period of time in a technologically intensive physical environment to complete an instrument or mission design. CACE is most often used to design space instruments and payloads or missions including orbital configuration; hardware such as spacecraft, landers, rovers, probes, or launchers; data and ground communication systems; other ground systems; and mission operations. But the CACE process applies beyond strict instrument and/or mission conceptual design.

Most NASA centers have a CACE facility. NASA CACE is built upon a people/process/tools/facility paradigm that enables the accelerated production of high-quality engineering design concepts in a concurrent, collaborative, rapid design environment. (See Figure 7.2-1.)

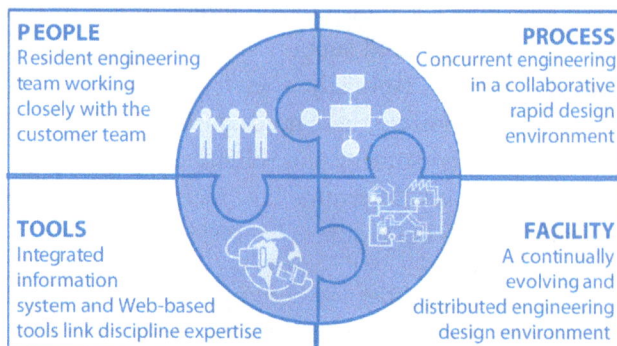

PEOPLE Resident engineering team working closely with the customer team	PROCESS Concurrent engineering in a collaborative rapid design environment
TOOLS Integrated information system and Web-based tools link discipline expertise	FACILITY A continually evolving and distributed engineering design environment

Figure 7.2-1 CACE people/process/tools/facility paradigm

Although CACE at NASA is based on a common philosophy and characteristics, specific CACE implementation varies in many areas. These variations include level of engineering detail, information infrastructure, knowledge base, areas of expertise, engineering staffing approach, administrative and engineering tools, type of facilitation, roles and responsibilities within CACE team, roles and responsibilities across CACE and stakeholder teams, activity execution approach, and duration of session. While primarily used to support early life-cycle phases such as pre-Formulation and Formulation, the CACE process has demonstrated applicability across the full project life cycle.

7.2.2 CACE Overview and Importance

CACE design techniques can be an especially effective and efficient method of generating a rapid articulation of concepts, architectures, and requirements.

The CACE approach provides an infrastructure for brainstorming and bouncing ideas between the engineers and stakeholder team representatives, which routinely results in a high-quality product that directly maps to the customer needs. The collaboration design paradigm is so successful because it enables a radical reduction in decision latency. In a non-CACE environment, questions, issues, or problems may take several days to resolve. If a design needs to be changed or a requirement reevaluated, significant time may pass before all engineering team members get the information or stakeholder team members can discuss potential requirement changes. These delays introduce the possibility, following initial evaluation, of another round of questions, issues, and

---

[1]From Miao and Haake "Supporting Concurrent Design by Integrating Information Sharing and Activity Synchronization."

changes to design and requirements, adding further delays.

The tools, data, and supporting information technology infrastructure within CACE provide an integrated support environment that can be immediately utilized by the team. The necessary skills and experience are gathered and are resident in the environment to synchronously complete the design. In a collaborative environment, questions can be answered immediately, or key participants can explore assumptions and alternatives with the stakeholder team or other design team members and quickly reorient the whole team when a design change occurs. The collaboration triggers the creativity of the engineers and helps them close the loop and rapidly converge on their ideas. Since the mid-1990s, the CACE approach has been successfully used at several NASA Centers as well as at commercial enterprises to dramatically reduce design development time and costs when compared to traditional methods.

CACE stakeholders include NASA programs and projects, scientists, and technologists as well as other Government agencies (civil and military), Federal laboratories, and universities. CACE products and services include:

- Generating mission concepts in support of Center proposals to science AO;
- Full end-to-end designs including system/subsystem concepts, requirements, and tradeoffs;
- Focused efforts assessing specific architecture sub-elements and tradeoffs;
- Independent assessments of customer-provided reports, concepts, and costs;
- Roadmapping support; and
- Technology and risk assessments.

As integrated design has become more accepted, collaborative engineering design efforts expanded from the participation of one or more Centers in a locally executed activity; to geographically distributed efforts across a few NASA Centers with limited scope and participation; to true OneNASA efforts with participation from many NASA integrated design teams addressing broad, complex architectures.

The use of geographically distributed CACE teams is a powerful engineering methodology to achieve lower risk and more creative solutions by factoring in the best skills and capabilities across the Agency. Using a geographically

distributed process must build upon common CACE elements while considering local CACE facility differences and the differences in the local Center cultures.

7.2.3 CACE Purpose and Benefits

The driving forces behind the creation of NASA's early CACE environments were increased systems engineering efficiency and effectiveness. More specifically, the early CACE environments addressed the need for:

- Generating more conceptual design studies at reduced cost and schedule,
- Creating a reusable process within dedicated facilities using well-defined tools,
- Developing a database of mission requirements and designs for future use,
- Developing mission generalists from a pool of experienced discipline engineers, and
- Infusing a broader systems engineering perspective across the organization.

Additional resulting strategic benefits across NASA included:

- Core competency support (e.g., developing systems engineers, maturing and broadening of discipline engineers, training environment, etc.);
- Sensitizing the customer base to end-to-end issues and implications of requirements upon design;
- Test-bed environment for improved tools and processes;
- Environment for forming partnerships;
- Technology development and roadmapping support;
- Improved quality and consistency of conceptual designs; and
- OneNASA environment that enables cooperative rather than competitive efforts among NASA organizations.

7.2.4 CACE Staffing

A management or leadership team, a multidisciplinary engineering team, a stakeholder team, and a facility support team are all vital elements in achieving a successful CACE activity.

A CACE team consists of a cadre of engineers, each representing a different discipline or specialty engineering area, along with a lead systems engineer and a team lead or facilitator. As required, the core engineering team is supplemented with specialty and/or nonstandard engineering skills to meet unique stakeholder needs. These supplementary engineering capabilities can be obtained either from the local Center or from an external source. The team lead coordinates and facilitates the CACE activity and interacts with the stakeholders to ensure that their objectives are adequately captured and represented. Engineers are equipped with techniques and software used in their area of expertise and interact with the team lead, other engineers, and the stakeholder team to study the feasibility of a proposed solution and produce a design for their specific subsystem.

A CACE operations manager serves as the Center advocate and manager, maintaining an operational capability, providing initial coordination with potential customers through final delivery of CACE product, and infusing continuous process and product improvement as well as evolutionary growth into the CACE environment to ensure its continued relevance to the customer base.

A CACE facility support team maintains and develops the information infrastructure to support CACE activities.

7.2.5 CACE Process

The CACE process starts with a customer requesting engineering support from CACE management. CACE management establishes that the customer's request is within the scope of the team capabilities and availability and puts together a multidisciplinary engineering team under the leadership of a team lead and lead systems engineer collaborating closely with the customer team. The following subsections briefly describe the three major CACE activity phases: (1) planning and preparation, (2) execution, and (3) wrap-up.

7.2.5.1 Planning and Preparation

Once a customer request is approved and team lead chosen, a planning meeting is scheduled. The key experts attending the planning meeting may include the CACE manager, a team lead, and a systems engineer as well as key representatives from the customer/stakeholder team. Interactions with the customer/stakeholder team and their active participation in the process are integral to the successful planning, preparation, and execution of a concurrent design session. Aspects addressed include establishing the activity scope, schedule, and costs; a general agreement on the type of product to be provided; and the success criteria and metrics. Agreements reached at the planning meeting are documented and distributed for review and comment.

Products from the planning and preparation phase include the identification of activities required by the customer/stakeholder team, the CACE team, or a combination of both teams, as well as the definition of the objectives, the requirements, the deliverables, the estimated budget, and the proposed schedule. Under some conditions, followup coordination meetings are scheduled that include the CACE team lead, the systems engineer(s), a subset of the remaining team members, and customer/stakeholder representatives, as appropriate. The makeup of participants is usually based on the elements that have been identified as the activity drivers and any work identified that needs to be done before the actual design activity begins.

During the planning and preparation process, the stakeholder-provided data and the objectives and activity plan are reviewed, and the scope of the activity is finalized. A discussion is held of what activities need to be done by each of the stakeholders and the design teams. For example, for planning a mission design study, the customer identifies the mission objectives by defining the measurement objectives and the instrument specifications, as applicable, and identifying the top-level requirements. A subset of the CACE engineering team may perform some preliminary work before the actual study (e.g., launch vehicle performance trajectory analysis; thrust and navigation requirements; the entry, descent, and landing profile; optical analysis; mechanical design; etc.) as identified in the planning meetings to further accelerate the concurrent engineering process in the study execution phase. The level of analysis in this phase is a function of many things, including the level of maturity of the incoming design, the stated goals and objectives of the engineering activity, engineer availability, and CACE scheduling.

7.2.5.2 Activity Execution Phase

A typical activity or study begins with the customer presentation of the overall mission concept and instrument concepts, as applicable, to the entire team. Additional information provided by the customer/stakeholders includes the team objectives, the science and technology goals, the initial requirements for payload, spacecraft and mission design, the task breakdown between providers of parts or functions, top challenges and concerns, and the approximate mission timeline. This information is often provided electronically in a format accessible to the engineering team and is presented by the customer/

stakeholder representatives at a high level. During this presentation, each of the subsystems engineers focuses on the part of the overall design that is relevant to their subsystem. The systems engineer puts the high-level system requirements into the systems spreadsheets and/or a database that is used throughout the process to track engineering changes. These data sources can be projected on the displays to keep the team members synchronized and the customer/stakeholders aware of the latest developments.

The engineering analysis is performed iteratively with the CACE team lead and systems engineer playing key roles to lead the process. Thus, issues are quickly identified, so consensus on tradeoff decisions and requirements redefinition can be achieved while maintaining momentum. The customer team actively participates in the collaborative process (e.g., trade studies, requirements relaxation, clarifying priorities), contributing to the rapid development of an acceptable product.

Often, there are breakout sessions, or sidebars, in which part of the team discusses a particular tradeoff study. Each subsystem has a set of key parameters that are used for describing its design. Because of the dependencies among the various subsystems, each discipline engineer needs to know the value of certain parameters related to other subsystems. These parameters are shared via the CACE information infrastructure. Often, there are conflicting or competing objectives for various subsystems. Many tradeoff studies, typically defined and led by the team systems engineer, are conducted among subsystem experts immediately as issues occur. Most of the communication among team members is face to face or live via video or teleconference. Additional subject matter experts are consulted as required. In the CACE environment, subsystems that need to interact extensively are clustered in close proximity to facilitate the communication process among the experts.

The team iterates on the requirements, and each subsystem expert refines or modifies design choices as schedule allows. This process continues until an acceptable solution is obtained. There may be occasions where it is not possible to iterate to an acceptable solution prior to the scheduled end of the activity. In those cases, the available iterated results are documented and form the basis of the delivered product.

In each iteration, activities such as the following take place, sometimes sequentially and other times in par-

allel. The subsystem experts of science, instruments, mission design, and ground systems collaboratively define the science data strategy for the mission in question. The telecommunications, ground systems, and command and data-handling experts develop the data-return strategy. The attitude control systems, power, propulsion, thermal, and structure experts iterate on the spacecraft design and the configuration expert prepares the initial concept. The systems engineer interacts with all discipline engineers to ensure that the various subsystem designs fit into the intended system architecture. Each subsystem expert provides design and cost information, and the cost expert estimates the total cost for the mission.

While design activity typically takes only days or weeks with final products available within weeks after study completion, longer term efforts take advantage of the concurrent, collaborative environment to perform more detailed analyses than those performed in the shorter duration CACE exercises.

7.2.5.3 Activity Wrap-Up

After the completion of a CACE study, the product is delivered to the customer. In some CACE environments, the wrap-up of the product is completed with minimal additional resources: the engineers respond to customer/stakeholder feedback by incorporating additional refinements or information emphasizing basic cleanup. In other CACE environments, significant time is expended to format the final report and review it with the customer/stakeholders to ensure that their expectations have been addressed adequately.

Some CACE environments have standardized their wrap-up activities to address the customer/stakeholder feedback and develop products that are structured and uniform across different ranges of efforts.

As part of activity followup, customer/stakeholder feedback is requested on processes, whether the product met their needs, and whether there are any suggested improvements. This feedback is factored back into the CACE environment as part of a continuous improvement process.

7.2.6 CACE Engineering Tools and Techniques

Engineering tools and techniques vary within and across CACE environments in several technical aspects (e.g., level of fidelity, level of integration, generally available commercial applications versus custom tools versus customized knowledge-based Excel spreadsheets, degree of parametric design and/or engineering analysis). For example, mechanical design tools range from white-board discussions to note pad translations to computer-aided design to 3D rapid design prototyping.

Important factors in determining which tools are appropriate to an activity include the purpose and duration of the activity, the engineers' familiarity or preference, the expected product, the local culture, and the evolution of the engineering environment. Factors to be considered in the selection of CACE tools and engineering techniques should also include flexibility, compatibility with the CACE environment and process, and value and ease of use for the customer after the CACE activities.

Engineering tools may be integrated into the CACE infrastructure, routinely provided by the supporting engineering staff, and/or utilized only on an activity-by-activity basis, as appropriate. As required, auxiliary engineering analysis outside of the scope of the CACE effort can be performed external to the CACE environment and imported for reference and incorporation into the CACE product.

7.2.7 CACE Facility, Information Infrastructure, and Staffing

Each CACE instantiation is unique to the Center, program, or project that it services. While the actual implementations vary, the basic character does not. Each implementation concentrates on enabling engineers, designers, team leads, and customer/stakeholders to be more productive during concurrent activities and communication. This subsection focuses on three aspects of this environment: the facility, the supporting information infrastructure, and the staff required to keep the facility operational.

7.2.7.1 Facility

The nature of communication among discipline specialists working together simultaneously creates a somewhat chaotic environment. Although it is the duty of the team lead to maintain order in the environment, the facility itself has to be designed to allow the participants to maintain order and remain on task while seeking to increase communication and collaboration. To do this effectively requires a significant investment in infrastructure resources.

The room needs sufficient space to hold active participants from the disciplines required, customer/stakeholder representatives, and observers. CACE managers encourage observers to show potential future CACE users the value of active CACE sessions.

It is also important to note that the room will get reconfigured often. Processes and requirements change, and the CACE facility must change with that. The facility could appear to an onlooker as a work in progress. Tables, chairs, computer workstations, network connections, electrical supplies, and visualization systems will continually be assessed for upgrades, modification, or elimination.

CACE requirements in the area of visualization are unique. When one subject matter expert wants to communicate to either a group of other discipline specialists or to the whole group in general, the projection system needs to be able to switch to different engineering workstations. When more than one subject matter expert wants to communicate with different groups, multiple projection systems need to be able to switch. This can typically require three to six projection systems with switching capability from any specific workstation to any specific projector. In addition, multiple projection systems switchable to the engineering workstations need to be mounted so that they can be viewed without impacting other activities in the room or so that the entire group can be refocused as required during the session. The ease of this reconfiguration is one measure of the efficacy of the environment.

7.2.7.2 Information Infrastructure

A CACE system not only requires a significant investment in the facility but relies heavily on the information infrastructure. Information infrastructure requirements can be broken down into three sections: hardware, software, and network infrastructure.

The hardware portion of the information infrastructure used in the CACE facility is the most transient element in the system. The computational resources, the communication fabric, servers, storage media, and the visualization capabilities benefit from rapid advances in technology. A CACE facility must be able to take advantage of the economy produced by those advances and must also be flexible enough to take advantage of the new capabilities.

One of the major costs of a CACE infrastructure is software. Much of the software currently used by engineering processes is modeling and simulation, usually produced by commercial software vendors. Infrastructure software to support exchange of engineering data; to manage the study archive; and to track, administer, and manage facility activities is integral to CACE success. One of the functions of the CACE manager is to determine how software costs can be paid, along with what software should be the responsibility of the participants and customers.

The network infrastructure of a CACE facility is critical. Information flowing among workstations, file servers, and visualization systems in real time requires a significant network infrastructure. In addition, the network infrastructure enables collaboration with outside consultants, external discipline experts, and intra-Center collaboration. The effective use of the network infrastructure requires a balance between network security and collaboration and, as such, will always be a source of modification, upgrade, and reconfiguration. A natural extension of this collaboration is the execution of geographically distributed CACE efforts; therefore it is essential that a CACE facility have the tools, processes, and communications capabilities to support such distributed studies.

7.2.7.3 Facility Support Staff Responsibilities

A core staff of individuals is required to maintain an operational CACE environment. The responsibilities to be covered include end-to-end CACE operations and the management and administration of the information infrastructure.

CACE information infrastructure management and administration includes computer workstation configuration; network system administration; documentation development; user help service; and software support to maintain infrastructure databases, tools, and Web sites.

7.2.8 CACE Products

CACE products are applicable across project life-cycle phases and can be clearly mapped to the various outputs associated with the systems engineering activities such as requirements definition, trade studies, decision analysis, and risk management. CACE products from a typical design effort include a requirements summary with driving requirements identified; system and subsystem

analysis; functional architectures and data flows; mass/power/data rackups; mission design and ConOps; engineering trades and associated results; technology maturity levels; issues, concerns, and risks; parametric and/or grassroots cost estimates; engineering analyses, models, and applicable tools to support potential future efforts; and a list of suggested future analyses.

CACE product format and content vary broadly both within and across CACE environments. The particular CACE environment, the goals/objectives of the supported activity, whether the activity was supported by multiple CACE teams or not, the customer's ultimate use, and the schedule requirements are some aspects that factor into the final product content and format. A primary goal in the identification and development of CACE products and in the packaging of the final delivery is to facilitate their use after the CACE activity.

Products include in-study results presentation, PowerPoint packages, formal reports and supporting computer-aided design models, and engineering analysis. Regardless of format, the CACE final products typically summarize the incoming requirements, study goal expectation, and study final results.

CACE environment flexibility enables support activities beyond that of a traditional engineering design study (e.g., independent technical reviews, cost validation, risk and technology assessments, roadmapping, and requirements review). Product contents for such activities might include feasibility assessment, technical recommendations, risk identification, recosting, technology infusion impact and implementation approach, and architectural options.

In addition to formal delivery of the CACE product to the customer team, the final results and planning data are archived within the CACE environment for future reference and for inclusion in internal CACE cross-study analyses.

7.2.9 CACE Best Practices

This subsection contains general CACE best practices for a successful CACE design activity. Three main topic areas—people, process, technologies—are applicable to both local and geographically distributed activities. Many lessons learned about the multi-CACE collaboration activities were learned through the NASA Exploration Design Team (NEDT) effort, a OneNASA multi-Center distributed collaborative design activity performed during FY05.

7.2.9.1 People

- **Training:** Individuals working in CACE environments benefit from specialized training. This training should equip individuals with the basic skills necessary for efficient and effective collaboration. Training should include what is required technically as well as orientation to the CACE environment and processes.

- **Characteristics:** Collaborative environment skills include being flexible, working with many unknowns, and willingness to take risks. Ability and willingness to think and respond in the moment is required as well as the ability to work as part of a team and to interact directly with customer representatives to negotiate requirements and to justify design decisions. Supporting engineers also need the ability to quickly and accurately document their final design as well as present this design in a professional manner. In addition, the CACE team leads or facilitators should have additional qualities to function well in a collaborative design environment. These include organizational and people skills, systems engineering skills and background, and broad general engineering knowledge.

7.2.9.2 Process and Tools

- **Customer Involvement:** Managing customer expectations is the number one factor in positive study outcome. It is important to make the customers continuously aware of the applications and limitations of the CACE environment and to solicit their active participation in the collaborative environment.

- **Adaptability:** The CACE environments must adapt processes depending on study type and objectives, as determined in negotiations prior to study execution. In addition to adapting the processes, engineers with appropriate engineering and collaborative environment skills must be assigned to each study.

- **Staffing:** Using an established team has the benefit of the team working together and knowing each other and the tools and processes. A disadvantage is that a standing army can get "stale" and not be fluent with the latest trends and tools in their areas of expertise. Supporting a standing army full time is also an expensive proposition and often not possible. A workable compromise is to have a full-time (or nearly full-time) leadership team complemented by an engineering team. This engineering team could be composed of engineers on rotational assignments or long-term detail to the team, as appropriate. An alternative paradigm is to partially staff the engineering team with personnel provided through the customer team.

- **Tools and Data Exchange:** In general, each engineer should use the engineering tools with which he or she is most familiar to result in an effective and efficient process. The CACE environment should provide an information infrastructure to integrate resulting engineering parameters.

- **Decision Process:** Capturing the decisionmaking and design rationale is of great interest and of value to CACE customers as well as being a major challenge in the rapid engineering environment. The benefit of this is especially important as a project progresses and makes the CACE product more valuable to the customer. Further along in the life cycle of a mission or instrument, captured decisions and design rationale are more useful than a point-design from some earlier time.

- **Communication:** CACE environments foster rapid communication among the team members. Because of the fast-paced environment and concurrent engineering activities, keeping the design elements "in synch" is a challenge. This challenge can be addressed by proactive systems engineers, frequent tag-ups, additional systems engineering support and the use of appropriate information infrastructure tools.

- **Standards Across CACE Environments:** Establishing minimum requirements and standard sets of tools and techniques across the NASA CACE environment would facilitate multi-Center collaborations.

- **Planning:** Proper planning and preparation are crucial for efficient CACE study execution. Customers wanting to forgo the necessary prestudy activity or planning and preparation must be aware of and accept the risk of a poor or less-than-desired outcome.

7.2.9.3 Facility

- **Communication Technologies:** The communication infrastructure is the backbone of the collaborative CACE environment. Certain technologies should be available to allow efficient access to resources external to a CACE facility. It is important to have "plug and play" laptop capability, for example. Multiple phones should be available to the team and cell phone access is desirable.

- **Distributed Team Connectivity:** Real-time transfer of information for immediate access between geographically distributed teams or for multi-Center activities can be complicated due to firewall and other networking issues. Connectivity and information transfer methods should be reviewed and tested before study execution.

7.3 Selecting Engineering Design Tools

NASA utilizes cutting-edge design tools and techniques to create the advanced analyses, designs, and concepts required to develop unique aerospace products, spacecraft, and science experiments. The diverse nature of the design work generated and overseen by NASA requires use of a broad spectrum of robust electronic tools such as computer-aided design tools and computer-aided systems engineering tools. Based on the distributed and varied nature of NASA projects, selection of a single suite of tools from only one vendor to accomplish all design tasks is not practical. However, opportunities to improve standardization of design policy, processes, and tools remain a focus for continuous improvement activities at all levels within the Agency.

These guidelines serve as an aid to help in the selection of appropriate tools in the design and development of aerospace products and space systems and when selecting tools that affect multiple Centers.

7.3.1 Program and Project Considerations

When selecting a tool to support a program or project, all of the upper level constraints and requirements must be identified early in the process. Pertinent information from the project that affects the selection of the tools will include the urgency, schedule, resource restrictions, extenuating circumstances, and constraints. A tool that does not support meeting the program master schedule or is too costly to be bought in sufficient numbers will not satisfy the project manager's requirements. For example, a tool that requires extensive modification and training that is inconsistent with the master schedule should not be selected by the technical team. If the activity to be undertaken is an upgrade to an existing project, legacy tools and availability of trained personnel are factors to be considered.

7.3.2 Policy and Processes

When selecting a tool, one must consider the applicable policies and processes at all levels, including those at the Center level, within programs and projects, and at other Centers when a program or project is a collaborative effort. In the following discussion, the term "organization" will be used to represent any controlling entity that establishes policy and/or processes for the use of tools in the design or development of NASA products. In other words,

"organization" can mean the user's Center, another collaborating Center, a program, a project, in-line engineering groups, or any combination of these entities.

Policies and processes affect many aspects of a tool's functionality. First and foremost, there are policies that dictate how designs are to be formally or informally controlled within the organization. These policies address configuration management processes that must be followed as well as the type of data object that will be formally controlled (e.g., drawings or models). Clearly this will affect the types of tools that will be used and how their designs will be annotated and controlled.

The Information Technology (IT) policy of the organization also needs to be considered. Data security and export control (e.g., International Traffic in Arms Regulations (ITAR)) policies are two important IT policy considerations that will influence the selection of a particular design tool.

The policy of the organization may also dictate requirements on the format of the design data that is produced by a tool. A specific format may be required for sharing information with collaborating parties. Other considerations are the organizations' quality processes, which control the versions of the software tools as well as their verification and validation. There are also policies on training and certifying users of tools supporting critical flight programs and projects. This is particularly important when the selection of a new tool results in the transition from a legacy tool to a new tool. Therefore, the quality of the training support provided by the tool vendor is an important consideration in the selection of any tool.

Also, if a tool is being procured to support a multi-Center program or project, then program policy may dictate which tool must be used by all participating Centers. If Centers are free to select their own tool in support of a multi-Center program or project, then consideration of the policies of all the other Centers must be taken into account to ensure compatibility among Centers.

7.3.3 Collaboration

The design process is highly collaborative due to the complex specialties that must interact to achieve a successful integrated design. Tools are an important part of a successful collaboration. To successfully select and integrate tools in this environment requires a clear understanding of the intended user community size, functionality required, nature of the data to be shared, and knowledge of tools to be used. These factors will dictate the number of licenses, hosting capacity, tool capabilities, IT security requirements, and training required. The sharing of common models across a broad group requires mechanisms for advancing the design in a controlled way. Effective use of data management tools can help control the collaborative design by requiring common naming conventions, markings, and design techniques to ensure compatibility among distributed design tools.

7.3.4 Design Standards

Depending on the specific domain or discipline, there may be industry and Center-specific standards that must be followed, particularly when designing hardware. This can be evident in the design of a mechanical part, where a mechanical computer-aided design package selected to model the parts must have the capability to meet specific standards, such as model accuracy, dimensioning and tolerancing, the ability to create different geometries, and the capability to produce annotations describing how to build and inspect the part. However, these same issues must be considered regardless of the product.

7.3.5 Existing IT Architecture

As with any new tool decision, an evaluation of defined Agency and Center IT architectures should be made that focuses on compatibility with and duplication of existing tools. Typical architecture considerations would include data management tools, middleware or integration infrastructure, network transmission capacity, design analysis tools, manufacturing equipment, approved hosting, and client environments.

While initial focus is typically placed on current needs, the scalability of the tools and the supporting IT infrastructure should be addressed too. Scalability applies to both the number of users and capacity of each user to successfully use the system over time.

7.3.6 Tool Interfaces

Information interfaces are ubiquitous, occurring whenever information is exchanged.

This is particularly characteristic of any collaborative environment. It is here that inefficiencies arise, infor-

mation is lost, and mistakes are made. There may be an organizational need to interface with other capabilities and/or analysis tools, and understanding the tools used by the design teams with which your team interfaces and how the outputs of your team drive other downstream design functions is critical to ensure compatibility of data.

For computer-aided systems engineering tools, users are encouraged to select tools that are compatible with the Object Management Group System Modeling Language (SysML) standard. SysML is a version of the Unified Modeling Language (UML) that has been specifically developed for systems engineering.

7.3.7 Interoperability and Data Formats

Interoperability is an important consideration when selecting tools. The tools must represent the designs in formats that are acceptable to the end user of the data. It is important that any selected tool include associative data exchange and industry-standard data formats. As the Agency increasingly engages in multi-Center programs and projects, the need for interoperability among different tools, and different versions of the same tool, becomes even more critical. True interoperability reduces human error and the complexity of the integration task, resulting in reduced cost, increased productivity, and a quality product.

When considering all end users' needs, it is clear that interoperability becomes a difficult challenge. Three broad approaches, each with their own strengths and weaknesses, are:

- Have all employees become proficient in a variety of different tool systems and the associated end use applications. While this provides a broad capability, it may not be practical or affordable.

- Require interoperability among whatever tools are used, i.e., requiring that each tool be capable of transferring model data in a manner that can be easily and correctly interpreted by all the other tools. Considerable progress has been made in recent years in the standards for the exchange of model data. While this would be the ideal solution for many, standard data formats that contain the required information for all end users do not yet exist.

- Dictate that all participating organizations use the same version of the same tool.

7.3.8 Backward Compatibility

On major programs and projects that span several years, it is often necessary to access design data that are more than 3 to 5 years old. However, access to old design data can be extremely difficult and expensive, either because tool vendors end their support or later versions of the tool can no longer read the data. Strategies for maintaining access include special contracts with vendors for longer support, archiving design data in neutral formats, continuous migration of archives into current formats, and recreating data on demand. Organizations should select the strategy that works best for them, after a careful consideration of the cost and risk.

7.3.9 Platform

While many tools will run on multiple hardware platforms, some perform better in specific environments or are only supported by specified versions of operating systems. In the case of open-source operating systems, many different varieties are available that may not fully support the intended tools. If the tool being considered requires a new platform, the additional procurement cost and administration support costs should be factored in.

7.3.10 Tool Configuration Control

Tool configuration control is a tradeoff between responsive adoption of the new capabilities in new versions and smooth operation across tool chain components. This is more difficult with heterogeneous (multiple vendor) tool components. An annual or biannual block upgrade strategy requires significant administrative effort. On the other hand, the desktop diversity resulting from user-managed upgrade timing also increases support requirements.

7.3.11 Security/Access Control

Special consideration should be given to the sensitivity and required access of all design data. Federal Government and Agency policy requires the assessment of all tools to ensure appropriate security controls are addressed to maintain the integrity of the data.

7.3.12 Training

Most of the major design tools have similar capabilities that will not be new concepts to a seasoned designer. However, each design tool utilizes different techniques to perform design functions, and each contains some unique tool sets that will require training. The more responsive vendors will provide followup access to instructors and onsite training with liberal distribution of training materials and worked examples. The cost and time to perform the training and time for the designer to become proficient can be significant and should be carefully factored in when making decisions on new design tools.

The disruptive aspect of training is an important consideration in adapting to a different tool. Before transitioning to a new tool, an organization must consider the schedule of deliverables to major programs and projects. Can commitments still be met in a timely fashion? It is suggested that organizations implement a phase-in approach to a new tool, where the old tool is retained for some time to allow people to learn the new tool and become proficient in its use. The transition of a fully functional and expert team using any one system, to the same team fully functional using another system, is a significant undertaking. Some overlap between the old tool and the new tool will ensure flexibility in the transition and ensure that the program and project work proceeds uninterrupted.

7.3.13 Licenses

Licenses provide and control access to the various modules or components of a product or product family. Consideration of the license scheme should be taken into account while selecting a tool package. Licenses are sometimes physical, like a hardware key that plugs into a serial or parallel port, or software that may or may not require a whole infrastructure to administer. Software licenses may be floating (able to be shared on many computers on a first-come, first-served basis) or locked (dedicated to a particular computer). A well-thought-out strategy for licenses must be developed in the beginning of the tool selection process. This strategy must take into consideration program and project requirements and constraints as well as other factors such as training and use.

7.3.14 Stability of Vendor and Customer Support

As in the selection of any support device or tool, vendor stability is of great importance. Given the significant investment in the tools (directly) and infrastructure (indirectly), it is important to look at the overall company stability to ensure the vendor will be around to support the tools. Maturity of company products, installed user base, training, and financial strength can all provide clues to the company's ability to remain in the marketplace with a viable product. In addition, a responsive vendor provides customer support in several forms. A useful venue is a Web-based user-accessible knowledge base that includes resolved issues, product documentation, manuals, white papers, and tutorials. Live telephone support can be valuable for customers who don't provide support internally. An issue resolution and escalation process involves customers directly in prioritizing and following closure of critical issues. Onsite presence by the sales team and application engineers, augmented by post-sales support engineers, can significantly shorten the time to discovery and resolution of issues and evolving needs.

7.4 Human Factors Engineering

The discipline of Human Factors (HF) is devoted to the study, analysis, design, and evaluation of human-system interfaces and human organizations, with an emphasis on human capabilities and limitations as they impact system operation. HF engineering issues relate to all aspects of the system life, including design, build, test, operate, and maintain, across the spectrum of operating conditions (nominal, contingency, and emergency).

People are critical components in complex aerospace systems: designers, manufacturers, operators, ground support, and maintainers. All elements of the system are influenced by human performance. In the world of human-system interaction, there are four avenues for improving performance, reducing error, and making systems more error tolerant: (1) personnel selection; (2) system, interface, and task design; (3) training; and (4) procedure improvement. Most effective performance improvement involves all four avenues. People can be highly selected for the work they are to perform and the environment they are to perform it in. Second, equipment and systems can be designed to be easy to use, error resistant, and quickly learned. Third, people can be trained to proficiency on their required tasks. Fourth, improving tasks or procedures can be an important intervention.

HF focuses on those aspects where people interface with the system. It considers all personnel who must interact with the system, not just the operator; deals with organizational systems as well as hardware; and examines all types of interaction, not just hardware or software interfaces. The role of the HF specialist is to advocate for the human component and to ensure that the design of hardware, software, tasks, and environment is compatible with the sensory, perceptual, cognitive, and physical attributes of those interacting with the system. The HF specialist should elucidate why human-related issues or features should be included in analyses, design decisions, or tests and explain how design options will affect human performance in ways that impact total system performance and/or cost. As system complexity grows, the potential for conflicts between requirements increases. Sophisticated human-system interfaces create conflicts such as the need to create systems that are easy for novices to learn while also being efficient for experts to use. The HF specialist recognizes these tradeoffs and constraints and provides guidance on balancing these competing requirements. The domain of application is anywhere there are concerns regarding human and organizational performance, error, safety, and comfort. The goal is always to inform and improve the design.

What distinguishes an HF specialist is the particular knowledge and methods used, the domain of employment, and the goal of the work. HF specialists have expertise in the knowledge of human performance, both general and specific. There are many academic specialties concerned with applying knowledge of human behavior. These include psychology, cognitive science, cognitive psychology, sociology, economics, instructional system development, education, physiology, industrial psychology, organizational behavior, communication, and industrial engineering. Project and/or process managers should consult with their engineering or SMA directorates to get advice and recommendations on specific HF specialists who would be appropriate for their particular activity.

It is recommended to consider having HF specialists on the team throughout all the systems engineering common technical processes so that they can construct the specific HF analysis techniques and tests customized to the specific process or project. Not only do the HF specialists help in the development of the end items in question, but they should also be used to make sure that the verification test and completeness techniques are compatible and accurate for humans to undertake. Participation early in the process is especially important. Entering the system design process early ensures that human systems requirements are "designed in" rather than corrected later. Sometimes the results of analyses performed later call for a reexamination of earlier analyses. For example, functional allocation typically must be refined as design progresses because of technological breakthroughs, unforeseen technical difficulties in design or programming, or task analysis may indicate that some tasks assigned to humans exceed human capabilities under certain conditions.

During requirements definition, HF specialists ensure that HF-related goals and constraints are included in the overall plans for the system. The HF specialist must identify the HF-related issues, design risks, and tradeoffs pertinent to each human-system component, and document these as part of the project's requirements so they are adequately addressed during the design phase. For stakeholder expectation definition from the HF perspective, the stakeholders include not only those who are specifying the system to be built, but also those who will be utilizing the system when it is put into operation. This approach yields requirements generated from the top down—what the system is intended to accomplish—and from the bottom up—how the system is anticipated to function. It is critical that the HF specialist contribute to the ConOps. The expectations of the role of the human in the system and the types of tasks the human is expected to perform underlie all the hardware and software requirements. The difference between a passive passenger and an active operator will drive major design decisions. The number of crewmembers will drive subsequent decisions about habitable volume and storage and about crew time available for operations and maintenance. HF specialists ensure appropriate system design that defines the environmental range in which the system will operate and any factors that impact the human components. Many of these factors will need to accommodate human, as well as machine, tolerances. The requirements may need to specify acceptable atmospheric conditions, including temperature, pressure, composition, and humidity, for example. The requirements might also address acceptable ranges of acoustic noise, vibration, acceleration, and gravitational forces, and the use of protective clothing. The requirements may also need to accommodate adverse or emergency conditions outside the range of normal operation.

7.4.1 Basic HF Model

A key to conducting human and organizational analysis, design, and testing is to have an explicit framework that relates and scopes the work in question. The following model identifies the boundaries and the components involved in assessing human impacts.

The HF interaction model (Figure 7.4-1) provides a reference point of items to be aware of in planning, analyzing, designing, testing, operating, and maintaining systems. Detailed checklists should be generated and customized for the particular system under development. The model presented in this module is adapted from David Meister's *Human Factors: Theory and Practice* and is one depiction of how humans and systems interact. Environmental influences on that interaction have been added. The model illustrates a typical information flow between the human and machine components of a system.

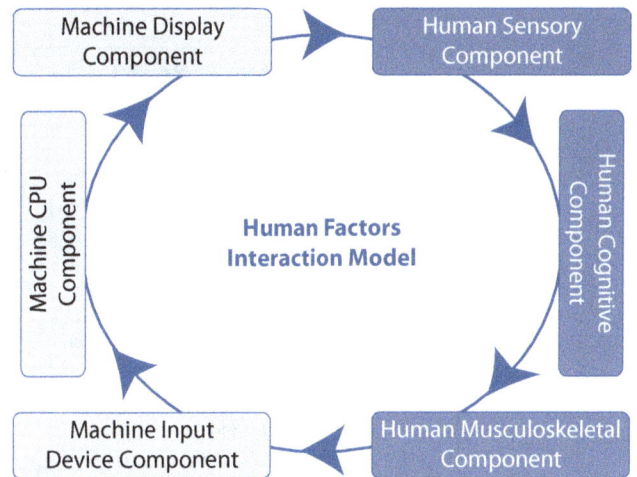

Figure 7.4-1 Human factors interaction model

Figure 7.4-2 provides a reference point of *human factors process phases* to be aware of in planning, analyzing, designing, testing, operating, and maintaining systems.

7.4.2 HF Analysis and Evaluation Techniques

Table 7.4-1 provides a set of techniques for human and organizational analysis and evaluation that can help to ensure that appropriate human and organizational factors have been considered and accommodated. These HF analysis methods are used to analyze systems, provide data about human performance, make predictions about human-system performance, and evaluate if the human-machine system performance meets design criteria. Most methods involve judgment and so are highly dependent on the skill and expertise of the analyst. In addition, both experienced and inexperienced operators provide valuable information about the strengths and weaknesses of old systems and how the new system might be used.

These methods are appropriate to all phases of system design with increasing specificity and detail as development progresses. While HF principles are often researched and understood at a generic level, their application is only appropriate when tailored to fit the design phase. Each type of analysis yields different kinds of information and so they are not interchangeable. The outputs or products of the analyses go into specification documents (operational needs document, ConOps, System Requirements Document (SRD), etc.) and formal review processes (e.g., ORR, SRR, SDR, PDR, CDR, PRR, PIR).

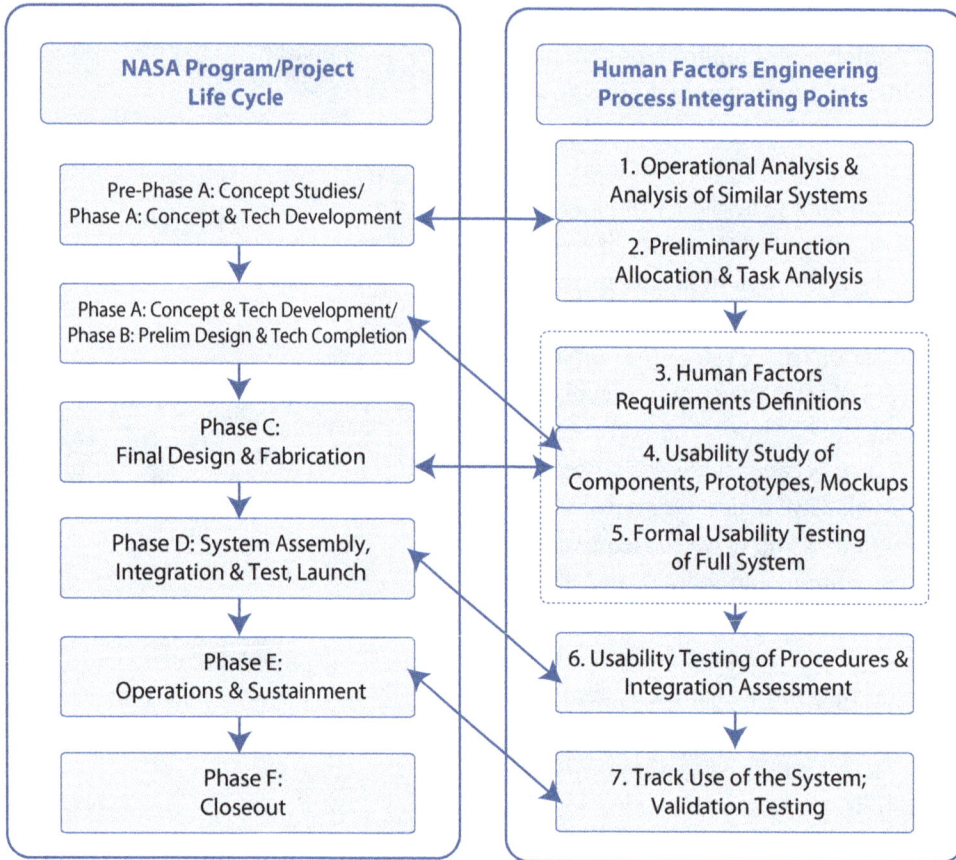

**Figure 7.4-2 HF engineering process and its links to the NASA program/
project life cycle**

The list shown in Table 7.4-1 is not exhaustive. The main point is to show examples that demonstrate the scope and usefulness of common methods used to evaluate system design and development.

Table 7.4-1 Human and Organizational Analysis Techniques

Process	Human/Individual Analysis	Additional Organizational Analysis
A. Operational Analysis		
Definition	When applied to HF, it is analysis of projected operations.	
Purpose	Obtain information about situations or events that may confront operators and maintainers using the new system. Systems engineers or operations analysts have typically done operational analyses. HF specialists should also be members of the analysis team to capture important operator or maintainer activities.	
Inputs	RFPs, planning documents, system requirements documents, and expert opinion.	
Process	Consult the systems engineer and projected users to extract implications for operators and maintainers.	Assess interactions and logistics between individuals and organizations. Evaluate operations under different types of organizations, structures, or distributions.
Outputs	Detailed scenarios (for nominal operations, hard and soft failures, and emergencies) including consequences; verbal descriptions of events confronting operators and maintainers; anticipated operations (list of feasible operations and those that may overstress the system); assumptions; constraints that may affect system performance; environments; list of system operation and maintenance requirements.	New or adjusted workflows to compensate for organizational impacts as appropriate.
B. Similar Systems Analysis		
Definition	When applied to HF, it examines previous systems or systems in use for information useful for the new system.	
Purpose	To obtain lessons learned and best practices useful in planning for a new system. Experiences gained from systems in use is valuable information that should be capitalized on.	
Inputs	Structured observations, interviews, questionnaires, activity analysis, accident/incident reports, maintenance records, and training records.	
Process	Obtain data on the operability, maintainability, and number of people required to staff the system in use. Identify skills required to operate and maintain the system and training required to bring operators to proficiency. Obtain previous data on HF design problems and problems encountered by previous users of the previous system or system in use.	Identify the existing system's organizational hierarchies and management distribution schemes (centralized versus decentralized).
Outputs	Identification of environmental factors that may affect personnel; preliminary assessments of workload and stress levels; assessment of skills required and their impact on selection, training, and design; estimates of future staffing and manpower requirements; identification of operator and maintainer problems to avoid; assessment of desirability and consequences of reallocation of systems functions.	Evaluation of the configuration's impact on performance and its potential risks.

(continued)

Table 7.4-1 Human and Organizational Analysis Techniques (continued)

Process	Human/Individual Analysis	Additional Organizational Analysis
C. Critical Incident Study		
Definition	When applied to HF, it identifies sources of difficulties for operators or maintenance or in the operational systems (or simulations of them).	
Purpose	To analyze and hypothesize sources of errors and difficulties in a system. This is particularly useful when a system has been operational and difficulties are observed or suspected, but the nature and severity of those difficulties is not known.	
Inputs	Operator/maintainer accounts of accidents, near-accidents, mistakes, and near-mistakes.	
Process	Interview large numbers of operators/maintainers; categorize incidents/accidents; use HF knowledge and experience to hypothesize sources of difficulty and how each one could be further studied; mitigate or redesign to eliminate difficulty.	Trace difficulties between individuals and organizations and map associated responsibilities and process assignments.
Outputs	Sources of serious HF difficulties in the operation of a system or its maintenance with suggested solutions to those difficulties.	Identification of potential gaps or disconnects based on the mapping.
D. Functional Flow Analysis		
Definition	When applied to HF, it is a structured technique for determining system requirements. Decomposes the sequence of functions or actions that a system must perform.	
Purpose	Provides a sequential ordering of functions that will achieve system requirements and a detailed checklist of system functions that must be considered in ensuring that the system will be able to perform its intended mission. These functions are needed for the solution of trade studies and determinations of their allocation among operators, equipment, software, or some combination of them. Decision-action analysis is often used instead of a functional flow analysis when the system requires binary decisions (e.g., software-oriented).	
Inputs	Operational analyses, analyses of similar systems, activity analyses.	
Process	Top-level functions are progressively expanded to lower levels containing more and more detailed information. If additional elaboration is needed about information requirements, sources of information, potential problems, and error-inducing features, for example, then an action-information analysis is also performed.	Map functional flows to associated organizational structures.
Outputs	Functional flow diagrams.	Identification of any logistics or responsibility gaps based on the integrated map.
E. Action-Information Analysis		
Definition	When applied to HF, it elaborates each function or action in functional flows or decision-action diagrams by identifying the information that is needed for each action or decision to occur. This analysis is often supplemented with sources of data, potential problems, and error-inducing features associated with each function or action.	

(continued)

Table 7.4-1 Human and Organizational Analysis Techniques (continued)

Process	Human/Individual Analysis	Additional Organizational Analysis
Purpose	Provides more detail before allocating functions to agents.	
Inputs	Data from the analysis of similar systems, activity analyses, critical incident studies, functional flow and decision-action analyses, and comments and data from knowledgeable experts.	
Process	Each function or action identified in functional flows or decision-action analyses is elaborated.	Map associated components (function, action, decisions) to the responsible organizational structures.
Outputs	Detailed lists of information requirements for operator-system interfaces, early estimates of special personnel provisions likely to be needed, support requirements, and lists of potential problems and probable solutions. Often produces suggestions for improvements in design of hardware, software, or procedures.	Identification of any logistics or responsibility gaps based on the integrated map.
F. Functional Allocation		
Definition	When applied to HF, it is a procedure for assigning each system function, action, and decision to hardware, software, operators, maintainers, or some combination of them.	
Purpose	To help identify user skill needs and provide preliminary estimates of staffing, training, and procedures requirements and workload assessments. Functional flows and decision-action analyses do not identify the agent (person or machine) that will execute the functions.	
Inputs	Functional flow analyses, decision-action analyses, action-information analyses, past engineering experience with similar systems, state-of-the-art performance capabilities of machines and software, and store of known human capabilities and limitations.	
Process	Identify and place to the side all those functions that must be allocated to personnel or equipment for reasons of safety, limitations of engineering technology, human limitations, or system requirements. List the remaining functions—those that could be either performed manually or by some combination of personnel and equipment. Prepare descriptions of implementation. Establish weighting criteria for each design alternative. Compare alternative configurations in terms of their effectiveness in performing the given function according to those criteria.	After initial assignment configuration is completed, evaluate allocations against relevant organizational norms, values, and organizational interfaces for logistics and management impacts.
Outputs	Allocations of system functions to hardware, software, operators, maintainers, or some combination of them. Task analyses are then performed on those functions allocated to humans.	List of potential impacts with recommended modifications in either functions or management or both.
G. Task Analysis		
Definition	When applied to HF, it is a method for producing an ordered list of all the things people will do in a system.	
Purpose	To develop input to all the analyses that come next. A subsequent timeline analysis chart can provide the temporal relationship among tasks—sequences of operator or maintainer actions, the times required for each action, and the time at which each action should occur.	

(continued)

Table 7.4-1 Human and Organizational Analysis Techniques (continued)

Process	Human/Individual Analysis	Additional Organizational Analysis
Inputs	Data from all the methods above supplemented with information provided by experts who have had experience with similar systems.	
Process	HF specialists and subject matter experts list and describe all tasks, subdividing them into subtasks with the addition of supplementary information.	Group all tasks assigned to a given organization and evaluate the range of skills, communications, and management capabilities required. Evaluate new requirements against existing organization's standard operating procedures, norms, and values.
Outputs	Ordered list of all the tasks people will perform in a system. Details on information requirements, evaluations and decisions that must be made, task times, operator actions, and environmental conditions.	Identify group-level workloads, management impacts, and training requirements.
H. Fault Tree Analysis		
Definition	When applied to HF, it determines those combinations of events that could cause specific system failures, faults, or catastrophes. Fault tree and failure mode and effects analysis are concerned with errors.	
Purpose	Anticipate mistakes that operators or maintainers might make and try to design against those mistakes. Limitation for HF is that each event must be described in terms of only two possible conditions, and it is extremely difficult to attach exact probabilities to human activities.	
Inputs	All outputs of the methods described above, supplemented with data on human reliability.	
Process	Construct a tree with symbols (logic gates) to represent events and consequences and describe the logical relationship between events.	Anticipate mistakes and disconnects that may occur in the workflow between individuals and organizations including unanticipated interactions between standard organization operating procedures and possible system events.
Outputs	Probabilities of various undesirable workflow-related events, the probable sequences that would produce them, and the identification of sensitive elements that could reduce the probability of a mishap.	Probabilities of various undesirable workflow-related events arranged by organizational interface points for the workflows, the probable sequences that would produce them, and the identification of sensitive elements that could reduce the probability of a mishap.
I. Failure Modes and Effects Analysis		
Definition	When applied to HF, it is a methodology for identifying error-inducing features in a system.	
Purpose	Deduce the consequences for system performance of a failure in one or more components (operators and maintainers) and the probabilities of those consequences occurring.	

(continued)

Table 7.4-1 Human and Organizational Analysis Techniques (continued)

Process	Human/Individual Analysis	Additional Organizational Analysis
Inputs	All outputs of methods described above, supplemented with data on human reliability.	
Process	Analyst identifies the various errors operators or maintainers may make in carrying out subtasks or functions. Estimates are made of the probabilities or frequencies of making each kind of error. The consequences of each kind of error are deduced by tracing its effects through a functional flow diagram to its final outcome.	Identify possible organizational entities and behaviors (i.e., political) involved with the system. Estimate the probabilities of occurrence and the impact of consequence.
Outputs	List of human failures that would have critical effects on system operation, the probabilities of system or subsystem failures due to human errors, and identification of those human tasks or actions that should be modified or replaced to reduce the probability of serious system failures.	List organizational behaviors that would have a critical effect on the system operation; the probabilities of the system or subsystem failures due to the organizational behaviors; and those organizational values, culture, or actions/standard operating procedures that should be modified or replaced to reduce the probability of serious system failures.
J. Link Analysis		
Definition	When applied to HF, it is an examination of the relationships between components, including the physical layout of instrument panels, control panels, workstations, or work areas to meet certain objectives.	
Purpose	To determine the efficiency and the effectiveness of the physical layout of the human-machine interface.	
Inputs	Data from activity and task analysis and observations of functional or simulated systems.	
Process	List all personnel and items. Estimate frequencies of linkages between items, operators, or items and operators. Estimate the importance of each link. Compute frequency-importance values for each link. Starting with the highest link values, successively add items with lower link values and readjust to minimize linkages. Fit the layout into the allocated space. Evaluate the new layout against the original objectives.	Assess links at individual versus organizational levels.
Outputs	Recommended layouts of panels, workstations, or work areas.	Adjusted layouts of panels, workstations, or work areas based on optimum individual and organizational performance priorities.
K. Simulation		
Definition	When applied to HF, it is a basic engineering or HF method to predict performance of systems. Includes usability testing and prototyping.	
Purpose	To predict the performance of systems, or parts of systems, that do not exist or to allow users to experience and receive training on systems, or parts of systems, that are complex, dangerous, or expensive.	
Inputs	Hardware, software, functions, and tasks elucidated in task analysis, operating procedures.	

(continued)

Table 7.4-1 Human and Organizational Analysis Techniques (continued)

Process	Human/Individual Analysis	Additional Organizational Analysis
Process	Users perform typical tasks on models or mockups prepared to incorporate some or all of the inputs.	Assess individual performance within possible organizational models.
Outputs	Predictions about system performance, assessment of workloads, evaluation of alternative configurations, evaluation of operating procedures, training, and identification of accident- or error-provocative situations and mismatches between personnel and equipment.	Predict system performance under varied organizational conditions, assess workloads, evaluate alternative configurations and operating procedures, train, and identify accident- or error-provocative situations and mismatches between personnel and equipment.
L. Controlled Experimentation		
Definition	When applied to HF, it is a highly controlled and structured version of simulation with deliberate manipulation of some variables.	
Purpose	To answer one or more hypotheses and narrow number of alternatives used in simulation.	
Inputs	From any or all methods listed thus far.	
Process	Select experimental design; identify dependent, independent, and controlled variables; set up test, apparatus, facilities, and tasks; prepare test protocol and instructions; select subjects; run tests; and analyze results statistically.	Scale to organizational levels where appropriate and feasible.
Outputs	Quantitative statements of the effects of some variables on others and differences between alternative configurations, procedures, or environments.	Quantitative statements of the effects of some organizational variables on others and differences between alternative organizational configurations, procedures, or environments.
M. Operational Sequence Analysis		
Definition	When applied to HF, it is a powerful technique used to simulate systems.	
Purpose	To permit visualization of interrelationships between operators and operators and equipment, identify interface problems, and explicitly identify decisions that might otherwise go unrecognized. Less expensive than mockups, prototypes, or computer programs that attempt to serve the same purpose.	
Inputs	Data from all above listed methods.	
Process	Diagram columns show timescale, external inputs, operators, machines, external outputs. Flow of events (actions, functions, decisions) is then plotted from top to bottom against the timescale using special symbology.	After initial analysis is complete, group results by responsible organizations.
Outputs	Time-based chart showing the functional relationships among system elements, the flow of materials or information, the physical and sequential distribution of operations, the inputs and outputs of subsystems, the consequences of alternative design configurations, potential sources of human difficulties.	Assessment of range of skills, communication processes, and management capabilities required. Evaluation of performance under various organizational structures.

(continued)

Table 7.4-1 Human and Organizational Analysis Techniques (continued)

Process	Human/Individual Analysis	Additional Organizational Analysis
N. Workload Assessment		
Definition	When applied to HF, it is a procedure for appraising operator and crew task loadings or the ability of personnel to carry out all assigned tasks in the time allotted or available.	
Purpose	To keep operator workloads at reasonable levels and to ensure that workloads are distributed equitably among operators.	
Inputs	Task time, frequency, and precision data are obtained from many of the above listed methods supplemented with judgments and estimates from knowledgeable experts.	
Process	DOD-HDBK-763 recommends a method that estimates the time required to perform a task divided by the time available or allotted to perform it. There are three classes of methods: performance measures, physiological measures, and subjective workloads either during or after an activity.	After initial analysis is complete, group results by responsible organizations.
Outputs	Quantitative assessments of estimated workloads for particular tasks at particular times.	Assessment of range of skills, communication processes, and management capabilities required. Evaluation of performance under various organizational structures.
O. Situational Awareness		
Definition	When applied to HF, it is a procedure for appraising operator and crew awareness of tasks and current situation.	
Purpose	To raise operator and maintainer awareness to maintain safety and efficiency.	
Inputs	All of the above listed analyses.	
Process	Different methods have been proposed including: situation awareness rating technique, situation awareness behaviorally anchored rating scale, situation awareness global assessment technique, situation awareness verification and analysis tool.	Collect organizational decisionmaking structures and processes and map the organization's situational awareness profiles.
Outputs	Quantitative estimates of situational awareness for particular tasks at particular times.	Identification of possible gaps, disconnects, and shortfalls.
P. Performance Modeling		
Definition	When applied to HF, it is a computational process for predicting human behavior based on current cognitive research.	
Purpose	To predict human limitations and capabilities before prototyping.	
Inputs	All of the above listed analyses.	
Process	Input results from above analyses. Input current relevant environmental and machine parameters. Can be interleaved with fast-time simulation to obtain frequency of error types.	Scale as appropriate to relevant organizational behaviors.
Outputs	Interrelationships between operators and operators and equipment and identification of interface problems and decisions that might otherwise go unrecognized.	Interrelationships between individuals and organizations and identification of organizational interface problems and decisions that might otherwise go unrecognized.

7.5 Environmental, Nuclear Safety, Planetary Protection, and Asset Protection Policy Compliance

7.5.1 NEPA and EO 12114

7.5.1.1 National Environmental Policy Act

The National Environmental Policy Act (NEPA) declares the basic national policy for protecting the human environment. NEPA sets the Nation's goals for enhancing and preserving the environment. NEPA also provides the procedural requirements to ensure compliance by all Federal agencies. NEPA compliance can be a critical path item in project or mission implementation. NEPA requires all Federal agencies to consider, before an action is taken, environmental values in the planning of actions and activities that may have a significant impact upon the quality of the human environment. NEPA directs agencies to consider alternatives to their proposed activities. In essence, NEPA requires NASA decisionmakers to integrate the NEPA process into early planning to ensure appropriate consideration of environmental factors, along with technical and economic ones. NEPA is also an environmental disclosure statute. It requires that available information be adequately addressed and made available to the NASA decisionmakers in a timely manner so they can consider the environmental consequences of the proposed action or activity before taking final action. Environmental information must also be made available to the public as well as to other Federal, state, and local agencies. NEPA does not require that the proposed action or activity be free of environmental impacts, be the most environmentally benign of potential alternatives, or be the most environmentally wise decision. NEPA requires the decisionmaker to consider environmental impacts as one factor in the decision to implement an action.

NASA activities are implemented through specific sponsoring entities, such as NASA HQ, NASA Centers (including component facilities, e.g., Wallops Flight Facility, White Sands Test Facility, and Michoud Assembly Facility), mission directorates, program, or mission support offices. The lead officials for these entities, the officials in charge, have the primary responsibility for ensuring that the NEPA process is integrated into their organizations' project planning activities before the sponsoring entities implement activities and actions. The sponsoring entities also are responsible for ensuring that records management requirements are met. NEPA func-

tions are not performed directly by lead officials. Each NASA Center has an Environmental Management Office (EMO), which is usually delegated the responsibility for implementing NEPA. The EMO performs the primary or working-level functions of the NEPA process, such as evaluating proposed activities, developing and/or reviewing and approving required documentation, advising project managers, and signing environmental decision documents on projects and programs having little or no environmental impact. However, ultimate responsibility for complying with NEPA and completing the process in a timely manner lies with the program or project manager. Since the EMO provides essential functional support to the sponsoring entity, and because its implementation responsibilities are delegated, the term "sponsoring entity" will be used throughout to include the implementing NEPA organization at any NASA facility. In cases where the sponsoring entity needs to be further defined, it will be specifically noted. For proposals made by tenants or entities using services or facilities at a NASA Center or component facility, the sponsoring entity shall be that Center or, if such authority is delegated to the component facility, the component facility.

NEPA compliance documentation must be completed before project planning reaches a point where NASA's ability to implement reasonable alternatives is effectively precluded (i.e., before hard decisions are made regarding project implementation). Environmental planning factors should be integrated into the Pre-Phase A concept study phase when a broad range of alternative approaches is being considered. In the Phase A concept development stage, decisions are made that could affect the Phase B preliminary design stage. At a minimum, an environmental evaluation should be initiated in the Phase A concept development stage. During this stage, the responsible project manager will have the greatest latitude in making adjustments in the plan to mitigate or avoid important environmental sensitivities and in planning the balance of the NEPA process to avoid unpleasant surprises later in the project cycle which may have schedule and/or cost implications. Before completing the NEPA process, no NASA official can take an action that would (1) affect the environment or (2) limit the choice of reasonable alternatives.

Accommodating environmental requirements early in project planning ultimately conserves both budget and schedule. Further detail regarding NEPA compliance requirements for NASA programs and projects can be found in *NPR 8580.1, Implementing The National Environmental Policy Act and Executive Order 12114.*

7.5.1.2 EO 12114 Environmental Effects Abroad of Major Federal Actions

Executive Order (EO) 12114 was issued "solely for the purpose of establishing internal procedures for Federal agencies to consider the significant effects of their actions on the environment outside the United States, its territories and possessions." The EO also specifically provided that its purpose is to enable the decisionmakers of the Federal agencies to be informed of pertinent environmental considerations, and factor such considerations in their decisions; however, such decisionmakers must still take into account considerations such as foreign policy, national security, and other relevant special circumstances.

The NASA Office of the General Counsel (OGC), or designee, is the NASA point of contact and official NASA representative on any matter involving EO 12114. Accordingly, any action by, or any implementation or legal interpretation of EO 12114 requires consultations with and the concurrence of the designee of the OGC. The sponsoring entity and local EMO contemplating an action that would have global environmental effects or effects outside the territorial jurisdiction of the United States must notify the NASA Headquarters/Environmental Management Division (HQ/EMD). The HQ/EMD will, in turn, coordinate with the Office of the General Counsel, the Assistant Administrator for External Relations, and other NASA organizations as appropriate; and assist the sponsoring entity to develop a plan of action. (Such a plan is subject to the concurrence of the OGC.) Further detail regarding EO 12114 compliance requirements for NASA programs and projects can be found in NPR 8580.1.

7.5.2 PD/NSC-25

NASA has procedural requirements for characterizing and reporting potential risks associated with a planned launch of radioactive materials into space, on launch vehicles and spacecraft, during normal or abnormal flight conditions. Procedures and levels of review and analysis required for nuclear launch safety approval vary with the quantity of radioactive material planned for use and potential risk to the general public and the environment.

Specific details concerning these requirements can be found in *NPR 8715.3, NASA General Safety Program Requirements.*

For any U.S. space mission involving the use of radioisotope power systems, radioisotope heating units, nuclear reactors, or a major nuclear source, launch approval must be obtained from the Office of the President per Presidential Directive/National Security Council Memorandum No. 25 (PD/NSC-25), "Scientific or Technological Experiments with Possible Large-Scale Adverse Environmental Effects and Launch of Nuclear Systems into Space," paragraph 9, as amended May 8, 1996. The approval decision is based on an established and proven review process that includes an independent evaluation by an ad hoc Interagency Nuclear Safety Review Panel (INSRP) comprised of representatives from NASA, the Department of Energy (DOE), the Department of Defense, and the Environmental Protection Agency, with an additional technical advisor from the Nuclear Regulatory Commission. The process begins with development of a launch vehicle databook (i.e., a compendium of information describing the mission, launch system, and potential accident scenarios including their environments and probabilities). DOE uses the databook to prepare a Preliminary Safety Analysis Report for the space mission. In all, three Safety Analysis Reports (SARs) are typically produced and submitted to the mission's INSRP—the PSAR, an updated SAR (draft final SAR), and a final SAR. The DOE project office responsible for providing the nuclear power system develops these documents.

The ad hoc INSRP conducts its nuclear safety/risk evaluation and documents their results in a nuclear Safety Evaluation Report (SER). The SER contains an independent evaluation of the mission radiological risk. DOE uses the SER as its basis for accepting the SAR. If the DOE Secretary formally accepts the SAR-SER package, it is forwarded to the NASA Administrator for use in the launch approval process.

NASA distributes the SAR and SER to the other cognizant Government agencies involved in the INSRP, and solicits their assessment of the documents. After receiving responses from these agencies, NASA conducts internal management reviews to address the SAR and SER and any other nuclear safety information pertinent to the launch. If the NASA Administrator recommends proceeding with the launch, then a request for nuclear safety launch approval is sent to the director of the Office of Science and Technology Policy (OSTP) within the Office of the President.

NASA HQ is responsible for implementing this process for NASA missions. It has traditionally enlisted the Jet Propulsion Laboratory (JPL) to assist in this activity. DOE supports the process by analyzing the response of redundant power system hardware to the different accident scenarios identified in the databook and preparing a probabilistic risk assessment of the potential radiological consequences and risks to the public and the environment for the mission. KSC is responsible for overseeing development of databooks and traditionally uses JPL to characterize accident environments and integrate databooks. Both KSC and JPL subcontractors provide information relevant to supporting the development of databooks. The development team ultimately selected for a mission would be responsible for providing payload descriptions, describing how the nuclear hardware integrates into the spacecraft, describing the mission, and supporting KSC and JPL in their development of databooks.

Mission directorate associate administrators, Center Directors, and program executives involved with the control and processing of radioactive materials for launch into space must ensure that basic designs of vehicles, spacecraft, and systems utilizing radioactive materials provide protection to the public, the environment, and users such that radiation risk resulting from exposures to radioactive sources are as low as reasonably achievable. Nuclear safety considerations must be incorporated from the Pre-Phase A concept study stage throughout all project stages to ensure that the overall mission radiological risk is acceptable. All space flight equipment (including medical and other experimental devices) that contain or use radioactive materials must be identified and analyzed for radiological risk. Site-specific ground operations and radiological contingency plans must be developed commensurate with the risk represented by the planned launch of nuclear materials. Contingency planning, as required by the National Response Plan, includes provisions for emergency response and support for source recovery efforts. *NPR 8710.1, Emergency Preparedness Program* and *NPR 8715.2, NASA Emergency Preparedness Plan Procedural Requirements—Revalidated* address the NASA emergency preparedness policy and program requirements.

7.5.3 Planetary Protection

The United States is a signatory to the United Nations' Treaty of Principles Governing the Activities of States in the Exploration and Use of Outer Space, Including the Moon and Other Celestial Bodies. Known as the Outer Space Treaty, it states in part (Article IX) that exploration of the Moon and other celestial bodies shall be conducted "so as to avoid their harmful contamination and also adverse changes in the environment of the Earth resulting from the introduction of extraterrestrial matter." NASA policy (*NPD 8020.7, Biological Contamination Control for Outbound and Inbound Planetary Spacecraft*) specifies that the purpose of preserving solar system conditions is for future biological and organic constituent exploration. This NPD also establishes the basic NASA policy for the protection of the Earth and its biosphere from planetary and other extraterrestrial sources of contamination. The general regulations to which NASA flight projects must adhere are set forth in *NPR 8020.12, Planetary Protection Provisions for Robotic Extraterrestrial Missions*. Different requirements apply to different missions, depending on which solar system object is targeted or encountered and the spacecraft or mission type (flyby, orbiter, lander, sample return, etc.). For some bodies (such as the Sun, Moon, and Mercury), there are minimal planetary protection requirements. Current requirements for the outbound phase of missions to Mars and Europa, however, are particularly rigorous. Table 7.5-1 shows the current planetary protection categories, while Table 7.5-2 provides a brief summary of their associated requirements.

At the core, planetary protection is a project management responsibility and a systems engineering activity. The effort cuts across multiple WBS elements, and failure to adopt and incorporate a viable planetary protection approach during the early planning phases will add cost and complexity to the mission. Planning for planetary protection begins in Phase A, during which feasibility of the mission is established. Prior to the end of Phase A, the project manager must send a letter to the Planetary Protection Officer (PPO) stating the mission type and planetary targets and requesting that the mission be assigned a planetary protection category.

Prior to the PDR, at the end of Phase B, the project manager must submit to the NASA PPO a planetary protection plan detailing the actions that will be taken to meet the requirements. The project's progress and completion of the requirements are reported in a planetary protection pre-launch report submitted to the NASA PPO for approval. The approval of this report at the FRR con-

Table 7.5-1 Planetary Protection Mission Categories

Planet Priorities	Mission Type	Category	Example
Not of direct interest for understanding the process of chemical evolution. No protection of such planets is warranted (no requirements).	Any	I	Lunar missions
Of significant interest relative to the process of chemical evolution, but only a remote chance that contamination by spacecraft could jeopardize future exploration.	Any	II	Stardust (outbound) Genesis (outbound) Cassini
Of significant interest relative to the process of chemical evolution and/or the origin of life or for which scientific opinion provides a significant chance of contamination which could jeopardize a future biological experiment.	Flyby, Orbiter	III	Odyssey Mars Global Surveyor Mars Reconnaissance Orbiter
	Lander, Probe	IV	Mars Exploration Rover Phoenix Europa Explorer Mars Sample Return (outbound)
Any solar system body.	Unrestricted Earth return[a]	V	Stardust (return) Genesis (return)
	Restricted Earth return[b]	V	Mars Sample Return (return)

a. No special precautions needed for returning material/samples back to Earth.

b. Special precautions need to be taken for returning material/samples back to Earth. See NPR 8020.12.

Table 7.5-2 Summarized Planetary Protection Requirements

Mission Category	Summarized Requirements
I	Certification of category.
II	Avoidance of accidental impact by spacecraft and launch vehicle. Documentation of final disposition of launched hardware.
III	Stringent limitations on the probability of impact. Requirements on orbital lifetime or requirements for microbial cleanliness of spacecraft.
IV	Stringent limitations on the probability of impact and/or the contamination of the object. Microbial cleanliness of landed hardware surfaces directly established by bioassays.
V	Outbound requirements per category of a lander mission to the target. Detailed restricted Earth return requirements will depend on many factors, but will likely include sterilization of any hardware that contacted the target planet before its return to Earth, and the containment of any returned sample.

stitutes the final planetary protection approval for the project and must be obtained for permission to launch. An update to this report, the planetary protection post-launch report, is prepared to report any deviations from the planned mission due to actual launch or early mission events. For sample return missions only, additional reports and reviews are required: prior to launch toward the Earth, prior to commitment to Earth reentry, and prior to the release of any extraterrestrial sample to the scientific community for investigation. Finally, at the formally declared End of Mission (EOM), a planetary protection EOM report is prepared. This document reviews the entire history of the mission in comparison to the original planetary protection plan and documents the degree of compliance with NASA's planetary protection requirements. This document is typically reported on by the NASA PPO at a meeting of the Committee on Space Research (COSPAR) to inform other spacefaring nations of NASA's degree of compliance with international planetary protection requirements.

7.5.4 Space Asset Protection

The terrorist attacks on the World Trade Center in New York and on the Pentagon on September 11, 2001 have created an atmosphere for greater vigilance on the part of Government agencies to ensure that sufficient security is in place to protect their personnel, physical assets, and information, especially those assets that contribute to the political, economic, and military capabilities of the United States. Current trends in technology proliferation, accessibility to space, globalization of space programs and industries, commercialization of space systems and services, and foreign knowledge about U.S. space systems increase the likelihood that vulnerable U.S. space systems may come under attack, particularly vulnerable systems. The ability to restrict or deny freedom of access to and operations in space is no longer limited to global military powers. The reality is that there are many existing capabilities to deny, disrupt, or physically destroy orbiting spacecraft and the ground facilities that command and control them. Knowledge of U.S. space systems' functions, locations, and physical characteristics, as well as the means to conduct counterspace operations is increasingly available on the international market. Nations or groups hostile to the United States either possess or can acquire the means to disrupt or destroy U.S. space systems by attacking satellites in space, their communications nodes on the ground and in space, the ground nodes that command these satellites or process their data, and/or the commercial infrastructure that supports a space system's operations.

7.5.4.1 Protection Policy

The new National Space Policy authorized by the President on August 31, 2006, states that space capabilities are vital to the Nation's interests and that the United States will "take those actions necessary to protect its space capabilities." The policy also gives responsibility for Space Situational Awareness (SSA) to the Secretary of Defense. In that capacity the Secretary of Defense will conduct SSA for civil space capabilities and operations, particularly human space flight activities. SSA provides an indepth knowledge and understanding of the threats posed to U.S., allied, and coalition space systems by adversaries and the environment, and is essential in developing and employing protection measures. Therefore, NASA's space asset protection needs will drive the requirements that the NASA levies on DOD for SSA.

7.5.4.2 Goal

The overall space asset protection goal for NASA is to support sustained mission assurance through the reduction of susceptibilities and the mitigation of vulnerabilities, relative to risk, and within fiscal constraints.

7.5.4.3 Scoping

Space asset protection involves the planning and implementation of measures to protect NASA space assets from intentional or unintentional disruption, exploitation, or attack, whether natural or manmade. It is essential that protection is provided for all segments of a space system (ground, communications/information, space, and launch) and covers the entire life cycle of a project. Space asset protection includes aspects of personnel, physical, information, communications, information technology, and operational security, as well as counterintelligence activities. The role of the systems engineer is to integrate security competencies with space systems engineering and operations expertise to develop mission protection strategies consistent with payload classifications as defined in *NPR 8705.4, Risk Classification for NASA Payloads*.

7.5.4.4 Protection Planning

Systems engineers use protection planning processes and products (which include engineering trade studies and cost-benefit analyses) to meet NASA's needs for acquiring, fielding, and sustaining secure and uncompromised space systems. Project protection plans are single-source documents that coordinate and integrate protection efforts and prevent inadvertent or uncontrolled disclosure of sensitive program information. Protection plans provide project management personnel (project manager, project scientist, mission systems engineer, operations manager, user community, etc.) with an overall view of the valid threats to a space system (both hostile and environmental), identify infrastructure vulnerabilities, and propose security countermeasures to mitigate risks and enhance survivability of the mission. An outline for a typical protection plan can be found in Appendix Q.

7.6 Use of Metric System

The decision whether a project or program could or should implement the System Internationale (SI), often called the "metric system," requires consideration of a number of factors, including cost, technical, risk, and other programmatic aspects.

The Metric Conversion Act of 1975 (Public Law 94-168) amended by the Omnibus Trade and Competitiveness Act of 1988 (Public Law 100-418) establishes a national goal of establishing the metric system as the preferred system of weights and measures for U.S. trade and commerce. NASA has developed *NPD 8010.2, Use of the SI (Metric) System of Measurement in NASA Programs*, which implements SI and provides specific requirements and responsibilities for NASA.

However, a second factor to consider is that there are possible exceptions to the required implementation approach. Both EO 12770 and NPD 8010.2 allow exceptions and, because full SI implementation may be difficult, allow the use of "hybrid" systems. Consideration of the following factors will have a direct impact on the implementation approach and use of exceptions by the program or project.

Programs or projects must do analysis during the early life-cycle phases when the design solutions are being developed to identify where SI is feasible or recommended and where exceptions will be required. A major factor to consider is the capability to actually produce or provide metric-based hardware components. Results and recommendations from these analyses must be presented by SRR for approval.

In planning program or project implementation to produce metric-based systems, issues to be addressed should include the following:

- Interfaces with heritage components (e.g., valves, pyrotechnic devices, etc.) built to English-based units:
  - ▶ Whether conversion from English to SI and/or interface to English-based hardware is required.
  - ▶ The team should review design implementation to ensure there is no certification impact with heritage hardware or identify and plan for any necessary re-certification efforts.
- Dimensioning and tolerancing:
  - ▶ Can result in parts that do not fit.

- ▶ Rounding errors have occurred when converting units from one unit system to the other.
- ▶ The team may require specific additional procedures, steps, and drawing Quality Assurance (QA) personnel when converting units.
- Tooling:
  - ▶ Not all shops have full metric tooling (e.g., drill bits, taps, end mills, reamers, etc.).
  - ▶ The team needs to inform potential contractors of intent to use SI and obtain feedback as to potential impacts.
- Fasteners and miscellaneous parts:
  - ▶ High-strength fastener choices and availability are more limited in metric sizes.
  - ▶ Bearings, pins, rod ends, bushings, etc., are readily available in English with minimal lead times.
  - ▶ The team needs to ascertain availability of acceptable SI-based fasteners in the timeframe needed.
- Reference material:
  - ▶ Some key aerospace reference materials are built only in English units, e.g., MIL-HDBK-5 (metallic material properties), and values will need to be converted when used.
  - ▶ Other key reference materials or commercial databases are built only in SI units.
  - ▶ The team needs to review the reference material to be used and ensure acceptable conversion controls are in place, if necessary.
- Corporate knowledge:
  - ▶ Many engineers presently think in English units, i.e., can relate to pressure in PSI, can relate to material strength in KSI, can relate to a tolerance of 0.003 inches, etc.
  - ▶ However, virtually all engineers coming out of school in this day and era presently think in SI units and have difficulty relating to English-based units such as slugs (for mass) and would require retraining with attendant increase in conversion errors.
  - ▶ The team needs to be aware of their program- or project-specific knowledge in English and SI units and obtain necessary training and experience.

- Industry practices:
  - ▸ Certain industries work exclusively in English units, and sometimes have their own jargon associated with English material properties. The parachute industry falls in this category, e.g., "600-lb braided Kevlar line."
  - ▸ Other industries, especially international suppliers, may work exclusively in metric units, e.g., "30-mm-thick raw bar stock."
  - ▸ The team needs to be aware of these unique cases and ensure both procurement and technical design and integration have the appropriate controls to avoid errors.
- Program or project controls: The team needs to consider, early in the SE process, what program- or project-specific risk management controls (such as configuration management steps) are required. This will include such straightforward concerns as the conversion(s) between system elements that are in English units and those in SI units or other, more complex, issues.

Several NASA projects have taken the approach of using both systems, which is allowed by NPD 8010.2. For example, the Mars soil drill project designed and developed their hardware using English-based components, while accomplishing their analyses using SI-based units. Other small-scale projects have successfully used a similar approach.

For larger or more dispersed projects or programs, a more systematic and complete risk management approach may be needed to successfully implement an SI-based system. Such things as standard conversion factors (e.g., from pounds to kilograms) should be documented, as should standard SI nomenclature. Many of these risk management aspects can be found in such documents as the National Institute of Standards and Technology's *Guide for the Use of the International System of Units (SI)* and the DOD *Guide for Identification and Development of Metric Standards*.

Until the Federal Government and the aerospace industrial base are fully converted to an SI-based unit system, the various NASA programs and projects will have to address their own level of SI implementation on a case-by-case basis. It is the responsibility of each NASA program and project management team, however, to comply with all laws and executive orders while still maintaining a reasonable level of risk for cost, schedule, and performance.

Appendix A: Acronyms

ACS	Attitude Control Systems
ACWP	Actual Cost of Work Performed
AD2	Advancement Degree of Difficulty Assessment
AHP	Analytic Hierarchy Process
AIAA	American Institute of Aeronautics and Astronautics
AO	Announcement of Opportunity
ASME	American Society of Mechanical Engineers
BAC	Budget at Completion
BCWP	Budgeted Cost for Work Performed
BCWS	Budgeted Cost for Work Scheduled
C&DH	Command and Data Handling
CACE	Capability for Accelerated Concurrent Engineering
CAIB	Columbia Accident Investigation Board
CAM	Control Account Manager or Cost Account Manager
CCB	Configuration Control Board
CDR	Critical Design Review
CE	Concurrent Engineering
CERR	Critical Event Readiness Review
CI	Configuration Item
CM	Configuration Management
CMC	Center Management Council
CMMI	Capability Maturity Model® Integration
CMO	Configuration Management Organization
CNSI	Classified National Security Information
CoF	Construction of Facilities
ConOps	Concept of Operations
COSPAR	Committee on Space Research
COTS	Commercial Off the Shelf
CPI	Critical Program Information or Cost Performance Index
CRM	Continuous Risk Management
CSA	Configuration Status Accounting
CWBS	Contract Work Breakdown Structure
DCR	Design Certification Review
DGA	Designated Governing Authority
DLA	Defense Logistics Agency
DM	Data Management
DOD	Department of Defense
DOE	Department of Energy
DODAF	DOD Architecture Framework
DR	Decommissioning Review

DRM	Design Reference Mission
EAC	Estimate at Completion
ECP	Engineering Change Proposal
ECR	Environmental Compliance and Restoration or Engineering Change Request
EEE	Electrical, Electronic, and Electromechanical
EFFBD	Enhanced Functional Flow Block Diagram
EIA	Electronic Industries Alliance
EMC	Electromagnetic Compatibility
EMI	Electromagnetic Interference
EMO	Environmental Management Office
EO	Executive Order
EOM	End of Mission
EV	Earned Value
EVM	Earned Value Management
FAD	Formulation Authorization Document
FAR	Federal Acquisition Requirement
FCA	Functional Configuration Audit
FDIR	Failure Detection, Isolation, And Recovery
FFBD	Functional Flow Block Diagram
FMEA	Failure Modes and Effects Analysis
FMECA	Failure Modes, Effects, and Criticality Analysis
FMR	Financial Management Requirements
FRR	Flight Readiness Review
FS&GS	Flight Systems and Ground Support
GEO	Geostationary
GFP	Government-Furnished Property
GMIP	Government Mandatory Inspection Point
GPS	Global Positioning Satellite
HF	Human Factors
HQ	Headquarters
HQ/EMD	NASA Headquarters/Environmental Management Division
HWIL	Hardware in the Loop
ICA	Independent Cost Analysis
ICD	Interface Control Document/Drawing
ICE	Independent Cost Estimate
ICP	Interface Control Plan
IDD	Interface Definition Document
IEEE	Institute of Electrical and Electronics Engineers
ILS	Integrated Logistics Support
INCOSE	International Council on Systems Engineering

INSRP	Interagency Nuclear Safety Review Panel		PERT	Program Evaluation and Review Technique
IPT	Integrated Product Team		PFAR	Post-Flight Assessment Review
IRD	Interface Requirements Document		PHA	Preliminary Hazard Analysis
IRN	Interface Revision Notice		PI	Performance Index/Principal Investigator
ISO	International Organization for Standardization		PIR	Program Implementation Review
IT	Information Technology or Iteration		PIRN	Preliminary Interface Revision Notice
ITA	Internal Task Agreement.		PKI	Public Key Infrastructure
ITAR	International Traffic in Arms Regulation		PLAR	Post-Launch Assessment Review
I&V	Integration and Verification		P(LOC)	Probability of Loss of Crew
IV&V	Independent Verification and Validation		P(LOM)	Probability of Loss of Mission
IWG	Interface Working Group		PMC	Program Management Council
JPL	Jet Propulsion Laboratory		PPBE	Planning, Programming, Budgeting, and Execution
KDP	Key Decision Point			
KSC	Kennedy Space Center		PPO	Planetary Protection Officer
LCCE	Life-Cycle Cost Estimate		PQASP	Program/Project Quality Assurance Surveillance Plan
LEO	Low Earth Orbit or Low Earth Orbiting			
LLIL	Limited Life Items List		PRA	Probabilistic Risk Assessment
LLIS	Lessons Learned Information System		PRD	Project Requirements Document
M&S	Modeling and Simulation		PRR	Production Readiness Review
MAUT	Multi-Attribute Utility Theory		P/SDR	Program/System Definition Review
MCDA	Multi-Criteria Decision Analysis		PSR	Program Status Review
MCR	Mission Concept Review		P/SRR	Program/System Requirements Review
MDAA	Mission Directorate Associate Administrator		PTR	Periodic Technical Reviews
MDR	Mission Definition Review		QA	Quality Assurance
MOE	Measure of Effectiveness		R&T	Research and Technology
MOP	Measure of Performance		RF	Radio Frequency
MOU	Memorandum of Understanding		RFA	Requests for Action
NASA	National Aeronautics and Space Administration		RFI	Request for Information
			RFP	Request for Proposal
NEDT	NASA Exploration Design Team		RID	Review Item Discrepancy
NEPA	National Environmental Policy Act		SAR	System Acceptance Review or Safety Analysis Report
NFS	NASA FAR Supplement			
NODIS	NASA On-Line Directives Information System		SBU	Sensitive But Unclassified
			SDR	System Definition Review
NIAT	NASA Integrated Action Team		SE	Systems Engineering
NOAA	National Oceanic and Atmospheric Administration		SEE	Single-Event Effects
			SEMP	Systems Engineering Management Plan
NPD	NASA Policy Directive		SER	Safety Evaluation Report
NPR	NASA Procedural Requirements		SI	System Internationale (metric system)
OCE	Office of the Chief Engineer		SIR	System Integration Review
OGC	Office of the General Counsel		SMA	Safety and Mission Assurance
OMB	Office of Management and Budget		SOW	Statement of Work
ORR	Operational Readiness Review		SP	Special Publication
OSTP	Office of Science and Technology Policy		SPI	Schedule Performance Index
OTS	Off-the-Shelf		SRB	Standing Review Board
PAR	Program Approval Review		SRD	System Requirements Document
PBS	Product Breakdown Structure		SRR	System Requirements Review
PCA	Physical Configuration Audit or Program Commitment Agreement		SSA	Space Situational Awareness
			STI	Scientific and Technical Information
PD/NSC	Presidential Directive/National Security Council		STS	Space Transportation System
			SysML	System Modeling Language
PDR	Preliminary Design Review		T&E	Test and Evaluation

TA	Technology Assessment	TRAR	Technology Readiness Assessment Report
TBD	To Be Determined	TRL	Technology Readiness Level
TBR	To Be Resolved	TRR	Test Readiness Review
TDRS	Tracking and Data Relay Satellite	TVC	Thrust Vector Controller
TDRSS	Tracking and Data Relay Satellite System	UML	Unified Modeling Language
TLA	Timeline Analysis	USML	United States Munitions List
TLS	Timeline Sheet	V&V	Verification and Validation
TMA	Technology Maturity Assessment	VAC	Variance at Completion
TPM	Technical Performance Measure	WBS	Work Breakdown Structure

Appendix B: Glossary

Term	Definition/Context
Acceptable Risk	The risk that is understood and agreed to by the program/project, governing authority, mission directorate, and other customer(s) such that no further specific mitigating action is required.
Acquisition	The acquiring by contract with appropriated funds of supplies or services (including construction) by and for the use of the Government through purchase or lease, whether the supplies or services are already in existence or must be created, developed, demonstrated, and evaluated. Acquisition begins at the point when Agency needs are established and includes the description of requirements to satisfy Agency needs, solicitation and selection of sources, award of contracts, contract financing, contract performance, contract administration, and those technical and management functions directly related to the process of fulfilling Agency needs by contract.
Activity	(1) Any of the project components or research functions that are executed to deliver a product or service or provide support or insight to mature technologies. (2) A set of tasks that describe the technical effort to accomplish a process and help generate expected outcomes.
Advancement Degree of Difficulty Assessment (AD²)	The process to develop an understanding of what is required to advance the level of system maturity.
Allocated Baseline (Phase C)	The allocated baseline is the approved performance-oriented configuration documentation for a CI to be developed that describes the functional and interface characteristics that are allocated from a higher level requirements document or a CI and the verification required to demonstrate achievement of those specified characteristics. The allocated baseline extends the top-level performance requirements of the functional baseline to sufficient detail for initiating manufacturing or coding of a CI. The allocated baseline is controlled by the NASA. The allocated baseline(s) is typically established at the Preliminary Design Review. Control of the allocated baseline would normally occur following the Functional Configuration Audit.
Analysis	Use of mathematical modeling and analytical techniques to predict the compliance of a design to its requirements based on calculated data or data derived from lower system structure end product validations.
Analysis of Alternatives	A formal analysis method that compares alternative approaches by estimating their ability to satisfy mission requirements through an effectiveness analysis and by estimating their life-cycle costs through a cost analysis. The results of these two analyses are used together to produce a cost-effectiveness comparison that allows decisionmakers to assess the relative value or potential programmatic returns of the alternatives.
Analytic Hierarchy Process	A multi-attribute methodology that provides a proven, effective means to deal with complex decision-making and can assist with identifying and weighting selection criteria, analyzing the data collected for the criteria, and expediting the decisionmaking process.
Approval	Authorization by a required management official to proceed with a proposed course of action. Approvals must be documented.
Approval (for Implementation)	The acknowledgment by the decision authority that the program/project has met stakeholder expectations and formulation requirements, and is ready to proceed to implementation. By approving a program/project, the decision authority commits the budget resources necessary to continue into implementation.
As-Deployed Baseline	The as-deployed baseline occurs at the Operational Readiness Review. At this point, the design is considered to be functional and ready for flight. All changes will have been incorporated into the documentation.

Term	Definition/Context
Baseline	An agreed-to set of requirements, designs, or documents that will have changes controlled through a formal approval and monitoring process.
Bidirectional Traceability	An association among two or more logical entities that is discernible in either direction (i.e., to and from an entity).
Brassboard	A research configuration of a system, suitable for field testing, that replicates both the function and configuration of the operational systems with the exception of nonessential aspects such as packaging.
Breadboard	A research configuration of a system, generally not suitable for field testing, that replicates both the function but not the actual configuration of the operational system and has major differences in actual physical layout.
Component Facilities	Complexes that are geographically separated from the NASA Center or institution to which they are assigned.
Concept of Operations (ConOps) (sometimes Operations Concept)	The ConOps describes how the system will be operated during the life-cycle phases to meet stakeholder expectations. It describes the system characteristics from an operational perspective and helps facilitate an understanding of the system goals. It stimulates the development of the requirements and architecture related to the user elements of the system. It serves as the basis for subsequent definition documents and provides the foundation for the long-range operational planning activities.
Concurrence	A documented agreement by a management official that a proposed course of action is acceptable.
Concurrent Engineering	Design in parallel rather than serial engineering fashion.
Configuration Items	A Configuration Item is any hardware, software, or combination of both that satisfies an end use function and is designated for separate configuration management. Configuration items are typically referred to by an alphanumeric identifier which also serves as the unchanging base for the assignment of serial numbers to uniquely identify individual units of the CI.
Configuration Management Process	A process that is a management discipline that is applied over a product's life cycle to provide visibility into and to control changes to performance and functional and physical characteristics. It ensures that the configuration of a product is known and reflected in product information, that any product change is beneficial and is effected without adverse consequences, and that changes are managed.
Context Diagram	A diagram that shows external systems that impact the system being designed.
Continuous Risk Management	An iterative process to refine risk management measures. Steps are to analyze risk, plan for tracking and control measures, track risk, carry out control measures, document and communicate all risk information, and deliberate throughout the process to refine it.
Contract	A mutually binding legal relationship obligating the seller to furnish the supplies or services (including construction) and the buyer to pay for them. It includes all types of commitments that obligate the Government to an expenditure of appropriated funds and that, except as otherwise authorized, are in writing.
Contractor	An individual, partnership, company, corporation, association, or other service having a contract with the Agency for the design, development, manufacture, maintenance, modification, operation, or supply of items or services under the terms of a contract to a program or project.
Control Account Manager	The person responsible for controlling variances at the control account level, which is typically at the subsystem WBS level. The CAM develops work and product plans, schedules, and time-phased resource plans. The technical subsystem manager/lead often takes on this role as part of their subsystem management responsibilities.
Control Gate (or milestone)	See "Key Decision Point."
Cost-Benefit Analysis	A methodology to determine the advantage of one alternative over another in terms of equivalent cost or benefits. It relies on totaling positive factors and subtracting negative factors to determine a net result.
Cost-Effectiveness Analysis	A systematic quantitative method for comparing the costs of alternative means of achieving the same equivalent benefit for a specific objective.

Term	Definition/Context
Critical Design Review	A review that demonstrates that the maturity of the design is appropriate to support proceeding with full-scale fabrication, assembly, integration, and test, and that the technical effort is on track to complete the flight and ground system development and mission operations in order to meet mission performance requirements within the identified cost and schedule constraints.
Critical Event (or key event)	An event that requires monitoring throughout the projected life cycle of a product that will generate critical requirements that would affect system design, development, manufacture, test, and operations (such as with an MOE, MOP, or TPM).
Critical Event Readiness Review	A review that confirms the project's readiness to execute the mission's critical activities during flight operation.
Customer	The organization or individual that has requested a product and will receive the product to be delivered. The customer may be an end user of the product, the acquiring agent for the end user, or the requestor of the work products from a technical effort. Each product within the system hierarchy has a customer.
Data Management	DM is used to plan for, acquire, access, manage, protect, and use data of a technical nature to support the total life cycle of a system.
Decision Analysis Process	A process that is a methodology for making decisions. It also offers techniques for modeling decision problems mathematically and finding optimal decisions numerically. The methodology entails identifying alternatives, one of which must be decided upon; possible events, one of which occurs thereafter; and outcomes, each of which results from a combination of decision and event.
Decision Authority	The Agency's responsible individual who authorizes the transition at a KDP to the next life-cycle phase for a program/project.
Decision Matrix	A methodology for evaluating alternatives in which valuation criteria typically are displayed in rows on the left side of the matrix, and alternatives are the column headings of the matrix. Criteria "weights" are typically assigned to each criterion.
Decision Support Package	Documentation submitted in conjunction with formal reviews and change requests.
Decision Trees	A portrayal of a decision model that displays the expected consequences of all decision alternatives by making discreet all "chance" nodes, and, based on this, calculating and appropriately weighting the possible consequences of all alternatives.
Decommissioning Review	A review that confirms the decision to terminate or decommission the system and assess the readiness for the safe decommissioning and disposal of system assets. The DR is normally held near the end of routine mission operations upon accomplishment of planned mission objectives. It may be advanced if some unplanned event gives rise to a need to prematurely terminate the mission, or delayed if operational life is extended to permit additional investigations.
Deliverable Data Item	Consists of technical data–requirements specifications, design documents, management data–plans, and metrics reports.
Demonstration	Use of a realized end product to show that a set of stakeholder expectations can be achieved.
Derived Requirements	For a program, requirements that are required to satisfy the directorate requirements on the program. For a project, requirements that are required to satisfy the program requirements on the project.
Descope	Taken out of the scope of a project.
Design Solution Definition Process	The process by which high-level requirements derived from stakeholder expectations and outputs of the Logical Decomposition Process are translated into a design solution.
Designated Governing Authority	The management entity above the program, project, or activity level with technical oversight responsibility.
Doctrine of Successive Refinement	A recursive and iterative design loop driven by the set of stakeholder expectations where a strawman architecture/design, the associated ConOps, and the derived requirements are developed.
Earned Value	The sum of the budgeted cost for tasks and products that have actually been produced (completed or in progress) at a given time in the schedule.

Term	Definition/Context
Earned Value Management	A tool for measuring and assessing project performance through the integration of technical scope with schedule and cost objectives during the execution of the project. EVM provides quantification of technical progress, enabling management to gain insight into project status and project completion costs and schedules. Two essential characteristics of successful EVM are EVM system data integrity and carefully targeted monthly EVM data analyses (i.e., risky WBS elements).
Enabling Products	The life-cycle support products and services (e.g., production, test, deployment, training, maintenance, and disposal) that facilitate the progression and use of the operational end product through its life cycle. Since the end product and its enabling products are interdependent, they are viewed as a system. Project responsibility thus extends to responsibility for acquiring services from the relevant enabling products in each life-cycle phase. When a suitable enabling product does not already exist, the project that is responsible for the end product may also be responsible for creating and using the enabling product.
Technical Cost Estimate	The cost estimate of the technical work on a project created by the technical team based on its understanding of the system requirements and operational concepts and its vision of the system architecture.
Enhanced Functional Flow Block Diagram	A block diagram that represents control flows and data flows as well as system functions and flow.
Entry Criteria	Minimum accomplishments each project needs to fulfill to enter into the next life-cycle phase or level of technical maturity.
Environmental Impact	The direct, indirect, or cumulative beneficial or adverse effect of an action on the environment.
Environmental Management	The activity of ensuring that program and project actions and decisions that potentially impact or damage the environment are assessed and evaluated during the formulation and planning phase and reevaluated throughout implementation. This activity must be performed according to all NASA policy and Federal, state, and local environmental laws and regulations.
Establish (with respect to processes)	The act of developing policy, work instructions, or procedures to implement process activities.
Evaluation	The continual, independent (i.e., outside the advocacy chain of the program/project) evaluation of the performance of a program or project and incorporation of the evaluation findings to ensure adequacy of planning and execution according to plan.
Extensibility	The ability of a decision to be extended to other applications.
Flexibility	The ability of a decision to support more than one current application.
Flight Readiness Review	A review that examines tests, demonstrations, analyses, and audits that determine the system's readiness for a safe and successful flight/launch and for subsequent flight operations. It also ensures that all flight and ground hardware, software, personnel, and procedures are operationally ready.
Flight Systems and Ground Support	FS&GS is one of four interrelated NASA product lines. FS&GS projects result in the most complex and visible of NASA investments. To manage these systems, the Formulation and Implementation phases for FS&GS projects follow the NASA project life-cycle model consisting of Phases A (concept development) through F (closeout). Primary drivers for FS&GS projects are safety and mission success.
Float	Extra time built into a schedule.
Formulation Phase	The first part of the NASA management life cycle defined in NPR 7120.5 where system requirements are baselined, feasible concepts are determined, a system definition is baselined for the selected concept(s), and preparation is made for progressing to the Implementation phase.
Functional Analysis	The process of identifying, describing, and relating the functions a system must perform to fulfill its goals and objectives.
Functional Baseline (Phase B)	The functional baseline is the approved configuration documentation that describes a system's or top-level CIs' performance requirements (functional, interoperability, and interface characteristics) and the verification required to demonstrate the achievement of those specified characteristics.

Term	Definition/Context
Functional Configuration Audit (FCA)	Examines the functional characteristics of the configured product and verifies that the product has met, via test results, the requirements specified in its functional baseline documentation approved at the PDR and CDR. FCAs will be conducted on both hardware- or software-configured products and will precede the PCA of the configured product.
Functional Decomposition	A subfunction under logical decomposition and design solution definition, it is the examination of a function to identify subfunctions necessary for the accomplishment of that function and functional relationships and interfaces.
Functional Flow Block Diagram	A block diagram that defines system functions and the time sequence of functional events.
Gantt Chart	Bar chart depicting start and finish dates of activities and products in the WBS.
Goal	Quantitative and qualitative guidance on such things as performance criteria, technology gaps, system context, effectiveness, cost, schedule, and risk.
Government Mandatory Inspection Points	Inspection points required by Federal regulations to ensure 100 percent compliance with safety/mission-critical attributes when noncompliance can result in loss of life or loss of mission.
Heritage (or legacy)	Refers to the original manufacturer's level of quality and reliability that is built into the parts which have been proven by (1) time in service, (2) number of units in service, (3) mean time between failure performance, and (4) number of use cycles.
Human Factors Engineering	The discipline that studies human-system interfaces and provides requirements, standards, and guidelines to ensure the human component of an integrated system is able to function as intended.
Implementation Phase	The part of the NASA management life cycle defined in NPR 7120.5 where the detailed design of system products is completed and the products to be deployed are fabricated, assembled, integrated, and tested and the products are deployed to their customers or users for their assigned use or mission.
Incommensurable Costs	Costs that cannot be easily measured, such as controlling pollution on launch or mitigating debris.
Influence Diagram	A compact graphical and mathematical representation of a decision state.
Inspection	Visual examination of a realized end product to validate physical design features or specific manufacturer identification.
Integrated Logistics Support	Activities within the SE process that ensure the product system is supported during development (Phase D) and operations (Phase E) in a cost-effective manner. This is primarily accomplished by early, concurrent consideration of supportability characteristics, performing trade studies on alternative system and ILS concepts, quantifying resource requirements for each ILS element using best-practice techniques, and acquiring the support items associated with each ILS element.
Interface Management Process	The process to assist in controlling product development when efforts are divided among parties (e.g., Government, contractors, geographically diverse technical teams) and/or to define and maintain compliance among the products that must interoperate.
Iterative	Application of a process to the same product or set of products to correct a discovered discrepancy or other variation from requirements. (See "Recursive" and "Repeatable.")
Key Decision Point (or milestone)	The event at which the decision authority determines the readiness of a program/project to progress to the next phase of the life cycle (or to the next KDP).
Key Event	See "Critical Event."
Knowledge Management	Getting the right information to the right people at the right time without delay while helping people create knowledge and share and act upon information in ways that will measurably improve the performance of NASA and its partners.
Least-Cost Analysis	A methodology that identifies the least-cost project option for meeting the technical requirements.

Term	Definition/Context
Liens	Requirements or tasks not satisfied that have to be resolved within a certain assigned time to allow passage through a control gate to proceed.
Life-Cycle Cost	The total cost of ownership over the project's or system's life cycle from Formulation through Implementation. The total of the direct, indirect, recurring, nonrecurring, and other related expenses incurred, or estimated to be incurred, in the design, development, verification, production, deployment, operation, maintenance, support, and disposal of a project.
Logical Decomposition Models	Requirements decomposed by one or more different methods (e.g., function, time, behavior, data flow, states, modes, system architecture).
Logical Decomposition Process	The process for creating the detailed functional requirements that enable NASA programs and projects to meet the ends desired by Agency stakeholders. This process identifies the "what" that must be achieved by the system at each level to enable a successful project. It utilizes functional analysis to create a system architecture and to decompose top-level (or parent) requirements and allocate them down to the lowest desired levels of the project.
Logistics	The management, engineering activities, and analysis associated with design requirements definition, material procurement and distribution, maintenance, supply replacement, transportation, and disposal that are identified by space flight and ground systems supportability objectives.
Maintain (with respect to establishment of processes)	The act of planning the process, providing resources, assigning responsibilities, training people, managing configurations, identifying and involving stakeholders, and monitoring process effectiveness.
Maintainability	The measure of the ability of an item to be retained in or restored to specified conditions when maintenance is performed by personnel having specified skill levels, using prescribed procedures and resources, at each prescribed level of maintenance.
Margin	The allowances carried in budget, projected schedules, and technical performance parameters (e.g., weight, power, or memory) to account for uncertainties and risks. Margin allocations are baselined in the Formulation process, based on assessments of risks, and are typically consumed as the program/project proceeds through the life cycle.
Measure of Effectiveness	A measure by which a stakeholder's expectations will be judged in assessing satisfaction with products or systems produced and delivered in accordance with the associated technical effort. The MOE is deemed to be critical to not only the acceptability of the product by the stakeholder but also critical to operational/mission usage. An MOE is typically qualitative in nature or not able to be used directly as a design-to requirement.
Measure of Performance	A quantitative measure that, when met by the design solution, will help ensure that an MOE for a product or system will be satisfied. These MOPs are given special attention during design to ensure that the MOEs to which they are associated are met. There are generally two or more measures of performance for each MOE.
Metric	The result of a measurement taken over a period of time that communicates vital information about the status or performance of a system, process, or activity. A metric should drive appropriate action.
Mission	A major activity required to accomplish an Agency goal or to effectively pursue a scientific, technological, or engineering opportunity directly related to an Agency goal. Mission needs are independent of any particular system or technological solution.
Mission Concept Review	A review that affirms the mission need and examines the proposed mission's objectives and the concept for meeting those objectives. It is an internal review that usually occurs at the cognizant organization for system development.
Mission Definition Review	A review that examines the functional and performance requirements defined for the system and the preliminary program or project plan and ensures that the requirements and the selected concept will satisfy the mission.
NASA Life-Cycle Phases (or program life-cycle phases)	Consists of Formulation and Implementation phases as defined in NPR 7120.5.

Term	Definition/Context
Objective Function (sometimes Cost Function)	A mathematical expression that expresses the values of combinations of possible outcomes as a single measure of cost-effectiveness.
Operational Readiness Review	A review that examines the actual system characteristics and the procedures used in the system or product's operation and ensures that all system and support (flight and ground) hardware, software, personnel, procedures, and user documentation accurately reflects the deployed state of the system.
Optimal Solution	A feasible solution that minimizes (or maximizes, if that is the goal) an objective function.
Other Interested Parties (Stakeholders)	A subset of "stakeholders," other interested parties are groups or individuals who are not customers of a planned technical effort but may be affected by the resulting product, the manner in which the product is realized or used, or have a responsibility for providing life-cycle support services.
Peer Review	Independent evaluation by internal or external subject matter experts who do not have a vested interest in the work product under review. Peer reviews can be planned, focused reviews conducted on selected work products by the producer's peers to identify defects and issues prior to that work product moving into a milestone review or approval cycle.
Performance Index	An overall measure of effectiveness for each alternative.
Performance Standards	Common metrics for use in performance standards include cost and schedule.
Physical Configuration Audits (or configuration inspection)	The PCA examines the physical configuration of the configured product and verifies that the product corresponds to the build-to (or code-to) product baseline documentation previously approved at the CDR. PCAs will be conducted on both hardware- and software-configured products.
Post-Flight Assessment Review	A review that evaluates the activities from the flight after recovery. The review identifies all anomalies that occurred during the flight and mission and determines the actions necessary to mitigate or resolve the anomalies for future flights.
Post-Launch Assessment Review	A review that evaluates the status, performance, and capabilities of the project evident from the flight operations experience since launch. This can also mean assessing readiness to transfer responsibility from the development organization to the operations organization. The review also evaluates the status of the project plans and the capability to conduct the mission with emphasis on near-term operations and mission-critical events. This review is typically held after the early flight operations and initial checkout.
Precedence Diagram	Workflow diagram that places activities in boxes, connected by dependency arrows; typical of a Gantt chart.
Preliminary Design Review	A review that demonstrates that the preliminary design meets all system requirements with acceptable risk and within the cost and schedule constraints and establishes the basis for proceeding with detailed design. It will show that the correct design option has been selected, interfaces have been identified, and verification methods have been described.
Process	A set of activities used to convert inputs into desired outputs to generate expected outcomes and satisfy a purpose.
Producibility	A system characteristic associated with the ease and economy with which a completed design can be transformed (i.e., fabricated, manufactured, or coded) into a hardware and/or software realization.
Product	A part of a system consisting of end products that perform operational functions and enabling products that perform life-cycle services related to the end product or a result of the technical efforts in the form of a work product (e.g., plan, baseline, or test result).
Product Baseline (Phase D/E)	The product baseline is the approved technical documentation that describes the configuration of a CI during the production, fielding/deployment, and operational support phases of its life cycle. The product baseline describes detailed physical or form, fit, and function characteristics of a CI; the selected functional characteristics designated for production acceptance testing; the production acceptance test requirements.

Term	Definition/Context
Product Breakdown Structure	A hierarchical breakdown of the hardware and software products of the program/project.
Product Implementation Process	The first process encountered in the SE engine, which begins the movement from the bottom of the product hierarchy up toward the Product Transition Process. This is where the plans, designs, analysis, requirement development, and drawings are realized into actual products.
Product Integration Process	One of the SE engine product realization processes that make up the system structure. In this process, lower level products are assembled into higher level products and checked to make sure that the integrated product functions properly. It is the first element of the processes that lead from realized products from a level below to realized end products at a level above, between the Product Implementation, Verification, and Validation Processes.
Product Realization	The act of making, buying, or reusing a product, or the assembly and integration of lower level realized products into a new product, as well as the verification and validation that the product satisfies its appropriate set of requirements and the transition of the product to its customer.
Product Transition Process	A process used to transition a verified and validated end product that has been generated by product implementation or product integration to the customer at the next level in the system structure for integration into an end product or, for the top-level end product, transitioned to the intended end user.
Product Validation Process	The second of the verification and validation processes conducted on a realized end product. While verification proves whether "the system was done right," validation proves whether "the right system was done." In other words, verification provides objective evidence that every "shall" was met, whereas validation is performed for the benefit of the customers and users to ensure that the system functions in the expected manner when placed in the intended environment. This is achieved by examining the products of the system at every level of the structure.
Product Verification Process	The first of the verification and validation processes conducted on a realized end product. As used in the context of systems engineering common technical processes, a realized product is one provided by either the Product Implementation Process or the Product Integration Process in a form suitable for meeting applicable life-cycle phase success criteria.
Production Readiness Review	A review that is held for FS&GS projects developing or acquiring multiple or similar systems greater than three or as determined by the project. The PRR determines the readiness of the system developers to efficiently produce the required number of systems. It ensures that the production plans; fabrication, assembly, and integration-enabling products; and personnel are in place and ready to begin production.
Program	A strategic investment by a mission directorate (or mission support office) that has defined goals, objectives, architecture, funding level, and a management structure that supports one or more projects.
Program/System Definition Review	A review that examines the proposed program architecture and the flowdown to the functional elements of the system. The proposed program's objectives and the concept for meeting those objectives are evaluated. Key technologies and other risks are identified and assessed. The baseline program plan, budgets, and schedules are presented.
Program/System Requirements Review	A review that is used to ensure that the program requirements are properly formulated and correlated with the Agency and mission directorate strategic objectives.
Programmatic Requirements	Requirements set by the mission directorate, program, project, and PI, if applicable. These include strategic scientific and exploration requirements, system performance requirements, and schedule, cost, and similar nontechnical constraints.
Project	(1) A specific investment having defined goals, objectives, requirements, life-cycle cost, a beginning, and an end. A project yields new or revised products or services that directly address NASA's strategic needs. They may be performed wholly in-house; by Government, industry, academia partnerships; or through contracts with private industry. (2) A unit of work performed in programs, projects, and activities.
Project Plan	The document that establishes the project's baseline for implementation, signed by the cognizant program manager, Center Director, project manager, and the MDAA, if required.

Term	Definition/Context
Project Technical Team	The whole technical team for the project.
Solicitation	The vehicle by which information is solicited from contractors to let a contract for products or services.
Prototype	Items (mockups, models) built early in the life cycle that are made as close to the flight item in form, fit, and function as is feasible at that stage of the development. The prototype is used to "wring out" the design solution so that experience gained from the prototype can be fed back into design changes that will improve the manufacture, integration, and maintainability of a single flight item or the production run of several flight items.
Quality Assurance	An independent assessment needed to have confidence that the system actually produced and delivered is in accordance with its functional, performance, and design requirements.
Realized Product	The desired output from the application of the four product realization processes. The form of this product is dependent on the phase of the product-line life cycle and the phase success criteria.
Recursive	Value is added to the system by the repeated application of processes to design next lower layer system products or to realize next upper layer end products within the system structure. This also applies to repeating application of the same processes to the system structure in the next life-cycle phase to mature the system definition and satisfy phase exit criteria.
Relevant Stakeholder	See "Stakeholder."
Reliability	The measure of the degree to which a system ensures mission success by functioning properly over its intended life. It has a low and acceptable probability of failure, achieved through simplicity, proper design, and proper application of reliable parts and materials. In addition to long life, a reliable system is robust and fault tolerant.
Repeatable	A characteristic of a process that can be applied to products at any level of the system structure or within any life-cycle phase.
Requirement	The agreed-upon need, desire, want, capability, capacity, or demand for personnel, equipment, facilities, or other resources or services by specified quantities for specific periods of time or at a specified time expressed as a "shall" statement. Acceptable form for a requirement statement is individually clear, correct, feasible to obtain, unambiguous in meaning, and can be validated at the level of the system structure at which stated. In pairs of requirement statements or as a set, collectively, they are not redundant, are adequately related with respect to terms used, and are not in conflict with one another.
Requirements Allocation Sheet	Documents the connection between allocated functions, allocated performance, and the physical system.
Requirements Management Process	A process that applies to the management of all stakeholder expectations, customer requirements, and technical product requirements down to the lowest level product component requirements.
Risk	The combination of the probability that a program or project will experience an undesired event (some examples include a cost overrun, schedule slippage, safety mishap, health problem, malicious activities, environmental impact, or failure to achieve a needed scientific or technological breakthrough or mission success criteria) and the consequences, impact, or severity of the undesired event, were it to occur. Both the probability and consequences may have associated uncertainties.
Risk Assessment	An evaluation of a risk item that determines (1) what can go wrong, (2) how likely is it to occur, (3) what the consequences are, and (4) what are the uncertainties associated with the likelihood and consequences.
Risk Management	An organized, systematic decisionmaking process that efficiently identifies, analyzes, plans, tracks, controls, communicates, and documents risk and establishes mitigation approaches and plans to increase the likelihood of achieving program/project goals.
Risk-Informed Decision Analysis Process	A five-step process focusing first on objectives and next on developing decision alternatives with those objectives clearly in mind and/or using decision alternatives that have been developed under other systems engineering processes. The later steps of the process interrelate heavily with the Technical Risk Management Process.

Term	Definition/Context
Safety	Freedom from those conditions that can cause death, injury, occupational illness, damage to or loss of equipment or property, or damage to the environment.
Search Space (or Alternative Space)	The envelope of concept possibilities defined by design constraints and parameters within which alternative concepts can be developed and traded off.
Software	As defined in *NPD 2820.1, NASA Software Policy*.
Specification	A document that prescribes completely, precisely, and verifiably the requirements, design, behavior, or characteristics of a system or system component.
Stakeholder	A group or individual who is affected by or is in some way accountable for the outcome of an under-taking. The term "relevant stakeholder" is a subset of the term "stakeholder" and describes the people identified to contribute to a specific task. There are two main classes of stakeholders. See "Customers" and "Other Interested Parties."
Stakeholder Expectations	A statement of needs, desires, capabilities, and wants that are not expressed as a requirement (not expressed as a "shall" statement) is to be referred to as an "expectation." Once the set of expectations from applicable stakeholders is collected, analyzed, and converted into a "shall" statement, the expectation becomes a requirement. Expectations can be stated in either qualitative (nonmeasurable) or quantitative (measurable) terms. Requirements are always stated in quantitative terms. Expectations can be stated in terms of functions, behaviors, or constraints with respect to the product being engineered or the process used to engineer the product.
Stakeholder Expectations Definition Process	The initial process within the SE engine that establishes the foundation from which the system is designed and the product realized. The main purpose of this process is to identify who the stakeholders are and how they intend to use the product. This is usually accomplished through use-case scenarios, design reference missions, and operational concepts.
Standing Review Board	The entity responsible for conducting independent reviews of the program/project per the life-cycle requirements. The SRB is advisory and is chartered to objectively assess the material presented by the program/project at a specific review.
State Diagram	A diagram that shows the flow in the system in response to varying inputs.
Success Criteria	Specific accomplishments that must be satisfactorily demonstrated to meet the objectives of a technical review so that a technical effort can progress further in the life cycle. Success criteria are documented in the corresponding technical review plan.
Surveillance (or Insight or Oversight)	The monitoring of a contractor's activities (e.g., status meetings, reviews, audits, site visits) for progress and production and to demonstrate fiscal responsibility, ensure crew safety and mission success, and determine award fees for extraordinary (or penalty fees for substandard) contract execution.
System	(1) The combination of elements that function together to produce the capability to meet a need. The elements include all hardware, software, equipment, facilities, personnel, processes, and procedures needed for this purpose. (2) The end product (which performs operational functions) and enabling products (which provide life-cycle support services to the operational end products) that make up a system.
System Acceptance Review	A review that verifies the completeness of the specific end item with respect to the expected maturity level and to assess compliance to stakeholder expectations. The SAR examines the system, its end items and documentation, and test data and analyses that support verification and validation. It also ensures that the system has sufficient technical maturity to authorize its shipment to the designated operational facility or launch site.
System Definition Review	A review that examines the proposed system architecture/design and the flowdown to all functional elements of the system.
System Integration Review	A review that ensures that the system is ready to be integrated; segments, components, and subsystems are available and ready to be integrated; and integration facilities, support personnel, and integration plans and procedures are ready for integration. SIR is conducted at the end of the final design phase (Phase C) and before the systems assembly, integration, and test phase (Phase D) begins.
System Requirements Review	A review that examines the functional and performance requirements defined for the system and the preliminary program or project plan and ensures that the requirements and the selected concept will satisfy the mission.

Term	Definition/Context
System Safety Engineering	The application of engineering and management principles, criteria, and techniques to achieve acceptable mishap risk within the constraints of operational effectiveness and suitability, time, and cost, throughout all phases of the system life cycle.
System Structure	A system structure is made up of a layered structure of product-based WBS models. (See "Work Breakdown Structure.")
Systems Analysis	The analytical process by which a need is transformed into a realized, definitive product, able to support compatibility with all physical and functional requirements and support the operational scenarios in terms of reliability, maintainability, supportability, serviceability, and disposability, while maintaining performance and affordability. Systems analysis is responsive to the needs of the customer at every phase of the life cycle, from pre-Phase A to realizing the final product and beyond.
Systems Approach	The application of a systematic, disciplined engineering approach that is quantifiable, recursive, iterative, and repeatable for the development, operation, and maintenance of systems integrated into a whole throughout the life cycle of a project or program.
Systems Engineering Engine	The technical processes framework for planning and implementing the technical effort within any phase of a product-line life cycle. The SE engine model in Figure 2.1-1 shows the 17 technical processes that are applied to products being engineered to drive the technical effort.
Systems Engineering Management Plan	The SEMP identifies the roles and responsibility interfaces of the technical effort and how those interfaces will be managed. The SEMP is the vehicle that documents and communicates the technical approach, including the application of the common technical processes; resources to be used; and key technical tasks, activities, and events along with their metrics and success criteria.
Tailoring	The documentation and approval of the adaptation of the process and approach to complying with requirements underlying the specific program or project.
Technical Assessment Process	The crosscutting process used to help monitor technical progress of a program/project through periodic technical reviews. It also provides status information in support of assessing system design, product realization, and technical management decisions.
Technical Data Management Process	The process used to plan for, acquire, access, manage, protect, and use data of a technical nature to support the total life cycle of a system. This includes its development, deployment, operations and support, eventual retirement, and retention of appropriate technical data beyond system retirement as required by current NASA policies.
Technical Data Package	An output of the Design Solution Definition Process, it evolves from phase to phase, starting with conceptual sketches or models and ending with complete drawings, parts list, and other details needed for product implementation or product integration.
Technical Measures	An established set of measures based on the expectations and requirements that will be tracked and assessed to determine overall system or product effectiveness and customer satisfaction. Common terms for these measures are MOEs, MOPs, and TPMs.
Technical Performance Measures	The set of critical or key performance parameters that are monitored by comparing the current actual achievement of the parameters with that anticipated at the current time and on future dates. Used to confirm progress and identify deficiencies that might jeopardize meeting a system requirement. Assessed parameter values that fall outside an expected range around the anticipated values indicate a need for evaluation and corrective action. Technical performance measures are typically selected from the defined set of MOPs.
Technical Planning Process	The first of the eight technical management processes contained in the SE engine, the Technical Planning Process establishes a plan for applying and managing each of the common technical processes that will be used to drive the development of system products and associated work products. This process also establishes a plan for identifying and defining the technical effort required to satisfy the project objectives and life-cycle-phase success criteria within the cost, schedule, and risk constraints of the project.
Technical Requirements Definition Process	The process used to transform the stakeholder expectations into a complete set of validated technical requirements expressed as "shall" statements that can be used for defining a design solution for the PBS model and related enabling products.

Term	Definition/Context
Technical Risk	Risk associated with the achievement of a technical goal, criterion, or objective. It applies to undesired consequences related to technical performance, human safety, mission assets, or environment.
Technical Risk Management Process	The process for measuring or assessing risk and developing strategies to manage it, an important component of managing NASA programs under its charter to explore and expand knowledge. Critical to this process is the proactive identification and control of departures from the baseline program, project, or activity.
Technical Team	A group of multidisciplinary individuals with appropriate domain knowledge, experience, competencies, and skills assigned to a specific technical task.
Technology Readiness Assessment Report	A document required for transition from Phase B to Phase C/D demonstrating that all systems, subsystems, and components have achieved a level of technological maturity with demonstrated evidence of qualification in a relevant environment.
Technology Assessment	A systematic process that ascertains the need to develop or infuse technological advances into a system. The technology assessment process makes use of basic systems engineering principles and processes within the framework of the PBS. It is a two-step process comprised of (1) the determination of the current technological maturity in terms of technology readiness levels and (2) the determination of the difficulty associated with moving a technology from one TRL to the next through the use of the AD^2.
Technology Development Plan	A document required for transition from Phase A to Phase B identifying technologies to be developed, heritage systems to be modified, alternative paths to be pursued, fallback positions and corresponding performance descopes, milestones, metrics, and key decision points. It is incorporated in the preliminary project plan.
Technology Maturity Assessment	The process to determine a system's technological maturity via TRLs.
Technology Readiness Level	Provides a scale against which to measure the maturity of a technology. TRLs range from 1, Basic Technology Research, to 9, Systems Test, Launch, and Operations. Typically, a TRL of 6 (i.e., technology demonstrated in a relevant environment) is required for a technology to be integrated into an SE process.
Test	The use of a realized end product to obtain detailed data to validate performance or to provide sufficient information to validate performance through further analysis.
Test Readiness Review	A review that ensures that the test article (hardware/software), test facility, support personnel, and test procedures are ready for testing and data acquisition, reduction, and control.
Traceability	A discernible association among two or more logical entities such as requirements, system elements, verifications, or tasks.
Trade Study	A means of evaluating system designs by devising alternative means to meet functional requirements, evaluating these alternatives in terms of the measures of effectiveness and system cost, ranking the alternatives according to appropriate selection criteria, dropping less promising alternatives, and proceeding to the next level of resolution, if needed.
Trade Study Report	A report written to document a trade study. It should include: he system under analysis; system goals, objectives (or requirements, as appropriate to the level of resolution), and constraints; measures and measurement methods (models) used; all data sources used; the alternatives chosen for analysis; computational results, including uncertainty ranges and sensitivity analyses performed; the selection rule used; and the recommended alternative.
Trade Tree	A representation of trade study alternatives in which each layer represents some system aspect that needs to be treated in a trade study to determine the best alternative.
Transition	The act of delivery or moving of a product from the location where the product has been implemented or integrated, as well as verified and validated, to a customer. This act can include packaging, handling, storing, moving, transporting, installing, and sustainment activities.
Utility	A measure of the relative value gained from an alternative. The theoretical unit of measurement for utility is the util.

Term	Definition/Context
Validated Requirements	A set of requirements that are well formed (clear and unambiguous), complete (agree with customer and stakeholder needs and expectations), consistent (conflict free), and individually verifiable and traceable to a higher level requirement or goal.
Validation	Testing, possibly under simulated conditions, to ensure that a finished product works as required.
Validation (of a product)	Proof that the product accomplishes the intended purpose. Validation may be determined by a combination of test, analysis, and demonstration.
Variance	In program control terminology, a difference between actual performance and planned costs or schedule status.
Verification	The process of proving or demonstrating that a finished product meets design specifications and requirements.
Verification (of a product)	Proof of compliance with specifications. Verification may be determined by test, analysis, demonstration, or inspection.
Waiver	A documented agreement intentionally releasing a program or project from meeting a requirement. (Some Centers use deviations prior to Implementation and waivers during Implementation).
WBS Model	Model that describes a system that consists of end products and their subsystems (which perform the operational functions of the system), the supporting or enabling products, and any other work products (plans, baselines) required for the development of the system.
Work Breakdown Structure (WBS)	A product-oriented hierarchical division of the hardware, software, services, and data required to produce the program/project's end product(s) structured according to the way the work will be performed, reflecting the way in which program/project costs, schedule, technical, and risk data are to be accumulated, summarized, and reported.
Workflow Diagram	A scheduling chart that shows activities, dependencies among activities, and milestones.

Appendix C: How to Write a Good Requirement

Use of Correct Terms

- Shall = requirement
- Will = facts or declaration of purpose
- Should = goal

Editorial Checklist

Personnel Requirement

1. The requirement is in the form "responsible party shall perform such and such." In other words, use the active, rather than the passive voice. A requirement must state who shall (do, perform, provide, weigh, or other verb) followed by a description of what must be performed.

Product Requirement

1. The requirement is in the form "product ABC shall XYZ." A requirement must state "The product shall" (do, perform, provide, weigh, or other verb) followed by a description of what must be done.

2. The requirement uses consistent terminology to refer to the product and its lower level entities.

3. Complete with tolerances for qualitative/performance values (e.g., less than, greater than or equal to, plus or minus, 3 sigma root sum squares).

4. Is the requirement free of implementation? (Requirements should state WHAT is needed, NOT HOW to provide it; i.e., state the problem not the solution. Ask, "Why do you need the requirement?" The answer may point to the real requirement.)

5. Free of descriptions of operations? (Is this a need the product must satisfy or an activity involving the product? Sentences like "The operator shall…" are almost always operational statements not requirements.)

Example Product Requirements

- The system shall operate at a power level of…
- The software shall acquire data from the…
- The structure shall withstand loads of…
- The hardware shall have a mass of…

General Goodness Checklist

1. The requirement is grammatically correct.

2. The requirement is free of typos, misspellings, and punctuation errors.

3. The requirement complies with the project's template and style rules.

4. The requirement is stated positively (as opposed to negatively, i.e., "shall not").

5. The use of "To Be Determined" (TBD) values should be minimized. It is better to use a best estimate for a value and mark it "To Be Resolved" (TBR) with the rationale along with what must be done to eliminate the TBR, who is responsible for its elimination, and by when it must be eliminated.

6. The requirement is accompanied by an intelligible rationale, including any assumptions. Can you validate (concur with) the assumptions? Assumptions must be confirmed before baselining.

7. The requirement is located in the proper section of the document (e.g., not in an appendix).

Requirements Validation Checklist

Clarity

1. Are the requirements clear and unambiguous? (Are all aspects of the requirement understandable and not subject to misinterpretation? Is the requirement free from indefinite pronouns (this, these) and ambiguous terms (e.g., "as appropriate," "etc.," "and/or," "but not limited to")?)

2. Are the requirements concise and simple?

3. Do the requirements express only one thought per requirement statement, a standalone statement as opposed to multiple requirements in a single statement, or a paragraph that contains both requirements and rationale?

4. Does the requirement statement have one subject and one predicate?

Completeness

1. Are requirements stated as completely as possible? Have all incomplete requirements been captured as TBDs or TBRs and a complete listing of them maintained with the requirements?

2. Are any requirements missing? For example have any of the following requirements areas been overlooked: functional, performance, interface, environment (development, manufacturing, test, transport, storage, operations), facility (manufacturing, test, storage, operations), transportation (among areas for manufacturing, assembling, delivery points, within storage facilities, loading), training, personnel, operability, safety, security, appearance and physical characteristics, and design.

3. Have all assumptions been explicitly stated?

Compliance

1. Are all requirements at the correct level (e.g., system, segment, element, subsystem)?

2. Are requirements free of implementation specifics? (Requirements should state what is needed, not how to provide it.)

3. Are requirements free of descriptions of operations? (Don't mix operation with requirements: update the ConOps instead.)

Consistency

1. Are the requirements stated consistently without contradicting themselves or the requirements of related systems?

2. Is the terminology consistent with the user and sponsor's terminology? With the project glossary?

3. Is the terminology consistently used through out the document?

4. Are the key terms included in the project's glossary?

Traceability

1. Are all requirements needed? Is each requirement necessary to meet the parent requirement? Is each requirement a needed function or characteristic? Distinguish between needs and wants. If it is not necessary, it is not a requirement. Ask, "What is the worst that could happen if the requirement was not included?"

2. Are all requirements (functions, structures, and constraints) bidirectionally traceable to higher level requirements or mission or system-of-interest scope (i.e., need(s), goals, objectives, constraints, or concept of operations)?

3. Is each requirement stated in such a manner that it can be uniquely referenced (e.g., each requirement is uniquely numbered) in subordinate documents?

Correctness

1. Is each requirement correct?

2. Is each stated assumption correct? Assumptions must be confirmed before the document can be baselined.

3. Are the requirements technically feasible?

Functionality

1. Are all described functions necessary and together sufficient to meet mission and system goals and objectives?

Performance

1. Are all required performance specifications and margins listed (e.g., consider timing, throughput, storage size, latency, accuracy and precision)?

2. Is each performance requirement realistic?

3. Are the tolerances overly tight? Are the tolerances defendable and cost-effective? Ask, "What is the worst thing that could happen if the tolerance was doubled or tripled?"

Interfaces

1. Are all external interfaces clearly defined?

2. Are all internal interfaces clearly defined?

3. Are all interfaces necessary, sufficient, and consistent with each other?

Maintainability

1. Have the requirements for system maintainability been specified in a measurable, verifiable manner?

2. Are requirements written so that ripple effects from changes are minimized (i.e., requirements are as weakly coupled as possible)?

Reliability

1. Are clearly defined, measurable, and verifiable reliability requirements specified?

2. Are there error detection, reporting, handling, and recovery requirements?

3. Are undesired events (e.g., single event upset, data loss or scrambling, operator error) considered and their required responses specified?

4. Have assumptions about the intended sequence of functions been stated? Are these sequences required?

5. Do these requirements adequately address the survivability after a software or hardware fault of the system from the point of view of hardware, software, operations, personnel and procedures?

Verifiability/Testability

1. Can the system be tested, demonstrated, inspected, or analyzed to show that it satisfies requirements? Can this be done at the level of the system at which the requirement is stated? Does a means exist to measure the accomplishment of the requirement and verify compliance? Can the criteria for verification be stated?

2. Are the requirements stated precisely to facilitate specification of system test success criteria and requirements?

3. Are the requirements free of unverifiable terms (e.g., flexible, easy, sufficient, safe, ad hoc, adequate, accommodate, user-friendly, usable, when required, if required, appropriate, fast, portable, light-weight, small, large, maximize, minimize, sufficient, robust, quickly, easily, clearly, other "ly" words, other "ize" words)?

Data Usage

1. Where applicable, are "don't care" conditions truly "don't care"? ("Don't care" values identify cases when the value of a condition or flag is irrelevant, even though the value may be important for other cases.) Are "don't care" conditions values explicitly stated? (Correct identification of "don't care" values may improve a design's portability.)

Appendix D: Requirements Verification Matrix

When developing requirements, it is important to identify an approach for verifying the requirements. This appendix provides the matrix that defines how all the requirements are verified. Only "shall" requirements should be included in these matrices. The matrix should identify each "shall" by unique identifier and be definitive as to the source, i.e., document from which the requirement is taken. This matrix could be divided into multiple matrices (e.g., one per requirements document) to delineate sources of requirements depending on the project. The example is shown to provide suggested guidelines for the minimum information that should be included in the verification matrix.

Table D-1 Requirements Verification Matrix

Require ment No. [a]	Document[b]	Paragraph[c]	Shall Statement[d]	Verification Suc cess Criteria[e]	Verifi cation Method[f]	Facility or Lab[g]	Phase[h]	Acceptance Require ment?[i]	Preflight Accept ance?[j]	Performing Organiza tion[k]	Results[l]
P-1	xxx	3.2.1.1 Capability: Support Uplinked Data (LDR)	System X shall pro- vide a max. ground- to-station uplink of…	1. System X locks to forward link at the min and max data rate tolerances 2. System X locks to the forward link at the min and max operating frequency tolerances	Test	xxx	5			xxx	TPS xxxx
P-i	xxx	Other paragraphs	Other "shalls" in PTRS	Other criteria	xxx	xxx	xxx			xxx	Memo xxx
S-i or other unique designator	xxxxx (other specs, ICDs, etc.)	Other paragraphs	Other "shalls" in specs, ICDs, etc.	Other criteria	xxx	xxx	xxx			xxx	Report xxx

a. Unique identifier for each System X requirement.

b. Document number the System X requirement is contained within.

c. Paragraph number of the System X requirement.

d. Text (within reason) of the System X requirement, i.e., the "shall."

e. Success criteria for the System X requirement.

f. Verification method for the System X requirement (analysis, inspection, demonstration, or test).

g. Facility or laboratory used to perform the verification and validation.

h. Phase in which the verification and validation will be performed: (1) Pre-Declared Development, (2) Formal Box-Level Functional, (3) Formal Box-Level Environmental, (4) Formal System-Level Environmental, (5) Formal System-Level Functional, (6) Formal End-to-End Functional, (7) Integrated Vehicle Functional, (8) On-Orbit Functional.

i. Indicate whether this requirement is also verified during initial acceptance testing of each unit.

j. Indicate whether this requirement is also verified during any pre-flight or recurring acceptance testing of each unit.

k. Organization responsible for performing the verification

l. Indicate documents that contain the objective evidence that requirement was satisfied

Appendix E: Creating the Validation Plan (Including Validation Requirements Matrix)

When developing requirements, it is important to identify a validation approach for how additional validation evaluation, testing, analysis or other demonstrations will be performed to ensure customer/sponsor satisfaction.

This validation plan should include a validation matrix with the elements in the example below. The final column in the matrix below uses a display product as a specific example.

Table E-1 Validation Requirements Matrix

Validation Product #	Activity	Objective	Validation Method	Facility or Lab	Phase	Performing Organization	Results
Unique identifier for validation product	Describe evaluation by the customer/sponsor that will be performed	What is to be accomplished by the customer/sponsor evaluation	Validation method for the System X requirement (analysis, inspection, demonstration, or test)	Facility or laboratory used to perform the validation	Phase in which the verification/validation will be performed[a]	Organization responsible for coordinating the validation activity	Indicate the objective evidence that validation activity occurred
1	Customer/sponsor will evaluate the candidate displays	1. Ensure legibility is acceptable 2. Ensure overall appearance is acceptable	Test	xxx	Phase A	xxx	

a. Example: (1) during product selection process, (2) prior to final product selection (if COTS) or prior to PDR, (3) prior to CDR, (4) during box-level functional, (5) during system-level functional, (6) during end-to-end functional, (7) during integrated vehicle functional, (8) during on-orbit functional.

Appendix F: Functional, Timing, and State Analysis

Functional Flow Block Diagrams

Functional analysis can be performed using various methods, one of which is Functional Flow Block Diagrams (FFBDs). FFBDs define the system functions and depict the time sequence of functional events. They identify "what" must happen and do not assume a particular answer to "how" a function will be performed. They are functionally oriented, not solution oriented.

FFBDs are made up of functional blocks, each of which represents a definite, finite, discrete action to be accomplished. The functional architecture is developed using a series of leveled diagrams to show the functional decomposition and display the functions in their logical, sequential relationship. A consistent numbering scheme is used to label the blocks. The numbers establish identification and relationships that carry through all the diagrams and facilitate traceability from the lower levels to the top level. Each block in the first- (top-) level diagram can be expanded to a series of functions in the second-level diagram, and so on. (See Figure F-1.) Lines connecting functions indicate function flow and not lapsed time or intermediate activity. Diagrams are laid out so that the flow direction is generally from left to right. Arrows are often used to indicate functional flows. The diagrams show both input (transfer to operational orbit) and output (transfer to STS orbit), thus facilitating the definition of interfaces and control process.

Each diagram contains a reference to other functional diagrams to facilitate movement between pages of the diagrams. Gates are used: "AND," "OR," "Go or No-Go," sometimes with enhanced functionality, including exclusive OR gate (XOR), iteration (IT), repetition (RP), or loop (LP). A circle is used to denote a summing gate and is used when AND/OR is present. AND is used to indicate parallel functions and all conditions must be satisfied to proceed (i.e., concurrency). OR is used to indicate that alternative paths can be satisfied to proceed (i.e., selection). G and $\overline{G}$ are used to denote Go and No-Go conditions. These symbols are placed adjacent to lines leaving a particular function to indicate alternative paths. For examples of the above, see Figures F-2 and F-3.

Enhanced Functional Flow Block Diagrams (EFFBDs) provide data flow overlay to capture data dependencies. EFFBDs (shown in Figure F-4) represent: (1) functions, (2) control flows, and (3) data flows. An EFFBD specification of a system is complete enough that it is executable as a discrete event model, capable of dynamic, as well as static, validation. EFFBDs provide freedom to use either control constructs or data triggers or both to specify execution conditions for the system functions. EFFBDs graphically distinguish between triggering and nontriggering data inputs. Triggering data are required before a function can begin execution. Triggers are actually data items with control implications. In Figure F-4, the data input shown with a green background and double-headed arrows is a triggering data input. The nontriggering data inputs are shown with gray backgrounds and single-headed arrows. An EFFBD must be enabled by: (1) the completion of the function(s) preceding it in the control construct and (2) triggered, if trigger data are identified, before it can execute. For example, in Figure F-4, "1. Serial Function" must complete and "Data 3" must be present before "3. Function in Concurrency" can execute. It should be noted that the "External Input" data into "1. Serial Function" and the "External Output" data from "6. Output Function" should not be confused with the functional input and output for these functions, which are represented by the input and the output arrows respectively. Data flows are represented as elongated ovals whereas functions are represented as rectangular boxes.

Functional analysis looks across all life-cycle processes. Functions required to deploy a system are very different from functions required to operate and ultimately dispose of the system. Preparing FFBDs for each phase of the life cycle as well as the transition into the phases themselves is necessary to draw out all the requirements. These diagrams are used both to develop requirements and to identify profitability. The functional analysis also incorporates alternative and contingency operations, which improve the probability of mission success. The flow diagrams provide an understanding of total operation of the system, serve as a basis for development of operational and contingency procedures, and pinpoint

TOP LEVEL

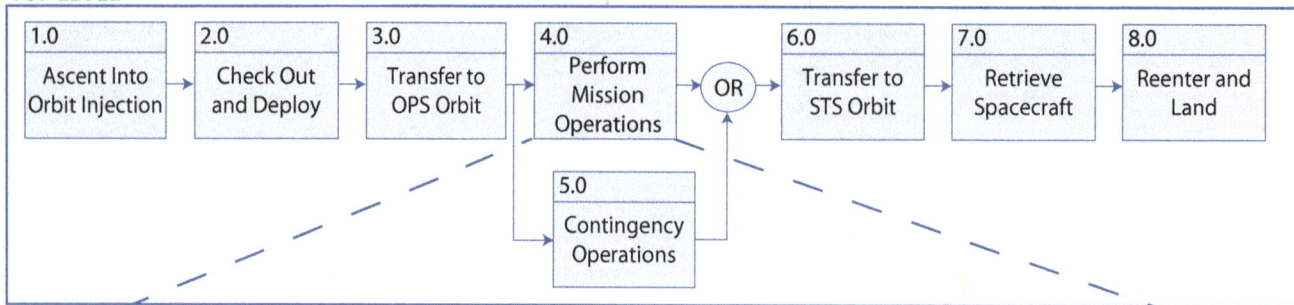

SECOND LEVEL

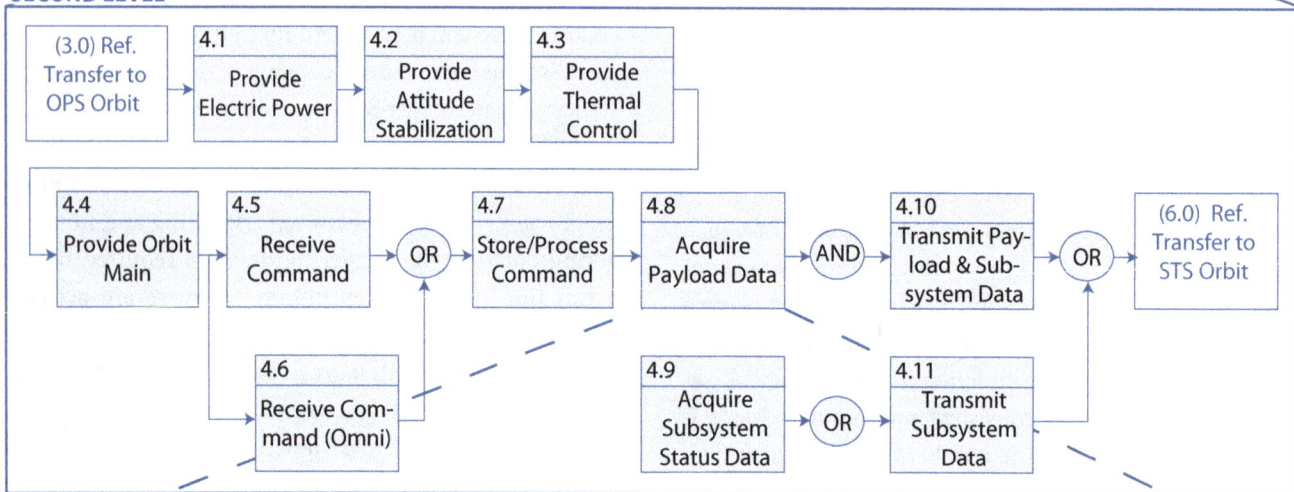

THIRD LEVEL

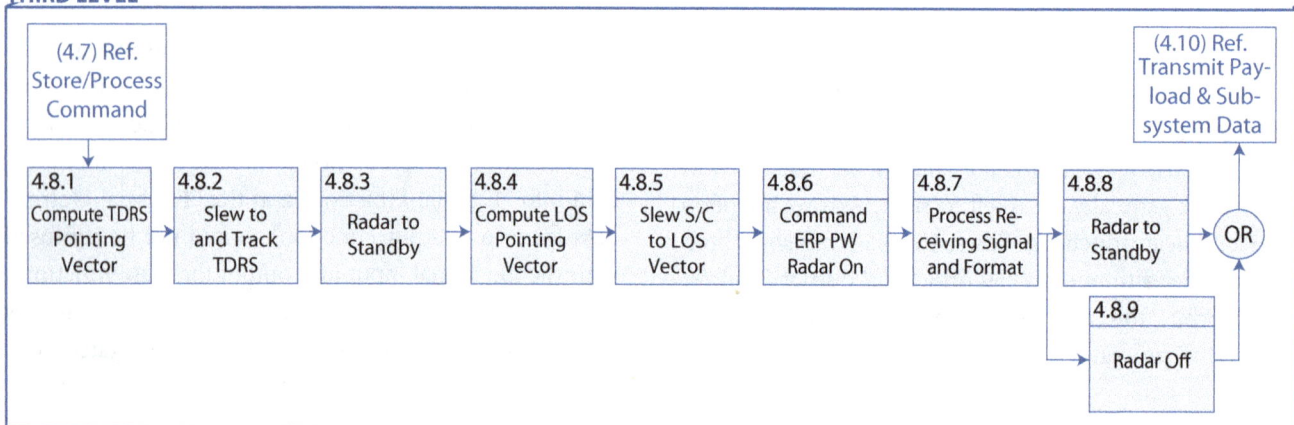

Figure F-1 FFBD flowdown

areas where changes in operational procedures could simplify the overall system operation. This organization will eventually feed into the WBS structure and ultimately drive the overall mission organization and cost. In certain cases, alternative FFBDs may be used to represent various means of satisfying a particular function until data are acquired, which permits selection among the alternatives. For more information on FFBDs and EFFBDs, see Jim Long's *Relationships between Common Graphical Representations in Systems Engineering*.

Requirements Allocation Sheets

Requirements allocation sheets document the connection between allocated functions, allocated performance, and the physical system. They provide traceability between Technical Requirements Definition functional analysis activities and Logical Decomposition and Design Solution Definition activities and maintain consistency between them, as well as show disconnects. Figure F-5 provides an example of a requirements allocation sheet. The

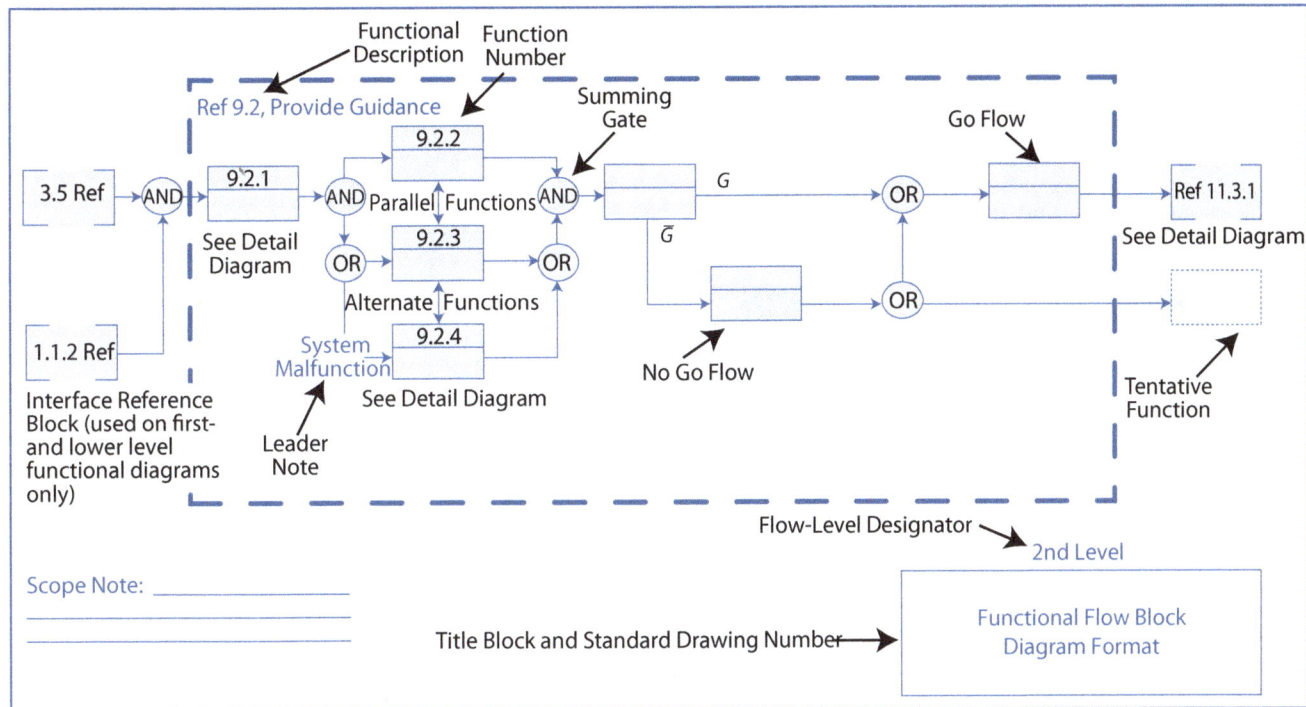

Figure F-2 FFBD: example 1

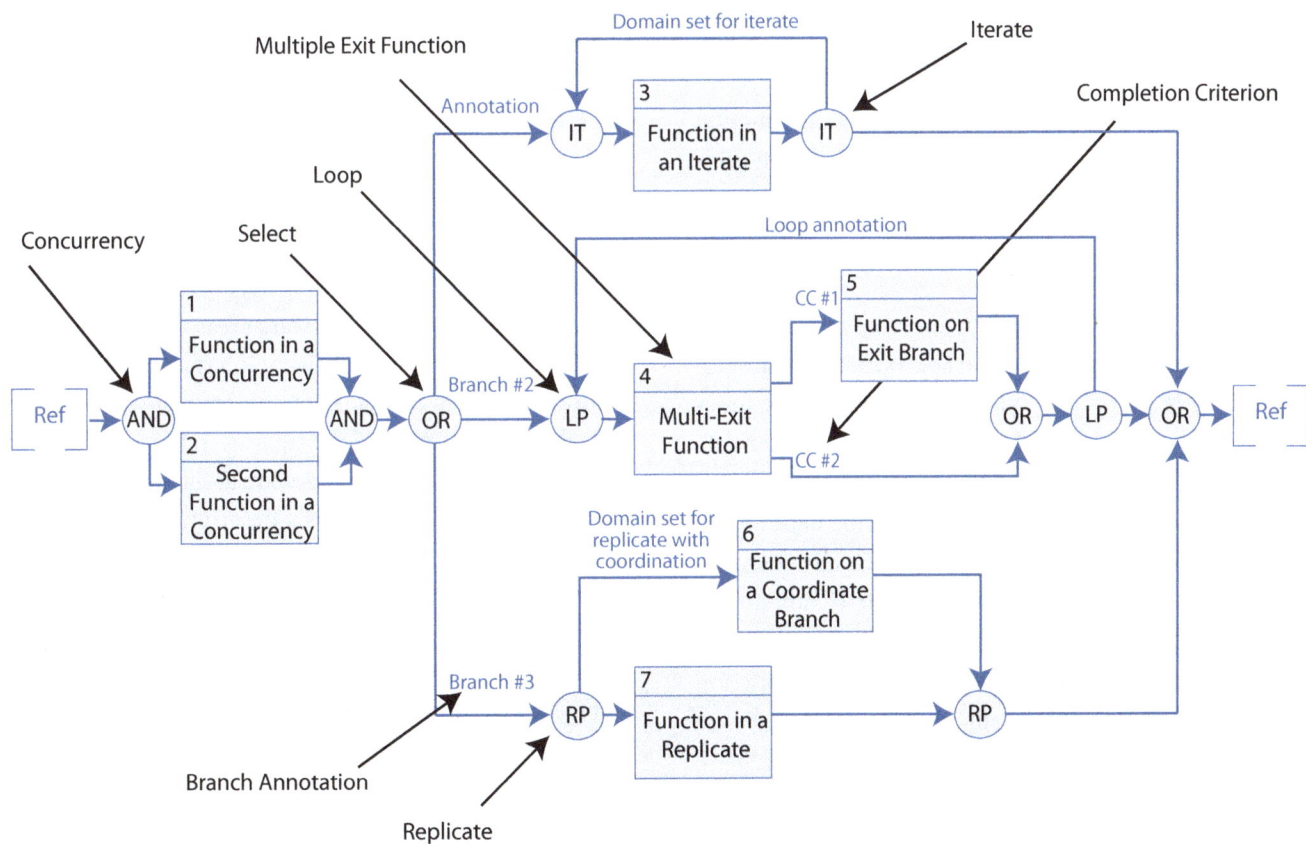

Figure F-3 FFBD showing additional control constructs: example 2

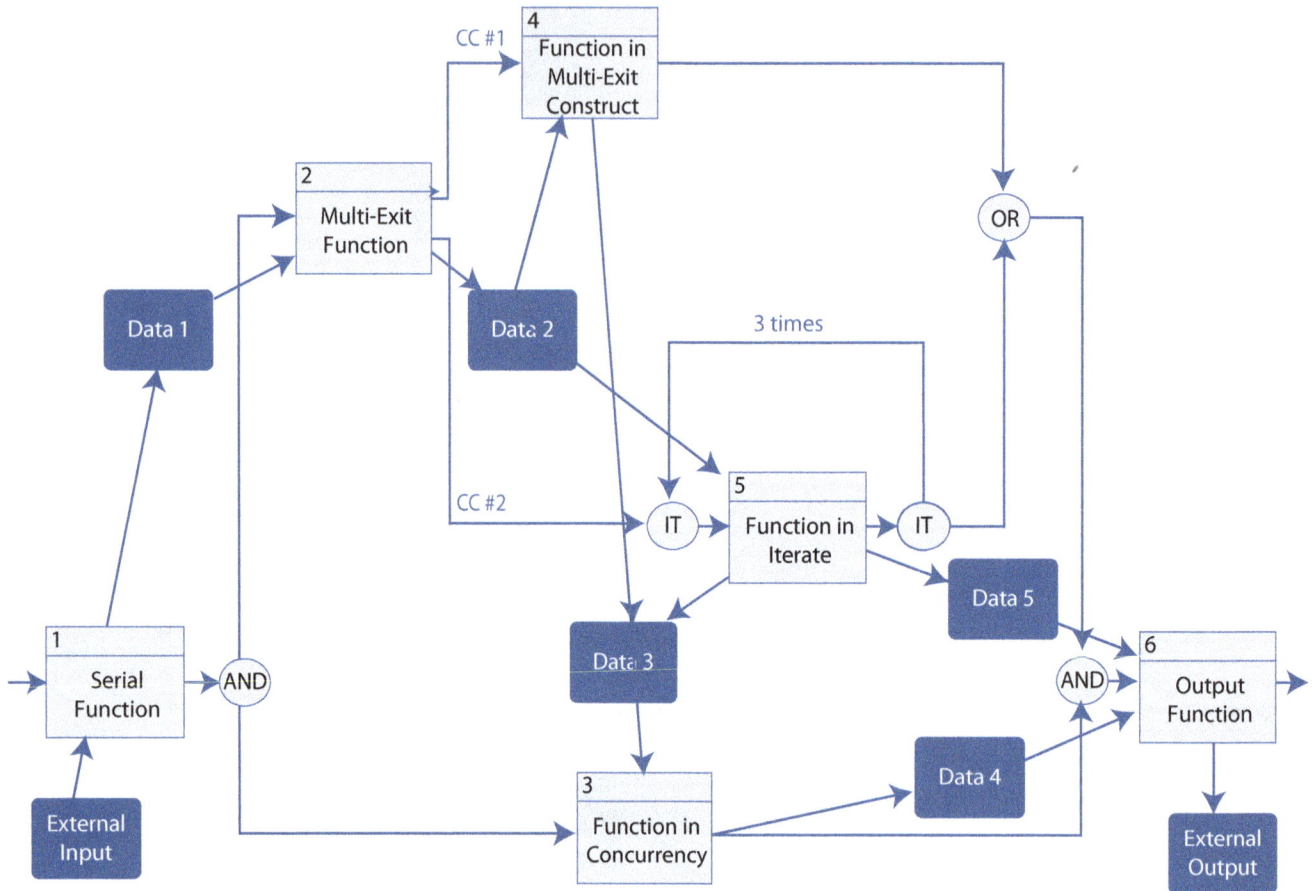

Figure F-4 Enhanced FFBD: example 3

reference column to the far right indicates the function numbers from the FFBDs. Fill in the requirements allocation sheet by performing the following:

1. Include the functions and function numbers from the FFBDs.

2. Allocate functional performance requirements and design requirements to the appropriate function(s) (many requirements may be allocated to one function, or one requirement may be allocated to many functions).

3. All system-level requirements must be allocated to a function to ensure the system meets all system requirements (functions without allocated requirements should be eliminated as unnecessary activities).

4. Allocate all derived requirements to the function that spawned the requirement.

5. Identify the physical equipment, configuration item, facilities, and specifications that will be used to meet the requirements.

(For a reference on requirements allocation sheets, see DOD's *Systems Engineering Fundamentals Guide*.)

N2 Diagrams

An N-squared (N2) diagram is a matrix representation of functional and/or physical interfaces between elements of a system at a particular hierarchical level. The N2 diagram has been used extensively to develop data interfaces, primarily in the software areas. However, it can also be used to develop hardware interfaces as shown in Figure F-6. The system components are placed on the diagonal. The remainder of the squares in the NxN matrix represent the interfaces. The square at the intersection of a row and a column contains a description of the interface between the two components represented on that row and that column. For example, the solar arrays have a mechanical interface with the structure and an electrical interface and supplied service interface with the voltage converters. Where a blank appears, there is no interface between the respective components.

The N2 diagram can be taken down into successively lower levels to the hardware and software component functional levels. In addition to defining the data that must be supplied across the interface, by showing the

ID	DESCRIPTION	REQUIREMENT	TRACED FROM	PERFORMANCE	MARGIN	COMMENTS	REF
M1	Mission Orbit	575 +/-15 km Sun-synchronous dawn-dusk orbit	S3, S11, P3	Complies	NA	Pegasus XL with HAPS provides required launch injection dispersion accuracy	F.2.c
M2	Launch Vehicle	Pegasus XL with HAPS	P2, P4	Complies	NA		F.2.c
M3	Observatory Mass	The observatory total mass shall not exceed 241 kg	M1, M2	192.5 kg	25.20%		F.5.b
M4	Data Acquisition Quality	The mission shall deliver 95% data with better than 1 in 100,000 BER	P1	Complies	NA	Standard margins and systems baselined; formal system analysis to be completed by PDR	F.7
M5	Communication Band	The mission shall use S-band SQPSK at 5 Mbps for spacecraft downlink and 2 kbps uplink	S12, P4	Complies	NA	See SC27, SC28, and G1, G2	F.3.f, F.7
M7	Tracking	MOC shall use NORAD two-line elements for observatory tracking	P4	Complies	NA		F.7
M8	Data Latency	Data latency shall be less than 72 hours	P12	Complies	NA		F.7
M9	Daily Data Volume	Accommodate average daily raw science data volume of 10.8 Gbits	P1, S12	Complies	12%	Margin based on funded ground contacts	F.3.e, F.7
M10	Ground Station	The mission shall be compatible with the Rutherford Appleton Laboratory Ground Station and the Poker Flat Ground Station	P1	Complies	NA		F.7
M11	Orbital Debris (Casualty Area)	Design observatory for demise upon reentry with <1/10,000 probability of injury	P3	1/51,000	400%	See Orbital Debris Analysis in Appendix M-6	F.2.e, App.6
M12	Orbital Debris (Lifetime)	Design observatory for reentry <25 years after end of mission	P3	<10 years	15 years	See Orbital Debris Analysis in Appendix M-6	F.2.e, App.6

Figure F-5 Requirements allocation sheet

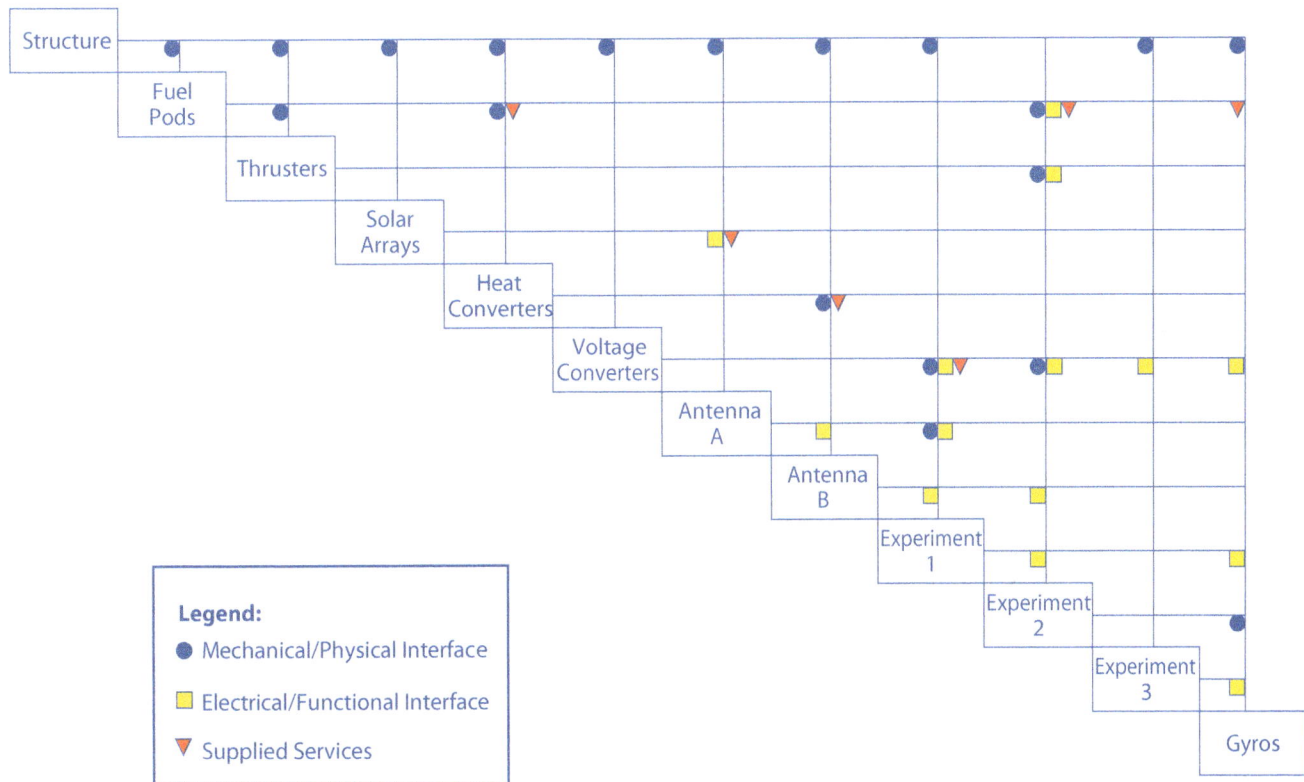

Figure F-6 N2 diagram for orbital equipment

Note: From NASA Reference Publication 1370, *Training Manual for Elements of Interface Definition and Control.*

data flows the N2 chart pinpoints areas where conflicts could arise in interfaces, and highlights input and output dependency assumptions and requirements.

Timing Analysis

There are several methods for visualizing the complex timing relationships in a system. Two of the more important ones are the timing diagram and the state transition diagram. The timing diagram (see Figure F-7) defines the behavior of different objects within a timescale. It provides a visual representation of objects changing state and interacting over time. Timing diagrams can be used for defining the behavior of hardware-driven and/or software-driven components. While a simple timeline analysis is very useful in understanding relationships such as concurrency, overlap, and sequencing, state diagrams (see Figure F-8) allow for even greater flexibility in that they can depict events such as loops and decision processes that may have largely varying timelines. Timing information can be added to an FFBD to create a timeline analysis. This is very useful for allocating resources and generating specific time-related design requirements. It also elucidates performance characteristics and design constraints. However, it is not complete.

State diagrams are needed to show the flow of the system in response to varying inputs.

The tools of timing analysis are rather straightforward. While some Commercial-Off-the-Shelf (COTs) tools are available, any graphics tool and a good spreadsheet will do. The important thing to remember is that timeline analysis is better for linear flows while circular, looping, multi-path, and combinations of these are best described with state diagrams. Complexity should be kept layered and track the FFBDs. The ultimate goal of using all these techniques is simply to force the thought process enough into the details of the system that most of the big surprises can be avoided.

State Analysis

State diagramming is another graphical tool that is most helpful for understanding and displaying the complex timing relationships in a system. Timing diagrams do not give the complete picture of the system. State diagrams are needed to show the flow of the system in response to varying inputs. State diagrams provide a sort of simplification of understanding on a system by breaking complex reactions into smaller and smaller known re-

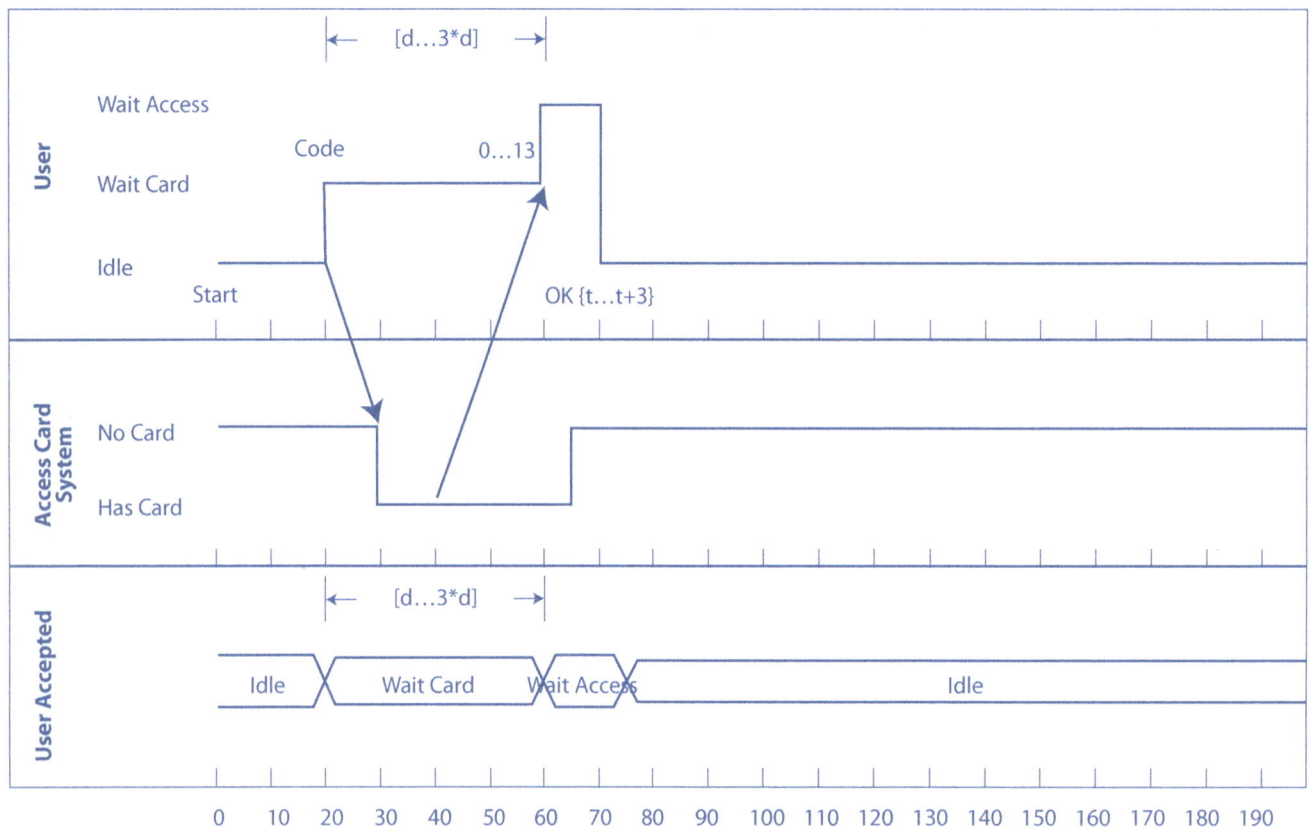

Figure F-7 Timing diagram example

sponses. This allows detailed requirements to be developed and verified with their timing performance.

Figure F-8 shows a slew command status state diagram from the James Webb Space Telescope. Ovals represent the system states. Arcs represent the event that triggers the state change as well as the action or output taken by the system in response to the event.

Self-loops are permitted. In the example in Figure F-8 the slew states can loop until they arrive at

the correct location, and then they can loop while they settle.

When it is used to represent the behavior of a sequential finite-state machine, the state diagram is called a state transition diagram. A sequential finite-state machine is one that has no memory, which means that the current output only depends on the current input. The state transition diagram models the event-based, time-dependent behavior of such a system.

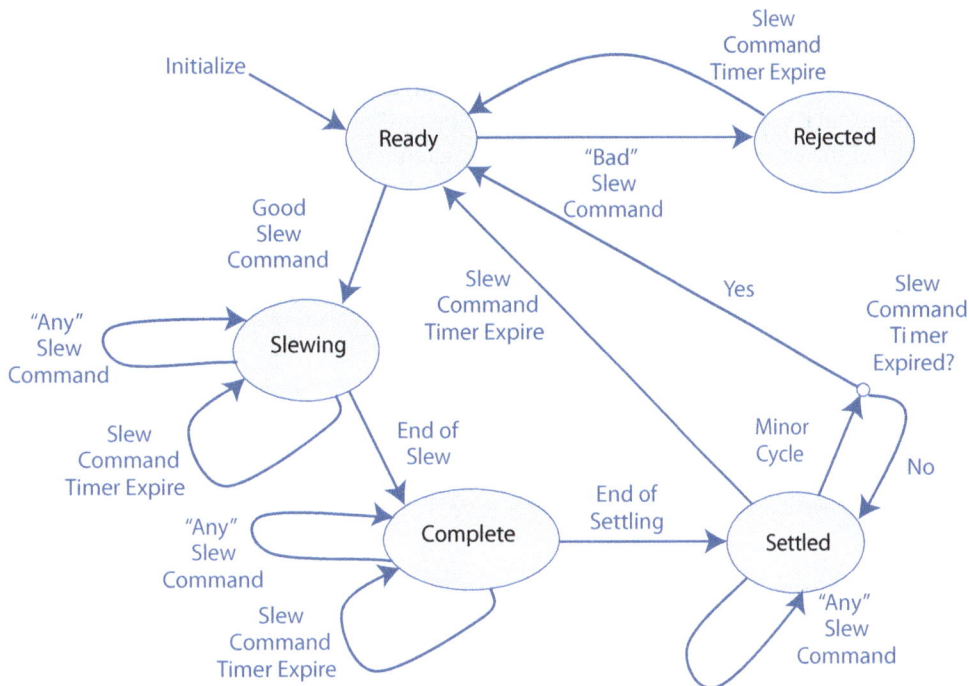

Figure F-8 Slew command status state diagram

Context Diagrams

When presented with a system design problem, the systems engineer's first task is to truly understand the problem. That means understanding the context in which the problem is set. A *context diagram* is a useful tool for grasping the system to be built and the *external domains* that are relevant to that system and which have interfaces to the system. The diagram shows the general structure of a context diagram. The system is shown surrounded by the external systems which have interfaces to the system. These systems are not part of the system, but they interact with the system via the system's external interfaces. The external systems can impact the system, and the system does impact the external systems. They play a major role in establishing the requirements for the system. Entities further removed are those in the system's context that can impact the system but cannot be impacted by the system. These entities in the system's context are responsible for some of the system's requirements.

Defining the boundaries of a system is a critical but often neglected task. Using an example from a satellite project, one of the external systems that is impacted by the satellite would be the Tracking and Data Relay Satellite System (TDRSS). The TDRSS is not part of the satellite system, but it defines requirements on the satellite and is impacted by the satellite since it must schedule contacts, receive and transmit data and commands, and downlink the satellite data to the ground. An example of an entity in the context of the satellite system that is not impacted by the satellite system is the Global Positioning Satellite (GPS) system. The GPS is not impacted in any way by the satellite, but it will levy some requirements on the satellite if the satellite is to use the GPS signals for navigation.

Reference: Diagram is from Buede, *The Engineering Design of Systems*, p. 38.

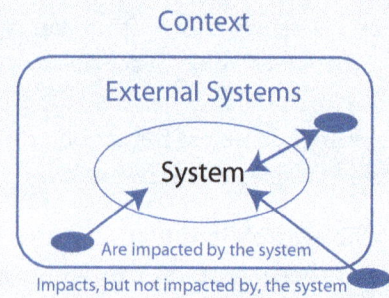

Appendix G: Technology Assessment/Insertion

Introduction, Purpose, and Scope

The Agency's programs and projects, by their very nature, frequently require the development and infusion of new technological advances to meet mission goals, objectives, and resulting requirements. Sometimes the new technological advancement being infused is actually a heritage system that is being incorporated into a different architecture and operated in different environment from that for which it was originally designed. In this latter case, it is often not recognized that adaptation of heritage systems frequently requires technological advancement and as a result, key steps in the development process are given short shrift—often to the detriment of the program/project. In both contexts of technological advancement (new and adapted heritage), infusion is a very complex process that has been dealt with over the years in an ad hoc manner differing greatly from project to project with varying degrees of success.

Frequently, technology infusion has resulted in schedule slips, cost overruns, and occasionally even to cancellations or failures. In post mortem, the root cause of such events has often been attributed to "inadequate definition of requirements." If such were indeed the root cause, then correcting the situation would simply be a matter of requiring better requirements definition, but since history seems frequently to repeat itself, this must not be the case—at least not in total.

In fact there are many contributors to schedule slip, cost overrun, and project cancellation and failure—among them lack of adequate requirements definition. The case can be made that most of these contributors are related to the degree of uncertainty at the outset of the project and that a dominant factor in the degree of uncertainty is the lack of understanding of the maturity of the technology required to bring the project to fruition and a concomitant lack of understanding of the cost and schedule reserves required to advance the technology from its present state to a point where it can be qualified and successfully infused with a high degree of confidence. Although this uncertainty cannot be eliminated, it can be substantially reduced through the early application of good systems engineering practices focused on understanding the technological requirements; the maturity of the required technology; and the technological advancement required to meet program/project goals, objectives, and requirements.

A number of processes can be used to develop the appropriate level of understanding required for successful technology insertion. The intent of this appendix is to describe a systematic process that can be used as an example of how to apply standard systems engineering practices to perform a comprehensive Technology Assessment (TA). The TA comprises two parts, a Technology Maturity Assessment (TMA) and an Advancement Degree of Difficulty Assessment (AD²). The process begins with the TMA which is used to determine technological maturity via NASA's Technology Readiness Level (TRL) scale. It then proceeds to develop an understanding of what is required to advance the level of maturity through AD². It is necessary to conduct TAs at various stages throughout a program/ project to provide the Key Decision Point (KDP) products required for transition between phases. (See Table G-1.)

The initial TMA provides the baseline maturity of the system's required technologies at program/project outset and allows monitoring progress throughout development. The final TMA is performed just prior to the Preliminary Design Review. It forms the basis for the Technology Readiness Assessment Report (TRAR), which documents the maturity of the technological advancement required by the systems, subsystems, and components demonstrated through test and analysis. The initial AD² assessment provides the material necessary to develop preliminary cost and to schedule plans and preliminary risk assessments. In subsequent assessment, the information is used to build the technology development plan in the process identifying alternative paths, fallback positions, and performance descope options. The information is also vital to preparing milestones and metrics for subsequent Earned Value Management (EVM).

Table G-1 Products Provided by the TA as a Function of Program/Project Phase

Gate	Product
KDP A—Transition from Pre-Phase A to Phase A	Requires an assessment of potential technology needs versus current and planned technology readiness levels, as well as potential opportunities to use commercial, academic, and other government agency sources of technology. Included as part of the draft integrated baseline.
KDP B—Transition from Phase A to Phase B	Requires a technology development plan identifying technologies to be developed, heritage systems to be modified, alternative paths to be pursued, fall-back positions and corresponding performance descopes, milestones, metrics, and key decision points. Incorporated in the preliminary project plan.
KDP C—Transition from Phase B to Phase C/D	Requires a TRAR demonstrating that all systems, subsystems, and components have achieved a level of technological maturity with demonstrated evidence of qualification in a relevant environment.

Source: NPR 7120.5.

The TMA is performed against the hierarchical breakdown of the hardware and software products of the program/project PBS to achieve a systematic, overall understanding at the system, subsystem, and component levels. (See Figure G-1.)

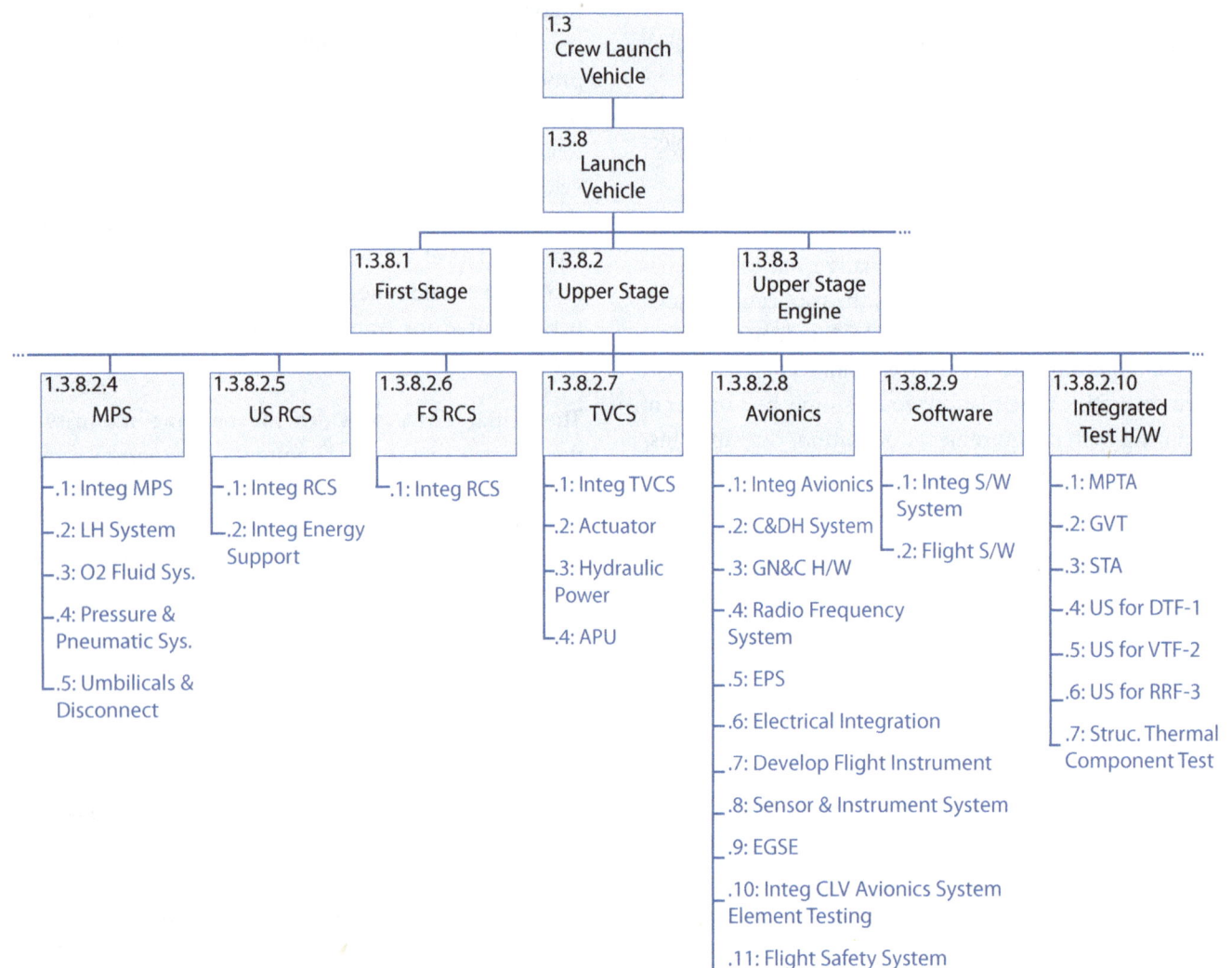

Figure G-1 PBS example

Inputs/Entry Criteria

It is extremely important that a TA process be defined at the beginning of the program/project and that it be performed at the earliest possible stage (concept development) and throughout the program/project through PDR. Inputs to the process will vary in level of detail according to the phase of the program/project, and even though there is a lack of detail in Pre-Phase A, the TA will drive out the major critical technological advancements required. Therefore, at the beginning of Pre-Phase A, the following should be provided:

- Refinement of TRL definitions.
- Definition of AD^2.
- Definition of terms to be used in the assessment process.
- Establishment of meaningful evaluation criteria and metrics that will allow for clear identification of gaps and shortfalls in performance.
- Establishment of the TA team.
- Establishment of an independent TA review team.

How to Do Technology Assessment

The technology assessment process makes use of basic systems engineering principles and processes. As mentioned previously, it is structured to occur within the framework of the Product Breakdown Structure (PBS) to facilitate incorporation of the results. Using the PBS as a framework has a twofold benefit—it breaks the "problem" down into systems, subsystems, and components that can be more accurately assessed; and it provides the results of the assessment in a format that can readily be used in the generation of program costs and schedules. It can also be highly beneficial in providing milestones and metrics for progress tracking using EVM. As discussed above, it is a two-step process comprised of (1) the determina-

tion of the current technological maturity in terms of TRLs and (2) the determination of the difficulty associated with moving a technology from one TRL to the next through the use of the AD^2. The overall process is iterative, starting at the conceptual level during program Formulation, establishing the initial identification of critical technologies and the preliminary cost, schedule, and risk mitigation plans. Continuing on into Phase A, it is used to establish the baseline maturity, the technology development plan and associated costs and schedule. The final TA consists only of the TMA and is used to develop the TRAR which validates that all elements are at the requisite maturity level. (See Figure G-2.)

Even at the conceptual level, it is important to use the formalism of a PBS to avoid having important technologies slip through the crack. Because of the preliminary nature of the concept, the systems, subsystems, and components will be defined at a level that will not permit detailed assessments to be made. The process of performing the assessment, however, is the same as that used for subsequent, more detailed steps that occur later in the program/project where systems are defined in greater detail.

Once the concept has been formulated and the initial identification of critical technologies made, it is nec-

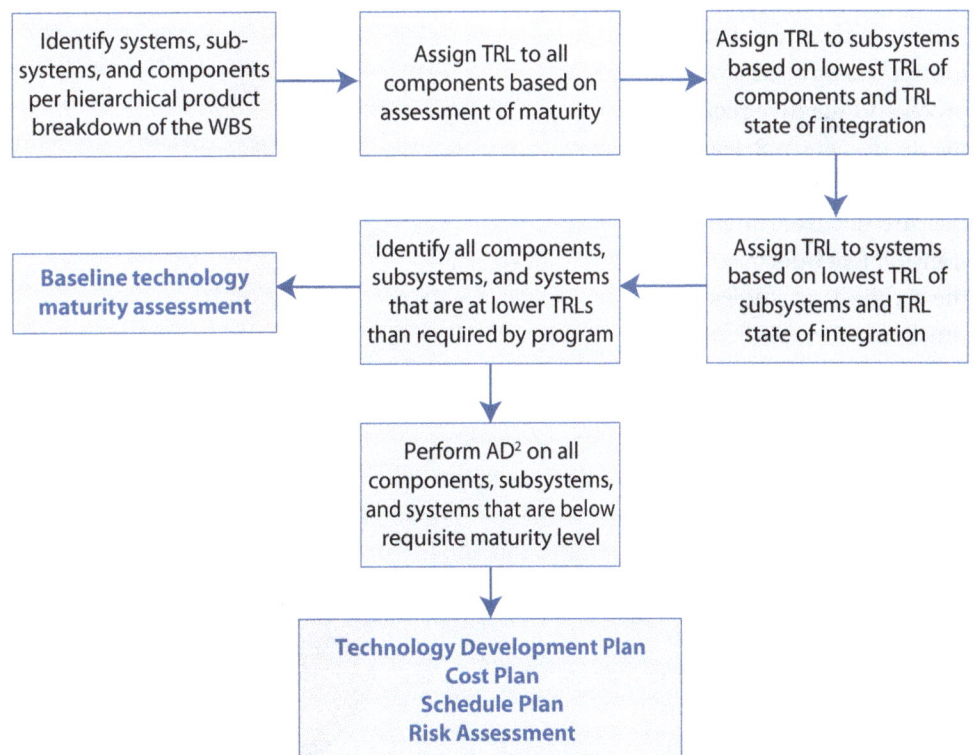

Figure G-2 Technology assessment process

essary to perform detailed architecture studies with the Technology Assessment Process intimately inter-

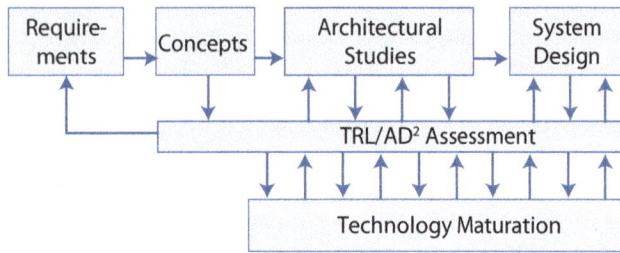

Figure G-3 Architectural studies and technology development

woven. (See Figure G-3.) The purpose of the architecture studies is to refine end-item system design to meet the overall scientific requirements of the mission. It is imperative that there be a continuous relationship between architectural studies and maturing technology advances. The architectural studies must incorporate the results of the technology maturation, planning for alternative paths and identifying new areas required for development as the architecture is refined. Similarly, it is incumbent upon the technology maturation process to identify requirements that are not feasible and development routes that are not fruitful and to transmit that information to the architecture studies in a timely manner. Similarly, it is incumbent upon the architecture studies to provide feedback to the technology development process relative to changes in requirements. Particular attention must be given to "heritage" systems in that they are often used in architectures and environments different from those in which they were designed to operate.

Establishing TRLs

TRL is, at its most basic, a description of the performance history of a given system, subsystem, or component relative to a set of levels first described at NASA HQ in the 1980s. The TRL essentially describes the state of the art of a given technology and provides a baseline from which maturity is gauged and advancement defined. (See Figure G-4.) Even though the concept of TRL has been around for almost 20 years, it is not well understood and frequently misinterpreted. As a result, we often undertake programs without fully understanding either the maturity of key technologies or what is needed to develop them to the required level. *It is impossible to understand the magnitude and scope of a development program without having a clear understanding of the baseline technological maturity of all elements of the system.* Establishing the TRL is a vital first step on the

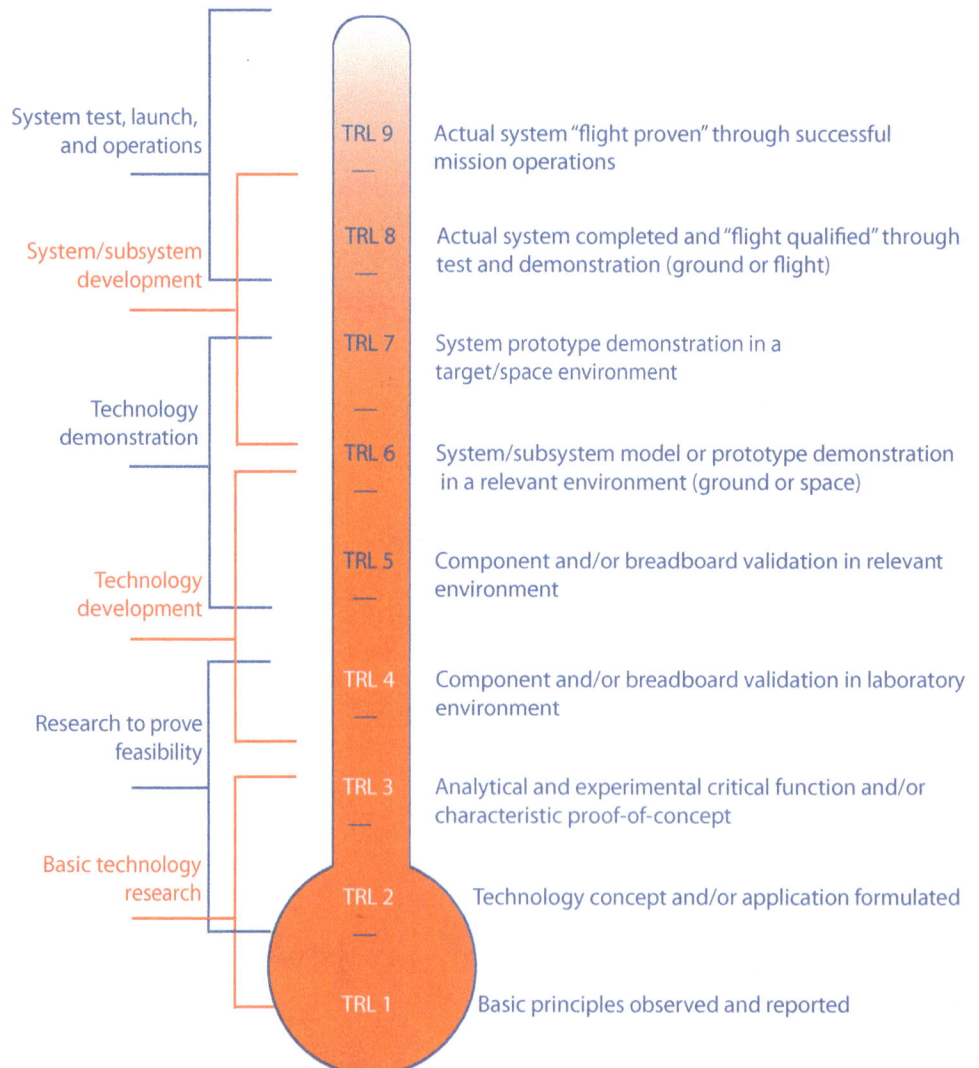

Figure G-4 Technology readiness levels

way to a successful program. A frequent misconception is that in practice it is too difficult to determine TRLs and that when you do it is not meaningful. On the contrary, identifying TRLs can be a straightforward systems engineering process of determining what was demonstrated and under what conditions was it demonstrated.

At first blush, the TRL descriptions in Figure G-4 appear to be straightforward. It is in the process of trying to assign levels that problems arise. A primary cause of difficulty is in terminology—everyone knows what a breadboard is, but not everyone has the same definition. Also, what is a "relevant environment"? What is relevant to one application may or may not be relevant to another. Many of these terms originated in various branches of engineering and had, at the time, very specific meanings to that particular field. They have since become commonly used throughout the engineering field and often take differences in meaning from discipline to discipline, some subtle, some not so subtle. "Breadboard," for example, comes from electrical engineering where the original use referred to checking out the functional design of an electrical circuit by populating a "breadboard" with components to verify that the design operated as anticipated. Other terms come from mechanical engineering, referring primarily to units that are subjected to different levels of stress under testing, i.e., qualification, protoflight, and flight units. The first step in developing a uniform TRL assessment (see Figure G-5) is to define the terms used. It is extremely important to develop and use a consistent set of definitions over the course of the program/project.

Having established a common set of terminology, it is necessary to proceed to the next step—quantifying "judgment calls" on the basis of past experience. Even with clear definitions there will be the need for judgment calls when it comes time to assess just how similar a given element is relative to what is needed (i.e., is it close enough to a prototype to be considered a prototype, or is it more like an engineering breadboard?). Describing what has been done in terms of form, fit, and function provides a means of quantifying an element based on its design intent and subsequent performance. The current definitions for software TRLs are contained in *NPR 7120.8, NASA Research and Technology Program and Project Management Requirements*.

A third critical element of any assessment relates to the question of who is in the best position to make judgment

calls relative to the status of the technology in question. For this step, it is extremely important to have a well-balanced, experienced assessment team. Team members do not necessarily have to be discipline experts. The primary expertise required for a TRL assessment is that the

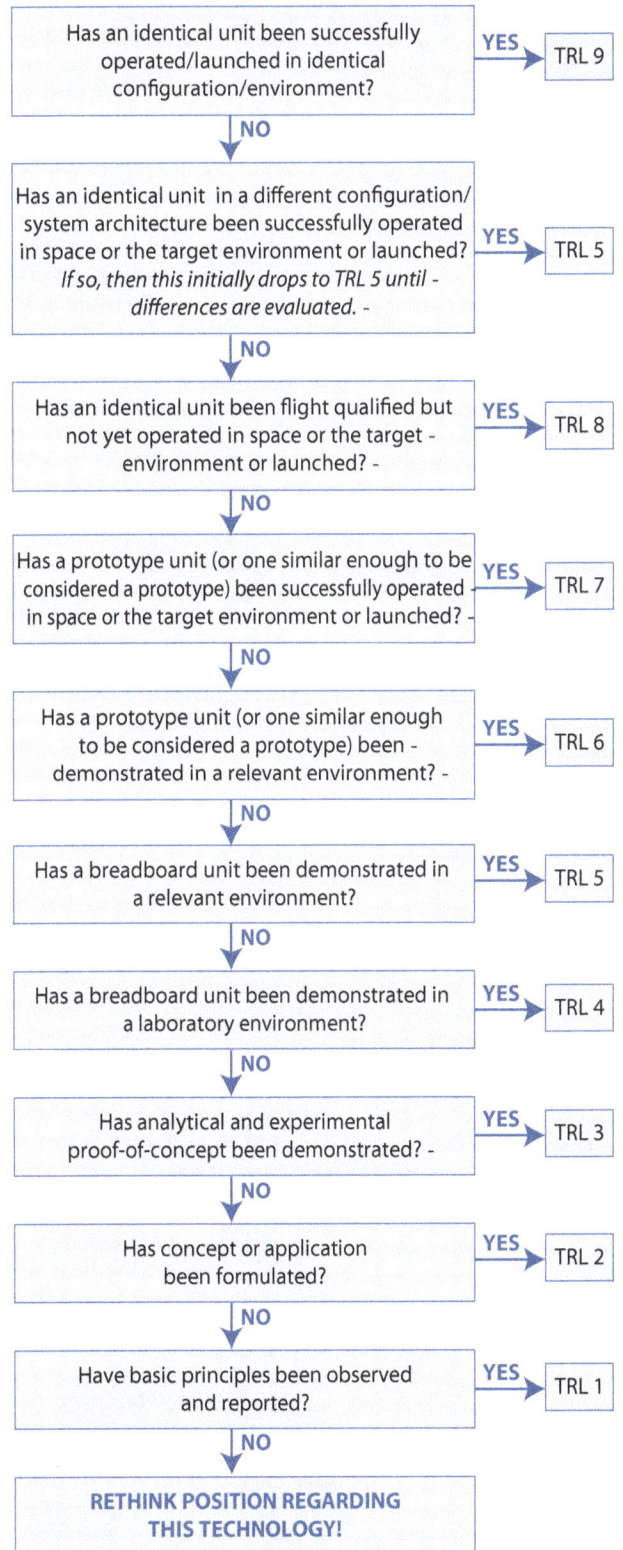

Figure G-5 The TMA thought process

systems engineer/user understands the current state of the art in applications. Having established a set of definitions, defined a process for quantifying judgment calls, and assembled an expert assessment team, the process primarily consists of asking the right questions. The flowchart depicted in Figure G-5 demonstrates the questions to ask to determine TRL at any level in the assessment.

Note the second box particularly refers to heritage systems. If the architecture and the environment have changed, then the TRL drops to TRL 5—at least intially. Additional testing may need to be done for heritage systems for the new use or new environment. If in subsequent analysis the new environment is sufficiently close to the old environment, or the new architecture sufficiently close to the old architecture then the resulting evaluation could be then TRL 6 or 7, but the most important thing to realize is that it is no longer at a TRL 9. Applying this process at the system level and then proceeding to lower levels of subsystem and component identifies those elements that require development and sets the stage for the subsequent phase, determining the AD^2.

A method for formalizing this process is shown in Figure G-6. Here, the process has been set up as a table: the rows identify the systems, subsystems, and components that are under assessment. The columns identify the categories that will be used to determine the TRL—i.e., what units have been built, to what scale, and in what environment have they been tested. Answers to these questions determine the TRL of an item under consideration. The TRL of the system is determined by the lowest TRL present in the system; i.e., a system is at TRL 2 if any single element in the system is at TRL 2. The problem of multiple elements being at low TRLs is dealt with in the AD^2 process. Note that the issue of integration affects the TRL of every system, subsystem, and component. All of the elements can be at a higher TRL, but if they have never been integrated as a unit, the TRL will be lower for the unit. How much lower depends on the complexity of the integration.

TRL ASSESSMENT

Red = Below TRL 3
Yellow = TRL 3,4 & 5
Green = TRL 6 and above
White = Unknown
X Exists

	Demonstration Units						Environment				Unit Description				
	Concept	Breadboard	Brassboard	Developmental Model	Prototype	Flight Qualified	Laboratory Environment	Relevant Environment	Space Environment	Space/Launch Operation	Form	Fit	Function	Appropriate Scale	Overall TRL
1.0 System															
1.1 Subsystem X															🟧
1.1.1 Mechanical Components															
1.1.2 Mechanical Systems															
1.1.3 Electrical Components				X					X		X	X	X		🟩
1.1.4 Electrical Systems															
1.1.5 Control Systems															
1.1.6 Thermal Systems								X			X	X			🟨
1.1.7 Fluid Systems		X													🟧
1.1.8 Optical Systems															
1.1.9 Electro-Optical Systems															
1.1.10 Software Systems															
1.1.11 Mechanisms	X														🟧
1.1.12 Integration															
1.2 Subsystem Y															🟨
1.2.1 Mechanical Components															

Figure G-6 TRL assessment matrix

Appendix H: Integration Plan Outline

Purpose

The integration plan defines the integration and verification strategies for a project interface with the system design and decomposition into the lower level elements.[1] The integration plan is structured to bring the elements together to assemble each subsystem and to bring all of the subsystems together to assemble the system/product. The primary purposes of the integration plan are: (1) to describe this coordinated integration effort that supports the implementation strategy, (2) to describe for the participants what needs to be done in each integration step, and (3) to identify the required resources and when and where they will be needed.

Questions/Checklist

- Does the integration plan include and cover integration of all of the components and subsystems of the project, either developed or purchased?

- Does the integration plan account for all external systems to be integrated with the system (for example, communications networks, field equipment, other complete systems owned by the government or owned by other government agencies)?

- Does the integration plan fully support the implementation strategy, for example, when and where the subsystems and system are to be used?

- Does the integration plan mesh with the verification plan?

- For each integration step, does the integration plan define what components and subsystems are to be integrated?

- For each integration step, does the integration plan identify all the needed participants and define what their roles and responsibilities are?

- Does the integration plan establish the sequence and schedule for every integration step?

- Does the integration plan spell out how integration problems are to be documented and resolved?

Integration Plan Contents

Table H-1 outlines the content of the integration plan by section.

---

[1] The material in this appendix is adapted from Federal Highway Administration and CalTrans, *Systems Engineering Guidebook for ITS, Version 2.0.*

Table H-1 Integration Plan Contents

Section	Contents
Title page	The title page should follow the NASA procedures or style guide. At a minimum, it should contain the following information: • INTEGRATION PLAN FOR THE [insert name of project] AND [insert name of organization] • Contract number • Date that the document was formally approved • The organization responsible for preparing the document • Internal document control number, if available • Revision version and date issued
1.0 Purpose of Document	A brief statement of the purpose of this document. It is the plan for integrating the components and subsystems of the project prior to verification.
2.0 Scope of Project	This section gives a brief description of the planned project and the purpose of the system to be built. Special emphasis is placed on the project's deployment complexities and challenges.
3.0 Integration Strategy	This section informs the reader what the high-level plan is for integration and, most importantly, why the integration plan is structured the way it is. The integration plan is subject to several, sometimes conflicting, constraints. Also, it is one part of the larger process of build, integrate, verify, and deploy, all of which must be synchronized to support the same project strategy. So, for even a moderately complex project, the integration strategy, based on a clear and concise statement of the project's goals and objectives, is described here at a high, but all-inclusive, level. It may also be necessary to describe the analysis of alternative strategies to make it clear why this particular strategy was selected. The same strategy is the basis for the build plan, the verification plan, and the deployment plan. This section covers and describes each step in the integration process. It describes what components are integrated at each step and gives a general idea of what threads of the operational capabilities (requirements) are covered. It ties the plan to the previously identified goals and objectives so the stakeholders can understand the rationale for each integration step. This summary-level description also defines the schedule for all the integration efforts.
4.0 Phase 1 Integration	This, and the following sections, define and explain each step in the integration process. The intent here is to identify all the needed participants and to describe to them what they have to do. In general, the description of each integration step should identify: • The location of the activities. • The project-developed equipment and software products to be integrated. Initially this is just a high-level list, but eventually the list must be exact and complete, showing part numbers and quantity. • Any support equipment (special software, test hardware, software stubs, and drivers to simulate yet-to-be-integrated software components, external systems) needed for this integration step. The same support equipment is most likely needed for the subsequent verification step. • All integration activities that need to be performed after installation, including integration with on-site systems and external systems at other sites. • A description of the verification activities, as defined in the applicable verification plan, that occur after this integration step. • The responsible parties for each activity in the integration step. • The schedule for each activity.
5.0 Multiple Phase Integration Steps (1 or N steps)	This, and any needed additional sections, follow the format for Section 3.0. Each covers each step in a multiple step integration effort.

Appendix I: Verification and Validation Plan Sample Outline

1. **Introduction**
   1.1 Purpose and Scope
   1.2 Responsibility and Change Authority
   1.3 Definitions

2. **Applicable and Reference Documents**
   2.1 Applicable Documents
   2.2 Reference Documents
   2.3 Order of Precedence

3. **System X Description**
   3.1 System X Requirements Flow Down
   3.2 System X Architecture
   3.3 End Item Architectures
       3.3.1   System X End Item A
       3.3.*n*   System X End Item *n*
   3.4 System X Ground Support Equipment
   3.5 Other Architecture Descriptions

4. **Verification and Validation Process**
   4.1 Verification and Validation Management Responsibilities
   4.2 Verification Methods
       4.2.1   Analysis
       4.2.2   Inspection
       4.2.3   Demonstration
       4.2.4   Test
           4.2.4.1 Qualification Testing
           4.2.4.2 Other Testing
   4.3 Validation Methods
   4.4 Certification Process
   4.5 Acceptance Testing

5. **Verification and Validation Implementation**
   5.1 System X Design and Verification and Validation Flow
   5.2 Test Articles
   5.3 Support Equipment
   5.4 Facilities

6. **System X End Item Verification and Validation**
   6.1 End Item A
       6.1.1   Developmental/Engineering Unit Evaluations
       6.1.2   Verification Activities
           6.1.2.1 Verification Testing
               6.1.2.1.1 Qualification Testing
               6.1.2.1.2 Other Testing

6.1.2.2 Verification Analysis

        6.1.2.2.1 Thermal Analysis

        6.1.2.2.2 Stress Analysis

        6.1.2.2.3 Analysis of Fracture Control

        6.1.2.2.4 Materials Analysis

        6.1.2.2.5 EEE Parts Analysis

6.1.2.3 Verification Inspection

6.1.2.4 Verification Demonstration

6.1.3   Validation Activities

6.1.4   Acceptance Testing

6.*n* End Item *n*

7. System X Verification and Validation

7.1 End-Item-to-End-Item Integration

    7.1.1   Developmental/Engineering Unit Evaluations

    7.1.2   Verification Activities

        7.1.2.1 Verification Testing

        7.1.2.2 Verification Analysis

        7.1.2.3 Verification Inspection

        7.1.2.4 Verification Demonstration

    7.1.3   Validation Activities

7.2 Complete System Integration

    7.2.1   Developmental/Engineering Unit Evaluations

    7.2.2   Verification Activities

        7.2.2.1 Verification Testing

        7.2.2.2 Verification Analysis

        7.2.2.3 Verification Inspection

        7.2.2.4 Verification Demonstration

    7.2.3   Validation Activities

8. System X Program Verification and Validation

8.1 Vehicle Integration

8.2 End-to-End Integration

8.3 On-Orbit V&V Activities

9. System X Certification Products

Appendix A: **Acronyms and Abbreviations**

Appendix B: **Definition of Terms**

Appendix C: **Requirement Verification Matrix**

Appendix D: **System X Validation Matrix**

Appendix J: SEMP Content Outline

SEMP Content

The SEMP is the foundation document for the technical and engineering activities conducted during the project. The SEMP conveys information on the technical integration methodologies and activities for the project within the scope of the project plan to all of the personnel. Because the SEMP provides the specific technical and management information to understand the technical integration and interfaces, its documentation and approval serves as an agreement within the project of how the technical work will be conducted. The technical team, working under the overall program/project plan, develops and updates the SEMP as necessary. The technical team works with the project manager to review the content and obtain concurrence. The SEMP includes the following three general sections:

- Technical program planning and control, which describes the processes for planning and control of the engineering efforts for the design, development, test, and evaluation of the system.
- Systems engineering processes, which includes specific tailoring of the systems engineering process as described in the NPR, implementation procedures, trade study methodologies, tools, and models to be used.
- Engineering specialty integration describes the integration of the technical disciplines' efforts into the systems engineering process and summarizes each technical discipline effort and cross references each of the specific and relevant plans.

Purpose and Scope

This section provides a brief description of the purpose, scope, and content of the SEMP. The scope encompasses the SE technical effort required to generate the work products necessary to meet the success criteria for the product-line life-cycle phases. The SEMP is a plan for doing the project technical effort by a technical team for a given WBS model in the system structure and to help meet life-cycle phase success criteria.

Applicable Documents

This section of the SEMP lists the documents applicable to this specific project and its SEMP implementation and describes major standards and procedures that this technical effort for this specific project needs to follow. Specific implementation of standardization tasking is incorporated into pertinent sections of the SEMP.

Provide the engineering standards and procedures to be used in the project. Examples of specific procedures could include any hazardous material handling, crew training for control room operations, special instrumentation techniques, special interface documentation for vehicles, and maintenance procedures specific to the project.

Technical Summary

This section contains an executive summary describing the problem to be solved by this technical effort and the purpose, context, and products of the WBS model to be developed and integrated with other interfacing systems identified.

System Description

This section contains a definition of the purpose/mission/objective of the system being developed, a brief description of the purpose of the products of the WBS models of the system structure for which this SEMP applies, and the expected scenarios for the system. Each WBS model includes the system end products and their subsystems and the supporting or enabling products and any other work products (plans, baselines) required for the development of the system. The description should include any interfacing systems and system products, including humans, with which the WBS model system products will interact physically, functionally, or electronically.

Identify and document system constraints, including cost, schedule, and technical (for example, environmental, design).

System Structure

This section contains an explanation of how the WBS models will be developed, how the resulting WBS model will be integrated into the project WBS, and how the overall system structure will be developed. This section contains a description of the relationship of the specification tree and the drawing tree with the products of the system structure and how the relationship and interfaces of the system end products and their life-cycle-enabling products will be managed throughout the planned technical effort.

Product Integration

This subsection contains an explanation of how the product will be integrated and will describe clear organizational responsibilities and interdependencies whether the organizations are geographically dispersed or managed across Centers. This includes identifying organizations—intra- and inter-NASA, other Government agencies, contractors, or other partners—and delineating their roles and responsibilities.

When components or elements will be available for integration needs to be clearly understood and identified on the schedule to establish critical schedule issues.

Planning Context

This subsection contains the product-line life-cycle model constraints (e.g., NPR 7120.5) that affect the planning and implementation of the common technical processes to be applied in performing the technical effort. The constraints provide a linkage of the technical effort with the applicable product-line life-cycle phases covered by the SEMP including, as applicable, milestone decision gates, major technical reviews, key intermediate events leading to project completion, life-cycle phase, event entry and success criteria, and major baseline and other work products to be delivered to the sponsor or customer of the technical effort.

Boundary of Technical Effort

This subsection contains a description of the boundary of the general problem to be solved by the technical effort. Specifically, it identifies what can be controlled by the technical team (inside the boundary) and what influences the technical effort and is influenced by the technical effort but not controlled by the technical team (outside the boundary). Specific attention should be given to physical, functional, and electronic interfaces across the boundary.

Define the system to be addressed. A description of the boundary of the system can include the following: definition of internal and external elements/items involved in realizing the system purpose as well as the system boundaries in terms of space, time, physical, and operational. Also, identification of what initiates the transitions of the system to operational status and what initiates its disposal is important. The following is a general listing of other items to include, as appropriate:

- General and functional descriptions of the subsystems,
- Document current and established subsystem performance characteristics,
- Identify and document current interfaces and characteristics,
- Develop functional interface descriptions and functional flow diagrams,
- Identify key performance interface characteristics, and
- Identify current integration strategies and architecture.

Cross References

This subsection contains cross references to appropriate nontechnical plans and critical reference material that interface with the technical effort. It contains a summary description of how the technical activities covered in other plans are accomplished as fully integrated parts of the technical effort.

Technical Effort Integration

This section contains a description of how the various inputs to the technical effort will be integrated into a coordinated effort that meets cost, schedule, and performance objectives.

The section should describe the integration and coordination of the specialty engineering disciplines into the systems engineering process during each iteration of the processes. Where there is potential for overlap of specialty efforts, the SEMP should define the relative responsibilities and authorities of each. This section should contain, as needed, the project's approach to the following:

- Concurrent engineering,
- The activity phasing of specialty engineering,
- The participation of specialty disciplines,
- The involvement of specialty disciplines,

- The role and responsibility of specialty disciplines,
- The participation of specialty disciplines in system decomposition and definition,
- The role of specialty disciplines in verification and validation,
- Reliability,
- Maintainability,
- Quality assurance,
- Integrated logistics,
- Human engineering,
- Safety,
- Producibility,
- Survivability/vulnerability,
- National Environmental Policy Act compliance, and
- Launch approval/flight readiness.

Provide the approach for coordination of diverse technical disciplines and integration of the development tasks. For example, this can include the use of integrated teaming approaches. Ensure that the specialty engineering disciplines are properly represented on all technical teams and during all life-cycle phases of the project. Define the scope and timing of the specialty engineering tasks.

Responsibility and Authority

This subsection contains a description of the organizing structure for the technical teams assigned to this technical effort and includes how the teams will be staffed and managed, including (1) what organization/panel will serve as the designated governing authority for this project and, therefore, will have final signature authority for this SEMP; (2) how multidisciplinary teamwork will be achieved; (3) identification and definition of roles, responsibilities, and authorities required to perform the activities of each planned common technical process; (4) planned technical staffing by discipline and expertise level, with human resource loading; (5) required technical staff training; and (6) assignment of roles, responsibilities, and authorities to appropriate project stakeholders or technical teams to ensure planned activities are accomplished.

Provide an organization chart and denote who on the team is responsible for each activity. Indicate the lines of authority and responsibility. Define the resolution authority to make decisions/decision process. Show how the engineers/engineering disciplines relate.

The systems engineering roles and responsibilities need to be addressed for the following: project office, user, Contracting Office Technical Representative (COTR),

systems engineering, design engineering, specialty engineering, and contractor.

Contractor Integration

This subsection contains a description of how the technical effort of in-house and external contractors is to be integrated with the NASA technical team efforts. This includes establishing technical agreements, monitoring contractor progress against the agreement, handling technical work or product requirements change requests, and acceptance of deliverables. The subsection will specifically address how interfaces between the NASA technical team and the contractor will be implemented for each of the 17 common technical processes. For example, it addresses how the NASA technical team will be involved with reviewing or controlling contractor-generated design solution definition documentation or how the technical team will be involved with product verification and product validation activities.

Key deliverables for the contractor to complete their systems and those required of the contractor for other project participants need to be identified and established on the schedule.

Support Integration

This subsection contains a description of the methods (such as integrated computer-aided tool sets, integrated work product databases, and technical management information systems) that will be used to support technical effort integration.

Common Technical Processes Implementation

Each of the 17 common technical processes will have a separate subsection that contains a plan for performing the required process activities as appropriately tailored. (See NPR 7123.1 for the process activities required and tailoring.) Implementation of the 17 common technical processes includes (1) the generation of the outcomes needed to satisfy the entry and success criteria of the applicable product-line life-cycle phase or phases identified in D.4.4.4 and (2) the necessary inputs for other technical processes. These sections contain a description of the approach, methods, and tools for:

- Identifying and obtaining adequate human and non-human resources for performing the planned process, developing the work products, and providing the services of the process.

- Assigning responsibility and authority for performing the planned process, developing the work products, and providing the services of the process.
- Training the technical staff performing or supporting the process, where training is identified as needed.
- Designating and placing designated work products of the process under appropriate levels of configuration management.
- Identifying and involving stakeholders of the process.
- Monitoring and controlling the process.
- Identifying, defining, and tracking metrics and success.
- Objectively evaluating adherence of the process and the work products and services of the process to the applicable requirements, objectives, and standards and addressing noncompliance.
- Reviewing activities, status, and results of the process with appropriate levels of management and resolving issues.

This section should also include the project-specific description of each of the 17 processes to be used, including the specific tailoring of the requirements to the system and the project; the procedures to be used in implementing the processes; in-house documentation; trade study methodology; types of mathematical and/or simulation models to be used; and generation of specifications.

Technology Insertion

This section contains a description of the approach and methods for identifying key technologies and their associated risks and criteria for assessing and inserting technologies, including those for inserting critical technologies from technology development projects. An approach should be developed for appropriate level and timing of technology insertion. This could include alternative approaches to take advantage of new technologies to meet systems needs as well as alternative options if the technologies do not prove appropriate in result or timing. The strategy for an initial technology assessment within the scope of the project requirements should be provided to identify technology constraints for the system.

Additional SE Functions and Activities

This section contains a description of other areas not specifically included in previous sections but that are essential for proper planning and conduct of the overall technical effort.

System Safety

This subsection contains a description of the approach and methods for conducting safety analysis and assessing the risk to operators, the system, the environment, or the public.

Engineering Methods and Tools

This subsection contains a description of the methods and tools not included in the technology insertion section that are needed to support the overall technical effort and identifies those tools to be acquired and tool training requirements.

Define the development environment for the project, including automation and software tools. If required, develop and/or acquire the tools and facilities for all disciplines on the project. Standardize when possible across the project, or enable a common output format of the tools that can be used as input by a broad range of tools used on the project. Define the requirements for information management systems and for using existing elements. Define and plan for the training required to use the tools and technology across the project.

Specialty Engineering

This subsection contains a description of engineering discipline and specialty requirements that apply across projects and the WBS models of the system structure. Examples of these requirement areas would include planning for safety, reliability, human factors, logistics, maintainability, quality, operability, and supportability. Estimate staffing levels for these disciplines and incorporate with the project requirements.

Integration with the Project Plan and Technical Resource Allocation

This section contains how the technical effort will integrate with project management and defines roles and responsibilities. It addresses how technical requirements will be integrated with the project plan to determinate the allocation of resources, including cost, schedule, and personnel, and how changes to the allocations will be coordinated.

This section describes the interface between all of the technical aspects of the project and the overall project management process during the systems engineering planning activities and updates. All activities to coordinate technical efforts with the overall project are included, such as technical interactions with the external stakeholders, users, and contractors.

Waivers

This section contains all approved waivers to the Center Director's Implementation Plan for SE NPR 7123.1 requirements for the SEMP. This section also contains a separate subsection that includes any tailored SE NPR requirements that are not related and able to be documented in a specific SEMP section or subsection.

Appendices

Appendices are included, as necessary, to provide a glossary, acronyms and abbreviations, and information published separately for convenience in document main-

tenance. Included would be: (1) information that may be pertinent to multiple topic areas (e.g., description of methods or procedures); (2) charts and proprietary data applicable to the technical efforts required in the SEMP; and (3) a summary of technical plans associated with the project. Each appendix should be referenced in one of the sections of the engineering plan where data would normally have been provided.

Templates

Any templates for forms, plans, or reports the technical team will need to fill out, like the format for the verification and validation plan, should be included in the appendices.

References

This section contains all documents referenced in the text of the SEMP.

SEMP Preparation Checklist

The SEMP, as the key reference document capturing the technical planning, needs to address some basic topics. For a generic SEMP preparation checklist, refer to *Systems Engineering Guidebook* by James Martin.

Appendix K: Plans

Appendix L: Interface Requirements Document Outline

1.0 Introduction

    1.1 Purpose and Scope. State the purpose of this document and briefly identify the interface to be defined. (For example, "This IRD defines and controls the interface(s) requirements between _____ and _____.")

    1.2 Precedence. Define the relationship of this document to other program documents and specify which is controlling in the event of a conflict.

    1.3 Responsibility and Change Authority. State the responsibilities of the interfacing organizations for development of this document and its contents. Define document approval authority (including change approval authority).

2.0 Documents

    2.1 Applicable Documents. List binding documents that are invoked to the extent specified in this IRD. The latest revision or most recent version should be listed. Documents and requirements imposed by higher–level documents (higher order of precedence) should not be repeated.

    2.2 Reference Documents. List any document that is referenced in the text in this subsection.

3.0 Interfaces

    3.1 General. In the subsections that follow, provide the detailed description, responsibilities, coordinate systems, and numerical requirements as they relate to the interface plane.

        3.1.1 Interface Description. Describe the interface as defined in the system specification. Use tables, figures, or drawings as appropriate.

        3.1.2 Interface Responsibilities. Define interface hardware and interface boundary responsibilities to depict the interface plane. Use tables, figures, or drawings as appropriate.

        3.1.3 Coordinate Systems. Define the coordinate system used for interface requirements on each side of the interface. Use tables, figures, or drawings as appropriate.

        3.1.4 Engineering Units, Tolerances, and Conversions. Define the measurement units along with tolerances. If required, define the conversion between measurement systems.

    3.2 Interface Requirements. In the subsections that follow, define structural limiting values at the interface, such as interface loads, forcing functions, and dynamic conditions.

        3.2.1 Interface Plane. Define the interface requirements on each side of the interface plane.

            3.2.1.1 Envelope

            3.2.1.2 Mass Properties. Define the derived interface requirements based on the allocated requirements contained in the applicable specification pertaining to that side of the interface. For example, this subsection should cover the mass of the element.

            3.2.1.3 Structural/Mechanical. Define the derived interface requirements based on the allocated requirements contained in the applicable specification pertaining to that side of the interface. For example, this subsection should cover attachment, stiffness, latching, and mechanisms.

            3.2.1.4 Fluid. Define the derived interface requirements based on the allocated requirements contained in the applicable specification pertaining to that side of the interface. For example, this subsection should cover fluid areas such as thermal control, O_2 and N_2, potable and waste water, fuel cell water, and atmospheric sampling.

3.2.1.5 Electrical (Power). Define the derived interface requirements based on the allocated requirements contained in the applicable specification pertaining to that side of the interface. For example, this subsection should cover various electric current, voltage, wattage, and resistance levels.

3.2.1.6 Electronic (Signal). Define the derived interface requirements based on the allocated requirements contained in the applicable specification pertaining to that side of the interface. For example, this subsection should cover various signal types such as audio, video, command data handling, and navigation.

3.2.1.7 Software and Data. Define the derived interface requirements based on the allocated requirements contained in the applicable specification pertaining to that side of the interface. For example, this subsection should cover various data standards, message timing, protocols, error detection/correction, functions, initialization, and status.

3.2.1.8 Environments. Define the derived interface requirements based on the allocated requirements contained in the applicable specification pertaining to that side of the interface. For example, cover the dynamic envelope measures of the element in English units or the metric equivalent on this side of the interface.

    3.2.1.8.1 Electromagnetic Effects

        3.2.1.8.1.a Electromagnetic Compatibility. Define the appropriate electromagnetic compatibility requirements. For example, end-item-1-to-end-item-2 interface shall meet the requirements [to be determined] of systems requirements for electromagnetic compatibility.

        3.2.1.8.1.b Electromagnetic Interference. Define the appropriate electromagnetic interference requirements. For example, end-item-1-to-end-item-2 interface shall meet the requirements [to be determined] of electromagnetic emission and susceptibility requirements for electromagnetic compatibility.

        3.2.1.8.1.c Grounding. Define the appropriate grounding requirements. For example, end-item-1-to-end-item-2 interface shall meet the requirements [to be determined] of grounding requirements.

        3.2.1.8.1.d Bonding. Define the appropriate bonding requirements. For example, end-item-1-to-end-item-2 structural/mechanical interface shall meet the requirements [to be determined] of electrical bonding requirements.

        3.2.1.8.1.e Cable and Wire Design. Define the appropriate cable and wire design requirements. For example, end-item-1-to-end-item-2 cable and wire interface shall meet the requirements [to be determined] of cable/wire design and control requirements for electromagnetic compatibility.

    3.2.1.8.2 Acoustic. Define the appropriate acoustics requirements. Define the acoustic noise levels on each side of the interface in accordance with program or project requirements.

    3.2.1.8.3 Structural Loads. Define the appropriate structural loads requirements. Define the mated loads that each end item must accommodate.

    3.2.1.8.4 Vibroacoustics. Define the appropriate vibroacoustics requirements. Define the vibroacoustic loads that each end item must accommodate.

3.2.1.9 Other Types of Interface Requirements. Define other types of unique interface requirements that may be applicable.

Appendix M: CM Plan Outline

A typical CM plan should include the following:

Table M-1 CM Plan Outline

Section	Description
1.0 Introduction	This section includes: • The purpose and scope of the CM plan and the program phases to which it applies • Brief description of the system or top-level configuration items
2.0 Applicable and Reference Documents	This section includes a list of the specifications, standards, manuals, and other documents, referenced in the plan by title, document number, issuing authority, revision, and as applicable, change notice, amendment, and issue date.
3.0 CM Concepts and Organization	This section includes: • CM objectives • Information needed to support the achievement of objectives in the current and future phases • Description and graphic portraying the project's planned organization with emphasis on the CM activities
4.0 CM Process • CM Management and Planning • Configuration Identification • Configuration Control • Configuration Status Accounting • Configuration Audits	This section includes a description of the project's CM process for accomplishing the five CM activities, which includes but is not limited to: • CM activities for the current and future phases • Baselines • Configuration items • Establishment and membership of configuration control boards • Nomenclature and numbering • Hardware/software identification • Functional configuration audits and physical configuration audits
5.0 Management of Configuration Data	This section describes the methods for meeting the CM technical data requirements.
6.0 Interface Management	This section includes a description on how CM will maintain and control interface documentation.
7.0 CM Phasing and Schedule	This section describes milestones for implementing CM commensurate with major program milestones.
8.0 Subcontractor/Vendor Control	This section describes methods used to ensure subcontractor/vendors comply with CM requirements.

Appendix N: Guidance on Technical Peer Reviews/Inspections

Introduction

The objective of technical peer reviews/inspections is to remove defects as early as possible in the development process. Peer reviews/inspections are a well defined review process for finding and fixing defects, conducted by a team of peers with assigned roles, each having a vested interest in the work product under review. Peer reviews/inspections are held within development phases, between milestone reviews, on completed products or completed portions of products. The results of peer reviews/inspections can be reported at milestone reviews. Checklists are heavily utilized in peer reviews/inspections to improve the quality of the review.

Technical peer reviews/inspections have proven over time to be one of the most effective practices available for ensuring quality products and on-time deliveries. Many studies have demonstrated their benefits, both within NASA and across industry. Peer reviews/inspections improve quality and reduce cost by reducing rework. The studies have shown that the rework effort saved not only pays for the effort spent on inspections, but also provides additional cost savings on the project. By removing defects at their origin (e.g., requirements and design documents, test plans and procedures, software code, etc.), inspections prevent defects from propagating through multiple phases and work products, and reduce the overall amount of rework necessary on projects. In addition, improved team efficiency is a side effect of peer reviews/inspections (e.g., by improving team communication, more quickly bringing new members up to speed, and educating project members about effective development practices).

How to Perform Technical Peer Reviews/Inspections

Figure N-1 shows a diagram of the peer review/inspection stages, and the text below the figure explains how to perform each of the stages. (Figure N-2, at the end of the

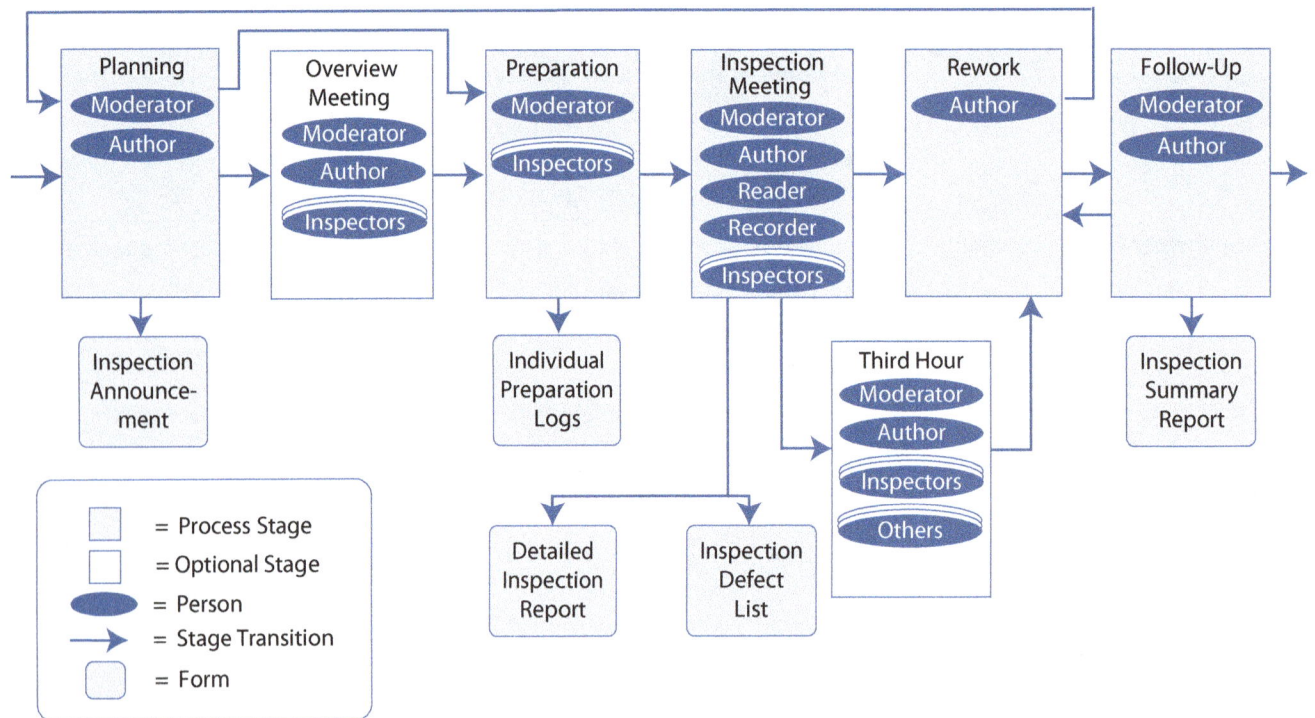

Figure N-1 The peer review/inspection process

appendix, summarizes the information as a quick reference guide.)

It is recommended that the moderator review the Planning Inspection Schedule and Estimating Staff Hours, Guidelines for Successful Inspections, and 10 Basic Rules of Inspections in Figure N-2 before beginning the planning stage. (Note: *NPR 7150.2, NASA Software Engineering Requirements* defines Agency requirements on the use of peer reviews and inspections for software development. NASA peer review/inspection training is offered by the NASA Office of the Chief Engineer.)

> Note: Where activities have an *, the moderator records the time on the inspection summary report.

A. Planning

The moderator of the peer review/inspection performs the following activities.[1]

1. Determine whether peer review/inspection entrance criteria have been met.
2. Determine whether an overview of the product is needed.
3. Select the peer review/inspection team and assign roles. For guidance on roles, see Roles of Participants in Figure N-2 at the end of this appendix. Reviewers have a vested interest in the work product (e.g., they are peers representing areas of the life cycle affected by the material being reviewed).
4. Determine if the size of the product is within the prescribed guidelines for the type of inspection. (See Meeting Rate Guidelines in Figure N-2 for guidelines on the optimal number of pages or lines of code to inspect for each type of inspection.) If the product exceeds the prescribed guidelines, break the product into parts and inspect each part separately. (It is highly recommended that the peer review/inspection meeting not exceed 2 hours.)
5. Schedule the overview (if one is needed).

_____
[1]Langley Research Center, *Instructional Handbook for Formal Inspections*. This document provides more detailed instructions on how to perform technical peer reviews/inspections. It also provides templates for the forms used in the peer review/inspection process described above: inspection announcement, individual preparation log, inspection defect list, detailed inspection report, and the inspection summary report.

6. Schedule peer review/inspection meeting time and place.
7. Prepare and distribute the inspection announcement and package. Include in the package the product to be reviewed and the appropriate checklist for the peer review/inspection.
8. Record total time spent in planning.*

B. Overview Meeting

1. Moderator runs the meeting, and the author presents background information to the reviewers.
2. Record total time spent in the overview.*

C. Peer Review/Inspection Preparation

1. Peers review the checklist definitions of defects.
2. Examine materials for understanding and possible defects.
3. Prepare for assigned role in peer review/inspection.
4. Complete and turn in individual preparation log to the moderator.
5. The moderator reviews the individual preparation logs and makes Go or No-Go decision and organizes inspection meeting.
6. Record total time spent in the preparation.*

D. Peer Review/Inspection Meeting

1. The moderator introduces people and identifies their peer review/inspection roles.
2. The reader presents work products to the peer review/inspection team in a logical and orderly manner.
3. Peer reviewers/inspectors find and classify defects by severity, category, and type. (See Classification of Defects in Figure N-2.)
4. The recorder writes the major and minor defects on the inspection defect list (for definitions of major and minor, see the Severity section of Figure N.2).
5. Steps 1 through 4 are repeated until the review of the product is completed.
6. Open issues are assigned to peer reviewers/inspectors if irresolvable discrepancies occur.
7. Summarize the number of defects and their classification on the detailed inspection report.
8. Determine the need for a reinspection or third hour. Optional: Trivial defects (e.g., redlined documents) can be given directly to the author at the end of the inspection.

9. The moderator obtains an estimate for rework time and completion date from the author, and does the same for action items if appropriate.

10. The moderator assigns writing of change requests and/or problem reports (if needed).

11. Record total time spent in the peer review/inspection meeting.*

E. Third Hour

1. Complete assigned action items and provide information to the author.

2. Attend third hour meeting at author's request.

3. Provide time spent in third-hour to the moderator.*

F. Rework

1. All major defects noted in the inspection defect list are resolved by the author.

2. Minor and trivial defects (which would not result in faulty execution) are resolved at the discretion of the author as time and cost permit.

3. Record total time spent in the rework on the inspection defect list.

G. Followup

1. The moderator verifies all major defects have been corrected and no secondary defects have been introduced.

2. The moderator ensures all open issues are resolved and verifies all success criteria for the peer review/inspection are met.

3. Record total time spent in rework and followup.*

4. File the inspection package.

5. The inspection summary report is distributed.

6. Communicate that the peer review/inspection has been passed.

Peer Reviews/Inspections

QUICK REFERENCE GUIDE

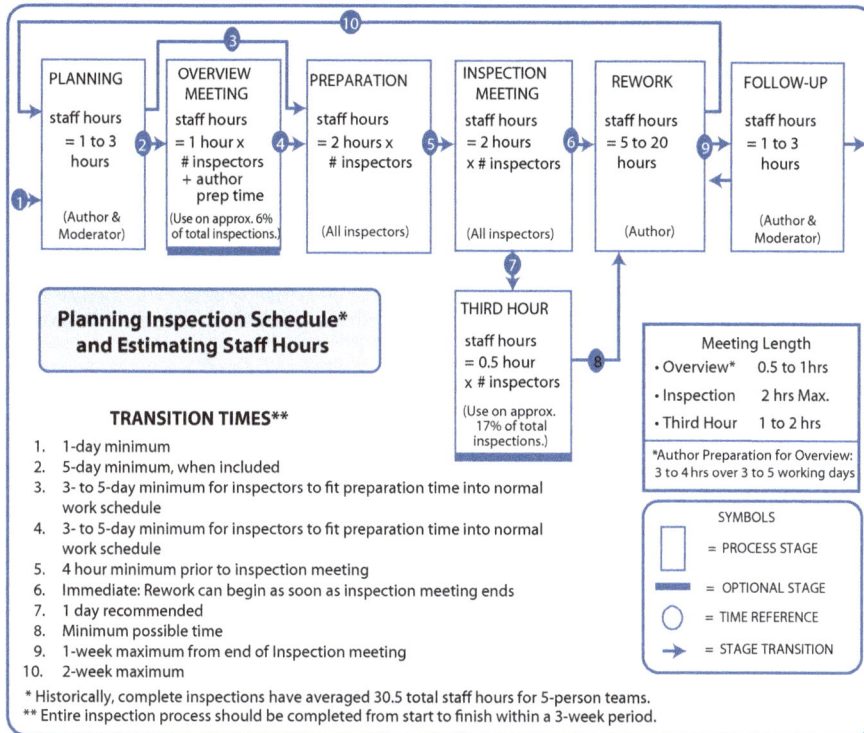

Process Flow Diagram

PLANNING	OVERVIEW MEETING	PREPARATION	INSPECTION MEETING	REWORK	FOLLOW-UP
staff hours = 1 to 3 hours	staff hours = 1 hour x # inspectors + author prep time	staff hours = 2 hours x # inspectors	staff hours = 2 hours x # inspectors	staff hours = 5 to 20 hours	staff hours = 1 to 3 hours
(Author & Moderator)	(Use on approx. 6% of total inspections.)	(All inspectors)	(All inspectors)	(Author)	(Author & Moderator)

THIRD HOUR

staff hours = 0.5 hour x # inspectors

(Use on approx. 17% of total inspections.)

Planning Inspection Schedule* and Estimating Staff Hours

TRANSITION TIMES**

1. 1-day minimum
2. 5-day minimum, when included
3. 3- to 5-day minimum for inspectors to fit preparation time into normal work schedule
4. 3- to 5-day minimum for inspectors to fit preparation time into normal work schedule
5. 4 hour minimum prior to inspection meeting
6. Immediate: Rework can begin as soon as inspection meeting ends
7. 1 day recommended
8. Minimum possible time
9. 1-week maximum from end of Inspection meeting
10. 2-week maximum

* Historically, complete inspections have averaged 30.5 total staff hours for 5-person teams.
** Entire inspection process should be completed from start to finish within a 3-week period.

Meeting Length
- Overview* 0.5 to 1hrs
- Inspection 2 hrs Max.
- Third Hour 1 to 2 hrs

*Author Preparation for Overview: 3 to 4 hrs over 3 to 5 working days

SYMBOLS
- ☐ = PROCESS STAGE
- ▬ = OPTIONAL STAGE
- ◯ = TIME REFERENCE
- → = STAGE TRANSITION

Types of Inspections

SY1	System Requirements
SY2	System Design
SU1	Subsystem Requirements
SU2	Subsystem Design
R1	Software Requirements
I0	Architectural Design
I1	Detailed Design
I2	Source Code
IT1	Test Plan
IT2	Test Procedures & Functions

Classification of Defects

Severity

Major
- An error that would cause a malfunction or prevents attainment of an expected or specified result.
- Any error that would in the future result in an approved change request or failure report.

Minor
- A violation of standards, guidelines, or rules that would not result in a deviation from requirements if not corrected, but could result in difficulties in terms of operations, maintenance, or future development.

Trivial
- Editorial errors such as spelling, punctuation, and grammar that do not cause errors or change requests. Recorded only as redlines. Presented directly to author.

Author is required to correct all major defects and should correct minor and trivial defects as time and cost permit.

Category

- Missing   - Wrong   - Extra

Type

Types of defects are derived from headings on checklist used for the inspection. Defect types can be standardized across inspections from all phases of the life cycle. A suggested standard set of defect types are:

- Clarity
- Completeness
- Compliance
- Consistency
- Correctness/Logic
- Data Usage
- Fault Tolerance
- Functionality
- Interface
- Level of Detail
- Maintainability
- Performance
- Reliability
- Testability
- Traceability
- Other

EXAMPLE

The following is an example of a defect classification that would be recorded on the Inspection Defect List:

Description	Classification
Line 169 – While counting the number of leading spaces in variable NAME, the wrong "I" is used to calculate "J."	Major Defect [X] Missing [ ] Minor Defect [ ] Wrong [X] Open Issue [ ] Extra [ ] Type: Data Usage Origin:

Based on JCK/LLW/SSP/HS: 10/92

Guidelines for Successful Inspections

- Train moderators, inspectors, and managers
- No more than 25% of developers' time should be devoted to inspections
- Inspect 100% of work product
- Be prepared
- Share responsibility for work product quality
- Be willing to associate and communicate
- Avoid judgmental language
- Do not evaluate author
- Have at least one positive and negative input
- Raise issues; don't resolve them
- Avoid discussions of style
- Stick to standard or change it
- Be technically competent
- Record all issues in public
- Stick to technical issues
- Distribute inspection documents as soon as possible
- Let author determine when work product is ready for inspection
- Keep accurate statistics

10 Basic Rules of Inspections

- Inspections are carried out at a number of points inside phases of the life cycle. Inspections are not substitutes for milestone reviews.
- Inspections are carried out by peers representing areas of life cycle affected by material being inspected (usually limited to 6 or fewer people). All inspectors should have a vested interest in the work product.
- Management is not present during inspections. Inspections are not to be used as a tool to evaluate workers.
- Inspections are led by a trained moderator.
- Trained inspectors are assigned roles.
- Inspections are carried out in a prescribed series of steps.
- Inspection meeting is limited to 2 hours.
- Checklists of questions are used to define task and to stimulate defect finding.
- Material is covered during inspection meeting within an optional page rate, which has been found to give maximum error-finding ability.
- Statistics on number of defects, types of defects, and time expended by engineers on inspections are kept.

Meeting* Rate Guidelines for Various Inspection Types

Type	Inspection Meeting	
	Target per 2 Hrs	Range per 2 Hrs
R0	20 pages	10 to 30 pages
R1	20 pages	10 to 30 pages
I0	30 pages	20 to 40 pages
I1	35 pages	25 to 45 Pages
I2	500 lines of source code**	400 to 600 lines of source code**
IT1	30 pages	20 to 40 pages
IT2	35 pages	25 to 45 pages

* Assumes a 2-hour meeting. Scale down planned meeting duration for shorter work products.
** Flight software and other highly complex code segments should proceed at about half this rate.

Roles of Participants

Moderator
Responsible for conducting inspection process and collecting inspection data. Plays key role in all stages of process except rework. Required to perform special duties during an inspection in addition to inspector's tasks.

Inspectors
Responsible for finding defects in work product from a general point of view, as well as defects that affect their area of expertise.

Author
Provides information about work product during all stages of process. Responsible for correcting all major defects and any minor and trivial defects that cost and schedule permit. Performs duties of an inspector.

Reader
Guides team through work product during inspection meeting. Reads or paraphrases work product in detail. Should be an inspector from same (or next) life-cycle phase as author. Performs duties of an inspector in addition to reader's role.

Recorder
Accurately records each defect found during inspection meeting on the Inspection Defect List. Performs duties of an inspector in addition to recorder's role.

Figure N-2 Peer reviews/inspections quick reference guide

Appendix O: Tradeoff Examples

Table O-1 Typical Tradeoffs for Space Systems

Development Related	Operations and Support Related
• Custom versus commercial-off-the-shelf	• Upgrade versus new start
• Light parts (expensive) versus heavy parts (less expensive)	• Manned versus unmanned
• On-board versus remote processing	• Autonomous versus remotely controlled
• Radio frequency versus optical links	• System of systems versus stand-alone system
• Levels of margin versus cost/risk	• One long-life unit versus many short-life units
• Class S versus non-class S parts	• Low Earth orbit versus medium Earth orbit versus geostationary orbit versus high Earth orbit
• Radiation-hardened versus standard components	• Single satellite versus constellation
• Levels of redundancy	• Launch vehicle type (e.g., Atlas versus Titan)
• Degrees of quality assurance	• Single stage versus multistage launch
• Built-in test versus remote diagnostics	• Repair in-situ versus bring down to ground
• Types of environmental exposure prior to operation	• Commercial versus Government assets
• Level of test (system versus subsystem)	• Limited versus public access
• Various life-cycle approaches (e.g., waterfall versus spiral versus incremental)	• Controlled versus uncontrolled reentry

Table O-2 Typical Tradeoffs in the Acquisition Process

Acquisition Phase	Trade Study Purpose
Mission needs analysis	Prioritize identified user needs
Concept exploration (concept and technology development)	1. Compare new technology with proven concepts 2. Select concepts best meeting mission needs 3. Select alternative system configurations 4. Focus on feasibility and affordability
Demonstration/validation	1. Select technology 2. Reduce alternative configurations to a testable number
Full-scale development (system development and demonstration	1. Select component/part designs 2. Select test methods 3. Select operational test and evaluation quantities
Production	1. Examine effectiveness of all proposed design changes 2. Perform make/buy, process, rate, and location decisions

Table O-3 Typical Tradeoffs Throughout the Project Life Cycle

Pre Phase A	Phase A	Phase B	Phases C&D	Phases D&E	Phases E&F
• Problem selection • Upgrade versus new start	• On-board versus ground processing • Low Earth orbit versus geostationary orbit	• Levels of redundancy • Radio frequency links versus optical links	• Single source versus multiple suppliers • Level of testing	• Platform STS-28 versus STS-3a • Launch go-ahead (Go or No-Go)	• Adjust orbit daily versus weekly • Deorbit now versus later

Appendix P: SOW Review Checklist

Editorial Checklist

1. Is the SOW requirement in the form "who" shall "do what"? An example is, "The Contractor shall (perform, provide, develop, test, analyze, or other verb followed by a description of what)."

    Example SOW requirements:

    - The Contractor shall design the XYZ flight software...
    - The Contractor shall operate the ABC ground system...
    - The Contractor shall provide maintenance on the following...
    - The Contractor shall report software metrics monthly ...
    - The Contractor shall integrate the PQR instrument with the spacecraft...

2. Is the SOW requirement a simple sentence that contains only one requirement?

    Compound sentences that contain more than one SOW requirement need to be split into multiple simple sentences. (For example, "The Contractor shall do ABC and perform XYZ" should be rewritten as "The Contractor shall do ABC" and "The Contractor shall perform XYZ.")

3. Is the SOW composed of simple, cohesive paragraphs, each covering a single topic? Paragraphs containing many requirements should be divided into subparagraphs for clarity.

4. Has each paragraph and subparagraph been given a unique number or letter identifier? Is the numbering or lettering correct?

5. Is the SOW requirement in the active rather than the passive voice? Passive voice leads to vague statements. (For example, state, "The Contractor shall hold monthly management review meetings" instead of "Management review meeting shall be held monthly.")

6. Is the SOW requirement stated positively as opposed to negatively? (Replace statements such as, "The Contractor shall not exceed the budgetary limits specified" with "The contractor shall comply with the budgetary limits specified.")

7. Is the SOW requirement grammatically correct?

8. Is the SOW requirement free of typos, misspellings, and punctuation errors?

9. Have all acronyms been defined in an acronym list or spelled out in the first occurrence?

10. Have the quantities, delivery schedules, and delivery method been identified for each deliverable within the SOW or in a separate attachment/section?

11. Has the content of documents to be delivered been defined in a separate attachment/section and submitted with the SOW?

12. Has the file format of each electronic deliverable been defined (e.g., Microsoft—Project, Adobe—Acrobat PDF, National Instruments—Labview VIs)?

Content Checklist

1. Are correct terms used to define the requirements?
    - **Shall** = requirement (binds the contractor)
    - **Should** = goal (leaves decision to contractor; avoid using this word)
    - **May** = allowable action (leaves decision to contractor; avoid using this word)

- **Will** = facts or declaration of intent by the Government (use only in referring to the Government)
- **Present tense** (e.g., "is") = descriptive text only (avoid using in requirements statements; use "shall" instead)
- **NEVER** use "must"

2. Is the scope of the SOW clearly defined? Is it clear what you are buying?

3. Is the flow and organizational structure of the document logical and understandable? (See LPR 5000.2 "Procurement Initiator's Guide," Section 12, for helpful hints.) Is the text compatible with the title of the section it's under? Are subheadings compatible with the subject matter of headings?

4. Is the SOW requirement clear and understandable?

   - Can the sentence be understood only one way?

   - Will all terminology used have the same meaning to different readers without definition? Has any terminology for which this is not the case been defined in the SOW? (E.g., in a definitions section or glossary.)

   - Is it free from indefinite pronouns ("this," "that," "these," "those") without clear antecedents (e.g., replace statements such as, "These shall be inspected on an annual basis" with "The fan blades shall be inspected on an annual basis")?

   - Is it stated concisely?

5. Have all redundant requirements been removed? Redundant requirements can reduce clarity, increase ambiguity, and lead to contradictions.

6. Is the requirement consistent with other requirements in the SOW, without contradicting itself, without using the same terminology with different meanings, without using different terminology for the same thing?

7. If the SOW includes the delivery of a product (as opposed to just a services SOW):

   - Are the technical product requirements in a separate section or attachment, apart from the activities that the contractor is required to perform? The intent is to clearly delineate between the technical product requirements and requirements for activities the contractor is to perform (e.g., separate SOW statements "The contractor shall" from technical product requirement statements such as "The system shall" and "The software shall").

   - Are references to the product and its subelements in the SOW at the level described in the technical product requirements?

   - Is the SOW consistent with and does it use the same terminology as the technical product requirements?

8. Is the SOW requirement free of ambiguities? Make sure the SOW requirement is free of vague terms (for example, "as appropriate," "any," "either," "etc.," "and/or," "support," "necessary," "but not limited to," "be capable of," "be able to").

9. Is the SOW requirement verifiable? Make sure the SOW requirement is free of unverifiable terms (for example, "flexible," "easy," "sufficient," "safe," "ad hoc," "adequate," "accommodate," "user-friendly," "usable," "when required," "if required," "appropriate," "fast," "portable," "lightweight," "small," "large," "maximize," "minimize," "optimize," "sufficient," "robust," "quickly," "easily," "clearly," other "ly" words, other "ize" words).

10. Is the SOW requirement free of implementation constraints? SOW requirements should state WHAT the contractor is to do, NOT HOW they are to do it (for example, "The Contractor shall design the XYZ flight software" states WHAT the contractor is to do, while "The Contractor shall design the XYZ software using object-oriented design" states HOW the contractor is to implement the activity of designing the software. In addition, too low a level of decomposition of activities can result in specifying how the activities are to be done, rather than what activities are to be done).

11. Is the SOW requirement stated in such a way that compliance with the requirement is verifiable? Do the means exist to measure or otherwise assess its accomplishment? Can a method for verifying compliance with the requirement be defined (e.g., described in a quality assurance surveillance plan)?

12. Is the background material clearly labeled as such (i.e., included in the background section of the SOW if one is used)?

13. Are any assumptions able to be validated and restated as requirements? If not, the assumptions should be deleted from the SOW. Assumptions should be recorded in a document separate from the SOW.

14. Is the SOW complete, covering all of the work the contractor is to do?

    - Are all of the activities necessary to develop the product included (e.g., system, software, and hardware activities for the following: requirements, architecture, and design development; implementation and manufacturing; verification and validation; integration testing and qualification testing.)?

    - Are all safety, reliability, maintainability (e.g., mean time to restore), availability, quality assurance, and security requirements defined for the total life of the contract?

    - Does the SOW include a requirement for the contractor to have a quality system (e.g., ISO certified), if one is needed?

    - Are all of the necessary management and support requirements included in the SOW (for example, project management; configuration management; systems engineering; system integration and test; risk management; interface definition and management; metrics collection, reporting, analysis, and use; acceptance testing; NASA Independent Verification and Validation (IV&V) support tasks.)?

    - Are clear performance standards included and sufficient to measure contractor performance (e.g., systems, software, hardware, and service performance standards for schedule, progress, size, stability, cost, resources, and defects)? See Langley's *Guidance on System and Software Metrics for Performance-Based Contracting* for more information and examples on performance standards.

    - Are all of the necessary service activities included (for example, transition to operations, operations, maintenance, database administration, system administration, and data management)?

    - Are all of the Government surveillance activities included (for example, project management meetings; decision points; requirements and design peer reviews for systems, software, and hardware; demonstrations; test readiness reviews; other desired meetings (e.g., technical interchange meetings); collection and delivery of metrics for systems, software, hardware, and services (to provide visibility into development progress and cost); electronic access to technical and management data; and access to subcontractors and other team members for the purposes of communication)?

    - Are the Government requirements for contractor inspection and testing addressed, if necessary?

    - Are the requirements for contractor support of Government acceptance activities addressed, if necessary?

15. Does the SOW only include contractor requirements? It should not include Government requirements.

16. Does the SOW give contractors full management responsibility and hold them accountable for the end result?

17. Is the SOW sufficiently detailed to permit a realistic estimate of cost, labor, and other resources required to accomplish each activity?

18. Are all deliverables identified (e.g., status, financial, product deliverables)? The following are examples of deliverables that are sometimes overlooked: management and development plans; technical progress reports that identify current work status, problems and proposed corrective actions, and planned work; financial reports that identify costs (planned, actual, projected) by category (e.g., software, hardware, quality assurance); products (e.g., source code, maintenance/user manual, test equipment); and discrepancy data (e.g., defect reports, anomalies). All deliverables should be specified in a separate document except for technical deliverables (e.g., hardware, software, prototypes), which should be included in the SOW.

19. Does each technical and management deliverable track to a paragraph in the SOW? Each deliverable should have a corresponding SOW requirement for its preparation (i.e., the SOW identifies the title of the deliverable in parentheses after the task requiring the generation of the deliverable).

20. Are all reference citations complete?

    - Are the complete number, title, and date or version of each reference specified?

    - Does the SOW reference the standards and other compliance documents in the proper SOW paragraphs?

- Is the correct reference document cited and is it referenced at least once?
- Is the reference document either furnished with the SOW or available at a location identified in the SOW?
- If the referenced standard or compliance document is only partially applicable, does the SOW explicitly and unambiguously reference the portion that is required of the contractor?

Appendix Q: Project Protection Plan Outline

The following outline will assist systems engineers in preparing a project protection plan. The plan is a living document that will be written and updated as the project progresses through major milestones and ultimately through end of life.

1. **Introduction**
   1.1 Protection Plan Overview
   1.2 Project Overview
   1.3 Acquisition Status

2. **References**
   2.1 Directives and Instructions
   2.2 Requirements
   2.3 Studies and Analyses

3. **References**
   3.1 Threats: Hostile Action
   - 3.1.1 Overview
   - 3.1.2 Threat Characterization
     - 3.1.2.1 Cyber Attack
     - 3.1.2.2 Electronic Attack
     - 3.1.2.3 Lasers
     - 3.1.2.4 Ground Attack
     - 3.1.2.5 Asymmetric Attack on Critical Commercial Infrastructure
     - 3.1.2.6 Anti-Satellite Weapons
     - 3.1.2.7 High-Energy Radio Frequency Weapons
     - 3.1.2.8 Artificially Enhanced Radiation Environment

   3.2 Threats: Environmental
   - 3.2.1 Overview
   - 3.2.2 Threat Characterization
     - 3.2.2.1 Natural Environment Storms
     - 3.2.2.2 Earthquakes
     - 3.2.2.3 Floods
     - 3.2.2.4 Fires
     - 3.2.2.5 Radiation Effects in the Natural Environment
     - 3.2.2.6 Radiation Effects to Spacecraft Electronics

4. **Protection Vulnerabilities**
   4.1 Ground Segment Vulnerabilities
   - 4.1.1 Command and Control Facilities
   - 4.1.2 Remote Tracking Stations
   - 4.1.3 Spacecraft Simulator(s)
   - 4.1.4 Mission Data Processing Facilities
   - 4.1.5 Flight Dynamic Facilities
   - 4.1.6 Flight Software Production/Verification/Validation Facilities

References

This appendix contains references that were cited in the sections of the handbook and sources for developing the material in the indicated sections. See the Bibliography for complete citations.

Section 2.0 Fundamentals of Systems Engineering

Griffin, Michael D., "System Engineering and the Two Cultures of Engineering." 2007.

Rechtin, Eberhardt. *Systems Architecting of Organizations: Why Eagles Can't Swim.* 2000.

Section 3.4 Project Phase A: Concept and Technology Development

NASA. *NASA Safety Standard 1740.14, Guidelines and Assessment Procedures for Limiting Orbital Debris.* 1995.

Section 4.1 Stakeholder Expectations Definition

ANSI. *Guide for the Preparation of Operational Concept Documents.* 1992.

Section 4.2 Technical Requirements Definition

NASA. *NASA Space Flight Human System Standard.* 2007.

Section 4.3 Logical Decomposition

Institute of Electrical and Electronics Engineers. *Standard Glossary of Software Engineering Terminology.* 1999.

Section 4.4 Design Solution

Blanchard, Benjamin S. *System Engineering Management.* 2006.

DOD. *MIL-STD-1472, Human Engineering.* 2003.

Federal Aviation Administration. *Human Factors Design Standard.* 2003.

International Organization for Standardization. *Quality Systems Aerospace—Model for Quality Assurance in Design, Development, Production, Installation, and Servicing.* 1999.

NASA. *NASA Space Flight Human System Standard.* 2007.

NASA. *Planning, Developing, and Maintaining and Effective Reliability and Maintainability (R&M) Program.* 1998.

U. S. Army Research Laboratory. *MIL HDBK 727, Design Guidance for Producibility.* 1990.

U.S. Nuclear Regulatory Commission. *Human-System Interface Design Review Guidelines.* 2002.

Section 5.1 Product Implementation

American Institute of Aeronautics and Astronautics. *AIAA Guide for Managing the Use of Commercial Off the Shelf (COTS) Software Components for Mission-Critical Systems.* 2006.

International Council on Systems Engineering. *Systems Engineering Handbook.* 2006.

NASA. *Off-the-Shelf Hardware Utilization in Flight Hardware Development.* 2004.

Section 5.3 Verification

Electronic Industries Alliance. *Processes for Engineering a System.* 1999.

Institute of Electrical and Electronics Engineers. *Standard for Application and Management of the Systems Engineering Process.* 1998.

International Organization for Standardization. *Systems Engineering—System Life Cycle Processes.* 2002.

NASA. *Project Management: Systems Engineering & Project Control Processes and Requirements.* 2004.

U.S. Air Force. *SMC Systems Engineering Primer and Handbook.* 2005.

Section 5.4 Validation

Electronic Industries Alliance. *Processes for Engineering a System.* 1999.

Institute of Electrical and Electronics Engineers. *Standard for Application and Management of the Systems Engineering Process.* 1998.

International Organization for Standardization. *Systems Engineering—System Life Cycle Processes*. 2002.

NASA. *Project Management: Systems Engineering & Project Control Processes and Requirements*. 2004.

U.S. Air Force. *SMC Systems Engineering Primer and Handbook*. 2005.

Section 5.5 Product Transition

DOD. *Defense Acquisition Guidebook*. 2004.

Electronic Industries Alliance. *Processes for Engineering a System*. 1999.

International Council on Systems Engineering. *Systems Engineering Handbook*. 2006.

International Organization for Standardization. *Systems Engineering—A Guide for the Application of ISO/IEC 15288*. 2003.

———. *Systems Engineering—System Life Cycle Processes*. 2002.

Naval Air Systems Command. *Systems Command SE Guide: 2003*. 2003.

Section 6.1 Technical Planning

American Institute of Aeronautics and Astronautics. *AIAA Guide for Managing the Use of Commercial Off the Shelf (COTS) Software Components for Mission-Critical Systems*. 2006.

Institute of Electrical and Electronics Engineers. *Standard for Application and Management of the Systems Engineering Process*. 1998.

Martin, James N. *Systems Engineering Guidebook: A Process for Developing Systems and Products*. 1996.

NASA. *NASA Cost Estimating Handbook*. 2004.

———. *Standard for Models and Simulations*. 2006.

Section 6.4 Technical Risk Management

Clemen, R., and T. Reilly. *Making Hard Decisions with DecisionTools Suite*. 2002.

Dezfuli, H. "Role of System Safety in Risk-Informed Decisionmaking." 2005.

Kaplan, S., and B. John Garrick. "On the Quantitative Definition of Risk." 1981.

Morgan, M. Granger, and M. Henrion. *Uncertainty: A Guide to Dealing with Uncertainty in Quantitative Risk and Policy Analysis*. 1990.

Stamelatos, M., H. Dezfuli, and G. Apostolakis. "A Proposed Risk-Informed Decisionmaking Framework for NASA." 2006.

Stern, Paul C., and Harvey V. Fineberg, eds. *Understanding Risk: Informing Decisions in a Democratic Society*. 1996.

U.S. Nuclear Regulatory Commission. *White Paper on Risk-Informed and Performance-Based Regulation*. 1998.

Section 6.5 Configuration Management

American Society of Mechanical Engineers. *Engineering Drawing Practices*. 2004.

———. *Types and Applications of Engineering Drawings*. 1999.

DOD. *Defense Logistics Agency (DLA) Cataloging Handbook*.

———. *MIL-HDBK-965, Parts Control Program*. 1996.

———. *MIL-STD-881B, Work Breakdown Structure (WBS) for Defense Materiel Items*. 1993.

DOD, U.S. General Services Administration, and NASA. *Acquisition of Commercial Items*. 2007.

———. *Quality Assurance, Nonconforming Supplies or Services*. 2007.

Institute of Electrical and Electronics Engineers. *EIA Guide for Information Technology Software Life Cycle Processes—Life Cycle Data*. 1997.

———. *IEEE Guide to Software Configuration Management*. 1987.

———. *Standard for Software Configuration Management Plans*. 1998.

International Organization for Standardization. *Information Technology—Software Life Cycle Processes Configuration Management*. 1998.

———. *Quality Management—Guidelines for Configuration Management*. 1995.

NASA. *NOAA-N Prime Mishap Investigation Final Report*. 2004.

National Defense Industrial Association. *Data Management*. 2004.

———. *National Consensus Standard for Configuration Management*. 1998.

Section 6.6 Technical Data Management

National Defense Industrial Association. *Data Management*. 2004.

——. *National Consensus Standard for Configuration Management*. 1998.

Section 6.8 Decision Analysis

Blanchard, Benjamin S. *System Engineering Management*. 2006.

Blanchard, Benjamin S., and Wolter Fabrycky. *Systems Engineering and Analysis*. 2006.

Clemen, R., and T. Reilly. *Making Hard Decisions with DecisionTools Suite*. 2002.

Keeney, Ralph L. *Value-Focused Thinking: A Path to Creative Decisionmaking*. 1992.

Keeney, Ralph L., and Timothy L. McDaniels. "A Framework to Guide Thinking and Analysis Regarding Climate Change Policies." 2001.

Keeney, Ralph L., and Howard Raiffa. *Decisions with Multiple Objectives: Preferences and Value Tradeoffs*. 1993.

Morgan, M. Granger, and M. Henrion. *Uncertainty: A Guide to Dealing with Uncertainty in Quantitative Risk and Policy Analysis*. 1990.

Saaty, Thomas L. *The Analytic Hierarchy Process*. 1980.

Section 7.1 Engineering with Contracts

Adams, R. J., S. Eslinger, P. Hantos, K. L. Owens, et al. *Software Development Standard for Space Systems*. 2005.

DOD, U.S. General Services Administration, and NASA. *Contracting Office Responsibilities*. 2007.

Eslinger, Suellen. *Software Acquisition Best Practices for the Early Acquisition Phases*. 2004.

Hofmann, Hubert F., Kathryn M. Dodson, Gowri S. Ramani, and Deborah K. Yedlin. *Adapting CMMI® for Acquisition Organizations: A Preliminary Report*. 2006.

International Council on Systems Engineering. *Systems Engineering Handbook: A "What To" Guide for all SE Practitioners*. 2004.

The Mitre Corporation. *Common Risks and Risk Mitigation Actions for a COTS-Based System*.

NASA. *Final Memorandum on NASA's Acquisition Approach Regarding Requirements for Certain Software Engineering Tools to Support NASA Programs*. 2006.

——. *The SEB Source Evaluation Process*. 2001.

——. *Solicitation to Contract Award*. 2007.

——. *Standard for Models and Simulations*. 2006.

——. *Statement of Work Checklist*.

——. *System and Software Metrics for Performance-Based Contracting*.

Naval Air Systems Command. *Systems Engineering Guide*. 2003.

Section 7.2 Integrated Design Facilities

Miao, Y., and J. M. Haake. "Supporting Concurrent Design by Integrating Information Sharing and Activity Synchronization." 1998.

Section 7.4 Human Factors Engineering

Blanchard, Benjamin S., and Wolter Fabrycky. *Systems Engineering and Analysis*. 2006.

Chapanis, A. "The Error-Provocative Situation: A Central Measurement Problem in Human Factors Engineering." 1980.

DOD. *Human Engineering Procedures Guide*. 1987.

——. *MIL-HDBK-46855A, Human Engineering Program Process and Procedures*. 1996.

Eggemeier, F. T., and G. F. Wilson. "Performance and Subjective Measures of Workload in Multitask Environments." 1991.

Endsley, M. R., and M. D. Rogers. "Situation Awareness Information Requirements Analysis for En Route Air Traffic Control." 1994.

Fuld, R. B. "The Fiction of Function Allocation." 1993.

Glass, J. T., V. Zaloom, and D. Gates. "A Micro-Computer-Aided Link Analysis Tool." 1991.

Gopher, D., and E. Donchin. "Workload: An Examination of the Concept." 1986.

Hart, S. G., and C. D. Wickens. "Workload Assessment and Prediction." 1990.

Huey, B. M., and C. D. Wickens, eds. *Workload Transition*. 1993.

Jones, E. R., R. T. Hennessy, and S. Deutsch, eds. *Human Factors Aspects of Simulation*. 1985.

Kirwin, B., and L. K. Ainsworth. *A Guide to Task Analysis*. 1992.

Kurke, M. I. "Operational Sequence Diagrams in System Design." 1961.

Meister, David. *Behavioral Analysis and Measurement Methods.* 1985.

———. *Human Factors: Theory and Practice.* 1971.

Price, H. E. "The Allocation of Functions in Systems." 1985.

Shafer, J. B. "Practical Workload Assessment in the Development Process." 1987.

Section 7.6 Use of Metric System

DOD. *DoD Guide for Identification and Development of Metric Standards.* 2003.

Taylor, Barry. *Guide for the Use of the International System of Units (SI).* 2007.

Appendix F: Functional, Timing, and State Analysis

Buede, Dennis. *The Engineering Design of Systems: Models and Methods.* 2000.

Defense Acquisition University. *Systems Engineering Fundamentals Guide.* 2001.

Long, Jim. *Relationships Between Common Graphical Representations in Systems Engineering.* 2002.

NASA. *Training Manual for Elements of Interface Definition and Control.* 1997.

Sage, Andrew, and William Rouse. *The Handbook of Systems Engineering and Management.* 1999.

Appendix H: Integration Plan Outline

Federal Highway Administration and CalTrans. *Systems Engineering Guidebook for ITS.* 2007.

Appendix J: SEMP Content Outline

DOD. *MIL-HDBK-881, Work Breakdown Structures for Defense Materiel Systems.* 2005.

DOD Systems Management College. *Systems Engineering Fundamentals.* 2001.

Martin, James N. *Systems Engineering Guidebook: A Process for Developing Systems and Products.* 1996.

NASA. *NASA Cost Estimating Handbook.* 2004.

The Project Management Institute®. *Practice Standards for Work Breakdown Structures.* 2001.

Bibliography

Adams, R. J., et al. *Software Development Standard for Space Systems*, Aerospace Report No. TOR—2004(3909)-3537, Revision B. March 11, 2005.

American Institute of Aeronautics and Astronautics. *AIAA Guide for Managing the Use of Commercial Off the Shelf (COTS) Software Components for Mission-Critical Systems*, AIAA G-118-2006e. Reston, VA, 2006.

American National Standards Institute. *Guide for the Preparation of Operational Concept Documents*, ANSI/AIAA G-043-1992. Washington, DC, 1992.

American Society of Mechanical Engineers. *Engineering Drawing Practices*, ASME Y14.100. New York, 2004.

———. *Types and Applications of Engineering Drawings*, ASME Y14.24. New York, 1999.

Blanchard, Benjamin S. *System Engineering Management*, 6th ed. New Dehli: Prentice Hall of India Private Limited, 2006.

Blanchard, Benjamin S., and Wolter Fabrycky. *Systems Engineering and Analysis*, 6th ed. New Dehli: Prentice Hall of India Private Limited, 2006.

Buede, Dennis. *The Engineering Design of Systems: Models and Methods*. New York: Wiley & Sons, 2000.

Chapanis, A. "The Error-Provocative Situation: A Central Measurement Problem in Human Factors Engineering." In *The Measurement of Safety Performance*. Edited by W. E. Tarrants. New York: Garland STPM Press, 1980.

Clemen, R., and T. Reilly. *Making Hard Decisions with DecisionTools Suite*. Pacific Grove, CA: Duxbury Resource Center, 2002.

Defense Acquisition University. *Systems Engineering Fundamentals Guide*. Fort Belvoir, VA, 2001.

Department of Defense. *DOD Architecture Framework, Version 1.5, Vol. 1*. Washington, DC, 2007.

———. *Defense Logistics Agency (DLA) Cataloging Handbook*, H4/H8 Series. Washington, DC.

———. *DoD Guide for Identification and Development of Metric Standards*, SD-10. Washington, DC: DOD, Office of the Under Secretary of Defense, Acquisition, Technology, & Logistics, 2003.

———. *DOD-HDBK-763, Human Engineering Procedures Guide*. Washington, DC, 1987.

———. *MIL-HDBK-965, Parts Control Program*. Washington, DC, 1996.

———. *MIL-HDBK-46855A, Human Engineering Program Process and Procedures*. Washington, DC, 1996.

———. *MIL-STD-881B, Work Breakdown Structure (WBS) for Defense Materiel Items*. Washington, DC, 1993.

———. *MIL-STD-1472, Human Engineering*. Washington, DC, 2003.

DOD, Systems Management College. *Systems Engineering Fundamentals*. Fort Belvoir, VA: Defense Acquisition Press, 2001.

DOD, U.S. General Services Administration, and NASA. *Acquisition of Commercial Items*, 14CFR1214–Part 1214–Space Flight 48CFR1814. Washington, DC, 2007.

———. *Contracting Office Responsibilities*, i 46.103(a). Washington, DC, 2007.

———. *Quality Assurance, Nonconforming Supplies or Services*, FAR Part 46.407. Washington, DC, 2007.

Dezfuli, H. "Role of System Safety in Risk-informed Decisionmaking." In *Proceedings, the NASA Risk Management Conference 2005*. Orlando, December 7, 2005.

Eggemeier, F. T., and G. F. Wilson. "Performance and Subjective Measures of Workload in Multitask Environments." In *Multiple-Task Performance*. Edited by D. Damos. London: Taylor and Francis, 1991.

Electronic Industries Alliance. *Processes for Engineering a System*, ANSI/EIA–632. Arlington, VA, 1999.

Endsley, M. R., and M. D. Rogers. "Situation Awareness Information Requirements Analysis for En Route Air Traffic Control." In *Proceedings of the Human Factors and Ergonomics Society 38th Annual Meeting*. Santa Monica: Human Factors and Ergonomics Society, 1994.

Eslinger, Suellen. *Software Acquisition Best Practices for the Early Acquisition Phases.* El Segundo, CA: The Aerospace Corporation, 2004.

Federal Aviation Administration. HF-STD-001, *Human Factors Design Standard.* Washington, DC, 2003.

Federal Highway Administration, and CalTrans. *Systems Engineering Guidebook for ITS*, Version 2.0. Washington, DC: U.S. Department of Transportation, 2007.

Fuld, R. B. "The Fiction of Function Allocation." *Ergonomics in Design* (January 1993): 20–24.

Glass, J. T., V. Zaloom, and D. Gates. "A Micro-Computer-Aided Link Analysis Tool." *Computers in Industry* 16, (1991): 179–87.

Gopher, D., and E. Donchin. "Workload: An Examination of the Concept." In *Handbook of Perception and Human Performance: Vol. II. Cognitive Processes and Performance.* Edited by K. R. Boff, L. Kaufman, and J. P. Thomas. New York: John Wiley & Sons, 1986.

Griffin, Michael D., NASA Administrator. "System Engineering and the Two Cultures of Engineering." Boeing Lecture, Purdue University, March 28, 2007.

Hart, S. G., and C. D. Wickens. "Workload Assessment and Prediction." In *MANPRINT: An Approach to Systems Integration.* Edited by H. R. Booher. New York: Van Nostrand Reinhold, 1990.

Hofmann, Hubert F., Kathryn M. Dodson, Gowri S. Ramani, and Deborah K. Yedlin. *Adapting CMMI® for Acquisition Organizations: A Preliminary Report*, CMU/SEI-2006-SR-005. Pittsburgh: Software Engineering Institute, Carnegie Mellon University, 2006, pp. 338–40.

Huey, B. M., and C. D. Wickens, eds. *Workload Transition.* Washington, DC: National Academy Press, 1993.

Institute of Electrical and Electronics Engineers. *EIA Guide for Information Technology Software Life Cycle Processes—Life Cycle Data*, IEEE Std 12207.1. Washington, DC, 1997.

———. *IEEE Guide to Software Configuration Management*, ANSI/IEEE 1042. Washington, DC, 1987.

———. *Standard for Application and Management of the Systems Engineering Process*, IEEE Std 1220. Washington, DC, 1998.

———. *Standard Glossary of Software Engineering Terminology*, IEEE Std 610.12-1990. Washington, DC, 1999.

———. *Standard for Software Configuration Management Plans*, IEEE Std 828. Washington, DC, 1998.

International Council on Systems Engineering. *Systems Engineering Handbook*, version 3. Seattle, 2006.

———. *Systems Engineering Handbook: A "What To" Guide for All SE Practitioners*, INCOSE-TP-2003-016-02, Version 2a. Seattle, 2004.

International Organization for Standardization. *Information Technology—Software Life Cycle Processes Configuration Management*, ISO TR 15846. Geneva, 1998.

———. *Quality Management—Guidelines for Configuration Management*, ISO 10007: 1995(E). Geneva, 1995.

———. *Quality Systems Aerospace—Model for Quality Assurance in Design, Development, Production, Installation, and Servicing*, ISO 9100/AS9100. Geneva: International Organization for Standardization, 1999.

———. *Systems Engineering—A Guide for the Application of ISO/IEC 15288*, ISO/IEC TR 19760: 2003. Geneva, 2003.

———. *Systems Engineering—System Life Cycle Processes*, ISO/IEC 15288: 2002. Geneva, 2002.

Jones, E. R., R. T. Hennessy, and S. Deutsch, eds. *Human Factors Aspects of Simulation.* Washington, DC: National Academy Press, 1985.

Kaplan, S., and B. John Garrick. "On the Quantitative Definition of Risk." *Risk Analysis* 1(1). 1981.

Keeney, Ralph L. *Value-Focused Thinking: A Path to Creative Decisionmaking.* Cambridge, MA: Harvard University Press, 1992.

Keeney, Ralph L., and Timothy L. McDaniels. "A Framework to Guide Thinking and Analysis Regarding Climate Change Policies." *Risk Analysis* 21(6): 989–1000. 2001.

Keeney, Ralph L., and Howard Raiffa. *Decisions with Multiple Objectives: Preferences and Value Tradeoffs.* Cambridge, UK: Cambridge University Press, 1993.

Kirwin, B., and L. K. Ainsworth. *A Guide to Task Analysis.* London: Taylor and Francis, 1992.

Kurke, M. I. "Operational Sequence Diagrams in System Design." *Human Factors* 3: 66–73. 1961.

Long, Jim. *Relationships Between Common Graphical Representations in Systems Engineering.* Vienna, VA: Vitech Corporation, 2002.

Martin, James N. *Processes for Engineering a System: An Overview of the ANSI/GEIA EIA-632 Standard and Its Heritage.* New York: Wiley & Sons, 2000.

———. *Systems Engineering Guidebook: A Process for Developing Systems and Products.* Boca Raton: CRC Press, 1996.

Meister, David. *Behavioral Analysis and Measurement Methods.* New York: John Wiley & Sons, 1985.

———. *Human Factors: Theory and Practice.* New York: John Wiley & Sons, 1971.

Miao, Y., and J. M. Haake. "Supporting Concurrent Design by Integrating Information Sharing and Activity Synchronization." In *Proceedings of the 5th ISPE International Conference on Concurrent Engineering Research and Applications (CE98).* Tokyo, 1998, pp. 165–74.

The Mitre Corporation. *Common Risks and Risk Mitigation Actions for a COTS-based System.* McLean, VA.

Morgan, M. Granger, and M. Henrion. *Uncertainty: A Guide to Dealing with Uncertainty in Quantitative Risk and Policy Analysis.* Cambridge, UK: Cambridge University Press, 1990.

NASA. *Final Memorandum on NASA's Acquisition Approach Regarding Requirements for Certain Software Engineering Tools to Support NASA Programs*, Assignment No. S06012. Washington, DC, NASA Office of Inspector General, 2006.

———. *NASA Cost Estimating Handbook.* Washington, DC, 2004.

———. *NASA-STD-3001, NASA Space Flight Human System Standard Volume 1: Crew Health.* Washington, DC, 2007.

———. *NASA-STD-(I)-7009, Standard for Models and Simulations.* Washington, DC, 2006.

———. *NASA-STD-8719.13, Software Safety Standard, NASA Technical Standard, Rev B.* Washington, DC, 2004.

———. *NASA-STD-8729.1, Planning, Developing, and Maintaining and Effective Reliability and Maintainability (R&M) Program.* Washington, DC, 1998.

———. *NOAA N-Prime Mishap Investigation Final Report.* Washington, DC, 2004.

———. *NPD 2820.1, NASA Software Policy.* Washington, DC, 2005.

———. *NPD 8010.2, Use of the SI (Metric) System of Measurement in NASA Programs.* Washington, DC, 2007.

———. *NPD 8010.3, Notification of Intent to Decommission or Terminate Operating Space Systems and Terminate Missions.* Washington, DC, 2004.

———. *NPD 8020.7, Biological Contamination Control for Outbound and Inbound Planetary Spacecraft.* Washington, DC, 1999.

———. *NPD 8070.6, Technical Standards.* Washington, DC, 2003.

———. *NPD 8730.5, NASA Quality Assurance Program Policy.* Washington, DC, 2005.

———. *NPR 1441.1, NASA Records Retention Schedules.* Washington, DC, 2003.

———. *NPR 1600.1, NASA Security Program Procedural Requirements.* Washington, DC, 2004.

———. *NPR 2810.1, Security of Information Technology.* Washington, DC, 2006.

———. *NPR 7120.5, NASA Space Flight Program and Project Management Processes and Requirements.* Washington, DC, 2007.

———. *NPR 7120.6, Lessons Learned Process.* Washington, DC, 2007.

———. *NPR 7123.1, Systems Engineering Processes and Requirements.* Washington, DC, 2007.

———. *NPR 7150.2, NASA Software Engineering Requirements.* Washington, DC, 2004.

———. *NPR 8000.4, Risk Management Procedural Requirements.* Washington, DC, NASA Office of Safety and Mission Assurance, 2007.

———. *NPR 8020.12, Planetary Protection Provisions for Robotic Extraterrestrial Missions.* Washington, DC, 2004.

———. *NPR 8580.1, Implementing The National Environmental Policy Act and Executive Order 12114.* Washington, DC, 2001.

———. *NPR 8705.2, Human-Rating Requirements for Space Systems.* Washington, DC, 2005.

———. *NPR 8705.3, Probabilistic Risk Assessment Procedures Guide for NASA Managers and Practitioners.* Washington, DC, 2002.

———. *NPR 8705.4, Risk Classification for NASA Payloads.* Washington, DC, 2004.

———. *NPR 8705.5, Probabilistic Risk Assessment (PRA) Procedures for NASA Programs and Projects.* Washington, DC, 2004.

———. *NPR 8710.1, Emergency Preparedness Program.* Washington, DC, 2006.

———. *NPR 8715.2, NASA Emergency Preparedness Plan Procedural Requirements—Revalidated.* Washington, DC, 1999.

———. *NPR 8715.3, NASA General Safety Program Requirements.* Washington, DC, 2007.

———. *NPR 8715.6, NASA Procedural Requirements for Limiting Orbital Debris.* Washington, DC, 2007.

———. *NPR 8735.2, Management of Government Quality Assurance Functions for NASA Contracts.* Washington, DC, 2006.

———. *NSS-1740.14, NASA Safety Standard Guidelines and Assessment Procedures for Limiting Orbital Debris.* Washington, DC, 1995.

———. *Off-the-Shelf Hardware Utilization in Flight Hardware Development,* MSFC NASA MWI 8060.1 Rev A. Washington, DC, 2004.

———. *Off-the-Shelf Hardware Utilization in Flight Hardware Development,* JSC Work Instruction EA-WI-016. Washington, DC.

———. *Project Management: Systems Engineering & Project Control Processes and Requirements,* JPR 7120.3. Washington, DC, 2004.

———. *The SEB Source Evaluation Process.* Washington, DC, 2001.

———. *Solicitation to Contract Award.* Washington, DC, NASA Procurement Library, 2007.

———. *Statement of Work Checklist.* Washington, DC.

———. *System and Software Metrics for Performance-Based Contracting.* Washington, DC.

———. *Systems Engineering Handbook,* SP-6105. Washington, DC, 1995.

———. *Training Manual for Elements of Interface Definition and Control,* NASA Reference Publication 1370. Washington, DC, 1997.

NASA Langley Research Center. *Instructional Handbook for Formal Inspections.*

———. *Guidance on System and Software Metrics for Performance-Based Contracting.*

National Defense Industrial Association. *Data Management,* ANSI/GEIA GEIA-859. Arlington, VA, 2004.

———. *National Consensus Standard for Configuration Management,* ANSI/GEIA EIA-649, Arlington, VA, 1998.

Naval Air Systems Command. *Systems Command SE Guide: 2003* (based on requirements of ANSI/EIA 632: 1998). Patuxent River, MD, 2003.

Nuclear Regulatory Commission. *NUREG-0700, Human-System Interface Design Review Guidelines, Rev. 2.* Washington, DC, Office of Nuclear Regulatory Research, 2002.

———. *Systems Engineering Guide.* Patuxent River, MD, 2003.

Price, H. E. "The Allocation of Functions in Systems." *Human Factors* 27: 33–45. 1985.

The Project Management Institute®. *Practice Standards for Work Breakdown Structures.* Newtown Square, PA, 2001.

Rechtin, Eberhardt. *Systems Architecting of Organizations: Why Eagles Can't Swim.* Boca Raton: CRC Press, 2000.

Saaty, Thomas L. *The Analytic Hierarchy Process.* New York: McGraw-Hill, 1980.

Sage, Andrew, and William Rouse. *The Handbook of Systems Engineering and Management.* New York: Wiley & Sons, 1999.

Shafer, J. B. "Practical Workload Assessment in the Development Process." In *Proceedings of the Human Factors Society 31st Annual Meeting.* Santa Monica: Human Factors Society, 1987.

Stamelatos, M., H. Dezfuli, and G. Apostolakis. "A Proposed Risk-Informed Decisionmaking Framework for NASA." In *Proceedings of the 8th International Conference on Probabilistic Safety Assessment and Management.* New Orleans, LA, May 14–18, 2006.

Stern, Paul C., and Harvey V. Fineberg, eds. *Understanding Risk: Informing Decisions in a Democratic Society.* Washington, DC: National Academies Press, 1996.

Taylor, Barry. *Guide for the Use of the International System of Units (SI)*, Special Publication 811. Gaithersburg, MD: National Institute of Standards and Technology, Physics Laboratory, 2007.

U.S. Air Force. *SMC Systems Engineering Primer and Handbook*, 3rd ed. Los Angeles: Space & Missile Systems Center, 2005.

U.S. Army Research Laboratory. *Design Guidance for Producibility*, MIL HDBK 727. Adelphi, MD: Weapons and Materials Research Directorate, 1990.

———. *White Paper on Risk-Informed and Performance-Based Regulation,* SECY-98-144. Washington, DC, 1998.

Index

To request print or electronic copies or provide comments,
contact the Office of the Chief Engineer via
SP6105rev1SEHandbook@nasa.gov

Electronic copies are also available from
NASA Center for AeroSpace Information
7115 Standard Drive
Hanover, MD 21076-1320
at
http://ntrs.nasa.gov/

www.ingramcontent.com/pod-product-compliance
Lightning Source LLC
Chambersburg PA
CBHW080906220326

41598CB00034B/5495